D0206179

# MODERN COMMUNICATION CIRCUITS

# McGraw-Hill Series in Electrical and Computer Engineering

SENIOR CONSULTING EDITOR

*Stephen W. Director, University of Michigan, Ann Arbor*

*Circuits and Systems*
*Communications and signal Processing*
*Computer Engineering*
*Control Theory*
*Electromagnetics*
*Electronics and VLSI Circuits*
*Introductory*
*Power and Energy*
*Radar and Antennas*

PREVIOUS CONSULTING EDITORS

Ronald N. Bracewell, Colin Cherry, James F. Gibbons,
Willis W. Harman, Hubert Heffner, Edward W. Harold,
John C. Linvill, Simon Ramo, Ronald A. Rohrer,
Anthony E. Siegman, Charles Susskind, Frederick E. Tehman,
John C. Truxal, Ernst Weber, and John R. Whinnery

# Communications and Signal Processing

SENIOR CONSULTING EDITOR

*Stephen W.* Director, University of Michigan, Ann Arbor

Also Available from McGraw-Hill

# Schaum's Outline Series in Electronics & Electrical Engineering

Most outlines include basic theory, definitions, and hundreds of example problems solved in step-by-step detail, and supplementary problems with answers.

Related titles on the current list include:

*Analog & Digital Communications*
*Basic Circuit Analysis*
*Basic Electrical Engineering*
*Basic Electricity*
*Basic Mathematics for Electricity & Electronics*
*Digital Principles*
*Electric Circuits*
*Electric Machines & Electromechanics*
*Electric Power Systems*
*Electromagnetics*
*Electronic Communication*
*Electronic Devices & Circuits*
*Feedback & Control Systems*
*Introduction to Digital Systems*
*Microprocessor Fundamentals*
*Signals & Systems*

## Schaum's Electronic Tutors

A Schaum's Outline plus the power of Mathcad®software. Use your computer to learn the theory and solve problems—every number, formula, and graph can be changed and calculated on screen.

Related titles on the current list include:

*Electric Circuits*
*Feedback & Control Systems*
*Electromagnetics*
*College Physics*

Available at most college bookstores, or for a complete list of titles and prices, write to: The McGraw-Hill Companies
        Schaum's
        11 West 19th Street
        New York, New York 10011-4285
        (212-337-4097)

# MODERN COMMUNICATION CIRCUITS

## SECOND EDITION

## JACK R. SMITH
Professor of Electrical Engineering
University of Florida

Boston, Massachusetts   Burr Ridge, Illinois   Dubuque, Iowa
Madison, Wisconsin   New York, New York   San Francisco, California   St. Louis, Missouri

# WCB/McGraw-Hill

*A Division of The **McGraw·Hill** Companies*

Modern Communication Circuits

Copyright © 1998, 1986 by The McGraw-Hill Companies, Inc. All rights reserved. Printed in the United States of America. Except as permitted under the United States Copyright Act of 1976, no part of this publication may be reproduced or distributed in any form or by any means, or stored in a data base or retrieval system, without the prior written permission of the publisher.

This book is printed on acid-free paper.

1 2 3 4 5 6 7 8 9 0 DOC DOC 9 0 0 9 8 7

ISBN 0-07-059283-7

Publisher: Thomas L. Casson
Sponsoring editor: Lynn B. Cox
Marketing manager: John T. Wannemacher
Project manager: Terri Wicks
Production supervisor: Rich DeVitto
Cover designer: Joe Piliero
Editorial assistant: Nina Kreiden
Compositor: Interactive Composition Corporation
Typeface: Times Roman
Printer: R. R. Donnelley & Sons Company

**Library of Congress Cataloging-in-Publication Data**

Smith, Jack, 1935-
　　Modern communication circuits / Jack Smith. — 2nd ed.
　　　　p.　　cr.
　　Includes bibliographical references and Index.
　　ISBN 0-07-059283-7
　　1. Radio circuits.　2. Telecommunication—Equipment and supplies.
　　3. Electronic circuit design.　I. Title.
　　TK6553.S5595　1997
　　621.384' 12—dc21
　　　　　　　　　　　　　　　　　　　　　97-28578
　　　　　　　　　　　　　　　　　　　　　CIP

http://www.mhhe.com

*To the descendants of*
*Laura Kornkven Smith*

# CONTENTS

There have been many advances in communication circuits since the publication of the first edition of this book. Yet most of the communication circuit fundamentals remain intact. Amplifiers, oscillators, tuned circuits, transformers, mixers, and power amplifiers are still among the fundamental building blocks of communication circuits. There is a shift toward the use of more broadband circuitry, but a few narrowband circuits are still required, particularly in oscillators. It is still not possible to realize an inductor in an integrated circuit, and inductors remain a necessary component of many communication circuits. The many new integrated circuits have markedly expanded the frequency range over which the material in this book is applicable. While the approach of the first edition has not changed, the second edition incorporates changes suggested by the advances in technology as well the tremendous improvements in, and availability of, computer simulation, particularly SPICE. Computer projects have been added to every chapter.

This book presents fundamental analysis and design techniques for modern communication circuits covering the frequency range up to several hundred megahertz. The coverage reflects the practice of modern communication systems design to use integrated circuits, to minimize tuning operations by using broadband circuits, and to use more field-effect transistors in high-frequency circuits. Much of the information presented here appears in book form for the first time. Some of it is original, much of it has been gathered from a diverse literature on communication electronics, and some has been adapted from early publications on vacuum-tube electronics. Practical approximations are emphasized rather than theoretical derivations based on complex circuit models. The approximations provide more insight into the design process than do lengthy derivations. Computer simulation is frequently used to establish the accuracy of the approximations.

The book is intended for use as a textbook in senior-level and beginning graduate-level courses in electrical engineering and as a reference book for practicing engineers. It is assumed that the reader has the basic background in linear transistor circuit theory obtained from a junior-level electrical engineering course, but not necessarily in the subjects which are becoming restricted to specialized courses in high-frequency electronics, such as tuned-circuit analysis. Chapter 2 presents a review of the necessary electronic circuit background.

A characteristic of almost all realistic communication circuits is that they are too complex for a complete analysis, and approximations must be made. The judicious use of approximations is also required in the design process. A goal of this book is to illustrate many of the approximations convenient and effective for the analysis and design of communication circuits. The circuit models are often incomplete, but more accurate and complex models are usually best treated with computer-aided analysis methods available to most electrical engineering students and electrical engineers. Computer-aided analysis is frequently used in the examples.

Most, if not all, communication circuits are available as integrated circuits, but there are several reasons why designers must still be familiar with the discrete circuit techniques. First, there are many communication systems that cannot yet be realized in integrated-circuit form, because circuit flexibility is compromised and an inductor-capacitor tuned circuit cannot be fabricated on a chip. Second, the performance of discrete communication circuits is superior. It is necessary for the communication circuit designer to consider the tradeoffs among size, cost, output power, power consumption, noise, and distortion. Third, even when integrated circuits are used, a basic understanding of communication circuitry is required. Also, monolithic power amplifiers create many additional problems, which result in the complete amplifier's occupying more space than would be required of a discrete power amplifier. This book incorporates the impact that integrated circuits are having on the design of communication systems. Two chapters are devoted to the phase-locked loop. The importance of this device in modern systems is largely due to its realization as an integrated circuit. Phase-locked loop applications are discussed in Chap. 8, and the analysis of phase-locked loops is presented in detail in Chap. 9.

It is not possible to completely cover a subject as broad as communication electronics in a single text. Much of the methodology presented here is applicable to communication systems in all frequency ranges, but distributed-parameter circuit analysis techniques, not covered in this book, are usually more accurate at frequencies above 100 MHz. Digital circuits are playing an increasingly important role in communication systems. Many applications of digital circuits are described in this text, but conventional logic circuits are not considered, as the subject is well covered in many texts.

Chapter 1 presents an introduction to communication circuits and discusses recent trends in receiver design. Chapter 2 reviews linear small-signal analysis of bipolar and field-effect transistor amplifiers. Operational amplifiers are included, since there are now devices available with gain-bandwidth products sufficiently large that they can be used in high-frequency communication circuits. Chapter 3 defines the noise and distortion specifications used to describe communication systems. The fundamentals of low-noise amplifier design are included in this chapter. The simple parallel and series tuned circuits still so important in modern communication systems are discussed in Chap. 4. Methods for analyzing the high-frequency performance of transistor amplifiers are given in Chap. 5, together with models for automatic gain control systems. The transformers, which are widely used in modern communication systems, are discussed in Chap. 6. The transmission-line transformer is still a suitable lumped impedance-matching network for the realization of untuned, broadband, interstage, and output-matching networks. Chapter 7 contains an in-depth treatment of the analysis and design of high-performance transistor oscillator circuits, including crystal and voltage-controlled oscillators. Chapter 8 is devoted to applications of the phase-locked loop, and Chap. 9 to its analysis. The phase-locked loop is one of the most versatile and widely applied circuits in communication systems. Its availability as an inexpensive integrated circuit implies that it will continue to find application in virtually all communication systems. Integrated circuits have also resulted in the design of frequency synthesizers, which have altered the design of most new communication systems using frequency tuning. The frequency synthesizer has also allowed the practical implementation of modern communication techniques such as frequency hopping. Chapter 10 provides a detailed treatment of frequency synthesizers. Methods for frequency shifting, modulation, and demodulation are discussed in Chap. 11, including methods for frequency- and phase-shift keying and other methods for handling digital signals. Chapter 12 discusses the design and analysis of power amplifiers and includes several curves useful for the design of class C power amplifiers.

The book contains enough material for a two-semester course, provided it is supplemented with articles from the current literature. Chapters 1, 3 through 8, and 10 have been used for a one-semester course at the University of Florida. Each chapter contains problems that emphasize particular points; several chapters include problems that extend the material covered. Component and integrated specification sheets contained in the appendixes are used in some of the problems.

The writing of the first edition of this book resulted from many discussions with my friend Dr. Ulrich Rohde, president of Compact Software. Rob Bruckner, now of Intel Corporation, assisted with the writing of this edition. The suggestions, corrections, and circuit designs of my students have also markedly improved the quality of this book.

Jack Smith

**JACK R. SMITH** is Professor Emeritus of Electrical Engineering at the University of Florida. He is founder and president of Neurotronics Incorporated. He also founded Microtronics Incorporated and served as its president until its sale to Oxford Instruments, Ltd. Dr. Smith has been a visiting scientist at the French National Center for Scientific Research, the Medical School of Geneva Switzerland, and the Tokyo Metropolitan Center for Neurological Research. He continues to consult for industry, most recently assisting the Land Mobile Productions Division of Motorola Incorporated.

# MODERN COMMUNICATION CIRCUITS

# Introduction to Radio Communication Systems

## 1.1

**INTRODUCTION**

Communication systems transmit information from one place to another by means of electric energy. The frequency used for the information transmission varies from the very low frequencies used in direct telephone communication to optical frequencies, also used for telephone communication. This book describes the analysis and design of electronic circuits used in radio-frequency communication systems covering the frequency range up through several hundred megahertz. The actual frequency limit depends upon whether the circuit is realized with discrete components or as an integrated circuit. The material is directly applicable to many other systems, including television and spectrum analyzers where the design of low-noise, high frequency receivers is of paramount importance. For very high frequency circuits, different circuit models, particularly distributed-parameter circuits, are more accurate. Low-frequency circuitry is discussed, but the emphasis here is on the radio-frequency circuits.

The following chapters treat the analysis and design of fundamental circuits of communication receivers and transmitters. Integrated circuits have simplified the system design, but the communication system designer still needs to be familiar with many circuit techniques and the simplifying approximations which apply in this frequency range. The designer is often faced with a choice between using an integrated circuit (IC) or a discrete component version of the same circuit. The decision is based on many factors, including cost, size, power consumption, noise, and distortion. Chapter 3 presents quantitative criteria for evaluating a circuit's noise and distortion performance. The application of integrated circuits in a communication system requires a knowledge of electronic circuit theory to properly interface the IC with the rest of the system. We will study the electronic circuits of the various subsystems, including oscillators, amplifiers, transformers, modulators, and demodulators, which make up a communication system. The mathematical analysis of the many modulation methods is not considered, as it is well described in the many good texts on communications theory.

## ▨ 1.2
### NETWORK THEORY

This section briefly reviews the concepts of network theory that are applied in the following chapters. The usual variables in an electronic circuit are the voltages and currents measured at various points in the circuit. The excitation and response can be described in the time domain, but determining the response in the time domain involves the solution of integrodifferential equations and rarely provides insight into the design process. For linear time-invariant systems, it is usually easier to obtain the system response using the Laplace transform. The Laplace transform of the time variable $v(t)$ is

$$V(s) = \int_0^\infty v(t)e^{-st}\,dt \tag{1.1}$$

where $s$ has the dimensions of frequency and is known as the *complex-frequency variable*. A linear system transfer function $H(s)$ is defined as

$$H(s) = \frac{R(s)}{V(s)} \tag{1.2}$$

where $R(s)$ is the Laplace transform of the response to an excitation $V(s)$. Linear circuit transfer functions can easily be obtained by interpreting an inductor as having a complex impedance $sL$ and a capacitor as having a complex impedance $(sC)^{-1}$. The method is illustrated by the following example.

> **EXAMPLE 1.1.** Determine the transfer function $V_o(s)/V_i(s)$ of the circuit shown in Fig. 1.1.
>
> **Solution.** In this circuit the inductor has been modeled as a complex impedance $sL$ and the capacitor as a complex impedance $(sC)^{-1}$. By using the voltage-divider rule of circuit analysis, we find that the transfer function $H(s)$ is
>
> $$H(s) = \frac{V_o(s)}{V_i(s)} = \frac{R/(RsC+1)}{R/(RsC+1)+sL} = \frac{R}{s^2RLC+sL+R} \tag{1.3}$$

For simplicity, the excitation and response are often written as $V_o$ and $V_i$, respectively, when there is no possibility of confusion.

Another advantage of describing the system response by the linear transfer function $H(s)$ is that the frequency response of the network can be obtained by setting $s = j\omega$. That is, if the linear system is stable and time-invariant and the excitation is a sinusoid, the steady-state response (after the transients have decayed to a

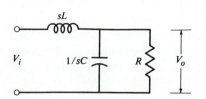

**FIGURE 1.1**
A low-pass filter.

negligible value) will also be a sinusoid of the same frequency. If

$$v_i(t) = V \sin \omega t$$

the steady-state response is

$$r(t) = R \sin(\omega t + \phi)$$

where
$$|H(j\omega)| = \frac{V}{R} \qquad \text{and} \qquad \arg H(j\omega) = \phi$$

**EXAMPLE 1.2.** Determine the frequency response of the network shown in Fig. 1.1 for $L = 0.5$ H, $C = 2$ F, and $R = 1\ \Omega$.

***Solution.*** The frequency response is obtained by substituting $s = j\omega$ in the transfer function. In this case, Eq. (1.3) becomes

$$H(j\omega) = [(j\omega)^2 + 0.5 j\omega + 1]^{-1}$$

The magnitude of the response is the frequency-dependent function

$$|H(j\omega)| = \left\{ \left[ (1 - \omega^2)^2 + \left(\frac{\omega}{2}\right)^2 \right]^{1/2} \right\}^{-1}$$

and the phase shift is also frequency-dependent

$$\arg H(j\omega) = -\tan^{-1} \frac{0.5\omega}{1 - \omega^2}$$

A linear transfer function without ideal delay elements will have the form

$$H(s) = \frac{A(s)}{B(s)} \tag{1.4}$$

where $A(s)$ and $B(s)$ are polynomials in $s$. The zeros of $A(s)$ are referred to as *zeros of the transfer function,* and the zeros of $B(s)$ are referred to as *poles of the transfer function.* The poles are the values of $s$ for which the magnitude of the transfer function is infinite. In order for the transfer function to be stable, all the poles must lie in the left half of the $s$ plane (the real part of the pole must be negative). The stability problem is considered in detail in Chap. 9.

**EXAMPLE 1.3.** Calculate the poles and zeros of the transfer function given in Example 1.2:

$$H(s) = (s^2 + 0.5s + 1)^{-1}$$

***Solution.*** The transfer function has no finite zeros. Since the order of the denominator polynomial is 2 higher than that of the numerator, the transfer function has two zeros at infinity. The poles

$$s_1, s_2 = -0.25 \pm j \frac{(3.75)^{1/2}}{2}$$

are located in the left half-plane. The real part of each pole is $-0.25$, and the imaginary parts are $\pm j (3.75)^{1/2}/2$.

## ■ 1.3

### MODULATION

For a signal to contain information, some feature of the signal must be varied in accordance with the information to be transmitted. Early radio communications conveyed information by the presence, or absence, of the signal. This method was soon surpassed by amplitude modulation of the radio wave by an audio signal. The amplitude modulation process provides a means of transmitting voice communications, and its development led to the rapid establishment of the radio broadcasting industry.

Angle modulation is another method of transmitting information widely used in high-frequency communication systems. An angle-modulated signal is described by the equation

$$S(t) = A \sin(\omega_o t + \phi)$$

The amplitude remains constant, and the angle $\phi$ is varied in response to the modulating signal. Both phase and frequency modulation can be used. One function of the receiver is to recover (demodulate) the original from the modulated signal. The circuitry for implementing the various types of modulation and demodulation is described in greater detail in Chap. 12. Today digital modulation techniques are being more frequently employed, particularly in satellite and telephone communication systems. Digital modulation implies that a parameter of the signal is varied in response to a digital signal. Amplitude, phase, or frequency modulation can be varied in response to a digital signal. That is, digital modulation is an extension of one or more of the conventional amplitude or angle modulation methods.

Figure 1.2 illustrates a simplified block diagram of a digital single-channel-per-carrier (SCPC) satellite communications channel. Each voice channel is sampled and then converted to a digitally encoded signal that modulates a low-frequency (baseband) carrier signal. The modulated signal uses a different carrier frequency sufficiently separated so that the signals can be combined without frequency overlay. This process is known as *frequency multiplexing*. The frequency-multiplexed signal is shifted up in frequency by mixing with a local oscillator signal, then amplified and transmitted. The transmission channel contains several voice channels. This illustration is only one of many possible methods of simultaneously transmitting several data channels. It is the function of the receiver to recover the original voice channels from the received signal.

## ■ 1.4

### RECEIVERS

The modulated signal is transmitted to a receiver where the signal is amplified and the information extracted. It is usually the case that many different signals are simultaneously present at the receiver input, and it is necessary for the receiver to be able to select the desired signal. This selection is made on the basis of the

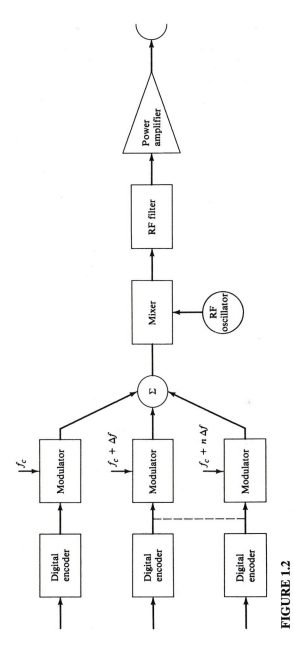

**FIGURE 1.2**
A satellite communications channel.

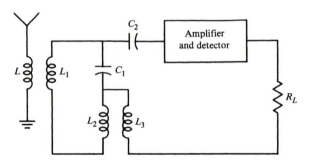

**FIGURE 1.3**
Schematic diagram of an
early regenerative receiver.

frequency of the incident signals—the receiver's ability to discriminate between signals of different frequencies is called *receiver selectivity*. Other functions of the receiver are to detect (demodulate) the information contained in the signal and perhaps to reconstruct and amplify the original waveform. Receivers vary in kind from telephone, radio, television, radar, and navigation to satellite communications models. The complexity of each type varies with the complexity of the transmitted signal, the frequency of operation, and the number and amplitude of the unwanted signals in the same frequency band. All receivers have the common problems of selectivity, rejecting input noise, and detecting the desired signal.

The name "radio" originated around 1910 to distinguish receivers that received voice transmission from the earlier receivers which received only code (pulse transmission). The first receivers did not have the capability of signal amplification, but this deficiency was soon overcome with the invention of the vacuum tube (triode). The first vacuum-tube circuits did not provide much gain, but E. H. Armstrong soon invented the "regenerative receiver," which used positive feedback from the output to the input. A simplified schematic of a regenerative receiver is illustrated in Fig. 1.3. A vacuum tube in the circuit serves as both an amplifier and a detector. A demodulator which recovers (detects) the modulating signal is known as a *detector*. The ac output signal is fed back in phase with the input signal (regenerated), increasing the loop gain. The regenerative receiver was probably the first application of electronic feedback; it quickly led to the invention of the electronic oscillator, since the circuit was susceptible to oscillations. The electronic oscillator greatly improved transmitter design.

The regenerative receiver was soon replaced by the tuned radio-frequency (TRF) receiver. A block diagram of a typical TRF receiver is illustrated in Fig. 1.4; it consists of three tuned RF amplifiers in cascade followed by a detector and power

**FIGURE 1.4**
Block diagram of an early tuned radio-frequency receiver.

amplifier. This receiver was hard to operate because of the difficulty of tuning all the RF amplifiers to the same frequency. The TRF receiver became obsolete with the invention of the superheterodyne receiver by E. H. Armstrong. The "superhet" eliminates the need for tuning all the RF amplifiers to the frequency of the input signal by shifting the input signal frequency to that of the fixed frequency of the receiver filter. It is possible to build fixed-frequency filters and amplifiers that are far superior to variable-frequency filters. The superheterodyne principle is still used in virtually all receivers. It consists of multiplying, or beating (heterodyne is from the Greek *heteros,* "other," and *dynamis,* "force"), the input signal with a signal generated in the local oscillator. If a sine wave of frequency $\omega_c$ is multiplied by a sine wave of frequency $\omega_L$, the resultant signal consists of the sinusoids of frequencies $\omega_c \pm \omega_L$. That is,

$$\sin \omega_c t \sin \omega_L t = \frac{\cos (\omega_c - \omega_L)t - \cos (\omega_c + \omega_L)t}{2}$$

A simplified block diagram of a superheterodyne receiver is shown in Fig. 1.5. In this type of receiver, the incoming signal is converted to an intermediate frequency by the first local oscillator and then is reduced to a low-frequency signal by the second mixer and low-pass filter. If the input signal consists of a carrier $f_c$ and an audio component $f_a$ and the first local oscillator frequency is $f_o$, the output of the first mixer consists of the two frequencies $f_c + f_a + f_o$ and $f_c + f_a - f_o$. The local oscillator frequency $f_o$ is selected so that one of these frequencies is equal to the center frequency of the intermediate-frequency (IF) filter ($f_{\text{IF}}$). Since $f_a$ is usually much less than $f_c$, for all practical purposes

$$f_o = |f_{\text{IF}} - f_c| \qquad \text{or} \qquad f_o = f_{\text{IF}} + f_c$$

in order that the mixer output frequency be at the center of the IF filter bandwidth.

One advantage of this form of detection is that the same high-quality filter can be used for all input frequencies. The frequency selection is obtained by varying the local oscillator frequency $f_o$. The IF filter output $f_{\text{IF}} + f_a$ is then reduced to $f_a$ in the second mixer, which mixes the IF output with the second oscillator frequency (which is fixed at $f_{\text{IF}}$). A problem with this form of detection occurs when there are a large number of signals of different frequencies present at the input. Consider, for

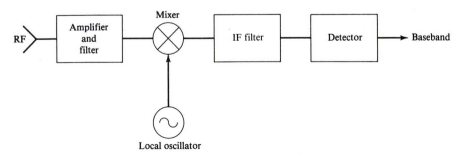

**FIGURE 1.5**
A superheterodyne receiver.

example, the receiver designed to select the difference frequency at the output of the first mixer. That is,

$$f_{IF} = |f_o - f_c| \qquad \text{down conversion}$$

or
$$f_{IF} = f_o + f_c \qquad \text{up conversion.}$$

There exists another signal frequency $f_{IM}$, referred to as the *image frequency,* which when mixed with the local oscillator frequency $f_o$ will produce a signal at the IF frequency. If $f_{IF} = |f_o - f_c|$, then $f_{IM} = f_o + f_{IF}$, or

$$f_{IM} = f_{IF} - f_o = 2f_{IF} + f_c$$

EXAMPLE 1.4. Consider a receiver with the IF filter centered at 455 kHz. If it is desired to receive a 1-MHz input signal, the local oscillator is tuned to 1.455 MHz. Then an image frequency

$$f_{IM} = f_{IF} + f_o \qquad \text{or} \qquad f_{IM} = f_o + 2f_{IF} = 1.91 \text{ MHz}$$

if present at the input, would also pass through the IF filter.

There is no way to separate the desired signal from a signal at the image frequency after they have entered the mixer. The image frequency signal must be removed before it arrives at the mixer. This can be accomplished by adding an image suppression filter (called a *preselector*) before the mixer. For a receiver designed to cover a band of frequencies, the preselector must be tunable. Tunable filters tend to be complex and represent a significant portion of the cost of receiver construction. The majority of receivers do include a preselector.

In Example 1.4, the filter center intermediate frequency was lower than the input signal frequency, and the input frequency was shifted down to the intermediate frequency. The intermediate center frequency can also be selected above the input signal frequency (up conversion). The following example illustrates the advantages of up conversion.

EXAMPLE 1.5. Consider again the receiver that is designed to cover the frequency band of 1 to 30 MHz, but which uses an IF filter centered at 40 MHz. For an input signal frequency $f_s$ of 1 MHz, there are two local oscillator frequencies $f_o$ (41 and 39 MHz) that will result in a mixer product at 40 MHz. If the local oscillator frequency is 41 MHz, the image frequency will be 81 MHz; and if the local oscillator frequency is 39 MHz, the image frequency will be 79 MHz. Table 1.1 lists the local oscillator frequency and corresponding image frequency for several input frequencies spanning the frequency band to be covered. Either set of local oscillator frequencies

TABLE 1.1
**Local oscillator frequency and corresponding image frequency, MHz**

| $f_s$ | $f_o$ | $f_{IM}$ | $f_o'$ | $f_{IM}'$ |
|-------|-------|----------|--------|-----------|
| 1     | 41    | 81       | 39     | 79        |
| 2     | 42    | 82       | 38     | 78        |
| 10    | 50    | 90       | 30     | 70        |
| 30    | 70    | 110      | 10     | 50        |

could be selected, but normally the first set, $f_o$, would be used, since the ratio of the highest to the lowest frequency, $70:41$, is lower than that used for $f'_o$, which is $39:10$. In the design and construction of variable-frequency oscillators, the ratio of highest to lowest frequencies is an important factor (the lower the ratio, the simpler the design).

One important feature of the up conversion technique is that all image frequencies lie above the frequency band to be covered. This implies that all image frequencies can be suppressed by adding a low-pass filter to the input (30-MHz bandwidth); a tunable bandpass filter is not needed with up conversion. Until recently it was not possible to use up conversion in this frequency range because high-quality bandpass filters were not readily available in the 30- to 50-MHz region; however, recent improvements in the manufacturing technology now provide for high-quality crystal filters in this frequency range.

Another advantage of up conversion is that the oscillator-tuning ratio $f_{max}/f_{min}$ is less than that for down conversion. If a down conversion receiver with a 455-kHz IF is used to cover the same frequency band, the local oscillator frequency will need to vary from $f_{min} = 1.455$ MHz to $f_{max} = 30.455$ MHz, a tuning range of $20.93:1$.

### A Modern Communications Receiver

A block diagram of the high-frequency section of a modern radio receiver is essentially the same as that of the superheterodyne receiver illustrated in Fig. 1.5, but the circuits differ from the earlier models. One difference is that up conversion is often used in high-quality receivers so that the input filter can remain a relatively simple low-pass filter that need not be tuned. Whether an amplifier is required before the mixer depends on the particular application and the receiver specifications. It will be shown in Chap. 3 that the absence of this amplifier can actually improve receiver performance in many applications.

Modern receivers differ from older receivers in many ways. A main difference is the frequency synthesizer used to generate the frequencies needed from the variable-frequency oscillator. The frequency synthesizer is capable of generating a large number of relatively precise frequencies from a single reference frequency. Although the receiver does contain at least two oscillators, it has less frequency drift and noise than the older, conventional variable-frequency oscillators. Today the output frequency of frequency synthesizers can be precisely controlled using digital circuitry. This makes possible the microprocessor control of radio receivers and spectrum analyzers. Frequency synthesizers are described in detail in Chap. 10. Since most synthesizers now incorporate a phase-locked loop, a thorough understanding of phase-locked loop characteristics is important for frequency synthesizer design. This information is presented in Chaps. 8 and 9.

### Direct Conversion Receivers

An immediate extension of the superheterodyne receiver is the direct conversion receiver in which the IF section is eliminated by converting the input signal directly

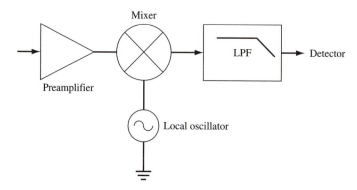

**FIGURE 1.6**
A phase-coherent direct conversion receiver.

to direct current baseband. Such a receiver is shown in Fig. 1.6. The local oscillator is set at the same frequency as the input carrier frequency. The mixer output then contains the baseband signal and a signal at twice the carrier frequency. The higher-frequency signal is then removed with a low-pass filter. An advantage of the direct conversion receiver is that it is much easier to build a low-pass filter than a narrow-band intermediate-frequency filter. Narrowband IF filters require more components and consume more power than do low-pass filters. The elimination of the intermediate frequency results in the direct conversion receiver's frequently being referred to as a zero IF or ZIFF receiver. One of the problems limiting the application of direct conversion receivers is local oscillator drift. And dc offsets are another problem. A problem for direct conversion receivers is local oscillator leakage back into the input stage from which it again mixes with the local oscillator output, creating an erroneous dc output. These problems are being reduced with improvements in modern components, and the direct conversion receiver is finding application in many battery-operated systems.

A major problem with the direct conversion method is created by the phase uncertainty $\Theta$ between the received signal and the local oscillator. (The phase difference is known in coherent detection schemes, but this requires some means of alerting the receiver of this phase difference.) If a constant-amplitude signal is sent, the dc component at the mixer output will be $K \cos \Theta$, where $\theta$ is the phase difference. This phase difference creates an error in the estimation of the input waveform amplitude. If the local oscillator is in phase synchronization with the input signal, the receiver is known as a homodyne receiver.

Figure 1.7 shows a modern direct conversion receiver used for the more common application in which the local oscillator is not phase-synchronized to the input signal. The phase uncertainty problem is solved by using an in-phase and a quadrature channel created by using the local oscillator signal and the local oscillator signal shifted by $90°$. The outputs of the two quadrature channels are then combined to recover the input signal (another problem). This topology is rather easily implemented in integrated circuits, and the ZIFF receiver is being used in many battery-operated applications, particularly those using some variant of digital modulation such as QPSK. The integrated-circuit realization minimizes the

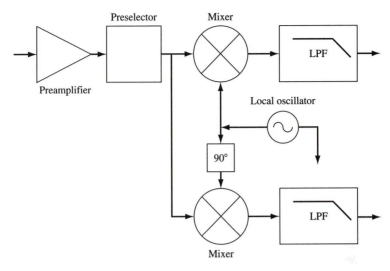

**FIGURE 1.7**
A modern direct conversion receiver.

problems associated with amplitude and phase imbalances between the two receiver sections. It is relatively easy to build such a phase shifter at a particular frequency, but most receivers must cover a band of frequencies. It is difficult to realize a 90° phase shifter over a wide bandwidth. An oscillator that does generate the in-phase and quadrature components of the oscillator signal is referred to as a *quadrature oscillator*. Quadrature oscillators are an important topic in contemporary receiver research and development. Any components of the ZIFF receiver, particularly the sections after the mixers, are easily implemented using a digital signal processor. A digital receiver usually refers not to a receiver of digital signals, but to one in which a significant part of the reception and signal extraction is done digitally.

There is no image frequency in the direct conversion receiver, and this is perhaps the major advantage of this receiver topology. AM receivers have an intermediate frequency of 455 kHz; broadcast FM uses a frequency of 10.7 MHz. This intermediate frequency is selected so that the image frequencies lie outside the broadcast band; NTSC television uses a 43.5-MHz intermediate frequency, and analog cellular telephones use a 90-MHz intermediate frequency.

## An Integrated-Circuit FM Receiver

Integrated-circuit technology has made it possible to design a radio with a minimum of external components. Figure 1.8 contains a simplified block diagram of the Siemans 5469, a high-frequency integrated circuit with a high level of system integration for FM communication. This bipolar device contains a complete FM receiver for input frequencies up to 50 MHz. The device includes (1) an RF input amplifier,

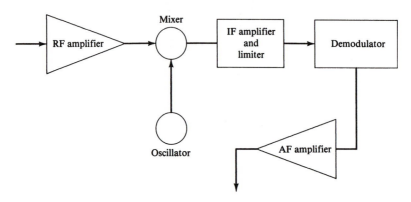

**FIGURE 1.8**
Block diagram of an integrated-circuit FM receiver.

(2) an oscillator, (3) a mixer, (4) an intermediate-frequency (IF) amplifier and limiter, (5) a coincidence demodulator, and (6) audio-frequency (AF) amplifiers with mute and volume control.

A complete FM system can be constructed with this one integrated circuit plus a crystal, a minimum of two inductors, and several resistors and capacitors used for frequency trimming and gain adjustment. The external resistors and capacitors provide the circuit with additional flexibility. Although the resistors and capacitors could be included in the IC, an inductor-capacitor tuned circuit cannot be fabricated in an IC chip; so a complete, self-contained IC receiver or transceiver (combined transceiver and receiver) is not possible with the current IC technology.

## ■ 1.5

### TRANSMITTERS

The transmitter modulates the information to be communicated onto a carrier, amplifies the waveform to the desired power level, and delivers it to the transmitting antenna. The elements of a transmitter are illustrated in Fig. 1.9. The transmitter includes a radio-frequency oscillator that is modulated by the message signal. The modulated signal is then multiplied in frequency up to the desired transmitting frequency and is amplified to the desired power level in the power amplifier. The first radio transmitters worked by charging two electrodes separated by a gap. When the charge became sufficiently large, a spark was created across the gap and electric energy was radiated. However, the spark-gap transmitter was slow, and it was difficult to accurately control the waveform and frequency of oscillation. The spark-gap transmitter became obsolete with the development of the electronic oscillator and vacuum tubes that could handle large amounts of power.

The transmitter topology illustrated in Fig. 1.9 is only one of many types. The modulation can actually take place in the power amplifier. Transmitter topology depends on the type of modulation used and the necessary power level. Narrowband transmitters usually employ pulse, amplitude, or frequency modulation. Wideband

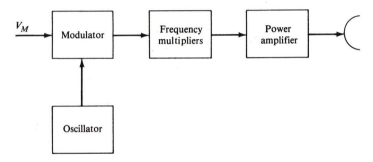

**FIGURE 1.9**
Block diagram of a transmitter.

transmitters use single-sided or multimode modulation and are used for long-range military, marine, aircraft, and amateur communications. Many transmitters and receiver circuits are similar; both require low-noise amplifiers and oscillators. Receivers are designed for the minimum detectable signal, while output power is of prime importance in the transmission of signals. The RF power amplifier is described in Chap. 11.

## ◼ 1.6
## PROBLEMS

**1.1** Determine the transfer function $V_o(s)/I(s)$ of the circuit shown in Fig. P1.1.

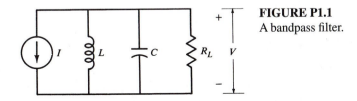

**FIGURE P1.1**
A bandpass filter.

**1.2** Calculate the poles and zeros of the circuit in Fig. P1.1 for $L = 1$ H, $C = 15$ F, and $R = 1\ \Omega$.

**1.3** A receiver is to be designed to cover the frequencies of 20 to 40 MHz using an IF filter centered at 10 MHz. Specify two different local oscillator frequencies for each input frequency, and determine the corresponding image frequency for each input frequency.

**1.4** Select the local oscillator frequencies and specify the preselector frequency response for an up conversion receiver covering the 2- to 30-MHz frequency range. The center frequency of the IF filter is 50 MHz.

**1.5** Figure P1.5 illustrates a direct conversion receiver in which the input signal is converted directly to an audio signal. Specify the local oscillator frequencies and the corresponding image frequencies for a direct conversion receiver covering the 2- to 30-MHz range.

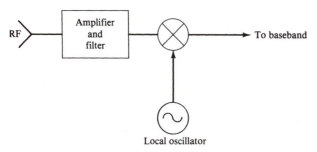

**FIGURE P1.5**
A direct conversion receiver.

**1.6** The double-conversion receiver in Fig. P1.6 employs two IF filters. Specify the required local oscillator frequencies ($f_1$ and $f_2$) for a receiver covering the 2- to 30-MHz range. The center frequency of the first IF filter is 50 MHz, and that of the second IF filter is 10 MHz.

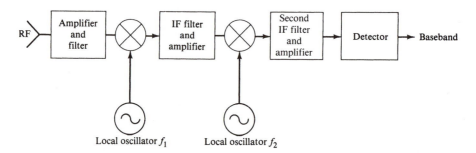

**FIGURE P1.6**

**1.7** NTSC television uses a 41.25-MHz intermediate frequency for the sound carrier and 45.75 MHz for the picture carrier. The local oscillator operates at 45.75 MHz above the desired incoming picture frequency. The VHF band has channels 2 to 6 covering 54 to 88 MHz and channels 7 to 13 covering 174 to 216 MHz. Determine the local oscillator frequency and image frequency for each VHF channel. The bandwidth of each channel is 6 MHz.

**\*1.8** For the circuit shown in Fig. 1.1 plot the input impedance and voltage transfer function as a function of frequency for $L = 10$ mH, $C = 0.01$ $\mu$F, and $R = 1$ k$\Omega$. What is the circuit's $-3$-dB bandwidth? Determine the circuit's rise time and peak overshoot to a unit step input of voltage.

---

*Problems with an asterisk are intended for solution using computer simulation (SPICE or PSPICE).

## ■ 1.7

### ADDITIONAL READING

Abidi, A. A.: "Direct-Conversion Radio Transceivers for Digital Communications," 1995
IEEE International Solid State Circuits Conference, 1995, p. 186.

Armstrong, F. H.: A New System of Short-Wave Amplification, *Proc. Radio Eng.* **9:**3–27,
1919.

Brannon, B., Wide-Dynamic-Range A/D Converters Pave the Way for Wideband Digital-
Radio Receivers, *EDN,* November 7, 1966, pp. 187–205.

Fessenden, R. H.: Wireless Telephony, *Am. Inst. Elect. Eng.,* **27:**553–629, 1908.

Lessing, L.: *Man of High Fidelity: Edwin Howard Armstrong,* Lippincott, Philadelphia,
1956.

Rohde, U. L., and T. T. N. Bucher: *Communication Receivers: Principles and Design,*
McGraw-Hill, New York, 1988.

Terman, F. F.: *Electronic and Radio Engineering,* 4th ed., McGraw-Hill, New York, 1955.

Tucker, D. G.: The History of the Homodyne and Synchrodyne, *J. British Inst. of Radio
Engineers,* **14**(4): 143–154, 1954.

# 2

# Small-Signal Amplifiers

## INTRODUCTION

Most amplifiers used in communications circuits can be considered small-signal amplifiers. These are amplifiers in which the input and output signals are sufficiently small so that the amplifier performance is described with linear equations. In this chapter we first discuss the low-frequency small-signal models for bipolar and field-effect transistor (FET) amplifiers. The hybrid-$\pi$ model is used to describe the bipolar transistor since it is the easiest to analyze. It will be shown that the same small-signal model can be used for the FET. (It will be shown in Chap. 5 that this model can be readily extended for the evaluation of high-frequency amplifier performance.) The importance of push-pull and operational amplifiers will be outlined here, as will the importance of the versatile differential amplifier used in the majority of linear integrated circuits. Operational-amplifier circuits will be discussed, and we will look at recent advances in integrated-circuit fabrication that have extended the gain-bandwidth product (a term defined in this chapter) of the operational amplifier to the extent that operational amplifiers now find many applications in high-frequency circuits.

■ 2.2

## BIPOLAR TRANSISTOR AMPLIFIERS

### Equivalent Circuits

In the hybrid-$\pi$ low-frequency equivalent circuit for the bipolar transistor (Fig. 2.1), terminal $b'$ represents the base junction and $b$, the base terminal. The resistor $r_b'$ connected between these two terminals is usually considered a constant in the range of 10 to 50 $\Omega$. The resistor $r_\pi$ is the base-emitter junction resistance and is usually

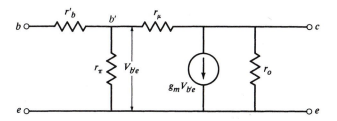

**FIGURE 2.1**
A small-signal, midfrequency equivalent-circuit model of a
bipolar transistor.

much larger than $r'_b$. A useful estimate of $r_\pi$ is given by the expression[1]

$$r_\pi = \frac{kT}{q} \frac{\beta}{I_C} = \frac{0.026\beta}{I_C} \qquad (2.1)$$

where    $\beta$ = transistor base-to-collector current gain
  $I_C$ = dc bias in collector
  $q$ = charge on an electron
  $k$ = Boltzmann's constant
  $T$ = the temperature

At room temperature $T = 290$ K, and $kT/q = 0.026$ V. The base-emitter resistance $r_\pi$ is inversely proportional to the collector bias current.

**EXAMPLE 2.1.** A 2N3904 transistor has a current gain $\beta$ of 100 and is biased so that the quiescent collector current is $10^{-3}$ A. What is the transistor base-emitter resistance?

*Solution.* From Eq. (2.1)

$$r_\pi = \frac{0.026(100)}{10^{-3}} = 2600 \ \Omega$$

The collector-to-emitter resistance $r_o$ and the collector-to-base resistance $r_u$ are also inversely proportional to the collector direct current. A typical value for $r_o$ is 15 k$\Omega$. And $r_u$ is a large resistance, on the order of megohms, used to model basewidth modulation effects in the transistor. Here $r_u$ is assumed to be an open circuit in all cases considered in this text. (It is of considerable importance in high-voltage-gain amplifiers with very large values of load impedance, but these applications are not normally found in high-frequency circuits.)

The transistor transconductance $g_m$ is determined from the formula

$$g_m r_\pi = \beta \qquad (2.2)$$

or

$$g_m = \frac{qI_C}{kT} \approx 40I_C \qquad (2.3)$$

The transconductance $g_m$ is directly proportional to the collector direct current.

Although the circuit appears rather complicated, under most conditions the resistor $r'_b$ can also be neglected. The equivalent circuit is then as shown in Fig. 2.2. The

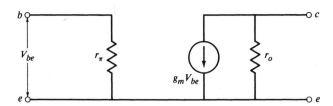

**FIGURE 2.2**
A simplified small-signal, midfrequency equivalent-circuit
model of a bipolar transistor.

model now consists of three independent parameters $g_m$, $r_\pi$, and $r_o$. And $g_m$ and $r_\pi$ are determined by the collector direct current and the current gain $\beta$ of the transistor.

**Common-Emitter Amplifier**

In order to analyze the small-signal behavior of a linear amplifier such as the common-emitter amplifier illustrated in Fig. 2.3a, the transistor is replaced by the small-signal model, and the complete equivalent circuit becomes as shown in Fig. 2.3b. If the coupling capacitor is assumed to be a short circuit, the base voltage $V$ is given by

$$V = \frac{R}{R + R_s} V_i \tag{2.4}$$

where $R$ is the equivalent resistance of $R_1$, $R_2$, and $r_\pi$ connected in parallel, or

$$R = \frac{R_1 R_2 r_\pi}{R_1 R_2 + r_\pi R_1 + r_\pi R_2} \tag{2.5}$$

The output voltage is given as

$$V_o = \frac{-g_m R_L r_o V}{r_o + R_L} = \frac{-g_m R_L r_o}{r_o + R_L} \frac{R V_i}{R + R_s} \tag{2.6}$$

and the midfrequency voltage gain is given as

$$A_v = \frac{V_o}{V_i} = \frac{-g_m R_L r_o}{r_o + R_L} \frac{R}{R + R_s} \tag{2.7}$$

The phase shift of the midfrequency voltage gain of the common-emitter amplifier is $180°$. The input impedance of the amplifier is by definition

$$Z_i = \frac{\Delta V_i}{\Delta I_i} \tag{2.8}$$

An increment of voltage is applied, and the change in input current is measured (assuming that all other independent sources remain constant) as shown in Fig. 2.4. In this figure, the source resistance is not included in determining the input impedance of the amplifier. For this circuit

$$Z_i = R_i = \frac{R_1 R_2 r_\pi}{R_1 R_2 + R_1 r_\pi + R_2 r_\pi} \tag{2.9}$$

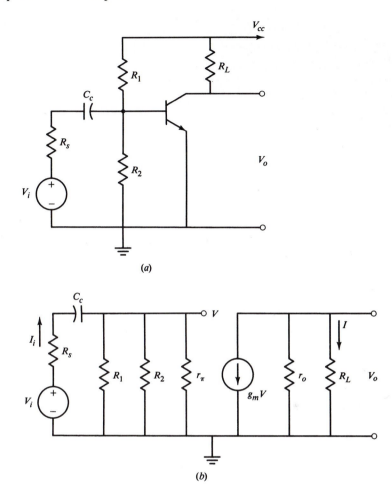

(a)

(b)

**FIGURE 2.3**
(a) A common-emitter amplifier; (b) small-signal, midfrequency equivalent circuit of the amplifier of Fig. 2.3a.

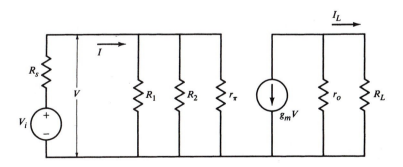

**FIGURE 2.4**
The input impedance of the amplifier circuit is defined as $V/I$.

Since $r_\pi$ depends on the bias direct current, the input impedance will depend on it also. The current gain is the load current $I_L$ divided by the input current, or

$$A_i = \frac{I_L}{I_i} = \frac{V_o/R_L}{V_i/(R_i + R_s)} = \frac{R_s + R_i}{R_L} A_v \tag{2.10}$$

The current gain from the base to the output is $-\beta$, provided that $r_o \gg R_L$. The midfrequency current gain also has a $180°$ phase shift.

Amplifier output impedance is determined by applying a voltage across the output terminals and measuring the change in the output current (with all other independent sources held constant). This is the Thévenin equivalent impedance seen by the load impedance:

$$Z_o = \frac{\Delta V_o}{\Delta I_o} \tag{2.11}$$

The load resistance is usually excluded from the definition, so

$$Z_o = r_o \tag{2.12}$$

for the common-emitter amplifier.

### Common-Base Amplifier

The same small-signal equivalent circuit can also be used for the common-base and common-collector amplifiers. Figure 2.5a shows a common-base amplifier, and Fig. 2.5b shows the midfrequency small-signal circuit. The direction of the dependent current source is now from emitter to collector, since the dependent voltage $V$

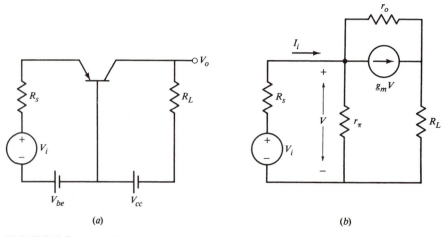

**FIGURE 2.5**
(a) A common-base amplifier; (b) a small-signal, midfrequency equivalent circuit of the common-base amplifier.

has been taken from emitter to base rather than from base to emitter. Since the sum of the currents leaving the emitter junction is zero,

$$\frac{V - V_i}{R_s} + \frac{V}{r_\pi} + \frac{V - V_o}{r_o} + g_m V = 0 \tag{2.13}$$

Also
$$\frac{V - V_o}{r_o} + g_m V = \frac{V_o}{R_L} \tag{2.14}$$

If the dependent voltage $V$ is eliminated from these two equations, we obtain an expression for the voltage gain of the common-base amplifier. This equation is rather complicated, and usually little accuracy is sacrificed by assuming that $r_o$ is large compared with $R_s$, $r_\pi$, and $R_L$. The voltage gain is then

$$A_v = \frac{V_o}{V_i} = \frac{g_m R_L r_\pi}{R_s + r_\pi + g_m R_s r_\pi} = \frac{\beta R_L}{r_\pi + R_s(1 + \beta)} \tag{2.15}$$

(Note that there is no phase inversion in the voltage gain of the common-base amplifier.) If the source impedance is small, $r_\pi \gg R_s(1 + \beta)$, the magnitude of the voltage gain will be the same as that of the common-emitter amplifier.

The input impedance of the common-base amplifier is determined using the simplified circuit of Fig. 2.6. Since

$$I_i = \frac{V_i}{r_\pi} + g_m V_i \tag{2.16}$$

then
$$Z_i = \frac{r_\pi}{1 + g_m r_\pi} = \frac{r_\pi}{1 + \beta} \approx \frac{1}{g_m} \tag{2.17}$$

The input impedance of the common-base amplifier is smaller than that of the common-emitter amplifier. Since it is inversely proportional to $g_m$, it is inversely proportional to the collector direct current. This property is found useful in setting the amplifier impedance to a desired level for impedance matching. Also, since

$$I_L = g_m V_i = g_m I_i Z_i \tag{2.18}$$

the current gain is

$$A_i = \frac{g_m r_\pi}{1 + g_m r_\pi} \approx \frac{\beta}{1 + \beta} \tag{2.19}$$

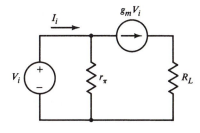

**FIGURE 2.6**
The input impedance of this common-base amplifier equivalent circuits is $V_i / I_i$.

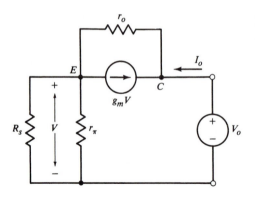

**FIGURE 2.7**
A simplified circuit model for calculating the output impedance of the common-base amplifier.

which is slightly less than 1. The output impedance can be determined from the two node equations (see Fig. 2.7)

$$-I_o = g_m V + \frac{V - V_o}{r_o} \tag{2.20}$$

and

$$\frac{V}{R} + \frac{V - V_o}{r_o} + g_m V = 0 \tag{2.21}$$

where

$$R = \frac{r_\pi R_s}{r_\pi + R_s} \tag{2.22}$$

Therefore, the output impedance is

$$Z_o = r_o + R(1 + g_m r_o) \tag{2.23}$$

which is larger than that of the common-emitter amplifier.

**EXAMPLE 2.2.** Calculate the midband voltage gain, the current gain, and the input impedance of the common-base amplifier shown in Fig. 2.8. The collector bias current is $10^{-3}$ A, and the transistor $\beta = 100$.

**Solution.** The small-signal, midfrequency model is shown in Fig. 2.9. The collector-emitter resistance is assumed to be infinite. Since $I_C = 1$ mA,

$$g_m = 40 I_C = 40 \times 10^{-3} S$$

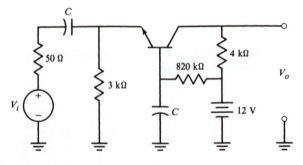

**FIGURE 2.8**
A common-base amplifier.

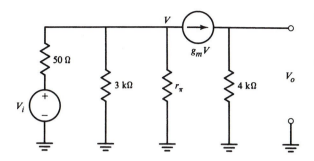

**FIGURE 2.9**
A small-signal equivalent circuit of the amplifier illustrated in Fig. 2.8.

and

$$r_\pi = \frac{\beta}{g_m} = 2500 \; \Omega$$

Therefore, from Eq. (2.15), the voltage gain is given by

$$A_v = \frac{100(4 \times 10^3)}{2.5 \times 10^3 + 50(101)} = 52.98$$

and the current gain is [using Eq. (2.19)]

$$A_i = \frac{100}{101} = 0.99$$

The input impedance is calculated using Eq. (2.17):

$$Z_i \approx \frac{1}{g_m} = 25 \; \Omega$$

The common-base amplifier has a voltage gain, but the current gain is less than unity. This amplifier is used in applications in which it is desired to build a non-inverting amplifier or an amplifier with a low-input or a high-output impedance. It does have much better high-frequency response than the common-emitter amplifier and is often used in high-frequency circuits. An application of this amplifier is described in the section on multistage amplifiers, and additional examples are provided in Chap. 5.

## Emitter-Follower

The common-collector amplifier, better known as an *emitter-follower,* has various applications. It will be shown that this amplifier has a noninverting voltage gain of less than 1, and a current gain approximately equal to the $\beta$ of the transistor used. Although it has a voltage gain less than 1, it can be combined with another amplifier stage, such as a common-emitter stage, to realize a greater combined voltage gain than could be achieved from the use of a common-emitter stage alone. This is particularly useful when a low-impedance load is used.

An emitter-follower amplifier is illustrated in Fig. 2.10, and the small-signal, midfrequency equivalent circuit is given in Fig. 2.11. It is normally the case that the base-biasing resistor $R_b$ is much larger than the source resistance $R_s$ and can

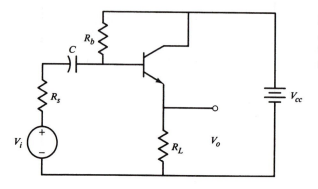

**FIGURE 2.10**
An emitter-follower
(common-collector
amplifier).

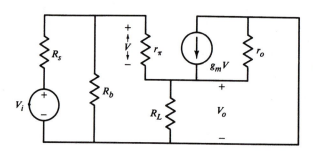

**FIGURE 2.11**
A small-signal equivalent circuit of the emitter-follower
illustrated in Fig. 2.10.

therefore be neglected. If this is the case, the equivalent circuit is as shown in
Fig. 2.12. The voltage gain is determined from three equations:

$$V_i = I_i(R_s + r_\pi) + V_o \tag{2.24}$$

$$V = I_i r_\pi \tag{2.25}$$

and
$$V_o = (I_i + g_m V)\frac{r_o R_L}{r_o + R_L} \tag{2.26}$$

If $V$ and $I_i$ are eliminated, the voltage gain is found to be

$$A_v = \frac{V_o}{V_i} = \frac{(1 + \beta)[r_o R_L/(r_o + R_L)]}{R_s + r_\pi + (1 + \beta)[r_o R_L/(r_o + R_L)]} \tag{2.27}$$

The voltage gain is noninverting (positive) and less than 1. As $R_L$ increases, the
gain approaches 1. The emitter-follower input impedance can be found from
Eqs. (2.24) and (2.27) by eliminating $V_o$ and setting $R_s = 0$. The input impedance is

$$Z_i = r_\pi + \frac{(1 + \beta)r_o R_L}{r_o + R_L} \tag{2.28}$$

If $r_o$ is very large, $Z_i \approx r_\pi + (1 + \beta)R_L$.

The input impedance of the emitter-follower is the largest of the three transistor
amplifier configurations. Actually, it is the same as that of the common-emitter amplifier if the same value of emitter resistance is used for both, but the additional resistance used for the common-emitter amplifier will create additional power dissipation.

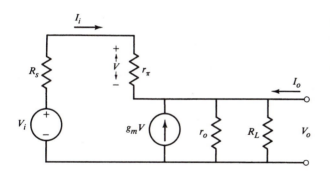

FIGURE 2.12
A simplified small-signal equivalent circuit of the emitter-follower.

The output impedance is determined by applying a voltage to the output terminals and measuring the current $I_o$. The output current is determined from the equation

$$I_o = \frac{V_o}{r_o} + g_m \frac{r_\pi}{r_\pi + R_s} V_o + \frac{V_o}{R_s + r_\pi} \tag{2.29}$$

The output impedance is

$$Z_o = \frac{V_o}{I_o} = \frac{r_o(r_\pi + R_s)}{r_\pi + R_s + r_o(1 + \beta)} \tag{2.30a}$$

Then

$$Z_o \approx \frac{r_\pi + R_s}{1 + \beta} \tag{2.30b}$$

The output impedance of the emitter-follower is the lowest of the three transistor amplifier configurations. Low output impedance is a requirement in many amplifier applications. The emitter-follower is used when a low output impedance is needed.

The emitter-follower current gain is

$$A_i = \frac{I_L}{I_i} = \frac{V_o/R_L}{I_i} = \frac{V_o}{V_i} \frac{R_s + Z_i}{R_L} \tag{2.31}$$

and by using Eqs. (2.27) and (2.28), the current gain is found to be

$$A_i = \frac{(1 + g_m r_\pi)r_o}{r_o + R_L} \approx 1 + \beta \tag{2.32}$$

Although the emitter-follower has a voltage gain less than 1, it has a large current gain. It is also frequently used as a power amplifier for low-impedance loads.

**EXAMPLE 2.3.** Calculate the power gain of the common-collector amplifier shown in Fig. 2.13$a$. The transistor is a 2N5901 with $\beta = 40$. The transistor output impedance is large and can be neglected. The collector direct current is 40 mA.

*Solution.* The small-signal, midfrequency equivalent circuit is shown in Fig. 2.13$b$ ($r_o$ has been neglected). Since the 100-$\Omega$ resistor, which is used for biasing, is in series with the dependent current source, it has no effect on the output signal. The collector

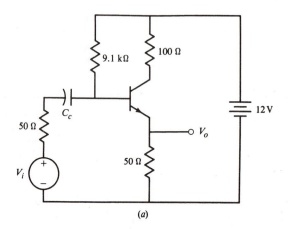

**FIGURE 2.13**
(*a*) An emitter-follower; (*b*) a
simplified equivalent circuit
of the emitter-follower.

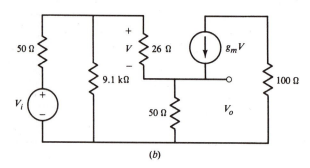

bias current $I_C = 40 \times 10^{-3}$, and $r_\pi$ is [Eq. (2.1)]

$$r_\pi \approx \frac{26(40)}{40} = 26 \ \Omega$$

The voltage gain is [using Eq. (2.27) and neglecting the 9.1-k$\Omega$ bias resistor]

$$A_v = \frac{\beta R_L}{R_s + r_\pi + (1 + \beta)R_L} = \frac{40(50)}{50 + 26 + 41(50)} = 0.94$$

The current gain is [Eq. (2.31)]

$$A_i \approx \beta + 1 = 41$$

The input and load impedances are real, so the voltages and currents are in phase and
the power gain is

$$A_p = A_i A_v = 38.54$$

The amplifier output impedance as seen by the 50-$\Omega$ resistor is

$$Z_o = \frac{26 + 50}{41} = 1.85 \ \Omega$$

## ◼ 2.3

### FIELD-EFFECT TRANSISTOR AMPLIFIERS

### Equivalent Circuits

Symbolic representations of three types of field-effect transistors (FETs) are given in Fig. 2.14. Figure 2.14a illustrates a JFET (junction field-effect transistor). The off-center location of the arrow is used if the drain-to-source channel of the particular device is asymmetric so that the drain and source cannot be interchanged. Many JFETs do have a symmetric channel. A depletion-type MOSFET (insulated-gate FET) is illustrated in Fig. 2.14b, and an enhancement-type MOSFET is illustrated in Fig. 2.14c. The additional terminal $U$ refers to the substrate (body) of the device and is usually connected to the source. If the gate arrow points toward the device, it is an $n$-channel device, while in $p$-channel FETs the arrow points out. The important signals are the gate, source, and drain currents ($I_g$, $I_s$, and $I_d$, respectively); the gate-to-source voltage $V_{gs}$; and the drain-to-source voltage $V_{ds}$. The small-signal, midfrequency model of this device (ignoring the dc components) is shown in Fig. 2.15.

For all practical purposes, the input impedance of an FET is so large that the input can be considered an open circuit in the midfrequency range, and the gate current $I_g = 0$. The small-signal model then consists only of a voltage-dependent current source whose value is proportional to the difference between the gate and

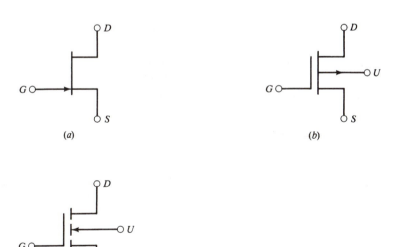

**FIGURE 2.14**
Circuit models for field-effect transistors: (a) a JFET; (b) a depletion-type MOSFET; (c) an enhancement-type MOSFET.

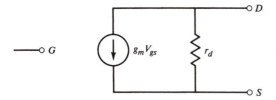

**FIGURE 2.15**
A small-signal low-frequency equivalent circuit for a field-effect transistor.

source small-signal voltages and a drain-to-source resistance $r_d$. The same model is used for both junction and insulated-gate FETs. Only the relation between $g_m$ and the dc biasing varies for the different devices. By definition, the transistor's transconductance relates the change in drain current to the change in gate-to-source voltage. That is,

$$g_m = \frac{dI_d}{dV_{gs}} \tag{2.33}$$

For JFETs it is really shown that

$$g_m = g_{mo} \left( \frac{I_D}{I_{DSS}} \right)^{1/2} \tag{2.34}$$

where $g_{mo}$ is the transconductance when gate-to-source bias voltage is zero.[2] $I_D$ is the drain direct current, and $I_{DSS}$ is the drain current when the gate-to-source voltage is zero. The transconductance $g_m$ is often referred to by the $y$ parameter symbol $y_{fs}$ on data sheets. For an MOS transistor, $g_m$ is given by

$$g_m = g_{mR} \left( \frac{I_D}{I_{DR}} \right)^{1/2} \tag{2.35}$$

where $g_{mR}$ is the transconductance at some specified drain bias current $I_{DR}$. In both transistor types, the drain current varies proportionally to the square root of the bias direct current.

## Common-Source Amplifier

The common-source amplifier is similar to the common-emitter amplifier. The midfrequency equivalent circuit of the common-source amplifier shown in Fig. 2.16a is given in Fig. 2.16b. Normally $R_g \gg R$, so $V_i \approx V_g$. The dependent current source $g_m V_{gs}$ depends on both the gate and the source voltages. The gate voltage is known ($V_g = V_i$), but the source voltage must be determined from the following equations:

$$\frac{V_o - V_s}{r_d} + \frac{V_o}{R_L} + g_m V_{gs} = 0 \tag{2.36}$$

and

$$\frac{V_s}{R_s} + \frac{V_s - V_o}{r_d} = g_m V_{gs} \tag{2.37}$$

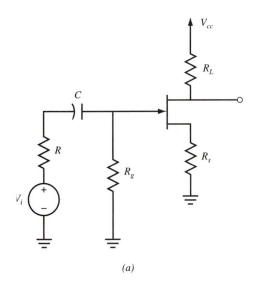

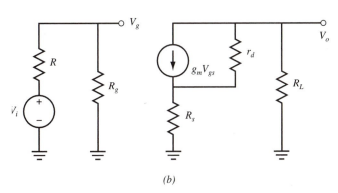

**FIGURE 2.16**
(*a*) A common-source
amplifier; (*b*) an equivalent
circuit for the common-
source amplifier.

(a)

(b)

The source voltage, in terms of the output voltage, is

$$V_s = \frac{-V_o R_s}{R_L} \tag{2.38}$$

and the voltage gain is

$$\frac{V_o}{V_i} = \frac{-g_m R_L r_d}{R_L + r_d + R_s(1 + g_m r_d)} \tag{2.39}$$

If the transistor output resistance $r_d$ is much larger than $R_s$ and $R_L$, as it normally will be, the voltage-gain expression simplifies to

$$\frac{V_o}{V_i} = \frac{-g_m R_L}{1 + g_m R_s} \cdot \tag{2.40}$$

and if $g_m R_s \gg 1$,

$$\frac{V_o}{V_i} = \frac{-R_L}{R_s} \tag{2.41}$$

The common-source amplifier with a source resistance can be used to design an inverting amplifier whose gain is independent of the transistor (provided $g_m R_s \gg 1$). This is a particularly valuable design procedure since it eliminates the costly process of having to carefully select a transistor.

Since the input impedance of the amplifier is very large, the current gain is also very large (for practical purposes, it is infinite). If $R_s = 0$, the output impedance is given by

$$Z_o = r_d \tag{2.42}$$

It is left as an exercise to calculate the output impedance for a nonzero $R_s$.

**EXAMPLE 2.4.** Figure 2.17$a$ contains an FET amplifier, and Fig. 2.17$b$ is the small-signal, midfrequency equivalent-circuit model. The coupling and bypass capacitors are assumed to act as short circuits in this frequency range. The manufacturer's data sheet for the 2N5486 gives the following minimum parameter values for $V_{DS} = 15$ V: $I_{DSS} = 8$ mA, $g_{mo} = 4 \times 10^{-3}$ S, and $r_d = 13$ kΩ. And $R_s$ is selected so that the amplifier is biased with $I_D = 2.0$ mA. What is the midband voltage gain of the amplifier?

**Solution.** From the equivalent circuit we see that

$$V_g = V_i \frac{10^6}{10^6 + 5 \times 10^3} = V_i$$

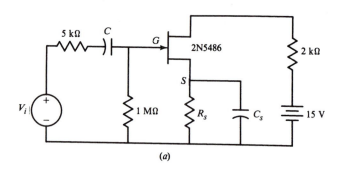

(a)

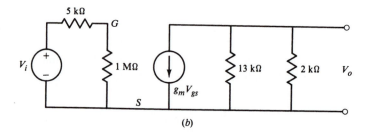

(b)

**FIGURE 2.17**
($a$) A common-source amplifier; ($b$) an equivalent circuit for the amplifier shown in ($a$).

Since the source is grounded, $V_s = 0$ and

$$V_{gs} = V_g - V_s = V_i$$

The equivalent load resistance is

$$R'_L = \frac{2\ \text{k}\Omega \times 13\ \text{k}\Omega}{2\ \text{k}\Omega + 13\ \text{k}\Omega} = 1.73\ \text{k}\Omega$$

and the output voltage is

$$V_o = -g_m V_{gs} R'_L = -g_m V_i R'_L$$

To determine the voltage gain, the transconductance must first be determined. From Eq. (2.34) the transconductance is

$$g_m = 4 \times 10^{-3} \left(\frac{2}{8}\right)^{1/2} = 2 \times 10^{-3}\ \text{S}$$

Therefore, the voltage gain is

$$\frac{V_o}{V_i} = -g_m R'_L = -3.46$$

### Source-Follower

The same equivalent circuit is used for the FET whether it is used in the common-source, common-gate, or common-drain configuration. Figure 2.18a illustrates a common-drain or source-follower circuit, and the small-signal, midfrequency equivalent circuit is given in Fig. 2.18b. If, as will usually be the case, $R_b \gg R_s$, then

$$V_g = V_i$$

and

$$V_{gs} = V_i - V_o \tag{2.43}$$

Also

$$V_o = g_m V_{gs} \frac{r_d R_L}{r_d + R_L} \tag{2.44}$$

so that

$$A_v = \frac{V_o}{V_i} = \frac{g_m[r_d R_L/(r_d + R_L)]}{1 + g_m[r_d R_L/(r_d + R_L)]} \tag{2.45}$$

which is a noninverting voltage gain less than 1. Another important parameter of this amplifier is the output impedance $Z_o$. By definition, from Eq. (2.11).

$$Z_o = \frac{\Delta V_o}{\Delta I_o}$$

with all over independent voltage and current sources held constant. For this amplifier, the output impedance can be determined by calculating the response $I_o$ to a voltage $V_o$, as illustrated in Fig. 2.19. Since $V_g = 0$, $V_{gs} = -V_o$, and

$$I_o = g_m V_o + \frac{V_o}{r_d} \tag{2.46}$$

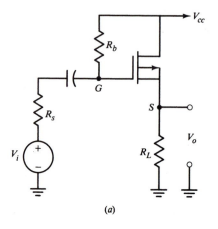

**FIGURE 2.18**
(*a*) A MOSFET source-follower;
(*b*) a low-frequency equivalent
circuit of the amplifier shown in (*a*).

(*a*)

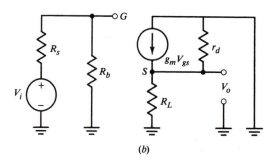

(*b*)

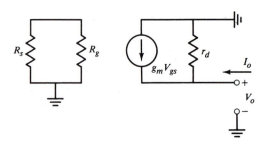

**FIGURE 2.19**
A simplified equivalent circuit for
measuring the output impedance of a
source-follower.

The output impedance is then

$$Z_o = \frac{r_d}{1 + g_m r_d} \tag{2.47}$$

The output impedance of the source-follower, like that of the emitter-follower, is much smaller than those of the other two FET amplifier configurations. This is the principal reason for using this amplifier configuration.

**EXAMPLE 2.5.** The circuit shown in Fig. 2.20 is used as an active antenna. Calculate the voltage gain for a 1-MHz input signal. The transconductance $g_m = 60 \times 10^{-3}$ S. The antenna source impedance is 50 $\Omega$.

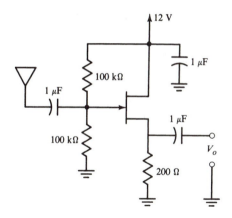

**FIGURE 2.20**
A source-follower used as an active antenna.

**Solution.** The midfrequency small-signal equivalent circuit for this amplifier is shown in Fig. 2.21. Since the source resistance is much smaller than the parallel combination of the two bias resistors shunting the gate, the gate voltage is

$$V_g = V_i$$

and
$$V_{gs} = V_i - V_o$$

A typical value for $r_d$ is 2.5 k$\Omega$, in which case the output impedance is

$$\frac{r_d}{1 + g_m r_d} = \frac{2.5 \times 10^3}{1 + 60(2.5)} = 16.6 \ \Omega$$

If the 200-$\Omega$ load is considered part of the amplifier, then the output impedance of the active antenna is

$$Z_o = \frac{200(16.6)}{216.6} = 15.3 \ \Omega$$

The voltage gain is found from Eq. (2.45) to be

$$A_v = \frac{(60 \times 10^{-3})(185)}{1 + (60 \times 10^{-3})(185)} = 0.92$$

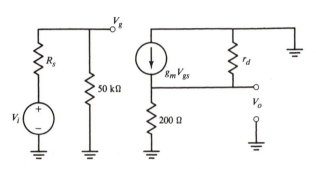

**FIGURE 2.21**
A small-signal equivalent circuit of the source-follower shown in Fig. 2.20.

**Common-Gate Amplifier**

The common-gate amplifier is often used in high-frequency applications. It will be shown in Chap. 5 that its bandwidth is much greater than that of the common-source amplifier. Another reason for its use is that the circuit configuration offers low input impedance, which is convenient for matching to transmission lines and other low-impedance sources. The basic circuit configuration is shown in Fig. 2.22a, and the small-signal, midfrequency equivalent circuit is given in Fig. 2.22b. Since the gate is grounded,

$$V_{gs} = -V_s \tag{2.48}$$

and the circuit can be redrawn as in Fig. 2.22c. If the currents are summed at the output node so that

$$\frac{V_o}{R_L} + \frac{V_o - V_s}{r_d} - g_m V_s = 0 \tag{2.49}$$

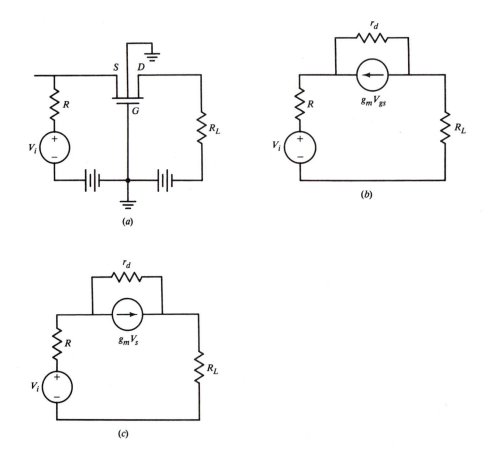

**FIGURE 2.22**
(a) A common-gate amplifier; (b) a simplified equivalent circuit for the common-gate amplifier; (c) another amplified circuit for the same amplifier.

we find that

$$\frac{V_o}{V_s} = \frac{R_L(1 + g_m r_d)}{r_d + R_L} \tag{2.50}$$

and

$$\frac{V_o}{V_s} \approx g_m R_L \tag{2.51}$$

for large $r_d \gg R_L$.

The voltage gain is noninverting, and for load resistances much less than the dynamic drain resistance, the magnitude of the voltage gain is approximately the same as that of the common-source amplifier. The current gain is close to 1.

**Input Impedance**

The input impedance at the source can be found by solving for the source current:

$$I_s = g_m V_s + \frac{V_s - V_o}{r_d} \tag{2.52}$$

After Eq. (2.50) is used to eliminate $V_o$, we obtain

$$Z_i = \frac{V_s}{I_s} = \frac{r_d + R_L}{1 + g_m r_d} \tag{2.53}$$

If $r_d \gg R_L$,

$$Z_i \approx (g_m)^{-1} \tag{2.54}$$

The common-gate amplifier has a low input impedance which is inversely proportional to the square root of the drain direct current.

**Voltage Gain**

Since

$$V_s = \frac{V_i Z_i}{Z_i + R} \tag{2.55}$$

the voltage gain is determined by combining Eqs. (2.50) and (2.53):

$$A_v = \frac{V_o}{V_i} = \frac{R_L}{R + (r_d + R_L)/(1 + g_m r_d)} \tag{2.56}$$

**Output Impedance**

The output impedance can be found by applying a voltage to the output terminals and determining the output current, as shown in Fig. 2.23:

$$I_o = \frac{V_o - V_s}{r_d} - g_m V_s \tag{2.57}$$

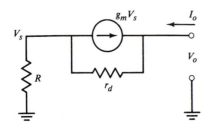

**FIGURE 2.23**
A circuit for determining the output impedance of the common-gate amplifier.

but $I_o$ must also be the current through the source resistance $R$, so

$$I_o = \frac{V_s}{R} \tag{2.58}$$

Combining these two equations, we find that

$$Z_o = \frac{V_o}{I_o} = r_d + (1 + g_m r_d)R \tag{2.59}$$

The common-gate amplifier has the largest output impedance of the three FET amplifier configurations.

## ▪ 2.4

### MULTISTAGE AMPLIFIERS

It is often necessary to use more than one amplifier stage for impedance matching or to obtain additional amplification. Power transistors possess a smaller gain-bandwidth product than low-power transistors; hence, the power amplification stage is often operated near unity voltage gain in order to maximize the bandwidth. The voltage amplification is carried out in the stages preceding the power amplification stage. The following example illustrates the use of an additional stage, with a voltage gain less than unity, to increase the overall voltage gain.

**EXAMPLE 2.6.** Design an amplifier using 2N3904 transistors to realize a voltage-gain magnitude of at least 10. The source resistance is 500 $\Omega$, and the load resistance is 50 $\Omega$.

**Solution.** The small value of the load resistance creates a design problem. If the transistor input impedance is large, the voltage gain will be

$$A_v \approx -g_m R_L$$

Since $R_L = 50$, a voltage gain of 10 can be realized with a common-emitter amplifier, provided $g_m \geq 0.2$ S. The transistor input resistance (assuming $\beta = 100$) will be

$$r_\pi = \frac{\beta}{g_m} \leq 500 \ \Omega$$

Therefore, with a 500-$\Omega$ source resistance and a 500-$\Omega$ amplifier input resistance, only one-half of the signal voltage will appear across the base and the overall amplification will be equal to 5. If $g_m$ is doubled, the transistor input impedance will be reduced to 250 $\Omega$, and only one-third of the applied signal will appear across the base. The overall gain will then be about $20/3$, but the gain specification cannot be met by increasing $I_C$. However, the problem of low load resistance can be solved by adding another stage, which simply increases the load impedance seen by the first stage. The input impedance of the second stage serves as the load impedance for the first stage. Since the input impedance of an emitter-follower from Eq. (2.28) is

$$Z_i \approx r_\pi + (1 + \beta)R_L$$

it is a logical choice for the second stage. For this transistor, $\beta$ is at least 100, so the input impedance will be more than 5 k$\Omega$. It is not a problem to realize a voltage gain of 10 with this large (5k $\Omega$) a load resistance. A complete two-stage amplifier is shown in Fig. 2.24. The resistors $R_{L_1}$, $R_{b_1}$, and $R_{b_2}$ are selected to complete the biasing. Typical values for use with a 12-V supply are

$$R_{L_1} = 2 \text{ k}\Omega \qquad R_{b_1} = 500 \text{ k}\Omega \qquad \text{and} \qquad R_{b_2} = 4.5 \text{ k}\Omega$$

With these values $I_{C_1} \approx 2$ mA and $I_{C_2} \approx 120$ mA. The load resistance seen by the first stage will be the 2-k$\Omega$ resistor in parallel with $R_{b_2}$ and the input impedance of the second stage (approximately 5 k$\Omega$). That is,

$$R'_{L_1} = 1.05 \text{ k}\Omega$$

The voltage gain of the second stage will be close to unity, and the voltage gain of the first stage will be

$$A_{v_1} \approx -g_{m_1} R'_L \frac{r_{\pi_1}}{r_{\pi_1} + R_s}$$

Since

$$g_{m_1} \approx 40 I_{C_1} = 0.08 \text{ S} \qquad \text{and} \qquad r_{\pi_1} = \frac{\beta}{g_{m_1}} = 1250 \ \Omega$$

$$A_{v_1} \approx \frac{-84(1250)}{1250 + 500} = -60$$

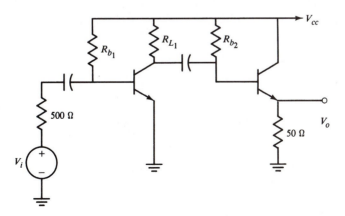

**FIGURE 2.24**
A two-stage amplifier consisting of a common-emitter amplifier followed by an emitter-follower.

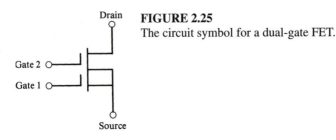

**FIGURE 2.25**
The circuit symbol for a dual-gate FET.

If less gain is desired, the gain can readily be reduced by adding a potentiometer in series with the source resistance.

## ■ 2.5

### DUAL-GATE FET

The FET cascode circuit has so many applications in high-frequency amplifiers that two FETs are often fabricated as a single transistor with two gates. The source of the one transistor is continuous with the drain of the other, so the device has one source, one drain, and two gates; it is referred to as a *dual-gate FET*. The schematic representation of a dual-gate MOSFET is shown in Fig. 2.25. The dual-gate MOSFET offers low noise and high gain in radio-frequency applications. It is a versatile device which can be used as a mixer or automatic gain control (AGC) amplifier, as described in Chap. 5. The construction is such that the capacitance from gate 2 to the drain is very small (approximately 1 percent of that of a single-gate MOSFET). This accounts for the excellent high-frequency performance of the device.

A linear amplifier circuit is shown in Fig. 2.26a, and the small-signal equivalent circuit is shown in Fig. 2.26b (where the dynamic drain resistances have been ignored). Drain 1 and source 2 are shown only as aids in understanding the circuit. Here gate 2 and the source are grounded for medium-frequency signals.

$$V_{D_1} = -g_{m_1} V_{gs_1} R_{L_1}$$

The load resistance of the first stage is the input resistance of the second stage, and the second stage is a common-gate amplifier. Therefore,

$$V_{D_1} = -g_{m_1} V_i \frac{1}{g_{m_2}}$$

since the input impedance of a common-gate amplifier is $-g_{m1}$. Since both transistors have the same drain current,

$$g_{m_1} = g_{m_2} \quad \text{and} \quad V_{D_1} = -V_1$$

Also, since gate 2 is grounded,

$$V_{gs_2} = -V_{s_2} = -V_{D_1} = V_i$$

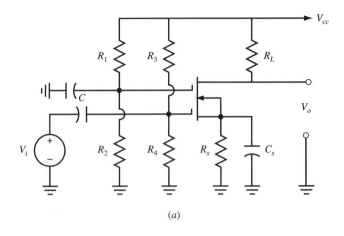

(a)

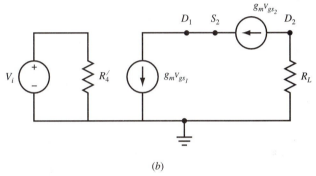

(b)

**FIGURE 2.26**
(a) An amplifier utilizing a dual-gate FET; (b) a simplified
equivalent circuit of the amplifier shown in Fig. 2.26a.

and the output voltage is

$$V_o = -g_m V_{gs_2} R_L = -g_m R_L V_i \tag{2.60}$$

The small-signal gain of the dual-gate MOSFET is equivalent to that of a single-gate MOSFET. The advantage of the dual-gate device for small-signal operation is that the much smaller gate-to-drain capacitance provides a much larger bandwidth than does the standard MOSFET. Also, the transconductance $g_m$, and hence the voltage gain, can be controlled by the bias voltage applied to gate 2.

## 2.6
### PUSH-PULL AMPLIFIERS

Transistors all exhibit a nonlinear characteristic that causes distortion of the input signal even at small-signal levels. Much of the distortion can be eliminated by an amplifier configuration known as the *push-pull amplifier.* An example of such an amplifier is illustrated in Fig. 2.27. This particular circuit uses two center-tapped transformers, one for separating the input signal into two signals 180° out of phase and one for summing the output currents of the two transistors. Circuit characteristics of center-tapped transformers are discussed in Chap. 6. Here it suffices to say that they are wound so that

$$V_1 = K_v V_i = -V_2 \qquad (2.61)$$

and

$$I_o = K_I(I_1 - I_2) \qquad (2.62)$$

Thus when the signal on the drain of $Q_1$ is positive, the signal on the drain of $Q_2$ is of equal magnitude but of opposite sign. The two drain signals are 180° out of phase. If the input signal is sinusoidal

$$V_i = A \cos \omega t$$

then the output currents of the two transistors are also periodic, and they can be expressed in a Fourier series:

$$I_1 = B_0 + B_1 \cos \omega t + B_2 \cos 2\omega t + B \cos 3\omega t + \cdots \qquad (2.63)$$

The higher-frequency components are created by the transistor nonlinearities. If the two transistors and associated components are identical, then $I_2$ and $I_1$ are identical except that $I_2$ lags $I_1$ by 180°. That is,

$$I_2(\omega t) = I_1(\omega t + \pi)$$

Therefore,

$$I_2 = B_0 - B_1 \cos \omega t + B_2 \cos 2\omega t - B_3 \cos 3\omega t + \cdots \qquad (2.64)$$

and the output current is

$$I_o = 2K_I(B_1 \cos \omega t + B_3 \cos 3\omega t + \cdots) \qquad (2.65)$$

The output does not contain any even harmonics since the even harmonics of the two transistors have been canceled by the push-pull arrangement. This is particularly

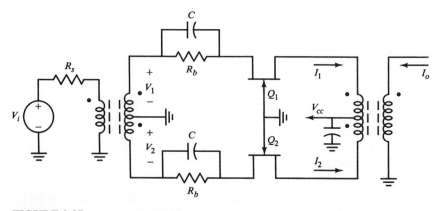

**FIGURE 2.27**
A balanced push-pull amplifier employing two common-gate amplifiers.

beneficial with FET amplifiers because they have a square-law characteristic that generates a relatively large second harmonic. Push-pull amplifiers are also used in power amplifiers. Several applications of their use in power amplifiers are provided in Chap. 11.

### ■ 2.7
### DIFFERENTIAL AMPLIFIER

A differential amplifier provides an output proportional to the difference between the signals present at the two input terminals; neither input need be grounded. Ideally, the output will be zero in response to a signal common to the two input terminals. A differential amplifier can be applied where it is desired to measure the voltage difference between two points, neither of which is grounded. Recent advances in transistor fabrication now make it possible to manufacture two transistors with closely matched characteristics on the same wafer. This has made the differential amplifier one of the most versatile building blocks at the designer's disposal, allowing for the relatively easy design of low-noise, low-drift, dc coupled amplifiers. Today the majority of analog integrated circuits use a differential amplifier input stage.

Consider the differential amplifier illustrated in Fig. 2.28. The positive terminal is referred to as the *noninverting terminal* since any voltage at this terminal will be amplified and will appear at the output without phase inversion, whereas any voltage on the negative (inverting) terminal will be amplified and will appear at the output with 180° phase shift (at least at low frequencies). The output may be either single-ended (one terminal grounded) or floating (neither output terminal grounded). Any combination of input signals can be decomposed into differential and common input signals. It will be shown that this decomposition greatly simplifies the analysis of differential amplifiers. The *differential input voltage* is defined as the difference between the two input signals, that is,

$$e_d = V_1 - V_2 \tag{2.66}$$

and the *common input voltage* is defined as

$$e_c = \frac{V_1 + V_2}{2} \tag{2.67}$$

Any two differential amplifier input signals, then, consist of a component in which the signals on the two input terminals are equal in magnitude and 180° out of phase and a component in which the two input terminal signals are equal in magnitude and in phase.

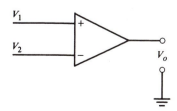

**FIGURE 2.28**
The circuit symbol for a differential amplifier.

**EXAMPLE 2.7.** Determine the differential and common input voltages of the two signals

$$V_1 = 5 + 3 \sin \omega t \qquad \text{and} \qquad V_2 = 3 - \sin \omega t$$

**Solution.** The differential signal is

$$e_d = V_1 - V_2 = 2 + 4 \sin \omega t$$

and the common signal is

$$e_c = \frac{V_1 + V_2}{2} = 4 + \sin \omega t$$

The amplification of a differential amplifier is characterized by its differential gain:

$$A_d = \frac{V_o}{e_d} \tag{2.68}$$

The *common-mode gain* is defined as

$$A_c = \frac{V_o}{V_1} \bigg|_{V_1 = V_2} = \frac{V_o}{e_c} \tag{2.69}$$

In an ideal differential amplifier, the common-mode gain is zero, but an actual differential amplifier always has a finite common-mode gain.

## Common-Mode Rejection Ratio

The ratio of differential gain to common-mode gain is known as the *common-mode rejection ratio (CMRR):*

$$\text{CMRR} = \frac{A_d}{A_c} \tag{2.70}$$

The actual amplifier output to any input signal is

$$V_o = A_d e_d + A_c e_c = A_d e_d + \frac{A_d}{\text{CMRR}} e_c \tag{2.71}$$

The term

$$\frac{A_d}{\text{CMRR}} e_c$$

is an error term caused by the finite CMRR. It is not possible to determine whether the output signal is due to a differential signal or a common signal. The larger the CMRR, the smaller the error term and the greater the amplifier accuracy.

**EXAMPLE 2.8.** Consider the problem of measuring the voltage across the 5-$\Omega$ resistor of the circuit shown in Fig. 2.29. This is a common problem in biomedical engineering and telemetry applications where the large impedances in series with the small resistors might represent the impedance of transducers or probe interfaces.

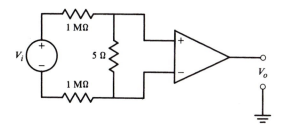

**FIGURE 2.29**
An application of a differential amplifier.

**Solution.** The voltage on the positive terminal is

$$V_1 = \frac{(5 + 10^6)V_i}{5 + 2 \times 10^6}$$

and the voltage on the inverting terminal is

$$V_2 = \frac{10^6 V_i}{5 + 2 \times 10^6}$$

So the differential voltage is

$$e_d = V_1 - V_2 = \frac{5V_i}{2 \times 10^6}$$

and the common-mode voltage is

$$e_c = \frac{V_1 + V_2}{2} = \frac{(5 + 2 \times 10^6)V_i}{2(5 + 2 \times 10^6)} = \frac{V_i}{2}$$

If the differential amplifier has a differential gain of $2 \times 10^3$ and a common-mode rejection ratio of 1 million (120 dB), the output voltage will be

$$V_o = \left[ \frac{5}{2 \times 10^6} (2 \times 10^3) \right] V_i + \frac{V_i}{2} (2) \frac{10^3}{10^6}$$

$$= (5 \times 10^{-3} + 1 \times 10^{-3})V_i$$

The second term represents a 20 percent error due to the finite CMRR. To realize a one percent measurement accuracy for this problem requires a differential amplifier with a minimum CMRR of $2 \times 10^7$ (146 dB).

**FET Differential Amplifier**

Consider the differential amplifier connection shown in Fig. 2.30. We will assume that the transistors are identical and at the same temperature. The small-signal, mid-frequency equivalent circuit is then as shown in Fig. 2.31. Although this circuit can be readily analyzed for arbitrary input voltages, we will first calculate the differential gain and then the common-mode gain, as this greatly simplifies the analysis.

Assume first that $V_{i_1} = -V_{i_2}$. If $V_{i_1}$ increases by $\Delta V$, then $V_{i_2}$ will decrease by the same amount; so the changes in the two dependent current sources in Fig. 2.31 will be equal in magnitude but opposite in sign. In this case it is readily shown that the voltage at the source terminal does not change, and so the current through $R_s$

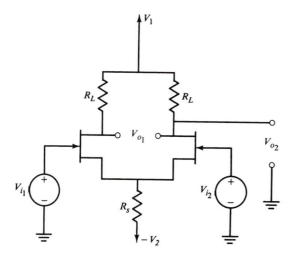

**FIGURE 2.30**
A differential amplifier realized
with two field-effect transistors.

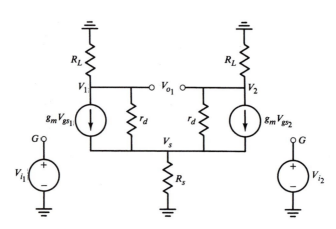

**FIGURE 2.31**
A small-signal equivalent circuit of the FET differential amplifier shown in Fig. 2.30.

does not change. Therefore, the small-signal equivalent circuit for differential
inputs can be simplified to that shown in Fig. 2.32. The output voltage is

$$V_{o_1} = \frac{-g_m r_d R_L (V_{i_1} - V_{i_2})}{r_d + R_L} = \frac{-e_d g_m r_d R_L}{r_d + R_L} \qquad (2.72)$$

and

$$V_{o_2} = \frac{e_d}{2} \frac{g_m r_d R_L}{r_d + R_L} \qquad (2.73)$$

The single-ended output differential gain is one-half of the differential output gain.
Also with the single-ended output as shown, input terminal 1 would be the noninverting input terminal.

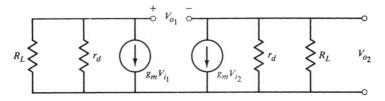

**FIGURE 2.32**
The small-signal equivalent circuit for determining the differential gain
of the FET differential amplifier.

To calculate the common-mode signal, we assume (with no loss of generality)
that the two input signals are the same, that is,

$$V_{i_1} = V_{i_2}$$

and the single-ended output voltage can be determined from an analysis of the circuit shown in Fig. 2.33a. Because of the circuit symmetry, the output can be derived
from the simplified circuit shown in Fig. 2.33b. Here

$$V_{gs} = V_i - V_s$$

The sum of the current at the source node is zero, so

$$\frac{V_s}{2R_s} + \frac{V_s - V_{o_2}}{r_d} = g_m V_{gs}$$

and the sum of the currents at the output node is zero, so

$$\frac{V_{o_2}}{R_L} + \frac{V_{o_2} - V_s}{r_d} + g_m V_{gs} = 0$$

If $V_s$ is eliminated from these equations, we obtain the common-mode gain
expression

$$A_c = \frac{V_{o_2}}{V_i} = \frac{V_{o_2}}{e_d} = \frac{g_m r_d R_L}{(1 + g_m r_d)2R_s + r_d + R_L}$$

$$\approx \frac{g_m R_L}{1 + 2g_m R_s} \qquad \text{for } r_d \gg R_L \tag{2.74}$$

The common-mode gain should ideally be zero. We see that the larger the
source resistance, the smaller the common-mode gain. Also, the differential gain
does not depend on $R_s$. Since an ideal current source has infinite impedance, the
source resistance can be replaced by a current source (which also provides a path
for the bias current). A current-source-biased differential amplifier is shown in
Fig. 2.34.

EXAMPLE 2.9. Calculate the output voltage of the circuit shown in Fig. 2.35 in response to an input signal $V_i = 5 \times 10^{-3}$ V. The equivalent output impedance of the current source is 150 kΩ. Assume that the SU2366 dual JFET is used. For the SU2366 dual
JFET, $g_m = 1.5 \times 10^{-3}$ S and $r_d = 500$ kΩ.

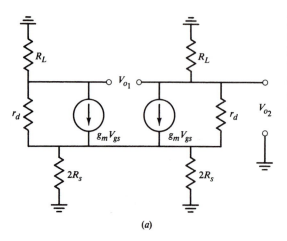

**FIGURE 2.33**
(*a*) A simplified circuit for calculating the common-mode gain of the FET differential amplifier shown in Fig. 2.30; (*b*) one-half of the symmetric equivalent circuit shown in Fig. 2.33*a*.

(*a*)

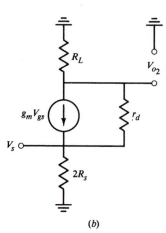

(*b*)

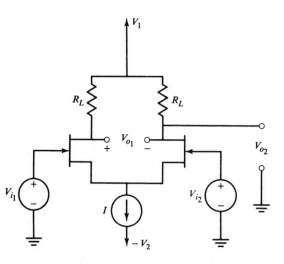

**FIGURE 2.34**
An FET differential amplifier with current-source biasing.

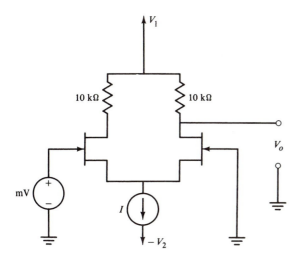

**FIGURE 2.35**
The amplifier discussed in Example 2.9.

**Solution.** Although it is possible to directly calculate the response to the single 5-mV input, it is far easier to consider the circuit to be a differential amplifier and to calculate the response to the differential and common-mode inputs. The differential input is

$$e_d = V_{i_1} - V_{i_2} = 5 \text{ mV}$$

and the common-mode input is

$$e_c = \frac{V_{i_1} + V_{i_2}}{2} = 2.5 \text{ mV}$$

(Note that this is equivalent to assuming $V_{i_1} = 2.5 \text{ mV} = -V_{i_2}$ for the differential input and $V_{i_1} = V_{i_2} = 2.5 \text{ mV}$ for the common-mode input.) Then by superposition

$$V_o = e_c A_c + e_d A_d$$

The common-mode gain is obtained using Eq. (2.74) (since $r_d \gg R_L$):

$$A_c = -\frac{1.5 \times 10^{-3} \times 10^4}{1 + 2(1.5 \times 10^{-3})(150 \times 10^3)} = -0.033$$

and the differential gain is [from Eq. (2.73)]

$$A_d = \frac{g_m R_L}{2} = \frac{1.5 \times 10^{-3} \times 10^4}{2} = 7.5$$

so that the output to the applied signal is

$$V_o = -0.033(2.5) + 7.5(5) = 37.42 \text{ mV}$$

### BJT Differential Amplifier

A bipolar junction transistor (BJT) differential amplifier is shown in Fig. 2.36a, and the small-signal equivalent circuit is shown in Fig. 2.36b (with $r_o$ and $r_b'$

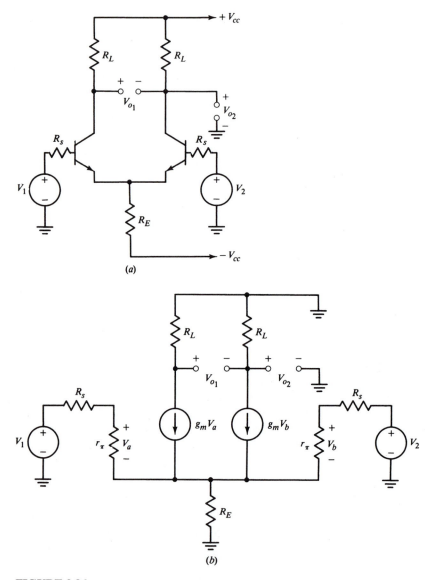

**FIGURE 2.36**
(*a*) A BJT differential amplifier; (*b*) amplifier shown in Fig. 2.36*a*.

neglected). The circuit can be analyzed with the method used for the FET differential amplifier. For differential inputs the emitter is at ground potential and

$$A_d = \frac{V_{o_2}}{e_d} = \frac{g_m R_L r_\pi}{2(R_s + r_\pi)} \tag{2.75}$$

The differential output gain is

$$\frac{V_{o_1}}{e_d} = \frac{-g_m R_L r_\pi}{R_s + r_\pi} \tag{2.76}$$

For common-mode input signals, the circuit is equivalent to two identical circuits in parallel, and

$$A_c = \frac{V_{o_2}}{e_c} = \frac{-g_m R_L r_\pi}{R_s + r_\pi + 2(\beta + 1)R_E} \tag{2.77}$$

The common-mode gain can be reduced by increasing $R_E$, so BJT differential amplifiers usually use emitter-current source biasing to achieve a large resistance to the emitters of the two transistors of the differential amplifier.

## ▓ 2.8

### OPERATIONAL AMPLIFIER

An operational amplifier (op amp) is a direct-coupled differential amplifier with high gain and large input impedance. The device is used with large amounts of negative feedback to obtain the desired gain-versus-frequency characteristic. It has been used mainly in low-frequency circuits, but integrated-circuit op amps are now available with gain-bandwidth products in the gigahertz ($10^9$ Hz) region. Op amps are currently being used in communication circuits as wideband or video amplifiers.

### Op-Amp Characteristics

The ideal operational amplifier is a differential amplifier with infinite gain, infinite input impedance, and infinite bandwidth. An actual operational amplifier has none of these characteristics, but the ideal characteristics are approximated over a limited frequency range. We will evaluate the accuracy of these approximations after considering the ideal case. Figure 2.37a contains a circuit representation of an op amp, and Fig. 2.37b contains a simplified small-signal equivalent circuit. The differential input impedance is represented by $Z_i$, *which can be greater than* $10^{12}$ Ω for some op amps with FET input stages, but can be as low as 10 kΩ for op amps with high gain-bandwidth products such as the HA5190. The output circuit contains a dependent voltage source where the dependent voltage is the differential input voltage $V_2 - V_1$ multiplied by $A_a$. The op-amp gain $A_a$ is often referred to as the *open-loop gain.* The output resistance $Z_o$ is on the order of 100 Ω, and it usually has a negligible effect on circuit operation, although it is a limiting factor in an op-amp's ability to drive capacitive loads.

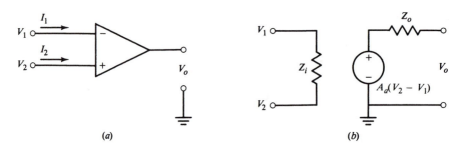

**FIGURE 2.37**
(*a*) A circuit symbol for an operational amplifier; (*b*) a small-signal equivalent circuit of
the operational amplifier.

### Ideal Inverting Amplifier

Figure 2.38 illustrates the most commonly used operational-amplifier configuration, the inverting amplifier. Here the noninverting input terminal is grounded, and
the output voltage is fed back to the inverting input terminal through the impedance
$Z_f$. For the ideal operational amplifier, the circuit analysis is simple. It is based on
the fact that no current will flow into the input terminals, and for any ideal op-amp
circuit the voltage on the positive terminal is equal to the voltage on the negative
input terminal:

$$V_1 = V_2$$

That the two voltages are equal follows from the fact that the output voltage is

$$V_o = A_a(V_2 - V_1) \tag{2.78}$$

In the ideal op amp, $A_a$ is infinite; therefore, the only way for the output voltage to
be finite is for $V_1$ to equal $V_2$. The two voltages are made equal through the feedback network connected around the op amp. It is, of course, possible that the feedback is such that the amplifier will be unstable, but it is assumed here that the
circuit configuration is such that the amplifier is stable. In the inverting amplifier

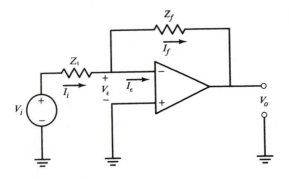

**FIGURE 2.38**
An inverting op-amp circuit.

shown in Fig. 2.38, if the amplifier is ideal, the voltage at the inverting terminal $V_\epsilon$ must be zero, since the positive terminal voltage $V_2 = 0$ and $V_1 = V_2$ in the ideal op-amp circuit. Also, in the ideal op amp, the input current $I_\epsilon = 0$, so

$$\frac{V_i}{Z_1} = I_i = I_f = \frac{-V_o}{Z_f}$$

That is, the transfer function of the inverting amplifier configuration, using an ideal op amp, is

$$A_v = \frac{V_o}{V_i} = \frac{-Z_f}{Z_1} \tag{2.79}$$

The closed-loop transfer function is independent of the operational amplifier and depends only on the two impedances used in the feedback network.

**EXAMPLE 2.10.** Design an amplifier with a voltage gain equal to $-50$.

**Solution.** If an ideal operational amplifier is used, the circuit configuration of Fig. 2.38 can be used, provided that $Z_f/Z_1 = 50$. Here one of the impedances can be arbitrarily chosen. For example, if $Z_1 = 1 \text{ k}\Omega$, then $Z_f = 50 \text{ k}\Omega$ will give the desired voltage gain.

The inverting amplifier configuration can also be used to realize a frequency-dependent transfer function. This concept is illustrated in the following example.

**EXAMPLE 2.11.** Design a low-pass amplifier with a dc voltage gain equal to $-50$ and a $-3$-dB bandwidth of $10^6$ rad/s.

**Solution.** Again there are many networks which can be used to meet the specifications, but one of the simplest is to select one such that

$$Z_f = \frac{R_f}{s R_f C + 1} \qquad \text{and} \qquad Z_1 = R_1$$

The voltage transfer function is then

$$A_v = \frac{-R_f/R_1}{sCR_f + 1} \tag{2.80}$$

which is a low-pass transfer function with a low-frequency gain equal to $-R_f/R_1$ and a $-3$-dB frequency:

$$\omega_L = \frac{1}{R_f C} \tag{2.81}$$

Therefore, we arbitrarily select $R_1 = 1 \text{ k}\Omega$. Then $R_f = 50 \text{ k}\Omega$ and $C = (10^6 \times 5 \times 10^4)^{-1} = 20 \text{ pF}$. The complete circuit is shown in Fig. 2.39.

An almost limitless number of different frequency-dependent transfer functions can be constructed using the inverting configuration.

Since the inverting terminal is also at ground potential, the input impedance of the amplifier shown in Fig. 2.38 is

$$Z_i = \frac{V_i}{I_i} = Z_1 \tag{2.82}$$

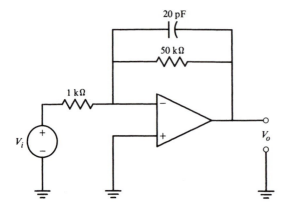

**FIGURE 2.39**
The low-pass amplifier discussed
in Example 2.11.

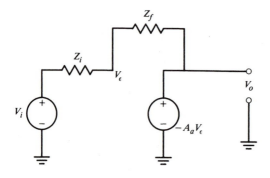

**FIGURE 2.40**
An equivalent circuit of the inverting
amplifier with finite open-loop gain
$(A_a)$.

The impedance is easily controlled by the selection of $Z_1$. The output impedance,
assuming an ideal op amp, is zero.

### Nonideal Inverting Amplifier: The Effect of Finite Loop Gain

The gain of an actual operational amplifier is never infinite and, in fact, is only large
over a limited frequency range. The effects of finite loop gain on the transfer function can be determined by analyzing Fig. 2.40, which is the small-signal equivalent
circuit for the operational amplifier with finite open-loop gain $A_a$. Since the amplifier input impedance is assumed to be infinite,

$$\frac{V_i - V_\epsilon}{Z_i} = \frac{V_\epsilon - V_o}{Z_f}$$

and if the op-amp output impedance is neglected,

$$V_o = -A_a V_\epsilon$$

If $V_\epsilon$ is eliminated from these two equations, the transfer function is

$$A_v = \frac{V_o}{V_i} = \frac{-Z_f/Z_i}{1 + (Z_i + Z_f)(A_a Z_i)} \tag{2.83}$$

If the operational-amplifier gain is known, the deviation of the ideal transfer function from its actual value can be calculated.

**EXAMPLE 2.12.** An inverting amplifier with a gain of 100 is designed by using an operational amplifier with an open-loop gain $A_a = 10^4$. The feedback resistor $R_f = 10^4$, and the input resistance $R_1 = 100$. What is the actual loop gain?

*Solution.* The actual gain is calculated by using Eq. (2.83):

$$A_v = \frac{-100}{1 + (100 + 1)/10^4} = \frac{-100}{1.01} = -99.0$$

The finite op-amp gain resulted in a 1 percent deviation in the closed-loop gain from the ideal value.

## Gain-Bandwidth Product

In addition to being finite, the amplifier gain $A_a$ is frequency-dependent. The majority of operational amplifiers are internally compensated so that the frequency dependence is of the form

$$A_a = \frac{A_o}{1 + j\omega/\omega_o} \tag{2.84}$$

where $A_o$ is the low-frequency open-loop gain and $\omega_o$ is the $-3$-dB frequency. This particular form of frequency dependence is selected because it enhances the stability characteristics of the closed-loop system. The magnitude of $A_a$ as a function of frequency is plotted in Fig. 2.41. The gain decreases at a rate of $-6$ dB per octave at high frequencies. At frequency $\omega_o$ the gain is 0.707 ($-3$ dB) of the dc value. This frequency is the $-3$-dB bandwidth of the amplifier. The frequency $\omega_T$ is defined as the frequency at which the magnitude of the gain is unity. That is,

$$A_a = 1 = \frac{A_o}{|1 + j\omega_T/\omega_o|} \approx \frac{A_o\omega_o}{\omega_T} \tag{2.85}$$

or $\qquad\qquad \omega_T = A_o\omega_o$

The frequency $\omega_T$ is equal to the low-frequency gain $A_o$ times the open-loop bandwidth $\omega_o$. For this reason, it is known as the *gain-bandwidth product*. The gain-bandwidth product is usually contained in the op-amp specification sheets.

**EXAMPLE 2.13.** The HA5190 operational amplifier has a gain-bandwidth product of 150 MHz and a low-frequency voltage gain of 90 dB. What is the transfer function for the open-loop voltage gain?

*Solution.* Since the low-frequency gain is 90 dB,

$$20 \log A_o = 90 \qquad \text{and} \qquad A_o = 3.16 \times 10^4$$

Since $\qquad\qquad \omega_o A_o = \omega_T = 2\pi(150 \times 10^6)$

the bandwidth is

$$\omega_o = 2\pi(4.74 \times 10^3) \qquad \text{rad/s}$$

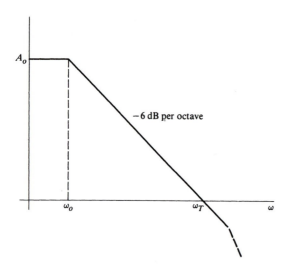

**FIGURE 2.41**
Freqeucny response of an internally compensated operational amplifier.

and the transfer function is

$$A_v = \frac{3.16 \times 10^4}{1 + j\omega/[2\pi(4.7 \times 10^3)]}$$

The open-loop gain of an op amp declines with increasing frequency so that the deviation of the actual transfer function from the ideal value will increase with frequency. At any frequency the actual value can be calculated by using Eq. (2.83) with the value of $A_a$ given by Eq. (2.84).

### Effect of Finite Input Impedance

For the case where the input impedance is finite, the equivalent circuit is as shown in Fig. 2.42a. This circuit is most easily analyzed by replacing the input circuit and $Z_i$ by the Thévenin equivalent (denoted by subscript th), as shown in Fig. 2.42b. Here

$$V_{th} = \frac{V_i Z_i}{Z_i + Z_1} \tag{2.86}$$

and

$$Z_{th} = \frac{Z_i Z_1}{Z_1 + Z_i} \tag{2.87}$$

A useful rule of thumb is that the effect of finite op-amp input impedance can be neglected, provided $Z_i > 100Z_1$. In either case, the output voltage will be

$$V_o = \frac{-Z_f/(Z_{th} V_{th})}{1 + (Z_{th} + Z_f)/(A_v Z_{th})} \tag{2.88}$$

where $V_{th}$ and $Z_{th}$ are given by Eqs. (2.86) and (2.87), respectively.

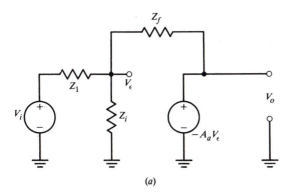

**FIGURE 2.42**
(a) A small-signal equivalent
circuit of an inverting amplifier
with finite op-amp input impedance $Z_i$ and op-amp gain $A_a$;
(b) a Thévenin equivalent circuit
of the amplifier shown in
Fig. 2.42a.

(a)

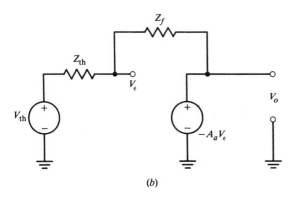

(b)

## Effect of Finite Input and Output Impedance

If neither $Z_i$ nor $Z_o$ is neglected, the small-signal equivalent circuit is as shown in
Fig. 2.43a. The circuit seen by the load $Z_L$ can be replaced by its Thévenin equivalent
circuit, as shown in Fig. 2.43b. To determine the Thévenin equivalent impedance
(which is the output impedance of the closed-loop amplifier), a voltage $V_o$ is applied
across the output terminals and the current $I_o$ is determined. The current $I_o$ is given by

$$I_o = \frac{V_o + A_a V_\epsilon}{Z_o} + \frac{V_o - V_\epsilon}{Z_f}$$

because

$$V_\epsilon = \frac{V_o Z_1'}{Z_1' + Z_f}$$

where

$$Z_1' = \frac{Z_i Z_1}{Z_i + Z_1}$$

The Thévenin equivalent impedance (the output impedance) is

$$Z_{th} = \frac{Z_o}{1 + (Z_f/Z_1' + A_a)/(2 + Z_f/Z_1')} \tag{2.89}$$

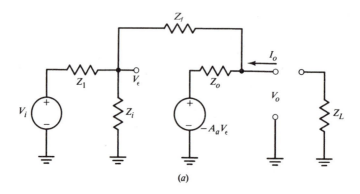

(a)

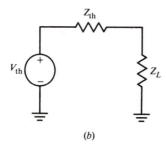

(b)

**FIGURE 2.43**
(a) An inverting amplifier equivalent circuit illustrating the
op-amp input impedance $Z_i$, output impedance $Z_o$, and gain $A_a$
(b) a Thévenin's equivalent circuit of the amplifier shown in
Fig. 2.43a.

The feedback circuit has the effect of reducing the output impedance from its open-circuit value $(Z_o)$; and the larger the op-amp gain $A_a$, the smaller the output impedance. The Thévenin equivalent voltage $V_{\text{th}}$ is the output voltage without the load $Z_L$ connected.

After eliminating $V_\epsilon$, we find the transfer function

$$\frac{V_o}{V_{\text{th}}} = \frac{-(Z_f - Z_o/A_a)}{Z_{\text{th}} + (Z_o + Z_f + Z_{\text{th}})/A_a} \tag{2.90}$$

This equation describes the inverting amplifier configuration with finite $Z_o$, $Z_i$, and $A_a$. Communication circuits frequently employ small $Z_i$ and $Z_f$, and the effects of nonzero $Z_o$ must be considered. Note that the significance of the output impedance increases as the open-loop gain $A_a$ decreases, as it does with increasing frequency. A finite output impedance can result in stability problems when the load is capacitive.[3]

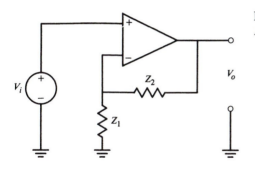

**FIGURE 2.44**
A noninverting amplifier.

### Noninverting Amplifier

The basic noninverting amplifier configuration is shown in Fig. 2.44. If the op-amp gain and input impedance are assumed to be infinite, the voltages at the plus and minus terminals must be equal. Therefore,

$$V_+ = V_i = V_- = \frac{V_o Z_1}{Z_1 + Z_2}$$

or

$$\frac{V_o}{V_i} = 1 + \frac{Z_2}{Z_1} \qquad (2.91)$$

This is the basic equation for the noninverting amplifier. If $Z_1$ and $Z_2$ are real, the gain is positive and greater than 1. The absence of a phase inversion from the input to the output is one reason that noninverting amplifiers are used. Another reason is that the input impedance is large (larger than the $Z_i$ of the device).

**EXAMPLE 2.14.** Design a noninverting amplifier with a voltage gain of 100, using an ideal op amp.

**Solution.** For the ideal op amp used in the amplifier configuration of Fig. 2.43a, the voltage gain is given by

$$A_v = \frac{V_o}{V_i} = 1 + \frac{Z_2}{Z_1}$$

Either $Z_2$ or $Z_1$ can be chosen arbitrarily. For example, if $Z_1 = 1 \text{ k}\Omega$, then $Z_2 = 99 \text{ k}\Omega$ will realize a voltage gain equal to 100.

The effects of finite voltage gain and input impedance are treated just as they were for the inverting amplifier. The calculations are left as an exercise.

## ▉ 2.9
### PROBLEMS

**2.1** What is the approximate midfrequency voltage gain $V_o/V_i$ of the circuit shown in Fig. P2.1? The transistor $\beta$ is 110, and the transistor output impedance is 50 k$\Omega$. What is the midfrequency power gain?

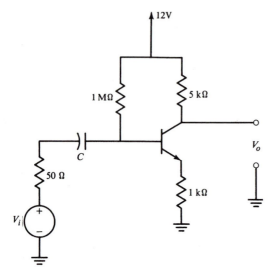

**FIGURE P2.1**
An emitter-loaded common-emitter amplifier.

**2.2** What are midfrequency voltage gain and input impedance $Z_i$ of the amplifier shown in Fig. P2.2? The transistor $\beta$ is 110, and the transistor output impedance is 100 kΩ.

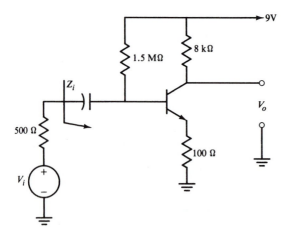

**FIGURE P2.2**
An emitter-loaded common-emitter amplifier.

**2.3** Design an amplifier, using one of the transistors specified in the appendixes, with a mid-frequency voltage gain of at least 26 dB. The source resistance is 600 Ω, and the load resistance is to be 5 kΩ. Show the complete circuit, including the biasing network.

**2.4** Calculate the approximate midfrequency voltage gain of the amplifier shown in Fig. P2.4. The $g_m$ of the transistor is $4 \times 10^{-3}$ S, and $r_d$ is 10 kΩ.

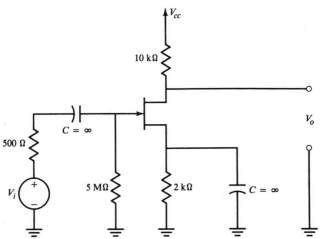

**FIGURE P2.4**
A common-source amplifier.

**2.5** Calculate the midfrequency voltage gain of the amplifier shown in Fig. P2.5. The transistor $g_m$ is $10^{-3}$ S, and the transistor output impedance $r_d$ is 100 kΩ.

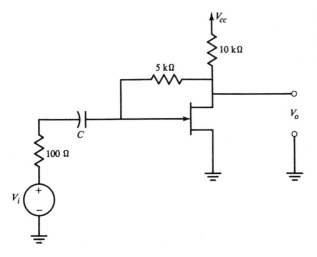

**FIGURE P2.5**
A common-source amplifier with current feedback.

**2.6** What are the midfrequency voltage gain and input impedance of the amplifier shown in Fig. P2.6?

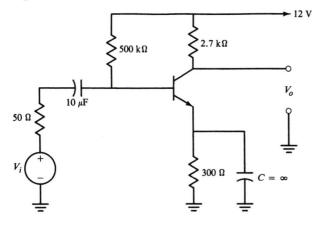

**FIGURE P2.6**
A common-emitter amplifier.

**2.7** Calculate the midfrequency voltage, power, and current gains of the amplifier shown in Fig. P2.7. The transistor $\beta = 100$, and the transistor output impedance is $20\ k\Omega$.

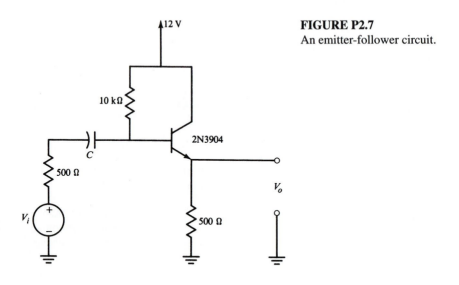

**FIGURE P2.7**
An emitter-follower circuit.

**2.8** For the two-stage amplifier shown in Fig. P2.8, calculate the midfrequency voltage and current gains. The $g_m$ of both transistors is $3 \times 10^{-3}$ S, and $r_d$ is $20\ k\Omega$.

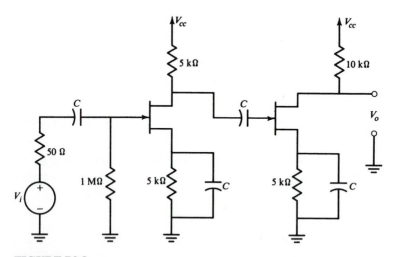

**FIGURE P2.8**
An amplifier consisting of two common-source amplifiers in cascade.

**2.9** Determine the approximate midfrequency voltage gain of the two-stage amplifier shown in Fig. P2.9. (All capacitors can be assumed to be short circuits.) The $\beta$ of the BJT is 100, and $g_m$ is $3 \times 10^{-3}$ S for the FET. The transistor output impedances are large enough to be neglected.

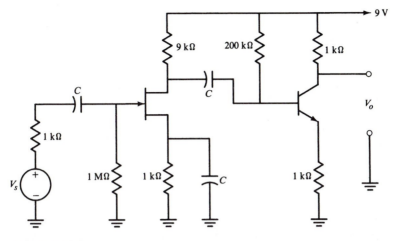

**FIGURE P2.9**
A two-stage amplifier.

**2.10** An amplifier configuration often referred to as a *Darlington amplifier* is shown in Fig. P2.10. Calculate the voltage gain and input resistance of this circuit. The transistors have identical $\beta$'s and their output impedances can be neglected, but the $g_m$ of the two devices will not be the same.

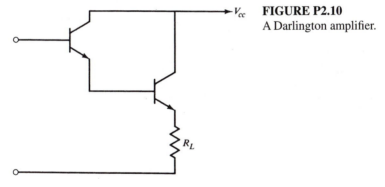

**FIGURE P2.10**
A Darlington amplifier.

**2.11** The amplifier circuit shown in Fig. P2.11 employs an operational amplifier with ideal characteristics except that its open-loop gain $A_a$ equals 20,000. How much does the actual closed-loop gain differ from the ideal value?

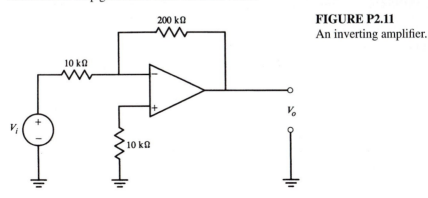

**FIGURE P2.11**
An inverting amplifier.

**2.12** Determine the gain of the amplifier shown in Fig. P2.12. The operational amplifier has ideal characteristics.

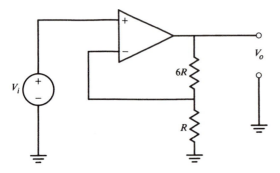

**FIGURE P2.12**
A noninverting amplifier.

**2.13** The inverting amplifier shown in Fig. P2.13 uses a compensated op amp that has a dc gain of 140 dB and an open-loop bandwidth of 10 Hz. What is the op-amp gain-bandwidth product? What are the gain and bandwidth of the closed-loop amplifier?

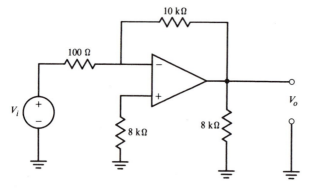

**FIGURE P2.13**
An inverting amplifier.

**2.14** What is the gain of the amplifier of Fig. P2.11 if it has an input impedance of 20 k$\Omega$ from the inverting terminal to ground?

**2.15** Calculate the voltage gain and input impedance of the amplifier shown in Fig. P2.13 under the conditions that the op amp is ideal. What is the gain if the op amp has an open-loop gain of 80 dB?

**2.16** Derive the transfer characteristic for the amplifier shown in Fig. P2.16. Assume the op amp is ideal.

**2.17** Derive an expression for $I_o$ in terms of $V_i$ for the amplifier shown in Fig. P2.17. Assume the op amp is ideal.

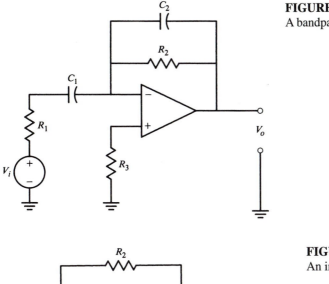

**FIGURE P2.16**
A bandpass amplifier.

**FIGURE P2.17**
An inverting amplifier.

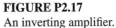

**2.18** Calculate the gain of the cascode circuit shown in Fig. P2.18. Assume the collector direct current is 1 mA.

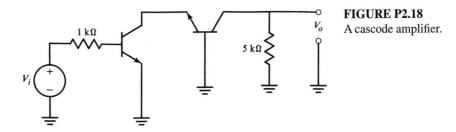

**FIGURE P2.18**
A cascode amplifier.

**2.19** What is the bandwidth of the amplifier illustrated in Fig. P2.13 if the Burr-Brown 3554 is used as the operational amplifier? See Appendix 5 for the 3554 specifications.

**2.20** Determine the output voltage by determining the common-mode and differential mode gains in Fig. P2.20.

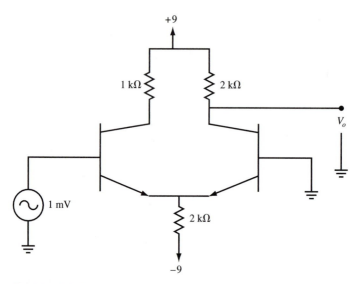

**FIGURE P2.20**
A BJT differential amplifier.

**2.21** Calculate the voltage $V_o$ in Fig. P2.21 after selecting $V_b$ so that the transistors are approximately in the middle of their linear operating region.

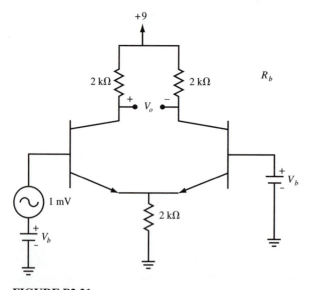

**FIGURE P2.21**
A BJT differential amplifier.

**2.22** Determine the output voltage of the circuit in Fig. P2.22 as a function of $e_i$ and $V_b$.

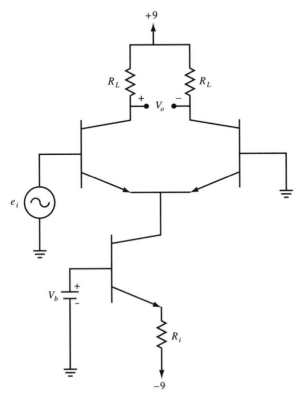

**FIGURE P2.22**
Current source biased differential amplifier.

**\*2.23** Design a differential voltage amplifier with a voltage gain of at least 30 dB. The source impedances is 50 $\Omega$, and the load impedance is not specified. 2N2222A transistors, resistors, and capacitors are available as are two 9-V supplies. The low-frequency response can be neglected (i.e., arbitrarily large capacitors can be used for the low-frequency response). Verify your design, using computer simulation, and determine the upper $-3$-dB frequency of the amplifier.

## ◼ 2.10

### REFERENCES

1. P. Horowitz and W. Hill, *The Art of Electronics,* 2d ed., Cambridge University Press, New York, 1989.

2. A. D. Evans, (ed.), *Designing with Field Effect Transistors,* McGraw-Hill, New York, 1981.
3. J. K. Roberge, *Operational Amplifiers: Theory and Practice,* Wiley, New York, 1975.

## ▒ 2.11

**ADDITIONAL READING**

P. R. Gray and R. G. Meyer, *Analysis and Design of Analog Integrated Circuits,* 3d ed., Wiley, New York, 1993.

# Network Noise and
# Intermodulation Distortion

## ▪ 3.1
### INTRODUCTION

One of the most important factors to consider in evaluating the performance of a communications network is the network's ability to process low-amplitude signals. Every system creates noise, which limits its ability to process weak signals; the principal noise sources are (1) random noise generated in the resistors and transistors, (2) mixer noise created by the nonideal properties of the mixers, (3) the undesired cross-coupling of signals between two sections of the receiver, and (4) power-supply noise. Except for random noise, all these sources of noise can be eliminated, at least in theory, by proper design and construction.

Random noise is inherent in all resistors and transistors. It is a critical factor in the performance of communication receivers since it determines the minimum signal level that can be detected. A measure of receiver performance, referred to as the *noise figure,* has long been used to quantitatively describe the noise generated in a network. This chapter discusses the nature of random noise and how it affects receiver performance and design of low-noise amplifiers.

At one time the noise figure was considered sufficient for characterizing a receiver's performance, but today large-amplitude, unwanted signals are often at the receiver input. It is often necessary to be able to detect low-amplitude signals that are adjacent in frequency to large-amplitude, unwanted signals, and the noise figure is not adequate to completely describe a receiver's performance. For this reason, the term "dynamic range" is introduced in this chapter to more completely describe receiver performance.

## ▪ 3.2
### NOISE

All signals, whether at the network input or output, are contaminated by noise which degrades the system performance. The "noisiness" of a signal is usually specified in terms of the signal-to-noise ratio $S/N$, which is, in general, a function of frequency. The *signal-to-noise ratio* can be defined as

$$\frac{S(f)}{N(f)} = \frac{\text{rms signal voltage}}{\text{rms noise voltage}}$$

or

$$= \frac{\text{peak signal power}}{\text{average noise power}}$$

or

$$= \frac{\text{average signal power}}{\text{average noise power}}$$

Unless otherwise stated, the last definition will be used here for the signal-to-noise ratio, that is,

$$\frac{S}{N} = \frac{\text{average signal power}}{\text{average noise power}}$$

In order to define the average noise power, the origins and characteristics of network noise will first be discussed. An appropriate definition of noise is that it is "any unwanted input." Noise consists of both nonrandom, or periodic, components and random components. Types of nonrandom, or periodic, noise include power-supply noise and noise due to the unwanted cross-coupling of large signals such as that from a local oscillator. Note that the oscillator signal is considered to be a noise if it occurs at a point in the system where it is not desired. Noise of human origin is usually the dominant factor in receiver noise. Most of this type of noise is deterministic and can (at least theoretically) be eliminated through proper circuit design, layout, and shielding.

Random noise, by its very nature, cannot be eliminated. It places a lower theoretical limit, much like the uncertainty principle in physics, on the receiver noise level. An understanding of the properties of random noise allows one to control, by system design, its effect on receiver performance.

Random noise is described in terms of its statistical properties. At any instant, noise amplitude cannot be predicted exactly, but it can be expressed in terms of a probability density function. For system design and evaluation it suffices to describe the noise in terms of its mean-square or root-mean-square (rms) values. The mean-square noise voltages (currents) are often referred to as the *noise power* since they are directly proportional to the power dissipated in a resistor. The mean-square power is normally frequency-dependent and is usually expressed as a power spectral density function (unit of power per hertz). The total noise power $P$ is

$$P = \int_{f_1}^{f_2} p(f)\, df$$

In the following sections lowercase letters denote spectral density functions (volts squared per hertz, etc.), and uppercase letters denote mean-square values (volts squared, etc.).

Random noise can be subdivided into noise occurring external to the receiver, such as atmospheric and interstellar noise over which the receiver designer has no control, and random noise occurring in the receiver. The most common form of random noise originating in the receiver is thermal noise.

**Thermal Noise**

J. B. Johnson discovered *thermal noise*—the minute currents caused by the thermal motion of the conduction electrons in a resistor that constitute a random noise. H. Nyquist was able to demonstrate, using statistical thermodynamics, that the thermal noise generated in an impedance $Z(f)$ in a frequency interval $\Delta f$ is given by (Fig. 3.1)

$$E_n^2 = 4kTR(f) \, \Delta f \tag{3.1}$$

where      $E_n$ = rms value of thermal noise voltage
$R(f)$ = resistive component of impedance $Z(f)$, $\Omega$
$T$ = absolute temperature, K
$k$ = Boltzmann's constant, $1.38 \times 10^{-23}$ W · s/K

Since the real part of the impedance $R$ will in general be a function of frequency, the thermal noise voltage will also be frequency-dependent. However, for a resistor $R$, Eq. (3.1) states that the thermal noise voltage squared will be proportional to the frequency interval $\Delta f$. This implies that if the interval is infinite, the noise power contributed by the resistor is also infinite. Actually, Eq. (3.1) must be modified at very high frequencies (above 100 MHz), but it is sufficiently accurate for the purposes of this text.

For a linear network the total mean-square voltage density appearing across any two terminals is given by

$$E_n^2 = 4kTR(f) \, df \tag{3.2}$$

where $R(f)$ is the real part of the impedance $Z(f)$. Impedance $Z(f)$ is the impedance looking into the two terminals between which $E_n$ is measured. The total mean-square noise is then obtained by integrating this expression over all frequencies.

An alternative but equivalent interpretation is to find the noise due to each resistor. If a resistor is connected to a frequency-dependent network as shown in Fig. 3.2, then the total noise at the output due to $R$ will be given by

$$E_o^2 = \int_0^\infty 4kTRG(f) \, df \tag{3.3}$$

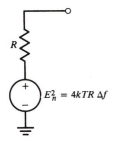

**FIGURE 3.1**
A resistor together with the mean-square thermal noise voltage of the resistance.

$E_n^2 = 4kTR \, \Delta f$

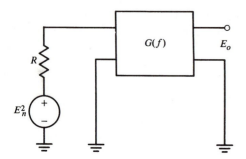

**FIGURE 3.2**
A resistor connected to a linear network with a frequency-dependent transfer function $G(f)$.

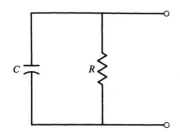

**FIGURE 3.3**
A simple frequency-dependent network.

where $G(f)$ is the magnitude squared of the frequency-dependent transfer function between the input and the output voltages, or

$$G(f) = \left| \frac{E_o(f)}{E_n(f)} \right|^2 \tag{3.4}$$

Since in this case $R$ is not a function of frequency, Eq. (3.3) is equivalent to

$$E_o^2 = 4kTR \int_0^\infty G(f)\, df \tag{3.5}$$

The integral of the magnitude squared of the transfer function (normalized for unity gain) is referred to as the *noise bandwidth* $B_n$ of the system. The noise bandwidth differs from the system's 3-dB bandwidth in that the noise bandwidth is the area under the curve $G(f)$. A system can have a narrow 3-dB bandwidth and yet have a large noise bandwidth.

**EXAMPLE 3.1.** The impedance of the parallel combination of a resistor $R$ and capacitor $C$ shown in Fig. 3.3 is given by

$$Z(\omega) = \frac{R}{1 + j\omega RC} \tag{3.6}$$

The real part of $Z(\omega) = R(\omega)$ is given by

$$R(\omega) = \frac{R}{1 + \omega^2 R^2 C^2} \tag{3.7}$$

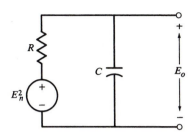

**FIGURE 3.4**
A simple frequency-dependent network including the thermal noise source.

so the total output noise power generated by the impedance will be

$$E_o^2 = 4kTR \int_0^\infty \frac{df}{1 + \omega^2 R^2 C^2} = \frac{kT}{C} \tag{3.8}$$

which is independent of the size of the resistor and depends only on the capacitance $C$ and temperature $T$.

**EXAMPLE 3.2.** The circuit of Example 3.1 could also be interpreted as shown in Fig. 3.4. Here the noise is represented as the equivalent thermal noise of the resistor $R$. The transfer function is given by

$$G(f) = \left| \frac{E_o}{E_n} \right|^2 = (1 + \omega^2 R^2 C^2)^{-1} \tag{3.9}$$

Thus, using Eq. (3.4), the output noise voltage squared is found to be

$$E_o^2 = 4kTR \int_0^\infty \frac{df}{1 + \omega^2 R^2 C^2} = \frac{kT}{C} \tag{3.10}$$

which is the same result as obtained in Example 3.1. Since

$$E_o^2 = 4kTRB_n$$

the circuit noise bandwidth $B_n = 1/(4RC)$ Hz. The 3-dB bandwidth is $2\pi/(RC)$ Hz.

In this example the same result was obtained by interpreting the noise to originate in the real part of the impedance $Z(f)$ or by interpreting the noise to originate in the resistor. It makes no difference whether the thermal noise voltage is interpreted to arise in the real part $R(f)$ of the impedance $Z(f)$ or whether the resistive elements are taken as the ultimate source of the thermal noise.

### Resistors in series

Equation (3.1) states that the noise voltage is the rms value of a randomly varying signal. If two resistors are added in series, as shown in Fig. 3.5, it is the noise voltages squared, not the noise voltages, which are added.

$$E_n^2 = E_1^2 + E_2^2 = 4kT(R_1 + R_2)\,\Delta f \tag{3.11}$$

Similarly, for two resistors in parallel the equivalent noise voltage is

$$E_n^2 = 4kT\frac{R_1 R_2}{R_1 + R_2}\,\Delta f \tag{3.12}$$

$R_1 + R_2$

$E_n^2 = 4kT(R_1 + R_2)\,\Delta f$

$R_1$

$E_1^2 = 4kTR_1\,\Delta f$

$R_2$

$E_2^2 = 4kTR_2\,\Delta f$

**FIGURE 3.5**
Equivalent mean-square noise voltage representation of two resistors in series.

The noise sources described here all refer to rms quantities, and the noise source has no polarity associated with it. In order to keep the notation simple, the noise sources are expressed in terms of the square of the voltage or current, and the values referred to are always the mean-square values.

### Current-source representation

Equation (3.1) states that the thermal noise can be represented by a voltage source in series with a noiseless resistor. Norton's theorem shows that the voltage noise source illustrated in Fig. 3.6b can also be represented by a current generator in parallel with a noiseless resistor as shown in Fig. 3.6a.

### Excess resistor noise

The thermal-noise power density generated in resistors does not vary with frequency, but many resistors also generate a frequency-dependent noise referred to as

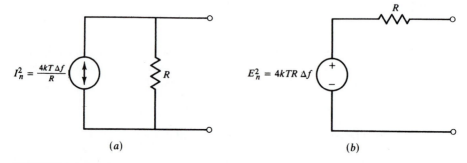

$I_n^2 = \dfrac{4kT\,\Delta f}{R}$          $R$

$E_n^2 = 4kTR\,\Delta f$          $R$

(a)                    (b)

**FIGURE 3.6**
Equivalent mean-square noise source representations of a resistor: (a) current generator with noiseless resistor; (b) voltage noise source.

*excess noise.* The excess noise power has a $1/f$ spectrum; the excess noise voltage is inversely proportional to the square root of the frequency. Noise that exhibits a $1/f$-power spectral characteristic at low frequencies is often referred to as *pink noise.*

The amount of excess noise generated in a resistor depends upon the resistor's composition. Carbon resistors generate the largest amount of excess noise, whereas the amount generated in wire-wound resistors is usually negligible. However, the inductance inherent in wire-wound resistors restricts them to low-frequency applications. Metal film resistors are usually the best choice for high-frequency communications circuits, where low noise and constant resistance are required.

**Active Device Noise**

Besides the thermal noise of resistors, the other sources of random noise of importance in network design are the active devices—integrated circuits, diodes, and transistors. The two main types of device noise are $1/f$, or flicker, noise and shot noise. *Flicker noise* is a low-frequency phenomenon in which the noise power density follows a $1/f^\alpha$ curve; the value of $\alpha$ is close to unity.

An electric current composed of discrete charge carriers flows through an active device. The discreteness of the charge-carrier fluctuations is present in the current crossing a barrier where the charge carriers pass independently of one another. Examples of such barriers are the semiconductor *pn* junction, in which the passage takes place by diffusion, and the cathode of a vacuum tube, where electron emission occurs as a result of thermal motion. The current fluctuations represent a noise component referred to as *shot noise,* which can be represented by an appropriate current source in parallel with the dynamic resistance of the barrier across which the noise originates. The spectral density of this shot noise is given by

$$i_{n_o}^2 = qkI_o \qquad \text{A}^2/\text{Hz} \qquad (3.13)$$

where $q$ is the charge on an electron, $I_o$ is the direct current, and $k$ is a constant that varies from device to device and also depends on how the junction is biased. In a junction transistor $k$ is equal to 2. Figure 3.7 illustrates the shot noise equivalent circuit for a forward-biased *pn* junction. Lowercase letters are used to denote the noise density as mean square (volts or amps) per hertz. Shot noise, like thermal noise, has

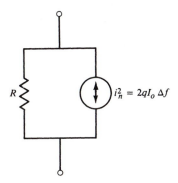

**FIGURE 3.7**
Network including the shot noise current source of a *pn* junction (lowercase $i^2$ denotes current spectral density).

$i_n^2 = 2qI_o\,\Delta f$

a uniform power spectral density, and the total noise current squared is proportional to the bandwidth. That is,

$$I_n^2 = i_{n_o}^2 \, \Delta f \tag{3.14}$$

The current source represented in Fig. 3.7 denotes that no direction is associated with the current source since it is a mean-square value. If the additional $1/f$ noise is included, the total mean-square current noise density can be written as

$$i_n^2(f) = i_{n_o}^2 \left(1 + \frac{f_L}{f}\right) \qquad A^2/Hz \tag{3.15}$$

where $f_L$ is the frequency where the shot noise current is equal to the $1/f$ noise current. Frequency $f_L$ varies from device to device and is usually determined empirically.

Figure 3.8 illustrates the power density of the total noise current as a function of frequency. At frequencies below $f_L$ the noise power density increases at a rate of 6 dB per octave, while at frequencies much higher than $f_L$ the noise power is equal to the shot noise and is independent of frequency. If the noise current is connected to a frequency-dependent network, the mean-square current at the output will be

$$I_o^2 = \int_0^\infty A_i(f) i_n^2 \, df \tag{3.16}$$

where $A_i(f)$ is the magnitude squared of the current transfer function between input and output.

### Noise in Transistor Amplifiers [1]

The previous discussion has shown that any amplifier must generate noise, which consists of thermal noise generated in the resistors plus the short and $1/f$ noise generated in the active devices. An equivalent circuit of a transistor amplifier, which

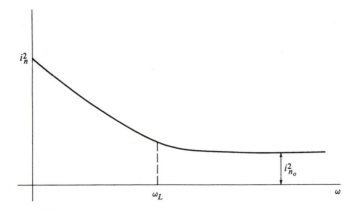

**FIGURE 3.8**
Spectral density function of the total noise current (including $1/f$ noise).

identifies the shot noise sources, is shown in Fig. 3.9. And $i_{n_2}$ represents the shot noise current density due to the bias current on the output of the device, and $i_{n_1}$ is the shot noise current density due to the input bias current. The other noise source is due to the load resistor $R_L$.

If the transistor output impedance is much larger than $R_L$, the output noise voltage due to $i_{n_2}$ will be

$$(e_o)_{n_2}^2 = i_{n_2}^2 R_L^2$$

It is convenient to refer all the noise sources to the input. The amplifier voltage gain (assuming $R_L$ is much less than the transistor output impedance) is approximately

$$\left| \frac{V_o}{V_i} \right| = g_m R_L$$

so the output noise current source can be replaced by an equivalent input noise voltage source

$$e_2^2 = \frac{i_{n_2}^2 R_L^2}{g_m^2 R_L^2} = \frac{i_{n_2}^2}{g_m^2} \tag{3.17}$$

The noise source $e_2^2$ can be interpreted as due to thermal noise of the transconductance $g_m$. Likewise, the thermal noise of the load can also be represented as a noise $e_3$ in series with the input, where

$$e_3^2 = \frac{4kTR_L}{g_m^2 R_L^2} \tag{3.18}$$

Normally $e_3 \ll e_2$ and can be neglected. The amplifier, with the noise sources referred to the input, can be designated as shown in Fig. 3.10. The amplifier is considered noiseless, and the noise is represented by the noise voltage and current sources. The model as represented has been simplified by assuming that the voltage

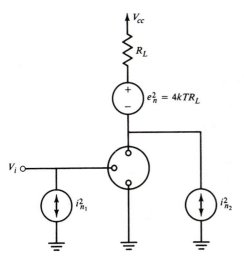

**FIGURE 3.9**
Model of a transistor amplifier including two shot noise sources and the noise source of the load resistance.

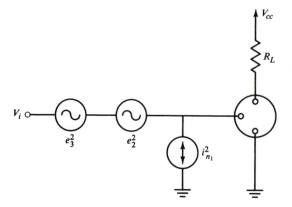

**FIGURE 3.10**
Another representation of the model shown in Fig. 3.9, but with the noise sources referenced to the input side of the amplifier.

gain and amplifier transadmittance are independent of frequency. In this case the total mean-square noise voltage will be proportional to frequency; but in the more general case, frequency-dependent transfer functions must be used, and the total mean-square noise voltage can only be obtained by integrating the instantaneous values over the frequency region of interest. The model also does not include any thermal noise present in the amplifier.

Transistor noise can originate from one of two sources.

### BJT noise

The principal noise sources in a bipolar transistor are the two shot noise sources and the thermal noise created in the base spreading resistor $r'_b$:

$$e_n^2 = 4kTr'_b \qquad \text{V}^2/\text{Hz} \tag{3.19}$$

$$i_{n_1}^2 = 2qI_B \qquad \text{A}^2/\text{Hz} \tag{3.20}$$

$$i_{n_2}^2 = 2qI_C \qquad \text{A}^2/\text{Hz} \tag{3.21}$$

The noise current source $i_{n_1}^2$ is connected between the base and emitter junctions, and the other current source is connected between the collector and emitter junctions. At high frequencies (above $f_\beta$ of the transistor) the noise currents increase with increasing frequencies, and the complete expression is

$$i_{n_o}^2 = i_n^2 \left(1 + \beta \frac{f^2}{f_T^2}\right) \tag{3.22}$$

Section 3.4 describes a more detailed model of the noise sources in a BJT amplifier.

### FET noise

The FET noise sources (excluding excess noise) are given by the following expressions:

$$e_n^2 = \frac{2.8kT}{g_m} \qquad \text{V}^2/\text{Hz} \tag{3.23}$$

and

$$i_n^2 = 2qI_g \qquad \text{A}^2/\text{Hz} \tag{3.24}$$

where $g_m$ is the mutual conductance and $I_g$ is the gate leakage current. The noise sources of MOSFETs and JFETs are the same, but $I_g$ is negligible for MOSFETs. The shot noise increases with frequency at very high frequencies; the total noise current is

$$i_n^2 = 2qI_g + \frac{2.8kT}{g_m} \, \omega^2 C_{gs'}^2 \qquad \text{A/Hz} \qquad (3.25)$$

where $C_{gs'}$ is approximately two-thirds of the transistor gate-to-source capacitance.

The transistor amplifier noise model will subsequently be used for the design of low-noise amplifiers, but first one of the parameters most often used to characterize the "noisiness" of a system, the noise figure, will be described.

## ▬ 3.3

### NOISE FIGURE, NOISE FACTOR, AND SENSITIVITY

Although the signal-to-noise ratio is the best measure of system input and output signal quality, a quantitative measure is also needed of how much noise is added by the circuit, whether the circuit is a passive filter, an amplifier, or an entire receiver. The noise factor $F$ has become a standard figure of merit of the amount of noise added in a circuit. According to the *IEEE Standards,* "The noise factor, at a specified input frequency, is defined as the ratio of (1) the total noise power per unit bandwidth available at the output port when the noise temperature of the input termination is standard (290 K) to (2) that portion of (1) engendered at the input frequency by the input termination."[2]

Available power refers to the maximum power that can be delivered from a generator with a source impedance $R_s$ to a load impedance $R_L$. For the network of Fig. 3.11 it is easily shown that the load receives maximum power when the load is matched to the source, that is, when

$$Z_L = Z_s^*$$

where $Z^*$ is the complex conjugate of the impedance $Z$. Under matched conditions the load power will be

$$P_o = \frac{E_s^2}{4R_s} \qquad (3.26)$$

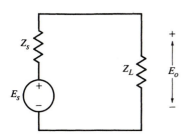

**FIGURE 3.11**
A simple circuit model for determining the maximum power transfer to the load impedance.

This is the maximum available power from the source $E_s$. Therefore the available noise power from a resistor $R$ is equal to $kT \, \Delta f$. That is, the available noise power is independent of the size of the resistor!

The IEEE definition for noise factor $F$ can be stated as

$$F = \frac{\text{available output noise power}}{\text{available output noise due to source}} \tag{3.27}$$

If $N_i$ is used to denote the noise power available from the source and $N$ is the total noise,

$$F = \frac{N_o}{(N_i)_o} \tag{3.28}$$

The $o$ subscripts indicate that the noise powers are specified at the output. It must be kept in mind that the symbols refer to the available noise powers. The noise factor depends upon the noise generated in a device and on its input termination, but not upon the output termination. Since the output noise power in a linear system is the sum of the noise due to the source plus the noise $N_a$ added in the system, the noise factor can be written as

$$F = \frac{N_i + N_a}{N_i} \tag{3.29}$$

The definition for available noise power used in Eqs. (3.26) to (3.29) must specify whether the noise is the total noise or the noise per unit bandwidth. Noise per unit bandwidth will be used unless otherwise noted. In Eq. (3.29) the values for noise can refer to the values at the input or output, and it is important only that one be consistent. We will use as standard notation

$N_i \; = \;$ available input noise per unit bandwidth

$N_a \, = \,$ available noise power added per unit bandwidth (referred to input)

Since the output power $S_o$ is the available input signal power $S_i$ times the power gain $G(f)$, Eq. (3.29) can be written as

$$F = \frac{N_i + N_a G(f) S_i}{N_i S_o} = \frac{S_i}{N_i} \frac{N_o}{S_o} \tag{3.30}$$

[The noise at the output $N_o$ is the noise at the input $N_i + N_a$ multiplied by the power gain $G(f)$.] Therefore the noise factor can also be written as

$$F = \frac{\text{input signal-to-noise ratio}}{\text{output signal-to-noise ratio}}$$

The noise factor $F$ is a measure of the degradation of the signal-to-noise ratio due to the noise added in the system. Note that the output signal-to-noise ratio is less than the input signal-to-noise ratio. This seeming contradiction is due to the noise figure's being specified at frequencies where both signal and noise are present.

Since the maximum available noise power is $E^2/(4R_g)$, the maximum available noise power per unit bandwidth from a source resistance is

$$N_i = \frac{4kTR_g}{4R_g} = kT \tag{3.31}$$

independent of the size of the source resistance. Hence the noise factor of a receiver is

$$F = 1 + \frac{N_a}{kT} \tag{3.32}$$

This is the noise factor measured in a unit bandwidth at a particular frequency and is often referred to as the *spot noise factor*. The source impedance does not appear in this expression for the noise factor, but it will subsequently be shown that the noise added depends on the source impedance, and hence so does the noise factor.

Note that in an ideal receiver no noise is added $(N_a = 0)$, so the receiver has a unity noise factor. Since the noise factor is always greater that 1, the output signal-to-noise ratio is always less than the input signal-to-noise ratio. That this does not agree with experience is a result of the definition of noise factor. A receiver will usually improve the signal-to-noise ratio through filtering of the input noise. Since the noise factor definition uses the same bandwidth in defining the two signal-to-noise ratios, the noise factor does not reflect the filtering quality of the receiver, and it is only one parameter to be considered in completely describing receiver performance.

**Average Noise Factor**

The noise performance of a communications system normally needs to be described over a range of frequencies. One method is to determine the spot noise factor at several frequencies. Another method found useful in noise measurements is to specify the average noise factor. The average noise factor is defined as the ratio of (1) the total noise power delivered into the output termination by the transducer when the noise temperature of the input termination is standard ($290°$ K) at all frequencies to (2) that portion of (1) engendered by the input termination.

The average noise factor $\bar{F}$ is

$$\bar{F} = \frac{\displaystyle\int F(f)G(f)\,df}{\displaystyle\int G(f)\,df} \tag{3.33}$$

where $G(f)$ is the system power (transducer) gain and $F(f)$ is the frequency-dependent noise factor. For a heterodyne system, the noise created by the input includes only that portion of the noise from the input termination that appears in the output via the principal frequency transformation of the system and does not include spurious contributions such as those from an image-frequency transformation.

## Noise Figure

The noise factor is often expressed in decibels. In this case it is called the *noise figure* (NF) and is defined as

$$\text{NF} = 10 \log F \tag{3.34}$$

Since the minimum value of $F = 1$, the noise figure of an ideal noiseless network is 0 dB.

## Noise Factor of Cascaded Networks

If the noise factor and power gain of individual networks are known, the noise factor of cascaded networks is readily determined. First consider the series combination of two networks with noise factors and power gains $F_1$, $G_1$ and $F_2$, $G_2$, respectively. If the available input noise power $N_i$ is equal to $kT$, the noise added by network 1 is

$$F_1 N_i - N_i = N_{a_1} = kT(F_1 - 1) \tag{3.35}$$

Likewise, the noise added by network 2 is

$$N_{a_2} = kT(F_2 - 1)$$

and the noise added in network 2, referred to the input, is

$$\frac{N_{a_2}}{G_1} = \frac{kT(F_2 - 1)}{G_1} \tag{3.36}$$

The overall noise factor is thus

$$F = \frac{\text{available input noise power} + \text{noise added}}{\text{available input noise power}}$$

$$= \frac{kT + (F_1 - 1)kT + (F_2 - 1)kT/G_1}{kT} \tag{3.37}$$

$$= F_1 + \frac{F_2 - 1}{G_1} \tag{3.38}$$

Equation (3.38) states that if the power gain of the first stage is large, the overall noise factor will be essentially that of the first stage. In other cases, the noise factor of the second stage, and even of succeeding stages, will be an important factor in the overall noise factor. Equation (3.38) is readily extended to $n$ stages. For an $n$-stage system

$$F = F_1 + \frac{F_2 - 1}{G_1} + \frac{F_3 - 1}{G_1 G_2} + \cdots + \frac{F_n - 1}{G_1 G_2 \cdots G_{n-1}} \tag{3.39}$$

EXAMPLE 3.3. For the system shown in Fig. 3.12, the first stage has a noise figure of 2 dB and a gain of 12 dB; the second stage has a noise figure of 6 dB and a power gain of

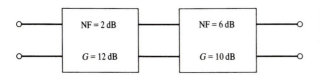

**FIGURE 3.12**
Numerical example of two cascaded, noisy networks.

10 dB; the second stage has a noise figure of 6 dB and a power gain of 10 dB. What is the overall noise figure?

**Solution.** Equation (3.38) expresses the noise factor $F$ in terms of the noise factors for each stage. Thus the noise figures must first be converted to noise factor values:

$$F_1 = 1.59 \qquad F_2 = 4$$

The corresponding gain values are

$$G_1 = 15.9 \qquad G_2 = 10$$

The overall noise factor is

$$F = 1.59 + \frac{4-1}{15.9} = 1.779$$

and the noise figure of the two-stage system is

$$NF = 10 \log 1.779 = 2.5 \text{ dB}$$

**EXAMPLE 3.4.** If $G_1$ and $G_2$ of Example 3.3 are independent of frequency, what will be the total output noise power of the cascaded system in a 3-kHz bandwidth? The operating temperature is $290°$ K.

**Solution.** Since $N_i + N_a = FkTB$,

$$kTB = 1.37 \times 10^{-23} \times 290 \times 3 \times 10^3$$

$$= 1.192 \times 10^{-17} \text{ W}$$

$$N_i + N_a = 1.779 kTB = 2.12 \times 10^{-17} \text{ W}$$

and the output noise

$$N_o = G_1 G_2 (N_i + N_a) = 159 \times 2.12 \times 10^{-17} = 337 \times 10^{-17} \text{ W}$$

**Noise Temperature**

The noise factor will normally lie between 1 and 10. For situations in which an expanded scale is needed, the system noise factor is usually expressed in terms of noise temperature. The noise factor is given by

$$F = 1 + \frac{N_a}{N_i} = 1 + \frac{N_a}{kT} \tag{3.40}$$

where $T$ is the reference noise temperature. The noise added can be interpreted as the available noise from a resistor whose temperature is $T_r$. That is,

$$F = 1 + \frac{T_r}{T} \tag{3.41}$$

or
$$T_r = (F - 1)T \tag{3.42}$$

where $T_r$ is referred to as the *system noise temperature.*

**EXAMPLE 3.5.** What is the variation in noise temperature as the noise factor varies from 1 to 1.6 (NF varies from 0 to 2 dB)? Assume the reference temperature is 290 K.

**Solution.** When the noise factor is 1, the noise temperature is 0. When the noise factor is 1.6,

$$T_r = (1.6 - 1)290 = 174 \text{ K}$$

Thus the change in the noise temperature is much greater than the change in the noise factor. This is a principal reason for using noise temperature to describe system noise.

## Sensitivity

The available input signal level $S_i$ for a given output signal-to-noise ratio $(S/N)_o$ is referred to as the *system sensitivity,* or *noise floor.* The input voltage level corresponding to $S_i$ is called the *minimum detectable signal.* Although the signal-to-noise ratio will depend on the system frequency response, we will assume for simplicity that the frequency response can be represented by the ideal characteristic shown in Fig. 3.13. Although this frequency response can never be realized in an actual receiver, it is closely approximated in many communication systems, especially those which include a narrow bandpass filter. When the frequency characteristic is ideal, Eq. (3.5) for the total available noise power from a resistor can be written as

$$\frac{E_n^2}{4R} = kTB \tag{3.43}$$

where $B$ is the bandwidth. Therefore Eq. (3.30) can be written as

$$S_i = F(kTB)\left(\frac{S}{N}\right)_o \tag{3.44}$$

where $N_o$ is now the total noise power at the output.

**EXAMPLE 3.6.** What minimum input signal will give an output signal-to-noise ratio of 0 dB in a system that has an input impedance equal to 50 $\Omega$, a noise figure (NF) of 8 dB, and a bandwidth of 2.1 kHz?

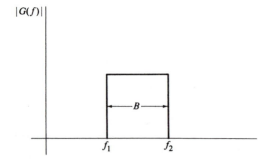

**FIGURE 3.13**
Frequency response of the magnitude of an ideal bandpass filter.

*Solution.* For a 0-dB output signal-to-noise ratio and a 290 K operating temperature, Eq. (3.44) can be written

$$10 \log S_i = \text{NF} - 144 + 10 \log B$$

where $S_i$ is in milliwatts and $B$ is in kilohertz. For a bandwidth of 2.1 kHz,

$$S_i = -133 \text{ dBm} \qquad 133 \text{ dB below 1-mW level}$$

where $S_i$ is the available input power and is related to the input signal voltage by Eq. (3.26). Thus

$$S_i = \frac{E_i^2}{4R_s} = 5.02 \times 10^{-17} \text{ W}$$

Since $R_s = 50 \text{ }\Omega$,

$$E_i = 0.10 \text{ } \mu\text{V}$$

That is, for these specifications the noise floor for an output signal-to-noise ratio of 1 is 0.10 $\mu$V.

**EXAMPLE 3.7.** What is the minimum detectable signal or noise floor of the system in the previous example for an output signal-to-noise ratio of 10 dB?

*Solution.* In this case Eq. (3.40) becomes

$$10 \log S_i = \text{NF} - 134 + 10 \log B = -123 \text{ dBm}$$

$$S_i = 5 \times 10^{-13} \times 10^{-3} \text{ W}$$

and the minimum detectable signal is

$$E_i = 0.32 \text{ } \mu\text{V}$$

Sensitivity is always specified for a given signal-to-noise ratio. Although the required output signal-to-noise ratio may not be the same as that used in the sensitivity specification, sensitivity does provide an objective measure for comparing receiver performance. The required signal-to-noise ratio at the receiver output will depend on the function of the receiver and on whether or not additional signal processing (such as correlation detection) is performed. An output signal-to-noise ratio between 0 and 10 dB is adequate for normal listening.

Receiver noise figure is a measure of how much noise is added by the system. A low noise figure is often desirable, but there are situations in which this is of little importance. This is particularly true when the input noise is much greater than the noise added by the system. Numerical examples will illustrate this point.

**EXAMPLE 3.8.** Consider a communications receiver with a 50-$\Omega$ input impedance, a $B$ of 3 kHz, and a 4-dB noise figure. The noise floor of this receiver for an output signal-to-noise ratio of 10 dB is found to be, using Eqs. (3.26) and (3.44),

$$S_i = -125 \text{ dBm} = 3 \times 10^{-16} \text{ W}$$

$$E_i = 0.245 \text{ } \mu\text{V}$$

An input signal of 0.245 $\mu$V will produce a 10-dB output signal-to-noise ratio. Now consider the performance of this receiver when it is connected to an antenna with a noise figure of 20 dB. Expressing the noise at the antenna in terms of the noise figure has become accepted practice, as it facilitates the numerical analysis. *Antenna noise factor* is defined to be

$$F_{\text{ant}} = \frac{\text{noise}_{\text{ant}} + \text{thermal noise}}{\text{thermal noise}} \qquad (3.45)$$

or $\text{noise}_{\text{ant}} = (F_a - 1) \times$ thermal noise. Antenna noise refers to the total noise picked up by the antenna, primarily from external sources. The antenna noise factor in this example is 100. Hence the antenna noise is seen to be

$$N_{\text{ant}} = 99 \times \text{thermal noise} = 99kTB$$

The total input noise is the antenna noise plus the source noise, or $100kTB$. The output noise (referred to the input) is

$$N_o = N_{\text{ant}} + N_i + N_a = (F_a - 1)kTB + F_r kTB$$

$$= N_{\text{ant}} + F_r kTB$$

where $F_r$ refers to the receiver noise factor. The output signal-to-noise ratio is thus

$$\frac{S_i}{N_i F} = \frac{S_o}{N_o}$$

$$\left( \frac{S}{N} \right)_o = \frac{S_i}{(F_a + F_r - 1)kTB} \qquad (3.46)$$

In this example the antenna noise figure is 20 dB which corresponds to a noise factor of 100. Since the receiver noise factor is 2.5 (NF $= 4$ dB), the input signal required for a 10-dB output signal-to-noise ratio is

$$S_i = \left( \frac{S}{N} \right)_o (100 + 2.5 - 1)kTB$$

$$= 10 \times 101.5 \times 397 \times 10^{-23} \times 3 \times 10^3$$

$$= 1.203 \times 10^{-14} \text{ W}$$

Thus the minimum detectable signal for a 10-dB output signal-to-noise ratio is

$$E_i = 1.56 \, \mu\text{V}$$

This is much larger than the $0.245 \, \mu\text{V}$ required if there were no antenna noise.

**EXAMPLE 3.9.** What will be the minimum detectable signal level in the previous example if a receiver with a noise figure of 10 dB is substituted?

*Solution.* Since the receiver noise factor is 10, Eq. (3.46) becomes for this system

$$S_i = \left( \frac{S}{N} \right)_o (100 + 10 - 1)kTB$$

$$= 1.29 \times 10^{-14} \text{ W}$$

and the minimum detectable signal is

$$E_i = 1.6 \, \mu\text{V}$$

A 6-dB increase in the receiver noise figure results in only a 0.3-dB reduction in the output signal-to-noise ratio because the noise added by the receiver is much less than the antenna noise.

Examples 3.8 and 3.9 illustrate that if the input noise is large, very little is gained by reducing the system noise figure below some acceptable level. For communications receivers operating below 30 MHz, 8 to 10 dB is usually taken as an acceptable receiver noise figure because of the large antenna noise figure. However, as the frequency is increased above 30 MHz, receivers with lower noise figures are desirable because the antenna noise is much less at the higher frequencies. When the input noise is large, not only the receiver bandwidth but also the actual passband must be selected for optimum performance by considering the frequency characteristics of the antenna noise.

Noise comparisons of two receivers must be used with care since the network with the lowest noise figure does not necessarily have the highest output signal-to-noise ratio. The following section on low-noise design proves this important point.

## ▉ 3.4

### DESIGN OF LOW-NOISE NETWORKS

### Network Noise Representation

Any linear noisy two-port network (or amplifier) can be represented by a noiseless network plus two noise generators $e_n$ and $i_n$, as shown in Fig. 3.14. Two noise sources are required to represent the network noise because noise may exist at the output with the input terminals short- or open-circuited. In general, $e_n$ and $i_n$ will be frequency-dependent. In the noise model for the common-emitter amplifier described in Sec. 3.2, the noise current $i_n$ was due to the input bias current shot noise, and the noise voltage $e_n$ was due to the output bias current shot noise plus the thermal noise due to the load resistor. The equivalent short-circuit input rms noise voltage $e_n$ is the noise voltage that would appear to originate at the input of

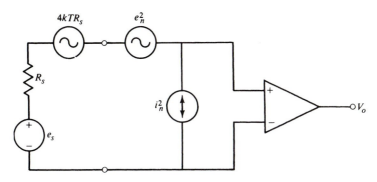

**FIGURE 3.14**
An amplifier noise source model.

the noiseless network if the input terminals were short-circuited. The rms noise current $i_n$ represents the remainder of the network noise. To determine $i_n$, a resistor $R_s$ is shunted across the input terminals and the output noise is measured.

### Network Noise Factors

In this section we will consider the more general problem in which the source and network input resistances are not necessarily matched.

The output noise power referred to the input in a unit bandwidth of the amplifier illustrated in Fig. 3.14 is then (assuming $i_n$ and $e_n$ are uncorrelated)

$$N = e_n^2 + i_n^2 R_s^2 + 4kTR_s \tag{3.47}$$

The noise voltage $e_n$ is determined by measuring the output voltage with the input terminals short-circuited. The output noise is then measured with a specified source resistance. From this measurement $i_n$ can be determined, provided the power gain is known. The noise added by the network is

$$N_a = e_n^2 + i_n^2 R_s^2 \tag{3.48}$$

The network noise factor in any unit bandwidth can be defined as

$$F = \frac{e_n^2 + i_n^2 R_s^2 + 4kTR_s}{4kTR_s} \tag{3.49}$$

which agrees with the previous definition when $R_s = R_i$.

EXAMPLE 3.10. Plots of the input noise voltage and input noise current for the Fairchild 741 operational amplifier are presented in Fig. 3.15. At 1 kHz the input noise voltage is approximately

$$e_n^2 \gtrsim 8 \times 10^{-16} \ \text{V}^2/\text{Hz}$$

and the input noise current is approximately

$$i_n^2 = 9 \times 10^{-25} \ \text{A}^2/\text{Hz}$$

If a 10-k$\Omega$ source impedance is used, the amplifier noise factor will be

$$F = 1 + \frac{8 \times 10^{-16} + 0.9 \times 10^{-16}}{4 \times 1.37 \times 10^{-23} \times 290 \times 10^4}$$

$$= 6.6$$

Equation (3.49) indicates that a network's noise factor depends upon the source resistance. The value of source resistance that minimizes the noise factor can be formed by differentiating Eq. (3.49) with respect to $R_s$:

$$\frac{dF}{dR_s} = \frac{1}{4kT} \left( i_n^2 - \frac{e_n^2}{r_s^2} \right) = 0$$

$$\tag{3.50}$$

or

$$R_s = \frac{e_n}{i_n}$$

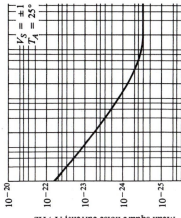

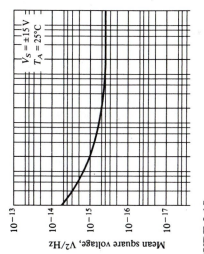

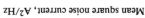

**FIGURE 3.15**

Mean-square voltage and noise sources of a 741 operational amplifier. (*Courtesy of Fairchild Semiconductor.*)

The value of source resistance that minimizes the noise factor is equal to the input noise voltage divided by the input noise current. If $i_n$ is large relative to $e_n$, a low value of source resistance is called for; if $i_n$ is small (such as in a FET input amplifier), a large value of $R_s$ is needed to minimize the noise factor. Since $e_n$ and $i_n$ are, in general, frequency-dependent, the source resistance required for minimization will be also; $R_s$ is usually chosen to minimize the spot noise factor at a specified frequency.

**EXAMPLE 3.11.** What will be the minimum noise figure of the 741 operational amplifier of the preceding example at 1 kHz?

**Solution.** The value of source resistance that minimizes the noise figure is found from Eq. (3.50) (and Fig. 3.15):

$$(R_s)_{\text{opt}} = \left(\frac{8 \times 10^{-16}}{9 \times 10^{-25}}\right)^{1/2} = 30 \times 10^3 \ \Omega$$

The minimum noise factor is found by using this value of source resistance in Eq. (3.49):

$$F = 1 + \frac{8 \times 10^{-16} + 9 \times 10^{-25} \times 9 \times 10^8}{4 \times 1.37 \times 10^{-23} \times 290 \times 30 \times 10^3}$$

$$= 4.35$$

and the minimum noise figure is

$$\text{NF} = 6.4 \text{ dB}$$

## Low-Noise Design

The concept of optimizing the source resistance to minimize the noise figure must be used with caution. A better criterion of a network's noise performance is the output signal-to-noise ratio. Minimizing a network's noise figure does not necessarily maximize the output signal-to-noise ratio, since changing the source resistance also changes the input signal-to-noise ratio. The signal-to-noise ratio is readily determined by replacing the input to the noiseless network by its Thévenin equivalent. If the noise sources are independent, the equivalent input circuit is as shown in Fig. 3.16. There are three noise sources due to $e_n$, $i_n$, and the thermal noise generated in the source resistor $R_s$. Since the voltage sources are in series with the signal $e_s$, they will have the same input-to-output transfer function, and the signal-to-noise ratio in any unit bandwidth is given by

$$\frac{S}{N} = \frac{e_s^2}{e_n^2 + i_n^2 R_s^2 + 4kTR_s} \tag{3.51}$$

If the applied signal is not a function of source resistance, then it is evident that the smaller the source resistance, the larger will be the signal-to-noise ratio. It would be detrimental to circuit performance to increase the source resistance in order to optimize the noise figure. Although this would result in the minimum noise figure, it would not maximize the output signal-to-noise ratio.

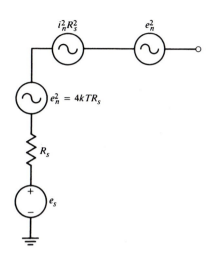

**FIGURE 3.16**
Amplifier equivalent noise voltage sources.

$$e_n^2 = 4kTR_s$$

$R_s$

$e_s$

**EXAMPLE 3.12.** The previous example found that a source resistance of 9.42 k$\Omega$ resulted in a minimum noise factor of 4.35. For this value of source resistance and the values of $e_n$ and $i_n$ specified in Example 3.10, the total noise is found to be

$$N = e_n^2 + i_n^2 R_s^2 + 4kTR_s = 17.48 \times 10^{-16} \text{ V}^2/\text{Hz}$$

If the source resistance is zero, the total noise is

$$N = e_n^2 = 8 \times 10^{-16} \text{ V}^2/\text{Hz}$$

which is less than 50 percent of the noise obtained when the noise factor is minimized.

If the signal voltage is a function of source resistance, then it is often possible to simultaneously minimize the noise figure and maximize the output signal-to-noise ratio. A most important case is that in which the source can be transformer-coupled to the amplifier input. In this case

$$R_s' = R_s N^2 \tag{3.52}$$

where $N$ is the transformer turns ratio and $R_s'$ is the reflected source impedance. The equivalent signal voltage will be

$$e_s' = e_s N \tag{3.53}$$

so the output signal-to-noise ratio (assuming the transformer does not add any noise) will be

$$\left(\frac{S}{N}\right)_o = \frac{N^2 e_s^2}{e_n^2 + i_n^2 N^4 R_s^2 + 4kTN^2 R_s} \tag{3.54}$$

The rate of change of the output signal-to-noise ratio as a function of $N$ is given by

$$\frac{d(S/N)_o}{dN} = \frac{2Ne_s^2(e_n^2 + i_n N^4 R_s^2 + 4kTN^2 R_s - 2N^4 i_n^2 R_s^2 - 4kTN^2 R_s)}{e_n^2 + i_n^2 N^4 R_s^2 + 4kTN^2 R_s}$$

which is equal to zero for

$$N^2 R_s = \frac{e_n}{i_n} \tag{3.55}$$

provided $e_s$ is not a function of $R_s$. This is also the value of source resistance that minimizes the noise figure. Therefore, if noiseless transformer coupling can be used to match the source to the amplifier input, then the turns ratio that minimizes the noise figure will also maximize the output signal-to-noise ratio.

## A Low-Noise Amplifier

Figure 3.17 illustrates a low-noise amplifier design using the concepts discussed in the previous section. This amplifier is a common-gate amplifier selected for its relatively low input impedance so that it can be matched to the source resistance $R_s$. The resistor $R_b$ serves to bias the circuit for linear operation, and capacitors $C_1$ and $C_2$ are short circuits in the frequency region of interest. The equivalent source impedance looking back from the gate to source terminals is found from the equivalent circuit, shown in Fig. 3.18, to be

$$Z'_s = R_s \left( \frac{N_1 + N_2}{N_1} \right)^2 \tag{3.56}$$

For the minimum noise factor and maximum signal-to-noise ratio, the turns ratio is selected so that

$$R_s \left( \frac{N_1 + N_2}{N_1} \right)^2 = \frac{e_n}{i_n} = R_n$$

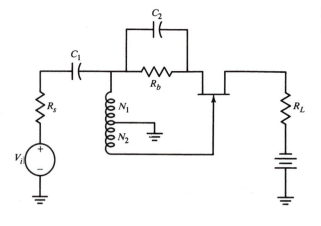

**FIGURE 3.17**
A low-noise FET amplifier.

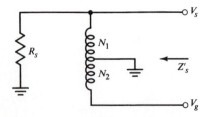

**FIGURE 3.18**
Input circuit of amplifier illustrated in Fig. 3.17.

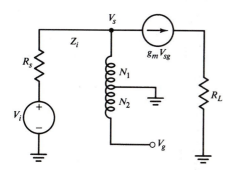

**FIGURE 3.19**
A small-signal equivalent circuit of amplifier shown in Fig. 3.17.

where $e_n$ and $i_n$ are the noise voltage and current, respectively, of the transistor.

The midfrequency small-signal equivalent circuit, with $r_d$ neglected, is shown in Fig. 3.19. If the reactance of the transformer is sufficiently large, little current will flow through the windings, and the input impedance will be

$$Z_i = \frac{V_s}{g_m V_{sg}} = \frac{V_s}{g_m[1 + (N_2/N_1)]V_s}$$

$$= \left[ g_m \left( 1 + \frac{N_2}{N_1} \right) \right]^{-1}$$

The circuit could also be realized with the gate grounded, as shown in Fig. 3.20a. In this case the small-signal equivalent circuit will be as shown in Fig. 3.20b The source impedance is

$$Z'_s = R_s \left( \frac{N_1 + N_2}{N_1} \right)^2$$

and

$$Z_i = \left[ g_m \left( \frac{N_1 + N_2}{N_1} \right)^2 \right]^{-1}$$

**Optimization of BJT Bias Current**

The total input noise of a common-emitter amplifier is

$$N_i = 4kTR_s + 4kTR'_b + \left( \frac{r_\pi + r'_b}{\beta} \right)^2 \left( i_{n_2}^2 + \frac{4kT}{R_L} \right) + i_{n_1}^2 R_s^2 \qquad \text{V}^2/\text{Hz} \qquad (3.57)$$

Since $R_\pi$, $i_{n_1}$, and $i_{n_2}$ are functions of the collector direct current, it is possible to adjust the bias current for minimum noise. This will maximize the signal-to-noise ratio and minimize the noise factor. The noise contributed by the load resistor will be assumed small compared to the other terms. (This approximation is valid as long as $R_L \gg r_\pi/\beta$, which is usually the case.) Since

$$i_{n_1}^2 = 2q I_B = 2q \frac{I_c}{\beta}$$

$$i_{n_2}^2 = 2q I_c$$

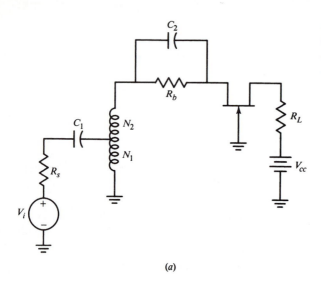

**FIGURE 3.20**
(*a*) A common-gate amplifi-er; (*b*) simplified equivalent circuit.

(*a*)

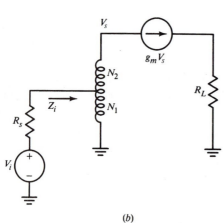

(*b*)

and
$$\frac{r_\pi}{\beta} = \frac{1}{g_m} = \frac{V_T}{I_c}$$

the input noise can be written as

$$N_i = 4kTR_s + 4kTR_b' + \frac{2qI_c}{\beta}R_s^2 + \frac{2qV_T^2}{I_c} + \frac{4qV_Tr_{b'}}{\beta} + \frac{2qI_cr_{b'}^2}{\beta^2} \quad \text{V}^2/\text{Hz} \quad (3.58)$$

The derivative of the noise with respect to collector current is

$$\frac{dN_i}{dI_c} = \frac{2qR_s^2}{\beta} + \frac{2qr_{b'}^2}{\beta^2} - \frac{2qV_T^2}{I_c^2}$$

The value of direct current, which minimizes the noise, is

$$I_c = \frac{V_T\beta}{(r_{b'}^2 + \beta R_s^2)^{1/2}} \quad (3.59)$$

## ▨ 3.5

## INTERMODULATION DISTORTION

The previous sections have considered the effects of low-level noise on receiver performance. In this section we will show that larger signals that are close in frequency to the desired signal can also affect receiver performance.

All communications receivers contain some degree of nonlinearity which can cause a change in the frequencies of the input signals and/or a change in the network gain. For these reasons the network nonlinearities need to be clearly delineated and considered during the design phase. The network nonlinearities can be described by the expansion

$$y(x) = k_1 f(x) + k_2 [f(x)]^2 + k_3 [f(x)]^3 + \text{higher-order terms} \qquad (3.60)$$

It is assumed that the nonlinearity is frequency-independent and can be adequately described by the first three terms; the higher-order terms will be ignored. Let $f(x)$ consist of two sinusoidal signals:

$$f(x) = A_1 \cos \omega_1 t + A_2 \cos \omega_2 t$$

If $\omega_1$ and $\omega_2$ are sufficiently close together, $k_i$ can be considered the same for both signals. Also, for simplicity we will assume that all the $k_i$ are real. If Eq. (3.60) describes the network's response to an input $f(x)$, the response will be

$$y = k_1 (A_1 \cos \omega_1 t + A_2 \cos \omega_2 t) + k_2 (A_1 \cos \omega_1 t + A_2 \cos \omega_2 t)^2$$

$$+ k_3 (A_1 \cos \omega_1 t + A_2 \cos \omega_2 t)^3$$

$$= k_1 (A_1 \cos \omega_1 t + A_2 \cos \omega_2 t)$$

$$+ k_2 \left[ A_1^2 \frac{1 + \cos 2\omega_1 t}{2} + A_2^2 \frac{1 + \cos 2\omega_2 t}{2} \right.$$

$$\left. + A_1 A_2 \frac{\cos(\omega_1 + \omega_2)t + \cos(\omega_1 - \omega_2)t}{2} \right] \qquad (3.61)$$

$$+ k_3 \left\{ \left[ A_1^3 \left( \frac{\cos \omega_1 t}{2} + \frac{\cos \omega_1 t}{4} + \frac{\cos 3\omega_1 t}{4} \right) + A_2^3 \left( \frac{3 \cos \omega_2 t}{4} + \frac{\cos 3\omega_2 t}{4} \right) \right] \right.$$

$$+ A_1^2 A_2 [\tfrac{3}{2} \cos \omega_2 t + \tfrac{3}{4} \cos(2\omega_1 + \omega_2)t + \tfrac{3}{4} \cos(2\omega_1 - \omega_2)t]$$

$$\left. + A_2^2 A_1 [\tfrac{3}{2} \cos \omega_1 t + \tfrac{3}{4} \cos(2\omega_2 + \omega_1)t + \tfrac{3}{4} \cos(2\omega_2 - \omega_1)t] \right\}$$

### Gain Compression

One effect of the nonlinearity that can be deduced from Eq. (3.61) is that the amplitude of the $\cos \omega_1 t$ signal has become

$$A_1' = k_1 A_1 + k_3 (\tfrac{3}{4} A_1^3 + \tfrac{3}{2} A_1 A_2^2) \qquad (3.62)$$

Normally $k_3$ will be negative, and a large signal $A_2 \cos \omega_2 t$ can effectively mask a smaller signal $A_1 \cos \omega_1 t$, since it results in a reduced gain because of the third-order coefficient $k_3$. To avoid the "gain compression," the third-order coefficient $k_3$ must be reduced. Also, multiple signals will result in a further reduction of the gain. If only one signal is present, the ratio of gain with distortion to the idealized (linear) gain is

$$\frac{A_1'}{A_1} = \frac{k_1 + k_3(\frac{3}{4}A_1^2)}{k_1} \tag{3.63}$$

and is referred to as the *single-tone gain compression factor.* Figure 3.21 illustrates how the $k_3$ term causes the gain to deviate from the idealized curve. The point at which the power gain is down 1 dB from the ideal is referred to as the *1-dB compression point.* Receivers must be operated below their gain compression point if the nonlinear gain region is to be avoided.

**Second Harmonic Distortion**

Second harmonics will occur at the receiver output because of the $k_2$ term. If a single signal is present at the receiver input, the amplitude of the second harmonic will be

$$\frac{k_2 A_1^2}{2} \tag{3.64}$$

**Intermodulation Distortion Ratio**

Another important effect of receiver nonlinearity is the intermodulation distortion caused by the cubic term in Eq. (3.61). Equation (3.61) shows that the cubic term

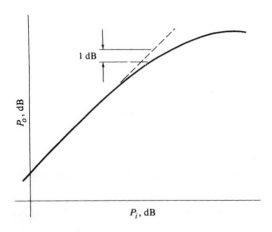

**FIGURE 3.21**
Idealized amplifier power-transfer characteristic illustrating the 1-dB compression point.

creates the intermodulation frequencies $2\omega_1 \pm \omega_2$ and $2\omega_2 \pm \omega_1$. If $\omega_1$ and $\omega_2$ are of approximately the same frequency, the higher frequencies $2\omega_1 + \omega_2$ and $2\omega_2 + \omega_1$ will normally be outside the passband and can be eliminated with filtering, but the two frequencies $2\omega_1 - \omega_2$ and $2\omega_2 - \omega_1$ can lie in the system passband and appear at the output as signal distortion. The *intermodulation distortion ratio (IMR)* is defined as the ratio of the amplitude of one of the intermodulation terms to the amplitude of the desired output signal. For the two-tone input signals, Eq. (3.62) yields

$$\text{IMR} = \frac{\frac{3}{4}k_3 A_1^2 A_2}{k_1 A_1} = \frac{3}{4}\frac{k_3}{k_1} A_1 A_2 \tag{3.65}$$

**Intercept point**

The *intermodulation distortion (IMD) power* is defined as

$$P_d = \frac{(\frac{3}{4}k_3 A_1^2 A_2)^2}{2} \tag{3.66}$$

If the two input amplitudes are the same, the intermodulation distortion power varies as the cube of the input power; that is, for every 1-dB change in input power there is a 3-dB change in the power of the intermodulation terms. In this case,

$$P_d = (k_d P_i)^3 \tag{3.67}$$

where $P_i = A_1^2/2$, the power in one signal component, and $k_d$ is the scale factor. The ratio ($P_{\text{IMR}}$) of the IMD power to the desired output power for the case where the two input signal amplitudes are the same is defined as

$$P_{\text{IMR}} = \frac{P_d}{P_o} \tag{3.68}$$

Since the distortion power is proportional to the cube of the input power and the output power is directly proportional to the input power,

$$P_{\text{IMR}} = (K_i P_i)^2 \tag{3.69}$$

A normalized plot of the desired output and intermodulation powers is shown in Fig. 3.22. On a logarithmic scale, the IMD power increases 3 times as fast as the desired output power. The value of input power for which the IMD power is equal to the output power contributed by the linear term $(k_1 A_1)^2/2$ is referred to as the *intercept point* $P_I$, a term which is finding increasing usage, especially in describing the distortion characteristics of frequency mixers. In order to express $P_I$ in terms of $P_{\text{IMR}}$ and $P_i$, note that when the output distortion power and the desired output power are equal (the intercept point), the IMR ratio is, by definition, unity, and thus Eq. (3.69) becomes

$$1 = (K_i P_i)^2$$

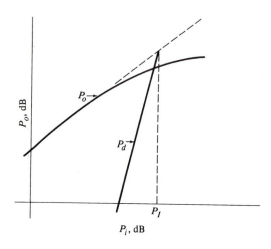

**FIGURE 3.22**
Power-transfer characteristic, including the third-order intermodulation distortion $P_d$ and the two-tone third-order intercept point $P_I$.

Since
$$P_i = P_I$$

at this signal level,

$$K_i = (P_I)^{-1}$$

and Eq. (3.69) can be written as

$$P_{\text{IMR}} = \left( \frac{P_i}{P_I} \right)^2 \tag{3.70}$$

where $P_i$ is the input power $A_1^2/2$.

EXAMPLE 3.13. If the system intercept point is $+20$ dBm, what is the IMR for an input signal power of 0 dBm?

*Solution.* To solve this problem Eq. (3.70) is used. Thus $P_{\text{IMR}} = 0 - 2 \times 20 = -40$ dB.

A receiver's intercept point is a measure of the distortion created in the receiver, and it is also a measure of its ability to reject large-amplitude signals that lie in close frequency proximity to a weak signal targeted for reception. The receiver intercept point is primarily determined by the intercept point of the input mixer. Double-balanced diode ring mixers with intercept points of $+15$ to $+27$ dBm are readily available and relatively inexpensive, but they are used in only the more expensive receivers because they require higher oscillator drive levels ($+7$ to $+23$ dBm) than do other types of mixers (such as integrated-circuit mixers which use field-effect transistors). Higher local oscillator levels will usually require additional shielding of the system components. Mixer specifications normally list either the two-tone third-order distortion at some level, from which the corresponding intercept point can be determined, or the 1-dB RF input compression level, which is much easier to measure. As a practical rule of thumb, the 1-dB compression point is approximately 15 dB below the two-tone third-order intercept point.

## Dynamic Range

The minimum detectable signal in a receiver is determined by the input thermal noise and the noise contributed by the receiver. At the other extreme, when the input signal is too large, the signal detection is limited by the distortion. The amount of distortion that can be tolerated will depend somewhat on the type of signals, but for purposes of an objective definition, the upper limit of signal detectability will be considered the signal level at which the intermodulation distortion is equal to the minimum detectable signal. The ratio of the minimum detectable signal to the signal power that causes the distortion power (in one frequency component) to be equal to the noise floor $N_f$ is referred to as the receiver's *dynamic range*. Since the ideal power out is

$$P_o = k_1^2 P_i \tag{3.71}$$

the intermodulation distortion ratio can be written as

$$P_{\text{IMR}} = \frac{P_d}{P_o} = \frac{P_d}{k_1^2 P_i} = \left(\frac{P_i}{P_I}\right)^2$$

Define $P_{di} = P_d/k_1^2$ (the distortion referred to the input); then

$$P_{\text{IMR}} = \frac{P_{di}}{P_i} = \left(\frac{P_i}{P_I}\right)^2 \tag{3.72}$$

When $P_{di}$ is equal to the noise floor $N_f$,

$$\frac{N_f}{P_i} = \frac{P_i^2}{P_I^2}$$

or

$$P_i = (P_I^2 N_f)^{1/3} \tag{3.73}$$

Therefore the dynamic range DR is

$$\text{DR} = \frac{(P_I^2 N_f)^{1/3}}{N_f} = \left(\frac{P_I}{N_f}\right)^{2/3} \tag{3.74}$$

It must be kept in mind that the intercept point and noise floor are measured at the same point in the system. Also, the noise floor depends upon the specified output signal-to-noise ratio, and thus so does the dynamic range.

**EXAMPLE 3.14.** The receiver of Example 3.6 has an intercept point of 20 dBm. What will be the dynamic range for an output signal-to-noise ratio of 10 dB?

*Solution.* From Example 3.7 it is known that the required available input signal power $S_i = -123$ dBm for a 10-dB output signal-to-noise ratio. The receiver's dynamic range [using Eq. (3.74)] is thus

$$\text{DR} = 0.67(20 + 123) = 95.3 \text{ dB}$$

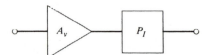

**FIGURE 3.23**
Circuit with a power intercept point $P_I$ preceded by a preamplifier with a voltage gain $A_v$.

If a linear preamplifier with a voltage gain $A_v$ is added before a network that has an intercept point $P_I$, as illustrated in Fig. 3.23, then the overall intercept point is $P_I/A_v^2$. The output power due to the linear term is

$$P_o = \frac{(k_1 A_1)^2}{2} A_v^2 \tag{3.75}$$

and the intermodulation distortion output will be

$$P_d = \frac{\frac{3}{4}(k_3 A_v^3 A_1^3)^2}{2} \tag{3.76}$$

and the IMD ratio is

$$P_{\text{IMR}} = \frac{\frac{3}{4}(k_3 A_1^3)^2}{(k_1 A_1)^2} A_v^4$$

$$= (k_i P_i)^2 A_v^4$$

when $P_{\text{IMR}} = 1$, $P_i = P_I$, so

$$P_I = (K_i A_v^2)^{-1} \tag{3.77}$$

The addition of a linear preamplifier reduces the intercept point. Unless the preamplifier can reduce the noise floor by the same amount the intercept point is reduced, the dynamic range will be decreased by the addition of the preamplifier.

## SINAD

Another figure of merit which is becoming widely used for commercial (particularly stereo) receivers is the SINAD ratio. SINAD is the ratio of signal plus noise plus distortion powers to noise and distortion powers.

$$\text{SINAD} = \frac{S + N + D}{N + D} \tag{3.78}$$

For many applications the distortion, even at low levels, is an important factor in describing performance. Also, the SINAD ratio is easy to measure using the method illustrated in Fig. 3.24. The measurement procedure consists of applying an RF signal modulated by an audio signal (usually 1 kHz) and measuring signal plus noise plus distortion. The audio signal is then filtered out, and the noise plus distortion is determined. A SINAD measurement is the same as a total harmonic distortion measurement. The SINAD ratio can also be used to define receiver sensitivity. One possibility is to define receiver sensitivity as the amount of RF signal needed to get a specified SINAD ratio.

**FIGURE 3.24**
SINAD-measuring network.

## 3.6

## PROBLEMS

**3.1** Determine an expression for the total noise voltage squared across the output terminals of the circuit shown in Fig. P3.1.

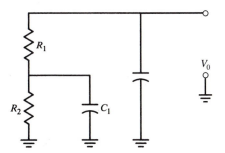

**FIGURE P3.1**
Frequency-dependent network including two noise sources.

**3.2** Calculate the 3-dB and noise bandwidths of the circuit shown in Fig. P3.2.

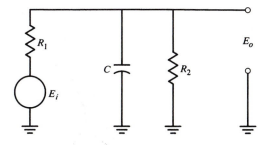

**FIGURE P3.2**
Frequency-dependent network with two noisy resistors.

**3.3** Show that the amount of excess noise $E_n^2 = (k/f)\,df$ generated in each decade of frequency is constant, independent of frequency.

**3.4** Derive the equation for the noise factor of $n$ cascaded networks, each with a noise factor $F_i$ and power gain $G_i$.

**3.5** Determine the equivalent input noise sources of a common-base amplifier in terms of the thermal and shot noise sources.

**3.6** A receiver has a 3-kHz bandwidth, a 50-$\Omega$ input impedance, and a 5-dB noise figure. It is connected to an antenna by means of a 50-$\Omega$ coaxial cable that has an equivalent gain (loss) of $-3$ dB. What is the overall noise figure?

**3.7** A receiver with an 8-dB noise figure, a 50-$\Omega$ input impedance, and a 3-kHz bandwidth is connected to an antenna that has a noise temperature of 2000 K. What is the minimum detectable signal for a 10-dB output signal-to-noise ratio? If a preamplifier with a gain of 10 dB, an NF of 5 dB, and bandwidth of 4 kHz (which overlaps the receiver's frequency response) is added between the antenna and receiver, what is the minimum detectable signal for a 10-dB output signal-to-noise ratio?

**3.8** A receiver is to be designed to have an overall noise figure of 4 dB. The input mixer has a noise figure of 8 dB, and the preamplifier (which is to be located at the input) has a noise figure of 3 dB. What must be the minimum preamplifier gain?

**3.9** An amplifier with a 10-dB noise figure and a 4-dB power gain is cascaded with a second amplifier which has 10-dB noise figure and a 10-dB power gain. What are the overall noise figure and power gain?

**3.10** Calculate the value of source resistance which will minimize the noise figure of a 741 operational amplifier at a frequency of 500 Hz. What is the minimum noise figure?

**3.11** A 741 operational amplifier is used with a source resistance of 20 k$\Omega$. If the input signal level is 1 mV, what is the output signal-to-noise ratio? Assume that the amplifier bandwidth is 1 Hz and that the center frequency is 1 kHz. What is the output signal-to-noise ratio with the source resistance that minimizes the noise figure?

**3.12** Consider an amplifier in which the source resistance is larger than that required to minimize the noise figure. Show that shunting the source with a resistor in order to minimize the noise figure will reduce the output signal-to-noise ratio.

**3.13** A receiver with a 3-kHz bandwidth and 50-$\Omega$ input impedance has a noise figure (NF) of 8 dB. What is the minimum detectable signal for an output signal-to-noise ratio of 10 dB? If the two-tone intercept point is $+20$ dBm, what is the receiver's dynamic range? What will be the dynamic range if a linear noiseless preamplifier with a voltage gain of 10 is added at the input?

**3.14** A receiver has a 10-dB noise figure, a 50-$\Omega$ input impedance, a $-5$-dBm two-tone intercept point ($P_I$), and 3.5-kHz bandwidth. What is the minimum detectable signal for a 0-dB output signal-to-noise ratio? What is the receiver's dynamic range?

**3.15** A linear preamplifier with a voltage gain of 5 and a 4-dB noise figure is inserted before the receiver of Prob. 3.14. What is the overall dynamic range?

**3.16** A receiver has a 3-kHz bandwidth, a 70-$\Omega$ input impedance, and a noise figure of 6 dB. It is connected to an antenna with a cable which has an equivalent loss of 6 dB and an NF of 3 dB. (The cable is matched to the input impedance.) What is the minimum detectable input signal for an output signal-to-noise ratio of 10 dB? If the antenna noise temperature is 3000 K, what is the minimum detectable signal for the same output $S/N$?

**3.17** Use a transistor with $\beta = 100$ to design a common-emitter amplifier to couple a 100-$\Omega$ source to a 1-k$\Omega$ load resistance. The base spreading resistance can be neglected. What is the optimum value of collector current for the lowest noise figure?

**3.18** A network consists of a voltage source with a 1-k$\Omega$ source resistance and a 1-k$\Omega$ load resistance. Determine the noise factor at 100 Hz and at 10 MHz. Also determine the total output noise power.

**3.19** Describe the noise characteristics of the amplifier you designed for the Chapter 2 Spice project. Determine the noise factor at 100 Hz and at 10 MHz. Also determine the total output noise power. The measurements should be made with the source matched to the amplifier input impedance. Can you determine the equivalent input noise generators? (*Note:* Add KF = 1E-13 to the transistor model. This is a flicker noise coefficient.)

## ▨ 3.7

### REFERENCES

1. Y. Netzer, The Design of Low-Noise Amplifiers, *Proc. IEEE,* **69:** 728–741 (1981).
2. IRE Standards on Methods of Measuring Noise in Linear Twoports, *Proc. IRE,* **48:** 60–68 (1960).

## ▨ 3.8

### ADDITIONAL READING

Carson, R. S.: *Radio Communications Concepts: Analog,* Wiley, New York, 1990.

Das, M. B.: FET Noise Sources and Their Effects on Amplifier Performance at Low Frequencies, *IEEE Trans. Electron Devices,* ED-19, 1972.

Fisk, J. R.: Receiver Noise Figure, Sensitivity and Dynamic Range—What the Numbers Mean, *Ham Radio,* October 1975, pp. 8–25.

Fris, H. T.: Noise Figures of Radio Receivers, *Proc. IRE,* **32:** 410–422 (1944).

Goldman, S.: *Frequency Analysis, Modulation and Noise,* McGraw-Hill, New York, 1948.

Motchenbacher, C. D., and F. C. Fitchen: *Low-Noise Electronic Design,* Wiley, New York, 1973.

North, D. D.: The Absolute Sensitivity of Radio Receivers, *RCA Review,* 1942.

Perlow, S. W.: Third Order Distortion in Amplifiers and Mixers, *RCA Review,* **37:** 234–265 (1976).

Representation of Noise in Linear Twoports, *Proc. IRE,* **48:** 69–74 (1960).

Sherwin, J.: Noise Specs Confusing, National Semiconductor Application Note #104 (May 1974).

# Frequency-Selective Networks and Transformers

## INTRODUCTION

Communication networks must frequently select a band of frequencies and attenuate other undesired frequencies. Modern filter theory now provides methods for designing such filters to meet virtually any specification, but the most commonly used frequency-selective circuits are still the rather simple series- and parallel-tuned resonant circuits. Even these simple circuits lead to complex equations when nonideal elements, such as the nonzero resistance of inductors, are considered. While standard procedures for circuit analysis can be used, the resulting equations are often too complex to provide insight into the design process. However, years of research have yielded many approximations that greatly facilitate the design and analysis of simple resonant circuits.

Resonant circuits are normally used as narrowband circuits. As the modern architecture of communication circuits shifts toward broader band systems, there are fewer narrowband circuits. Yet the importance of resonant circuits is increasing where they do find application. One of the main applications for resonant circuits is the design of low-noise oscillators. Modern communication receivers are requiring resonant circuits with very narrow bandwidths.

This chapter first describes the analysis of these circuits and provides approximations that allow one to design and analyze such circuits with a minimum of mathematics. We then show how transformers are incorporated into resonant circuits. Methods are provided here for analyzing frequency-selective circuits containing simple transformers, one of the most useful components of communication circuits. And finally we demonstrate how frequency-selective methods can be used for impedance matching and how a combination of inductors and capacitors can be used to alter the input impedance of the network over a limited frequency range.

## 4.2

### SERIES RESONANT CIRCUITS

One of the simplest frequency-selective circuits is the series resonant circuit illustrated in Fig. 4.1. At low frequencies the current is blocked by the capacitor, and the

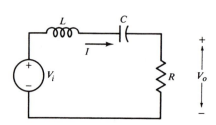

**FIGURE 4.1**
A series resonant circuit.

inductor blocks the current at high frequencies. At some intermediate frequency the impedance of the inductor is equal in magnitude and opposite in sign to the impedance of the capacitor. At this frequency, referred to as the *resonant frequency,* maximum current flows and is the phase with the applied voltage. This circuit will now be analyzed in order to quantitatively describe its frequency performance. The circuit current is

$$I(s) = \frac{V_i(s)}{sL + (sC)^{-1} + R}$$

and the output voltage $V_o(s)$ is

$$V_o(s) = \frac{RV_i(s)}{sL + (sC)^{-1} + R} = \frac{(R/L)sV_i(s)}{s^2 + Rs/L + (LC)^{-1}} \tag{4.1}$$

The voltage gain $V_o(s)/V_i(s)$ has a zero at the origin and two poles $s_1$ and $s_2$ located at

$$s_{1,2} = -\frac{R}{2L} \pm \left[\frac{(R/L)^2 - 4/(LC)}{4}\right]^{1/2} \tag{4.2}$$

Both poles may be real, or they can occur as a complex conjugate pair. If

$$R > 2\left(\frac{L}{C}\right)^{1/2}$$

the poles are both real, and the circuit is said to be *overdamped.* If the two poles are equal, that is,

$$s_1 = s_2 = -\frac{R}{2L}$$

the circuit is critically damped. In most frequency-selective networks, the circuit is underdamped, that is,

$$R < 2\left(\frac{L}{C}\right)^{1/2}$$

and the poles form a complex conjugate pair. The transfer function is frequently written as

$$A(s) = \frac{RCs}{s^2/\omega_o^2 + (2\zeta/\omega_o)s + 1} \tag{4.3}$$

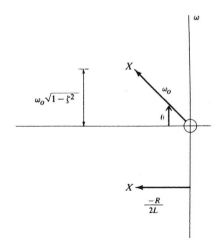

**FIGURE 4.2**
Pole-zero diagram of the series resonant circuit.

where $\omega_o = [(LC)^{1/2}]^{-1}$ is called the *undamped natural frequency* and

$$\zeta = \frac{R}{2} \left( \frac{C}{L} \right)^{1/2}$$

is referred to as the *damping ratio*. If $\zeta < 1$, the circuit is underdamped. The pole-zero plot of this transfer function for the underdamped case is given in Fig. 4.2. The real part of each pole is equal to $-R/(2L) = \zeta\omega_o$, and the magnitude of the imaginary part of each pole is

$$\frac{1}{2}\left[ \frac{4}{LC} - \left( \frac{R}{L} \right)^2 \right]^{1/2} = \omega_o \left( 1 - \frac{R^2 C}{4L} \right)^{1/2} = \omega_o(1 - \zeta^2)^{1/2}$$

Note that the distance of the pole from the origin is equal to

$$\left\{ \left( \frac{R}{2L} \right)^2 + \frac{1}{4}\left[ \frac{4}{LC} - \left( \frac{R}{L} \right)^2 \right] \right\}^{1/2} = [(LC)^{1/2}]^{-1} = \omega_o \tag{4.4}$$

and the angle $\theta$ is determined by

$$\cos \theta = \frac{R(LC)^{1/2}}{2L} = \zeta \tag{4.5}$$

where $\omega_o$ and $\zeta$ are convenient parameters for describing the transient response of the circuit. For a unit step input voltage $V_i(t) = U(t)$, it is readily shown (by taking the inverse Laplace transform) that if $\zeta$ is less than 1,

$$V_o(t) = \frac{2\zeta}{(1 - \zeta^2)^{1/2}} e^{-\zeta\omega_o t} \sin [\omega_o(1 - \zeta^2)^{1/2}t] U(t) \tag{4.6}$$

where $U(t)$ is the unit step function

$$U(t) = \begin{cases} 1 & t \geq 0 \\ 0 & t < 0 \end{cases}$$

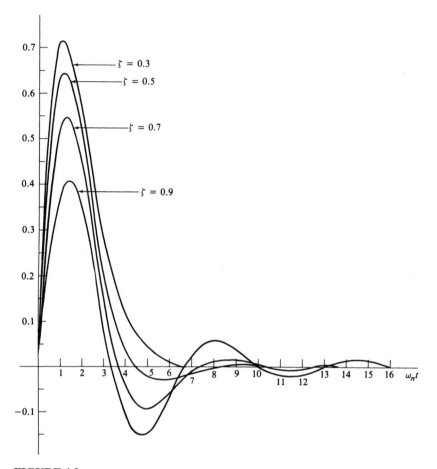

**FIGURE 4.3**
Transient response of underdamped series resonant circuits to a step input, for selected values of damping.

The step response (plotted in Fig. 4.3 for selected values of $\zeta$) is an exponentially damped sinusoid. The ringing frequency is

$$\omega_o(1 - \zeta^2)^{1/2} \qquad \text{rad/s}$$

which is equal to the imaginary part of each pole, and the envelope's damping factor $\zeta\omega_o = -R/(2L)$ is equal to the magnitude of the real part of each pole. The smaller the value of circuit resistance (for a fixed inductance), the smaller the damping factor and the longer it takes for the transient to die out.

The steady-state (frequency) analysis of the circuit is readily evaluated from Eq. (4.1) by setting $s = j\omega$. For convenience the equation is written in the form

$$
A(j\omega) = \left[1 + (j\omega RC)^{-1}\left(1 - \frac{\omega^2}{\omega_o^2}\right)\right]^{-1}
$$

$$
= \left[1 + \frac{j}{\omega_o RC}\left(\frac{\omega}{\omega_o} - \frac{\omega_o}{\omega}\right)\right]^{-1} = \left[1 + jQ\left(\frac{\omega}{\omega_o} - \frac{\omega_o}{\omega}\right)\right]^{-1}
$$

(4.7)

where the circuit $Q$ is defined by

$$Q = (\omega_o R C)^{-1} = \frac{\omega_o L}{R} = (2\zeta)^{-1} \tag{4.8}$$

The resonant frequency is defined as the frequency where the phase shift of the transfer function is equal to zero. (This is the frequency where the imaginary part of the transfer function is zero.) The imaginary part of the transfer function is zero when

$$\omega = \omega_o = [(LC)^{1/2}]^{-1} \tag{4.9}$$

This is the frequency at which the inductive reactance $j\omega L$ is equal in magnitude and opposite in sign to the capacitive reactance, or

$$-jX_c = (j\omega C)^{-1}$$

yielding a total series reactance of 0 Ω. This series circuit has only one frequency at which the phase shift is 0°. The magnitude of the transfer function at the resonant frequency is $|A(j\omega_o)| = 1$, which is its maximum value.

The corresponding phase shift of the transfer function arg $A(j\omega)$ is plotted in Fig. 4.4 for selected values of $Q$. The phase of the voltage gain changes from +90° to −90° as the frequency increases from below the resonant frequency to above the resonant frequency. The higher the circuit $Q$, the more abrupt is the transition in phase. This phenomenon is particularly important in oscillator design and is discussed further in Chap. 7.

The half-power frequencies $\omega_1$, $\omega_2$ at which $|A(j\omega)|$ is reduced to 0.707 times its maximum value (the −3-dB frequencies) can be found by solving

$$|A(j\omega)| = (2^{1/2})^{-1} = \left| \left[ 1 + jQ \left( \frac{\omega}{\omega_o} - \frac{\omega_o}{\omega} \right) \right]^{-1} \right|$$

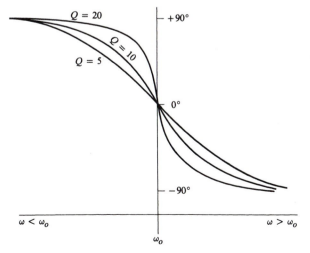

**FIGURE 4.4**
Phase response of under-damped series resonant circuits.

or
$$\frac{1}{2} = \left[1 + Q^2 \left(\frac{\omega}{\omega_o} - \frac{\omega_o}{\omega}\right)^2\right]^{-1}$$

which is equivalent to

$$Q\left(\frac{\omega}{\omega_o} - \frac{\omega_o}{\omega}\right) = 1 \tag{4.10}$$

The solution of this equation can be facilitated by noting that $|A(j\omega)|$ displays geometric symmetry. That is,

$$|A(j\omega)| = \left|A\left(j\frac{\omega_o^2}{\omega}\right)\right|$$

One of the half-power frequencies $\omega_1$ is determined by solving

$$\omega_1 - \frac{\omega_o^2}{\omega_1} = \frac{\omega_o}{Q}$$

From the symmetry argument it follows that the other half-power frequency is found from

$$\omega_2 = \frac{\omega_o^2}{\omega_1} \quad \text{and} \quad \omega_1 - \omega_2 = \frac{\omega_o}{Q}$$

The difference between the two half-power frequencies $\omega_1 - \omega_2$ is by definition the circuit bandwidth $B$. Therefore, the bandwidth is

$$B = \frac{\omega_o}{Q} \tag{4.11}$$

It will be shown in subsequent sections that this relationship among bandwidth, center frequency, and $Q$ also holds true for parallel resonant circuits with the bandwidth being inversely related to the circuit $Q$. The bandwidth can be specified independently of the center frequency $\omega_o$ since

$$B = \frac{\omega_o}{Q} = \omega_o^2 RC = \frac{R}{L} \tag{4.12}$$

and $R$ is not a function of $\omega_o$. The steady-state behavior of this circuit is completely described in terms of its center frequency $\omega_o$ and its $Q$.

**EXAMPLE 4.1.** Design a filter to couple a voltage source, with negligible source impedance, to a 50-$\Omega$ load resistance. The specifications are that the filter center frequency be 5 MHz and the bandwidth be 100 kHz.

*Solution.* A series resonant $LC$ circuit can be used to meet the specifications. Since the source impedance is negligible, the total series resistance is $50\,\Omega$. The inductance is determined using Eq. (4.12):

$$L = \frac{R}{B} = \frac{50}{2\pi(10^5)} = 79.6\,\mu\text{H}$$

The capacitance is determined from Eq. (4.8)

$$C = [(79.6 \times 10^{-6})(2\pi \times 5 \times 10^6)^2]^{-1} = 12.7\,\text{pF}$$

The complete filter, including the load resistor, is shown in Fig. 4.5.

The frequency response of the magnitude of Eq. (4.7) is plotted in Fig. 4.6 for selected values of $Q$. For the overdamped circuit the gain rolls off at $-6$ dB per octave at both high and low frequencies, but for the high-$Q$ case the attenuation rate is much greater near the resonant frequency. The frequency response becomes more selective (sharper) as the $Q$ is increased. The pole-zero diagram shown in Fig. 4.7 provides a means for estimating the roll-off characteristics of the high-$Q$ filter. In the high-$Q$ case the circuit poles are close to the $j\omega$ axis, and their distance from the origin is approximately $\omega_o$. The circuit gain at any frequency is the product of the distances from the zeros to the desired frequency, divided by the product of the distances from the poles to the desired frequency. Thus at any frequency $n\omega_o$ where $n$ is an integer, the zero distance is $n\omega_o$ and the pole distances are $(n-1)\omega_o$ and $(n+1)\omega_o$. Therefore, the magnitude of the gain is

$$|A(jn\omega_o)| = \frac{n\omega_o R/L}{(n-1)\omega_o(n+1)\omega_o} = \frac{n\omega_o R/L}{(n^2-1)\omega_o^2} \tag{4.13}$$

Since

$$\frac{R}{\omega_o L} = Q^{-1} \tag{4.14}$$

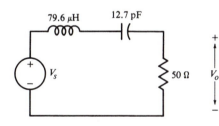

**FIGURE 4.5**
Series resonant circuit described in Example 4.1.

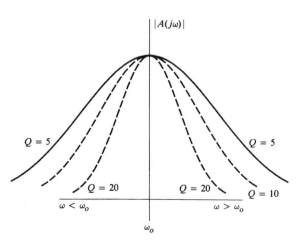

**FIGURE 4.6**
Magnitude of the series resonant circuit gain as a function of frequency.

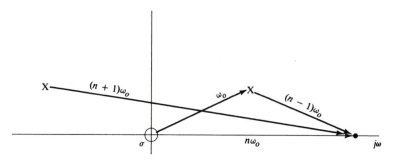

**FIGURE 4.7**
Pole-zero diagram used to evaluate the magnitude of the frequency
response at the $n$th harmonic of the resonant frequency.

the attenuation at any harmonic of the circuit's resonant frequency is given by

$$\left|\frac{A(jn\omega_o)}{A(j\omega_o)}\right| = \frac{n}{Q(n^2 - 1)} \tag{4.15}$$

As the $Q$ of the circuit increases, the attenuation at the harmonic frequencies
increases. This equation can be most useful in circuit design, as illustrated by the
following example.

> **EXAMPLE 4.2.** A series-tuned circuit is to be used to filter out the harmonics of a wave-
> form. What must be the minimum circuit $Q$ for the amplitude of the fifth harmonic to
> be 40 dB below the amplitude of the fundamental frequency?
>
> **Solution.** Forty decibels corresponds to a voltage ratio of 100:1. Therefore, since
> $|(A(j\omega_o)| = 1$,
>
> $$|A(j\omega_o 5)| = 0.01 = \frac{5}{Q(25 - 1)}$$
>
> or $\qquad\qquad Q_{min} = 20.83$

In some designs it may be necessary to keep $Q$ fixed and yet provide greater
out-of-band attenuation than can be realized using Eq. (4.15). For such problems a
higher-order filter will be necessary. Higher-order filter design is readily executed
by using a computer-aided design program such as FILSYN.[1]

Resonant circuits are usually used as narrowband circuits. As such, they are
used to filter out all but a narrow frequency range of the input signal. The transient performance of these circuits can cause difficulties. These resonant circuits
are bandpass circuits, and the steady-state response to low-frequency inputs will
be zero. Yet, as Fig. 4.3 shows, narrowband circuits can have a large transient
peak to step input signals. The lower the damping ration (the higher the $Q$), the
larger the transient response to step inputs. Many receivers use amplitude limiters before narrowband circuits to limit the ringing that can be caused by large-
amplitude noise spikes. Another difficulty is that the narrower the bandwidth, the
longer it takes for the circuit to reach steady state. Figure 4.8 illustrates the tran-

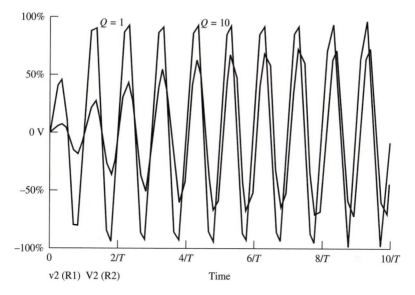

**FIGURE 4.8**

sient response of a series resonant circuit to a sinusoidal waveform whose frequency is the same as the circuit's resonant frequency. It is seen that the circuit with the higher $Q$ takes much longer to reach steady state than the lower-$Q$ circuit. Very high $Q$ circuits are best suited for steady-state applications, such as in oscillators.

### Effect of Source Resistance

The analysis up to this point assumed that the source resistance was zero. If the source resistance cannot be neglected, the equivalent circuit is as shown in Fig. 4.9; the transfer function for this circuit is

$$A(s) = \frac{V_o}{V_i} = \frac{R_L}{R_L + R_s + sL + (sC)^{-1}}$$

$$= \frac{R_L (R_L + R_s)sC}{R_L + R_s s^2 LC + (R_L + R_s)sC + 1} \qquad (4.16)$$

This equation is identical to Eq. (4.1) except that $R$ is replaced by $R_L + R_s$, and the transfer condition is multiplied by the frequency-independent attenuation factor

$$K = \frac{R_L}{R_L + R_s} \qquad (4.17)$$

The effect of the nonzero source resistance is to reduce the amplitude of the transfer function at all frequencies and to reduce the $Q$ of the circuit from $\omega_o L/R_L$ to

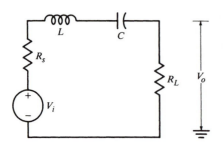

**FIGURE 4.9**
A series resonant circuit including a source resistance $R_s$.

$\omega_o L/(R_L + R_s)$, which is equivalent to widening the bandwidth by the same factor by which the gain is reduced. The analysis is obviously the same if the inductance $L$ is modeled as an ideal inductor $L$ in series with a resistance $R_s$. Since

$$Q = \frac{\omega_o L}{R_s + R_L}$$

any additional series resistance reduces the circuit $Q$ (and increases the bandwidth) without changing the resonant frequency.

### Voltage Application

Another property of the series resonant circuit is that the voltage across the reactive components can be much larger than the applied voltage. The voltage across the capacitor in the circuit shown in Fig. 4.1 is

$$V_c(s) = \frac{V_i(s)}{s^2 LC + RsC + 1}$$

At the resonant frequency, the magnitude of the capacitor voltage is

$$|V_c(j\omega_o)| = \frac{V_i}{\omega_o RC} = QV_i \tag{4.18}$$

At the resonant frequency, the voltage across the capacitor is $Q$ times the input voltage. Since the magnitude of the reactance of the inductor is the same as that of the capacitor at the resonant frequency, the voltage across the inductor will be the same.

### 4.3

## PARALLEL RESONANT CIRCUITS

In parallel resonance (actually antiresonance), two equal and opposite susceptances are added in parallel so that the admittance, instead of the impedance, is a minimum at the resonant frequency. For the parallel resonant circuit shown in Fig. 4.10, the

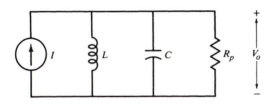

**FIGURE 4.10**
A parallel resonant circuit.

transfer function (transfer impedance) is

$$A(s) = \frac{V_o(s)}{I(s)} = [sC + (sL)^{-1} + (R_p)^{-1}]^{-1}$$

$$= \frac{sL/R_p}{s^2LC + sL/R_p + 1} R_p$$

(4.3a)

This equation is identical to Eq. (4.3), except for the scale factor, provided $L/R_p = RC$, where $R$ and $C$ denote the resistance and capacitance, respectively, of the series resonant circuit. Thus the results for the transient and steady-state responses of the series resonant circuit can be used for the parallel resonant circuit, provided

$$\frac{2\zeta}{\omega_o} = \frac{L}{R_p}$$

The only difference is that the magnitude of Eq. (4.3a) is equal to $R_p$ at the resonant frequency, whereas it is equal to unity for the series resonant circuit driven by a voltage source. The resonant frequency

$$\omega_o = [(LC)^{1/2}]^{-1}$$

(4.19)

is the same for both circuits. The $Q$ of the parallel resonant circuit is defined as

$$Q_p^{-1} = 2\zeta_p = R_p^{-1} \left(\frac{L}{C}\right)^{1/2} = R_p^{-1} \frac{L}{(LC)^{1/2}} = \frac{\omega_o L}{R_p}$$

or

$$Q_p = \frac{R_p}{\omega_o L}$$

(4.20)

Equations (4.19) and (4.20) permit one to determine the transient and steady-state responses for the parallel resonant circuit using the results derived for the series resonant circuit. Table 4.1 summarizes the relations for the parallel and series resonant circuits.

In the simple two-pole resonant circuits described here, $Q$ and $\omega_o$ completely describe the network. For the series resonant circuit, the higher the series resistance, the lower the $Q$. Just the opposite is true for the parallel resonant circuit. Any resistance added in parallel with the tuned circuit reduces the resistance across the circuit and hence reduces the $Q$.

Nonideal inductors always possess finite resistance in series with the inductance, and so a more accurate model of a parallel $LC$ circuit is as shown in Fig. 4.11. It will now be shown, with certain assumptions, that the analysis of this circuit can

■ **TABLE 4.1**
**Relation of series and parallel resonant circuits**

|           | Series | Parallel |
|-----------|:------:|:--------:|
| $\omega_o$ | $[(LC)^{1/2}]^{-1}$ | $[(LC)^{1/2}]^{-1}$ |
| $Q$ | $\dfrac{\omega_o L}{R}$ | $\dfrac{R_p}{\omega_o L}$ |
| $\zeta$ | $\dfrac{R}{2}\left(\dfrac{C}{L}\right)^{1/2}$ | $(2R_p)^{-1}\left(\dfrac{L}{C}\right)^{1/2}$ |

be reduced to that presented in the previous section for the parallel circuit with a lossless inductor. The transfer function for the circuit of Fig. 4.11 is

$$\frac{V_o(s)}{I(s)} = A(s) = \frac{sL + r_s}{s^2 LC + r_s Cs + 1} \tag{4.21}$$

This equation is of the same form as Eq. (4.3a) except that the zero is now located at $-r_s/L$ instead of at the origin. If this zero distance is small compared to $\omega$, that is, if $\omega \gg r_s/L$, then to a first approximation the zero can be assumed to be at the origin, and Eq. (4.21) can be accurately approximated by

$$A(s) \approx \frac{sL}{s^2 LC + sCr_s + 1}$$

which is identical to Eq. (4.3a), provided

$$\frac{L}{R_p} = r_s C$$

or
$$R_p = \frac{L}{Cr_s} = \frac{L}{r_s}\,\omega_o^2 L = \frac{(\omega_o L)^2}{r_s} \tag{4.22}$$

If $\omega L \gg r_s$, the parallel resonant circuit (Fig. 4.11) with a resistor in $r_s$ series with the inductor can be replaced by the parallel resonant circuit shown in Fig. 4.12. This approximation is almost always valid in the frequency region of interest since one would not select a low-$Q$ inductor for use in a high-$Q$ circuit.

The approximation is not valid at low frequencies where $\omega L \ll r_s$ and $\omega_o L/r_s$ is referred to as the *inductor Q*. Every inductor has a finite $Q$, so the overall $Q$

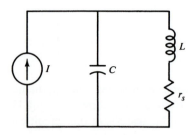

**FIGURE 4.11**
A parallel resonant circuit including a resistor in series with the inductor.

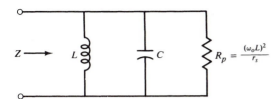

**FIGURE 4.12**
Approximate equivalent circuit of
the circuit illustrated in Fig. 4.11.

of the tuned circuit must also be finite. Note that the $Q$ of the circuit in
Fig. 4.12 is

$$Q = \frac{R_p}{\omega_o L} = \frac{\omega_o L}{r_s} \tag{4.23}$$

which is the same as the $Q$ of the coil. This is referred to as the *unloaded Q*, or $Q_u$,
of the circuit. Any additional load resistance added across the circuit will further
reduce the $Q$, so the circuit (or loaded) $Q$ will always be less than the inductor $Q$
unless active components are used.

EXAMPLE 4.3. The tuned circuit shown in Fig. 4.13 employs an inductor with a $Q_u$ of
100. If a 100-k$\Omega$ load resistor is added across the circuit, what is the loaded $Q_L$?

**Solution.** The resonant frequency of the circuit (ignoring the finite resistance) is

$$\omega_o = [(LC)^{1/2}]^{-1} = 10^8 \text{ rad/s}$$

The $Q$ of the coil is 100; therefore,

$$r_s = \frac{\omega_o L}{Q} = 10$$

Also, since $\omega_o L \gg r_s$ for frequencies near resonance, the series resistor can be replaced by a parallel resistor $R_p$, where

$$R_p = \frac{(\omega_o L)^2}{r_s} = 10^5 \ \Omega$$

The total parallel resistance, including $R_L$, is equal to two 100-k$\Omega$ resistors in parallel,
or 50 k$\Omega$, and the loaded $Q$ is

$$Q_L = \frac{50 \text{ k}\Omega}{\omega_o L} = 50$$

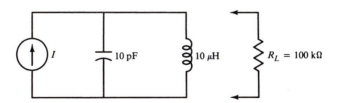

**FIGURE 4.13**
Parallel resonant circuit discussed in Example 4.3.

The parallel load resistor reduces the $Q$ of the circuit by a factor of 2 below that of the unloaded $Q$ of the inductor.

## Branch Currents

At the resonant frequency of the parallel $RLC$ circuit, the output voltage is $V_o = I_i R_p$. The magnitude of the current through the capacitor is then

$$I_c = V_o \omega_o C = I_i \omega_o C R_p$$

$$= I_i \frac{R_p}{\omega_o L} = Q I_i \tag{4.24}$$

This equation shows that at frequencies near resonance the branch currents through the inductor and capacitor can be much larger than the applied current.

## ■ 4.4

### PARALLEL RESONANT CIRCUITS INCLUDING TRANSFORMERS

Transformers are extensively used in resonant circuits to provide phase inversion, dc isolation, and impedance-level shifting. Since the transformers already contain an inductor, it is possible to form a parallel resonant circuit with a transformer by simply adding a capacitor. The parallel resonant circuit $Q$ is directly proportional to the parallel load resistance and inversely proportional to the magnitude of the inductive reactance. High-frequency circuits will usually have a low input imped-ance. For example, the equivalent load resistance of an antenna is small (on the order of 50 $\Omega$). These small impedance values make it difficult to realize high-$Q$ circuits unless some method is used to transform the load to a larger value. There are several methods available to realize this transformation; the magnetically cou-pled transformer is a frequently used technique. For the transformer circuit shown in Fig. 4.14 the equilibrium equations are

$$V_1(t) = L_1 \frac{di_1}{dt} + M \frac{di_2}{dt} \tag{4.25}$$

and
$$V_2(t) = M \frac{di_1}{dt} + L_2 \frac{di_2}{dt} \tag{4.26}$$

or
$$V_1(s) = sL_1 I_1(s) + sMI_2(s) \tag{4.25a}$$

and
$$V_2(s) = sMI_1(s) + sL_2 I_2(s) \tag{4.26a}$$

where $M$ is the mutual inductance between the input and secondary of the trans-former. The standard dot convention is used, which places a dot on one terminal of a coil and another on the terminal of a second coil such that if current is sent into the two dotted terminals, the magnetic fluxes linking the coils will reinforce each other. With this convention $M$ is always positive.

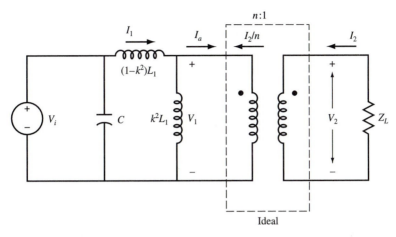

**FIGURE 4.14**
A magnetically coupled tuned circuit.

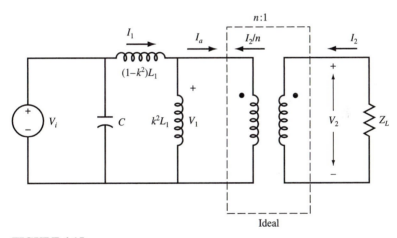

**FIGURE 4.15**
An equivalent circuit to the one shown in Fig. 4.14. Here the transformer
has been replaced by an ideal transformer plus two inductors.

There are many equivalent circuits for this transformer, but the one which has
the greatest utility in communication circuits is shown in Fig. 4.15. The ideal transformer contained in this circuit is a mathematical abstraction that facilitates the
analysis by simplifying the transformer circuit models. For the ideal transformer,

$$V_1 = nV_2$$

and
$$I_1 = -\frac{I_2}{n}$$

independent of frequency. Also, since no power is dissipated in an ideal transformer,

$$\frac{V_1 I_2}{n} = V_2 I_2$$

$$\frac{V_1}{I_a}(s) = Z_1 = -n^2 \frac{V_2}{I_2}(s) = n^2 Z_L$$

(4.27)

The circuit model for the transformer of Fig. 4.15 is equivalent to that of Fig. 4.14, provided the terminal voltage and current relations are the same as given by Eqs. (4.25a) and (4.26a). The equilibrium equations for the circuit of Fig. 4.15 are

$$V_i(s) = s(1 - k^2)L_1 I_1(s) + sk^2 L_1 \left[ I_1(s) + \frac{I_2(s)}{n} \right]$$

$$= sL_1 I_1(s) + \frac{sk^2 L_1 I_2(s)}{n}$$

(4.28)

and

$$V_2(s) = \frac{sk^2 L_1 I_1(s)}{n} + \frac{sk^2 L_1 I_2(s)}{n^2}$$

(4.29)

These equations are the same as Eqs. (4.25a) and (4.26a), provided

$$\frac{k^2 L_1}{n} = M$$

(4.30)

and

$$\frac{k^2 L_1}{n^2} = L_2$$

(4.31)

That is, the equivalent turns ratio is

$$n = k \left( \frac{L_1}{L_2} \right)^{1/2}$$

(4.32)

and $k$, referred to as the *coefficient of coupling,* is

$$k = \frac{M}{(L_1 L_2)^{1/2}}$$

(4.33)

One reason that this two-inductor model of the real transformer is so useful is that in narrowband circuits, transformers are used which have a coefficient of coupling $k$ near unity. If $k \approx 1$, the model can be simplified to that shown in Fig. 4.16; the transformer is represented by an inductor in parallel with an ideal transformer (which simply reflects the secondary impedance, multiplied by the turns ratio squared, to the primary). The model shown in Fig. 4.15 provides an easily analyzable circuit when the transformer coefficient of coupling is close to unity.

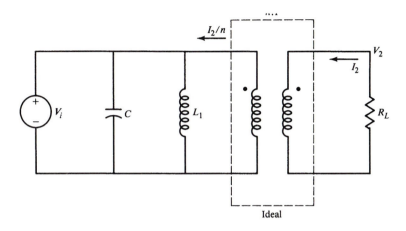

**FIGURE 4.16**
A simplified version of the circuit shown in Fig. 4.15, valid when the
coefficient of coupling $k \approx 1$.

**EXAMPLE 4.4.** A tightly coupled transformer with a primary inductance $L_1$ of 25 $\mu$H
and a secondary inductance $L_2$ of 400 $\mu$H is used in the circuit shown in Fig. 4.17a.
What is the overall frequency response of the circuit?

**Solution.** The tightly coupled transformer circuit can be replaced by the equivalent circuit shown in Fig. 4.16 (since $k \approx 1$). Therefore,

$$n = k \left( \frac{L_1}{L_2} \right)^{1/2} \approx \left( \frac{L_1}{L_2} \right)^{1/2} = \frac{1}{4}$$

and the total primary capacitance is

$$C_T = 8 + \frac{2}{n^2} = 8 + 2(4)^2 = 40 \text{ pF}$$

and the load resistance reflected to the primary is

$$n^2 R_L = \frac{400}{16} \times 10^3 = 25 \text{ k}\Omega$$

Thus, with the load reflected to the primary, the equivalent circuit is a parallel-tuned circuit as shown in Fig. 4.17b. The resonant frequency is

$$\omega_o = \{[(40 \times 10^{-12})(25 \times 10^{-6})]^{1/2}\}^{-1} = 31.6 \times 10^6 \text{ rad/s}$$

and

$$Q = \frac{R}{\omega_o L} = \frac{25 \times 10^3}{(31.6 \times 10^6)(25 \times 10^{-6})} = 31.68$$

Since the secondary in this circuit is coupled to the primary with an ideal transformer,
the output voltage is

$$V_o(t) = 4 V_i(t)$$

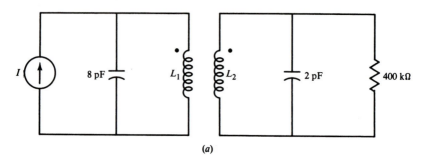

(a)

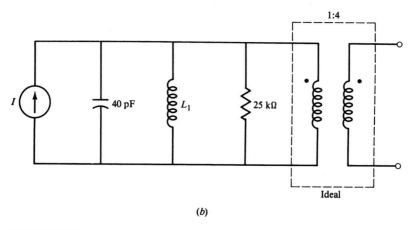

(b)

**FIGURE 4.17**
Equivalent circuits discussed in Example 4.4.

Transformers have a voltage gain equal to the turns ratio $1/n$ so, as in the preceding example, the transformer can provide voltage amplification. Transformers with a turns ratio greater than 1 are frequently used in amplifier design for circuits with a small load resistance. The transformer then increases the impedance seen at the amplifier input. If the amplifier gain without a transformer is $A_v = g_m R_L$, then the gain with the transformer is $A'_v = g_m R_L/n$.

### Transformers with Tuned Secondaries

Transformers with a turns ratio less than 1 are also used with voltage amplifiers. The reduced voltage gain results in a smaller Miller capacitance (see Chap. 5) and thus a wider bandwidth. Consider the amplifier with a tuned secondary circuit as illustrated in Fig. 4.18a. If the transformer is replaced by the equivalent circuit model, the equivalent circuit seen from the transistor collector is as shown in Fig. 4.18b.

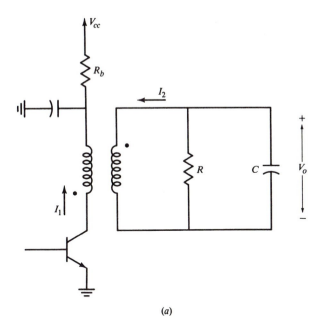

(a)

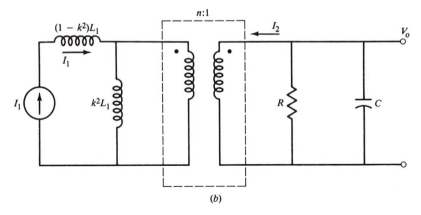

(b)

**FIGURE 4.18**
(a) A transistor amplifier including a transformer-coupled circuit with a tuned
secondary; (b) simplified equivalent circuit.

The equilibrium equations are

$$nV_o(s) = sk^2L_1\left[I_1(s) + \frac{I_2(s)}{n}\right] \tag{4.34}$$

and

$$V_o(s) = -I_2(s)\frac{R_L}{R_LsC + 1} \tag{4.35}$$

The transfer impedance $Z_{12}(s)$ can be found by eliminating $I_2$:

$$Z_{12}(s) = \frac{V_o}{I_1} = \frac{sk^2L_1/n}{s^2k^2L_1C/n^2 + sk^2L_1/(n^2R_L) + 1} \tag{4.36}$$

Since from Eq. (4.31)

$$\frac{k^2L_1}{n^2} = L_2$$

the transfer impedance is

$$Z_{12}(s) = \frac{V_o}{I_1} = \frac{nsL_2}{s^2L_2C + sL_2/R_L + 1} \tag{4.37}$$

which is also the equation of a parallel resonant circuit with

$$\omega_o = [(L_2C)^{1/2}]^{-1} \tag{4.38}$$

Equation (4.38) states that when the capacitor is placed in the transformer secondary, the resonant frequency is determined by this capacitance in parallel with the inductance of the transformer secondary. The input impedance seen by the current source $I_1$ is determined from

$$V_1(s) = sI_1(1 - k^2)L_1 + nV_o(s) \tag{4.39}$$

with $V_o$ replaced using Eq. (4.37)

$$V_1(s) = sI_1(1 - k^2)L_1 + nZ_{12}(s)I_1(s) \tag{4.40}$$

The circuit input impedance (at the collector of the transistor) is

$$Z_i = \frac{V_i(s)}{I_1(s)} = sL_1(1 - k^2) + nZ_{12}(s)$$

$$Z_i(s) = \frac{V_i(s)}{I_1(s)} = sL_1(1 - k^2) + \frac{n^2sL_2}{s^2L_2C + sL_2/R_L + 1} \tag{4.41}$$

At the resonant frequency $\omega_o$, the input impedance is

$$Z_i(j\omega_o) = j\omega_o L_1(1 - k^2) + n^2R_L \tag{4.42}$$

The transformer is tightly coupled if ($k \approx 1$)

$$Z_i(j\omega_o) = n^2R_L \tag{4.43}$$

At the resonant frequency the tightly coupled transformer reflects back to the collector of the transistor the load resistance $R_L$ amplified by $n^2$. If the turns ratio $n$ is less than 1, the collector voltage is less than the output voltage. This has the net effect of reducing the base-to-collector voltage gain and hence reducing the Miller capacitance.

If the capacitance is added to the primary, then

$$C_1 = (L_1\omega_o^2)^{-1}$$

and if it is added to the secondary, then

$$C_2 = (L_2\omega_o^2)^{-1}$$

So the ratio of possible capacitors is

$$\frac{C_1}{C_2} = \frac{L_2}{L_1} = (n^2)^{-1} \tag{4.44}$$

in a tightly coupled transformer. This equation shows that if $n > 1$, the transformer circuit can be tuned by a smaller capacitor across the input, but if $n < 1$, a tuned secondary will result in a smaller capacitor $C_2$.

## Double-Tuned Circuits

In addition to tuning either the primary or secondary of a transformer-coupled circuit, it is possible to include tuned circuits in both the primary and secondary (double-tuned circuits). A transformer-coupled circuit with parallel resonant circuits in the primary and secondary is shown in Fig. 4.19. The transadmittance of this network is

$$\frac{V_o}{I_i} = \frac{-k\omega_1\omega_2 s}{(1 - k^2)(C_1 C_2)^{1/2}(s^4 + a_3 s^3 + a_2 s^2 + a_1 s + a_0)} \tag{4.45}$$

Where $k$ is the coefficient of coupling, the primary resonant frequency is

$$\omega_1 = [(L_1 C_1)^{1/2}]^{-1} \tag{4.46}$$

and the resonant frequency of the secondary circuit is

$$\omega_2 = [(L_2 C_2)^{1/2}]^{-1} \tag{4.47}$$

The values of the coefficients of Eq. (4.45) are

$$a_3 = \frac{\omega_1}{Q_1} + \frac{\omega_2}{Q_2} \qquad a_2 = \frac{\omega_1\omega_2}{Q_1 Q_2} + \frac{\omega_1^2 + \omega_2^2}{1 - k^2}$$

$$a_1 = \frac{\omega_1^2\omega_2}{Q_2(1 - k^2)} + \frac{\omega_2^2\omega_1}{Q_1(1 - k^2)} \qquad a_0 = \frac{\omega_1^2\omega_2^2}{1 - k^2} \tag{4.48}$$

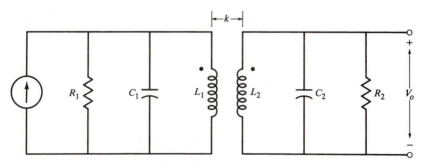

**FIGURE 4.19**
A double-tuned magnetically coupled circuit.

where $\qquad$ $Q_1 = \dfrac{R_1}{\omega_1 L_1}$ $\qquad$ and $\qquad$ $Q_2 = \dfrac{R_2}{\omega_2 L_2}$

This transformer-coupled double-tuned network is a bandpass network with four poles and a single zero at the origin. One method of double-tuned network design is to obtain numerical values for the coefficients that correspond to an approximation of a desired frequency response (such as the Butterworth or Chebyshev approximation) and then solve for the circuit values with Eq. (4.48).

For narrowband circuits the analysis can be simplified. If $Q_1 = Q_2$ and both circuits are tuned to the same frequency, $\omega_1 = \omega_2 = \omega_O$, then for the loosely coupled case ($k^2 \ll 1$) the approximate pole positions are found to be

$$s_1, \; s_1^*, \; s_2, \; s_2^* = \frac{-\omega_o}{2Q} \pm j\omega_o \left(1 \pm \frac{k}{2}\right) \qquad (4.49)$$

For this circuit with equal input and output $Q$'s, it is easy to find the desired circuit parameters for a maximally flat (Butterworth) filter (or any other filter requiring two pairs of complex-conjugate poles) where the real parts of the poles $-\omega_o/2Q$ are the same. For the Butterworth filter, the coefficient of coupling $k = 1/Q$. For this value of $k$, the circuit is said to be "critically coupled." The bandwidth is

$$B = \frac{2^{1/2}\omega_o}{Q} \qquad (4.50)$$

and the gain-bandwidth product is

$$GB = [(2C_1 C_2)^{1/2}]^{-1} \qquad (4.51)$$

For a single-tuned circuit, the gain-bandwidth product is

$$GB = (C_1 + C_2)^{-1} \qquad (4.52)$$

(Here $C_1$ and $C_2$ are the primary and secondary capacitances, respectively.) Therefore the double-tuned (lower-$Q$) method has a better gain-bandwidth product by at least a factor of $2^{1/2}$.

Double-tuned circuits can also be designed for wideband operation.[2]

## Autotransformers

The $Q$ of a parallel tuned circuit is directly proportional to the load resistance shunting the tuned circuit. In many applications this resistance is too small to realize the desired $Q$, and some method is needed to increase the load resistance shunting the tuned circuit. One method is to use a separate transformer, but a simple method which often suffices is to place the load resistance across only a portion of the inductor, as shown in Fig. 4.20. For analysis, the autotransformer can also be replaced by the equivalent circuit shown in Fig. 4.21. The inductor serves as an

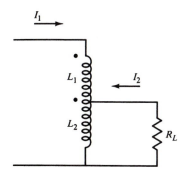

**FIGURE 4.20**
A step-up autotransformer.

autotransformer. For the two circuits to be equivalent the equilibrium equations must be the same. For the circuit of Fig. 4.20 the equations are

$$V_1 = s(L_1 + L_2 + 2M)I_1 + s(L_2 + M)I_2 \tag{4.53}$$

$$V_2 = s(L_2 + M)I_1 + sL_2I_2 \tag{4.54}$$

where $M$ is the mutual inductance between the two sections of the coil. For the circuit shown in Fig. 4.21, the equilibrium equations are

$$V_1 = sLI_1 + \frac{sk^2L}{n} I_2 \tag{4.55}$$

$$V_2 = \frac{sk^2L}{n} I_1 + \frac{sk^2L}{n^2} I_2 \tag{4.56}$$

These two sets of equations will be equivalent if

$$L = L_1 + L_2 + 2M \tag{4.57}$$

$$n = \frac{k^2L}{L_2 + M} \tag{4.58}$$

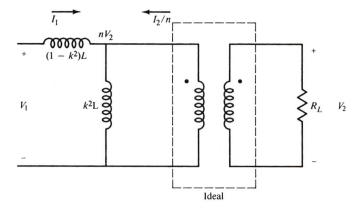

**FIGURE 4.21**
An equivalent circuit for the step-up autotransformer.

and
$$n^2 = \frac{k^2 L}{L_2} \tag{4.59}$$

Solving these two equations for $k$ and $n$, we obtain

$$k = \frac{L_2 + M}{\sqrt{LL_2}}$$

and
$$n = \frac{L_2 + M}{L_2}$$

Thus the equivalent turns ratio of the autotransformer is

$$n = k\sqrt{\frac{L}{L_2}} \tag{4.60}$$

where $L$ is the total inductance measured with the load open-circuited and $L_2$ is measured with the input open-circuited:

$$L_2 = \left. \frac{V_2}{sI_2} \right|_{I_1=0} \tag{4.61}$$

And $L$ and $L_2$ are proportional to the square of the corresponding number of turns $n \approx N/N_2$, where $N$ is the total number of turns on the coil and $N_2$ is the number of turns on the lower section. This autotransformer is often referred to as an *impedance step-up transformer* since the impedance seen by the primary in a closely coupled transformer is $n^2$ times the load impedance and $n$ is greater than 1. The derivation of the equivalent circuit for an impedance step-down transformer is left as an exercise (Prob. 4.8).

### Capacitive Transformers

Impedance level shifting can also be accomplished in narrowband circuits without incurring the cost of expensive and often bulky transformers. One simple method of increasing the impedance level is to use the capacitive transformer illustrated in Fig. 4.22a. To analyze this circuit, consider first the impedance $Z$ in parallel with the inductor $L$ (Fig. 4.22b). This impedance will be equal to the parallel combination of a resistor and capacitor if certain conditions, which will now be derived, hold.

The impedance of the circuit shown in Fig. 4.22b is

$$Z = (j\omega C_2)^{-1} + \frac{R}{Rj\omega C_1 + 1} = \frac{Rj\omega(C_1 + C_2) + 1}{j\omega C_2(Rj\omega C_1 + 1)} \tag{4.62}$$

and the admittance is

$$Y(\omega) = [Z(j\omega)]^{-1} = \frac{j\omega C_2(j\omega RC_1 + 1)}{1 + j\omega R(C_1 + C_2)} \tag{4.63}$$

$$= \frac{j\omega C_2(j\omega RC_1 + 1)[1 - j\omega R(C_1 + C_2)]}{1 + \omega^2 R^2(C_1 + C_2)^2}$$

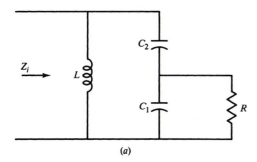

**FIGURE 4.22**
(*a*) A tuned circuit. (*b*) Impedance in parallel with the inductor containing a capacitive autotransformer.

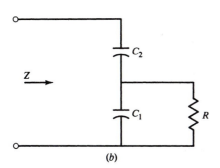

This circuit can be replaced (at some frequency) by the simpler parallel circuit shown in Fig. 4.23, provided the admittances are equal at that frequency. This requires that both the real and imaginary parts be equal. That is,

$$G_p = \frac{1}{R_p} = \frac{-\omega^2 RC_1C_2 + \omega^2 RC_2(C_1 + C_2)}{1 + \omega^2 R^2(C_1 + C_2)^2} = \frac{\omega^2 RC_2^2}{1 + \omega^2 R^2(C_1 + C_2)^2} \tag{4.64}$$

and

$$\omega C_p = \frac{\omega C_2 + \omega^3 R^2 C_1(C_1 + C_2)C_2}{1 + \omega^2 R^2(C_1 + C_2)^2} \tag{4.65}$$

The parallel resistance

$$R_p = \frac{1 + \omega^2 R^2(C_1 + C_2)^2}{\omega^2 RC_2^2} \tag{4.66}$$

is approximately

$$R_p \approx R \left(\frac{C_1 + C_2}{C_2}\right)^2 \tag{4.67}$$

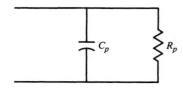

**FIGURE 4.23**
A narrowband equivalent circuit for the capacitive autotransformer.

provided that $\omega^2 R^2 (C_1 + C_2)^2 \gg 1$. Under the same conditions,

$$C_p = \frac{C_2 + \omega^2 R^2 C_1 (C_1 + C_2) C_2}{1 + \omega^2 R^2 (C_1 + C_2)^2} \approx \frac{C_1 C_2}{C_1 + C_2} \tag{4.68}$$

These equations show that the two capacitors have the effect of transforming the resistance up by the turns ratio squared where

$$n = 1 + \frac{C_1}{C_2} \tag{4.69}$$

provided $\omega^2 R^2 (C_1 + C_2)^2 \gg 1$. In this case the circuit of Fig. 4.22b can be replaced (for analysis) by the simpler parallel $RC$ circuit shown in Fig. 4.23.

If the approximation cannot be made that $\omega^2 R^2 (C_1 + C_2)^2 \gg 1$, then the circuit is best analyzed by using computer-aided techniques. Fortunately, narrowband capacitance transformers are normally designed so that the approximations are valid in the frequency region of interest.

**EXAMPLE 4.5.** Determine the response of the interstage coupling circuit shown in Fig. 4.24a. The output impedance of the first transistor amplifier can be assumed to be infinite, and the input impedance of the second stage is $R\Omega$.

**Solution.** The equivalent circuit is shown in Fig. 4.24b. If $\omega^2 R^2 (C_1 + C_2)^2 \gg 1$, the circuit is equivalent to that shown in Fig. 4.25, where

$$C = \frac{C_1 C_2}{C_1 + C_2} \tag{4.70}$$

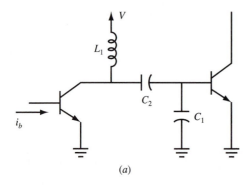

(a)

**FIGURE 4.24**
(a) A two-transistor amplifier with a capacitive transformer in the interstage coupling network; (b) a small-signal equivalent circuit for the interstage network.

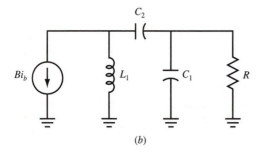

(b)

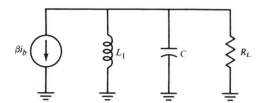

**FIGURE 4.25**
The interstage network of Fig. 4.24a modeled as a parallel resonant circuit.

and
$$R_L = \left(1 + \frac{C_1}{C_2}\right)^2 R \qquad (4.71)$$

The response of this equivalent parallel tuned circuit is now well known. The center frequency is

$$\omega_o = [(L_1 C)^{1/2}]^{-1} \qquad (4.72)$$

and the circuit

$$Q = \frac{R_L}{\omega_o L} = \frac{(C_1 + C_2)^2}{C_2^2 L^{1/2}} \left(\frac{C_1 C_2}{C_1 + C_2}\right)^{1/2} R \qquad (4.73)$$

At the resonant frequency the voltage at the collector of the first transistor is

$$V_c = -(\beta i_b) R_L = -\beta i_b \left(\frac{C_1 + C_2}{C_2}\right)^2 R \qquad (4.74)$$

and the input voltage to the second stage is

$$V_i = -(\beta i_b) R_L \frac{C_2}{C_2 + C_1} = -(\beta i_b) R \frac{C_1 + C_2}{C_2} \qquad (4.75)$$

## ■ 4.5

### IMPEDANCE MATCHING AND HARMONIC FILTERING USING REACTIVE NETWORKS

Besides the transformers previously discussed, reactive networks can be used to match impedances over a narrow frequency range. These networks, properly designed, also serve to filter out harmonics of the signal frequency. The filters described here are composed only of lossless reactive elements, since resistive components would result in power's being dissipated in the coupling network while no power is dissipated in a network consisting solely of inductors and capacitors. The simplest design method utilizes the fact that at any frequency any series combination of resistance and reactance can be converted to an equivalent parallel combination of similar elements (or vice versa).

The input impedances of the two networks illustrated in Fig. 4.26 are equal provided

$$Z_i = R_s \pm j X_s = \frac{\pm R_p j X_p}{R_p \pm j X_p} = \frac{R_p X_p^2}{R_p^2 + X_p^2} \pm j \frac{X_p R_p^2}{R_p^2 + X_p^2} \qquad (4.76)$$

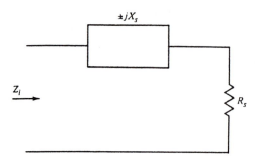

**FIGURE 4.26**
Any series combination of resistance and reactance is equivalent to an equivalent parallel combination at a given frequency.

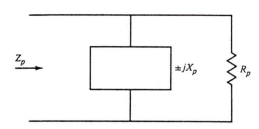

Since both the real and imaginary parts must be equal, Eq. (4.76) is equivalent to

$$R_s = \frac{R_p X_p^2}{R_p^2 + X_p^2} \tag{4.77}$$

and

$$X_s = \frac{X_p R_p^2}{R_p^2 + X_p^2} \tag{4.78}$$

It is seen that the series and parallel reactances will be of the same type (since they are of the same sign). If one is capacitive, the other will be also. The equations for converting from series to parallel impedances can be derived in a like manner (equating admittances). They are

$$R_p = \frac{R_s^2 + X_s^2}{R_s} \tag{4.79}$$

and

$$X_p = \frac{R_s^2 + X_s^2}{X_s} \tag{4.80}$$

The application of these relationships is best illustrated by an example.

**EXAMPLE 4.6.** The input impedance of a transistor amplifier is equal to 10 $\Omega$ in series with 0.2 $\mu$H. Design a matching network so that the input impedance is 50 $\Omega$ at 20 MHz.

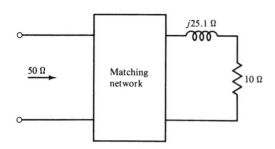

**FIGURE 4.27**
Network discussed in Example 4.6.

**Solution.** At 20 MHz the inductive reactance of a 0.2-$\mu$H inductor is 25.1 $\Omega$. The problem then is to convert the series impedance $10 + j25.1$ to a resistance of 50 $\Omega$, using a lossless matching network as illustrated in Fig. 4.27. Equation (4.79) shows that the equivalent parallel resistance is larger than the series resistance. In this case,

$$R_p = \frac{10^2 + 25.1^2}{10} = 73$$

which is larger than the desired value of 50 $\Omega$. The parallel resistance could be reduced by shunting the equivalent 73-$\Omega$ resistance with another resistor, but this would result in power loss in the coupling network. Equation (4.79) shows that if $X_s$ is smaller, $R_p$ will be smaller. In fact, if $X_s$ satisfies the equation

$$50 = \frac{10^2 + X_s^2}{10}$$

or $$jX_s = j400^{1/2} = j20$$

then the correct impedance level can be reached. The magnitude of $X_s$ can be reduced by adding a capacitive reactance of $-j5.1$ (0.00156 $\mu$F) in series with the load. If this is done, the equivalent parallel resistance is 50 $\Omega$, and the equivalent parallel reactance is

$$jX_p = j\frac{10^2 + 20^2}{20} = j25$$

The reactance can be canceled by adding a parallel capacitor of $-j25$ (318 pF) in parallel. The complete circuit is shown in Fig. 4.28.

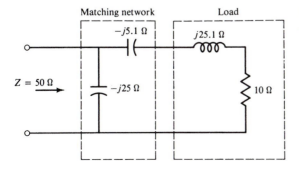

**FIGURE 4.28**
A lossless matching network solution to Example 4.6.

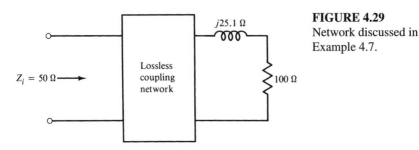

**FIGURE 4.29**
Network discussed in
Example 4.7.

The preceding example illustrates the matching of a load resistance to a larger source resistance. The same technique can be used for matching a load resistance to a smaller source resistance.

**EXAMPLE 4.7.** Design a lossless matching network to couple the impedance shown in Fig. 4.29 to a 50-$\Omega$ source impedance at 20 MHz.

**Solution.** Since the series-to-parallel transformation always results in a larger parallel resistance, there is no series reactance that will directly transform the 100-$\Omega$ resistor to a parallel equivalent of 50 $\Omega$. There are, nevertheless, many possible matching networks that can be used. One network would consist of adding a capacitive reactance of $-j25.1\ \Omega$ in series and then adding a reactance in parallel (as shown in Fig. 4.30) that will transfer the 100 $\Omega$ to a series resistance of 50 $\Omega$. Using this approach, $X_p$ is selected so that

$$R_s = \frac{R_p X_p^2}{R_p^2 + X_p^2} = 50 = \frac{100 X_p^2}{100^2 + X_p^2}$$

Therefore,

$$50 X_p^2 = 50(100)^2$$

or

$$X_p = 100$$

The magnitude of $X_p$ must be 100 $\Omega$, but it can be either capacitive or inductive. An inductor would be selected in cases where it is desired to filter the low-frequency components from the load, and a capacitor would be selected if it is desired to filter the high-frequency components. Once $X_p$ is selected, then a series reactance of the opposite sign must be added to cancel the equivalent series reactance (in this case

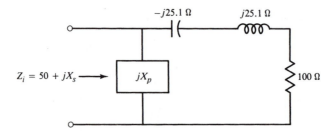

**FIGURE 4.30**
Intermediate solution to Example 4.7.

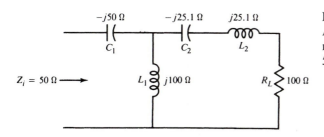

**FIGURE 4.31**
A lossless network for matching a 100-$\Omega$ load to a 50-$\Omega$ source.

$X_s = -j50$). A completed circuit is shown in Fig. 4.31. If the input impedance is to be real at 20 MHz, the corresponding component values are $C_1 = 159$ pF, $C_2 = 317$ pF, $L_1 = 0.8 \ \mu$H, and $L_2 = 0.2 \ \mu$H. If one preferred to filter the high-frequency components instead, the $-j50$-$\Omega$ capacitor would be replaced by a $+j50$-$\Omega$ inductor, and the shunt inductance would be replaced by a $-j100$-$\Omega$ capacitance.

In the design of the coupling networks in the preceding two examples, no consideration was given to the frequency response of the circuit. Circuit $Q$ is often specified to provide a quantitative measure of the attenuation. Circuit $Q$ is an ambiguous term, depending upon the transfer function under consideration. The transfer function of the circuit input impedance is generally different from that of the transimpedance (the latter is important when the circuit is driven by a current source), which is, in general, also different from the voltage-gain transfer function. In the previous example it was shown that the equivalent input impedance could be represented by a 50-$\Omega$ resistor in parallel with capacitive and inductive impedances of 100 $\Omega$. The circuit $Q$ is often interpreted as $Q = \frac{50}{100} = 0.5$. This is the $Q$ of the circuit input impedance, and it is only valid at one frequency. The transfer impedance of this circuit (shown in Fig. 4.31) can be written as

$$\frac{V_o(s)}{I_i(s)} = Z_{12}(s) = \frac{R_L L_1 C_2 s^2}{s^2(L_1 + L_2)C_2 + R_L C_2 s + 1} \tag{4.81}$$

Note that the transfer impedance describes a high-pass transfer function. The magnitude of the transimpedance as a function of frequency is plotted in Fig. 4.32. The $Q$ of a high-pass or low-pass network has no meaning in terms of bandwidth.

The transfer impedance represents a high-pass transfer function, and the $Q$ of such a filter cannot be interpreted as the center frequency divided by the bandwidth as in the bandpass filter. Also, if the source is a voltage rather than a current source, it is the voltage transfer function that is of interest. Since the voltage transfer function will be different from that of the transimpedance, the $Q$'s will not generally be the same. Whether the source is a voltage or current source must be specified before the $Q$ of the circuit can be determined. The network of Fig. 4.31 can be made to have a bandpass transfer impedance by adding a parallel resonant network across the input as shown in Fig. 4.33. If this network is antiresonant at 20 MHz, then the input impedance will still be 50 $\Omega$ at that frequency. If this is a high-$Q$ network, the $Q$ of the transfer impedance will approximate the $Q$ of the parallel network. If precise circuit impedance and frequency characteristics are required, it is better to use network synthesis techniques and computer-aided design programs.

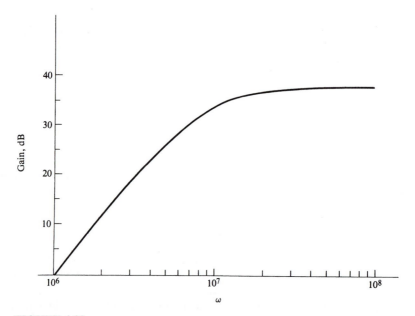

**FIGURE 4.32**
Frequency response of the magnitude of the transimpedance of the circuit shown in Fig. 4.31.

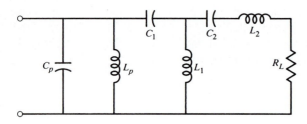

**FIGURE 4.33**
The high-pass transimpedance shown in Fig. 4.31 has been converted to a bandpass transfer function by the addition of the parallel resonant circuit at the input.

## Voltage Gain

Since no power is lost in a lossless network,

$$P_o = P_{in}$$

If the input impedance of the network is matched to the source resistance,

$$P_{in} = \frac{V_s^2}{4R_s} = \frac{V_o^2}{R_o}$$

so the magnitude of the voltage gain is

$$A_v = \frac{1}{2}\sqrt{\frac{R_o}{R_s}}$$

See Fig. 4.34.

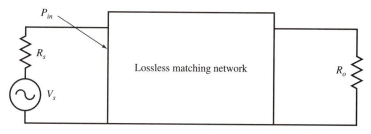

**FIGURE 4.34**

## ▪ 4.6

### FILTER DELAY AND SIGNAL DISTORTION

The design of filter frequency-response characteristics has been largely concerned with the magnitude of the frequency response. Filter phase characteristics are also of importance in modern communication systems. The effect of phase shift on the signal is seen by determining the response of a linear filter $F(s)$ to an input signal $V_i(s)$. The filter output

$$V_o(s) = V_i(s)F(s)$$

If the filter output signal is to be a delayed replica of the input signal (that is, undistorted in shape), then

$$V_o(t) = V_i(t - T)$$

and

$$V_o(s) = V_i(s)e^{-sT}$$

Therefore, the filter characteristics for this undistorted response must be of the form

$$F(s) = \frac{V_o(s)}{V_i(s)} = e^{-sT} \tag{4.82}$$

The magnitude of the filter response

$$|F(j\omega)| = |e^{-j\omega T}| = 1$$

must not vary with frequency, and the filter phase shift as a function of frequency

$$\arg F(j\omega) = \arg e^{-j\omega T} = -\omega T$$

must be linear if the output waveform is to be a delayed replica of the input signal. If the phase shift is not linear, then not all frequency components of the input signal are delayed to the same degree.

The filter delay has become an accepted method of describing the effect of phase nonlinearities on the input waveform. The common definitions for describing filter delay are group delay and phase delay. *Group delay* describes the delay of a group of frequencies; *phase delay* describes the delay of a single sinusoid. Group delay (also called *envelope delay*) is defined as

$$D(\omega) = -\frac{d \arg F(\omega)}{d\omega} \tag{4.83}$$

and phase delay is defined as

$$\phi_D(\omega) = -\frac{\arg F(\omega)}{\omega} \tag{4.84}$$

where $F(\omega)$ is the filter phase shift in radians. The group delay is the negative of the slope of the filter phase shift. The filter group delay and the phase delay will be equal at all frequencies if the filter phase shift is linear. That is, if $\arg F(\omega) = -\omega T$, then

$$D(\omega) = -\frac{d\omega T}{d\omega} = -T \quad \text{and} \quad \phi_D(\omega) = \frac{-\omega T}{\omega} = -T$$

For other filters the two delays are not the same.

**EXAMPLE 4.8.** Compare the group and phase delays of a first-order low-pass filter.

**Solution.** For a first-order low-pass filter, the filter response is

$$F(j\omega) = \left(\frac{j\omega}{\omega_L} + 1\right)^{-1}$$

where $\omega_L$ is the $-3$-dB frequency. The phase shift is

$$\arg F(j\omega) = -\tan^{-1}\frac{\omega}{\omega_L}$$

so the phase delay is

$$\phi_D(\omega) = \frac{\tan^{-1}(\omega/\omega_L)}{\omega} \tag{4.85}$$

and the group delay is

$$D(\omega) = \frac{\omega_L}{\omega^2 + \omega_L^2} \tag{4.86}$$

The two delays are plotted as a function of frequency in Fig. 4.35.

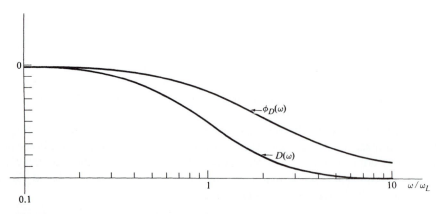

**FIGURE 4.35**
Phase delay $\phi_D(\omega)$ and group delay $D(\omega)$ of a first-order low-pass filter.

Phase-equalizer circuits, which have phase shifts but whose gain magnitudes do not change as a function of frequency, can be added to the filter in order to linearize the phase response over the filter passband.[3] Equalization can result in a phase-frequency diagram such as the one illustrated in Fig. 4.36. The phase is linear across the passband ($\omega_2 - \omega_1$) and can be approximated by

$$\arg F(\omega) = -T\omega - \theta_o$$

where $-\theta_o$ is the phase at $\omega = 0$. The nonzero value of the phase at $\omega = 0$ is known as *phase-intercept distortion*. This type of distortion can affect modulated signals whose carrier contains information. The problem will be described by showing the effect of phase-intercept distortion on an amplitude-modulated signal.

A carrier modulated by a sine wave is described by the equation

$$S(t) = (1 + m\cos\omega_m t)\cos\omega_c t$$

$$= \cos\omega_c t + \frac{m}{2}\cos(\omega_c - \omega_m)t + \frac{m}{2}\cos(\omega_c + \omega_m)t$$

where  $\omega_m$ = modulating frequency
$m$ = modulating index
$\omega_c$ = carrier frequency

and $1 + m\cos\omega_m t$ is known as the *signal envelope*. The modulated signal has a frequency component at the carrier frequency plus upper and lower sidebands at the sum and difference frequencies ($\omega_c \pm \omega_m$). If the modulated signal is passed through a linear phase filter that has phase-intercept distortion, the output signal (assuming the filter gain is constant) is

$$S_o = \cos[\omega_c(t - T) - \theta_o] + \frac{m}{2}\cos[(\omega_c - \omega_m)(t - T) - \theta_o]$$

$$+ \frac{m}{2}\cos[(\omega_c + \omega_m)(t - T) - \theta_o]$$

$$= [1 + m\cos\omega_m(t - T)]\cos\omega_c\left[t - \left(T + \frac{\theta_o}{\omega_c}\right)\right]$$

(4.87)

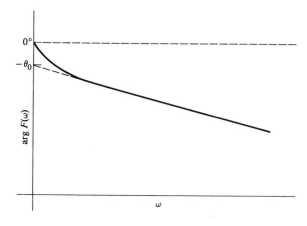

**FIGURE 4.36**
Phase response of a network that includes phase-equalization circuitry.

The envelope of the signal is delayed by the group delay $T$. If the group delay had not been constant, the two sideband signals would not have been delayed equally and the resulting envelope would also have been distorted. For the modulated signal the carrier component has been delayed by phase delay. If synchronous detection is used (see Chap. 12), the output amplitude can be reduced because of the phase-intercept distortion.

## ■ 4.7

### PROBLEMS

**4.1** (*a*) Determine the $Q$, bandwidth $B$, and resonant frequency of the circuit illustrated in Fig. P4.1.
  (*b*) Repeat (*a*) with a 50-$\Omega$ resistor added in series with the voltage source.

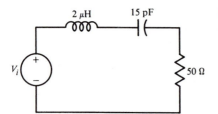

**FIGURE P4.1**
A series resonant circuit.

**4.2** For the circuit of Prob. 4.1 find the frequency at which the voltage across the capacitor is a maximum. Express the capacitor voltage at any frequency in terms of the circuit $Q$. Is this frequency the resonant frequency of the circuit? Explain.

**4.3** Derive Eq. (4.6).

**4.4** (*a*) Calculate the resonant frequency $Q$ and bandwidth of the circuit illustrated in Fig. P4.4.
  (*b*) What is the attenuation of the third harmonic compared to the fundamental frequency amplitude? Of the fifth harmonic?

**4.5** Repeat Prob. 4.4 for the case where a 5-$\Omega$ resistor is in series with the inductor.

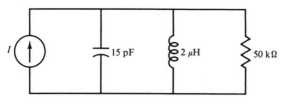

**FIGURE P4.4**
A parallel resonant circuit.

**4.6** A 0.1-$\mu$H inductor with a $Q_u$ of 100 at 1 MHz is used in a series resonant circuit with a 50-$\Omega$ load resistor. What must be the capacitance for the circuit to resonate at 1 MHz? What is the circuit $Q$?

**4.7** The output impedance of the transistor used in the amplifier shown in Fig. P4.7 is 50 kΩ. Does this affect the circuit bandwidth? What will the bandwidth be if the inductor has a $Q_u$ of 100? The capacitor $C = 160$ pF.

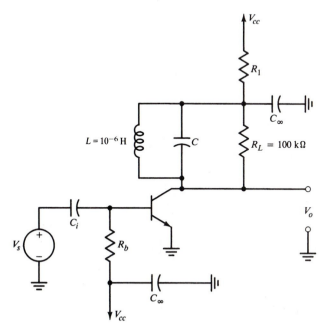

**FIGURE P4.7**
A transistor amplifier with a tuned circuit load.

**4.8** Derive the equivalent turns ratio for a closely coupled step-down autotransformer.

**4.9** If $C_2 = 100$ pF, for what value of $C_1$ will the input impedance of the circuit in Fig. P4.9 be real at 20 MHz? Are there any values of $C_2$ for which $C_1$ cannot be adjusted to make the input impedance real at this frequency?

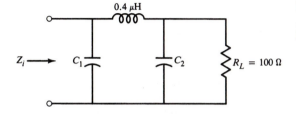

**FIGURE P4.9**
A load resistance and loss-less matching network.

**4.10** Design a lossless coupling network to match a 50-Ω load to a 20-Ω source resistance. What are the component values if the operating frequency is to be 20 MHz? Is the transfer impedance a low-pass bandpass or a high-pass circuit? What about the input impedance? What is the magnitude of the circuit's voltage gain?

**4.11** Design a lossless coupling network that matches a $10 + 5j$-$\Omega$ load to a 50-$\Omega$ source impedance.

**4.12** Design a lossless coupling network that matches a 10-$\Omega$ load in parallel with a $+5j$-$\Omega$ reactance to a 50-$\Omega$ source.

**4.13** Design a lossless coupling network that matches a 50-$\Omega$ load impedance to a 15-$\Omega$ source impedance. Do the $Q$ values of the input impedance and the transfer impedance have practical significance? What is the bandwidth of the transimpedance? What is the magnitude of the circuit's voltage gain?

**4.14** Determine the value of $C_2$ for the circuit shown in Fig. P4.14 to be resonant at $\omega_o = 4 \times 10^6$ rad/s. Determine the circuit input impedance and also the output voltage at the resonant frequency if $I = 10^{-2} \sin(4 \times 10^6 t)$. The transformer is tightly coupled. If the unloaded $Q$ of the secondary coil $Q_2$ is 70, what is the circuit bandwidth? Also, $L_1$ is 0.30 $\mu$H, and $M$ is 3 $\mu$H.

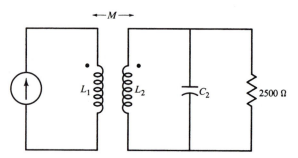

**FIGURE P4.14**
A transformer circuit with a tuned secondary.

**4.15** Find (approximate) the half-power bandwidth of the circuit shown in Fig. P4.15. Here $C$ is 1000 pF, $L$ is 10 $\mu$H, $r$ is 1 $\Omega$, and $R$ is 10 k$\Omega$.

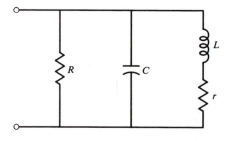

**FIGURE P4.15**
A parallel resonant circuit.

**4.16** In Prob. 4.15, if the circuit is excited by a 1-A peak sinusoidal current source with a frequency of 5 times the resonant frequency of the circuit, what will be the amplitude of the output?

**4.17** Design a lossless coupling network that matches a load impedance of $100 + 5j$ $\Omega$ to a 50-$\Omega$ source impedance.

**4.18** The transistor in the circuit shown in Fig. P4.18 has an input impedance of 1 k$\Omega$ and an output impedance of 80 k$\Omega$. The unloaded $Q$ of the inductor is 100. The transistor current gain $\beta$ is 100.

(a) What is the resonant frequency of the amplifier?
(b) What is the amplifier bandwidth?
(c) What is the radio of the third harmonic output to the fundamental output?
(d) What is the voltage gain at resonance?

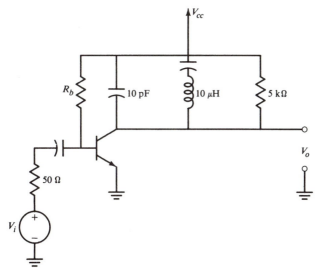

**FIGURE P4.18**
A voltage amplifier with a tuned circuit load.

**4.19** Show that the gain-bandwidth product of the single-tuned inductively coupled circuit is given by Eq. (4.52).

**4.20** For the circuit shown in Fig. P4.20, $L_1 = L_2 = 20$ $\mu$H and $M = 8$ $\mu$H. What value of $C$ is required for the input impedance to be resonant at 1 MHz? What will be the circuit $Q$?

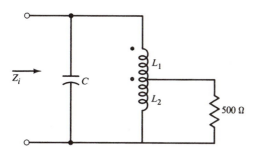

**FIGURE P4.20**
An autotransformer.

**4.21** Derive expressions for the voltage gain as a function of frequency, center frequency, and bandwidth of the circuit shown in Fig. P4.21.

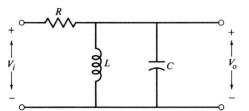

**FIGURE P4.21**
A bandpass filter.

**4.22** A low-pass filter consists of two first-order low-pass filters connected in cascade. If the −3-dB bandwidth of each filter section is 100 kHz, what are the phase and group delays of the composite filter at 1000 kHz?

**4.23** Calculate the output voltage of the amplifier in Fig. P4.23.

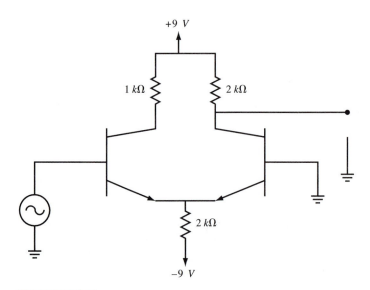

**FIGURE P4.23**

**4.24** Calculate the output voltage of the amplifier in Fig. P4.24.

*4.25** (a) Design a parallel-tuned circuit which resonates at 1 MHz, using a 0.1 $\mu$H inductor with a $Q$ of 100. Determine the accuracy (or inaccuracy) of Eq. (4.22). Evaluate the circuit's response, using steady-state analysis and by determining the transient response to a 10-MHz sine wave input. The two responses should agree. (Note that the analytical solution to this problem is relatively simple. Some of the parameters used in the computer transient simulation will need to be optimized in

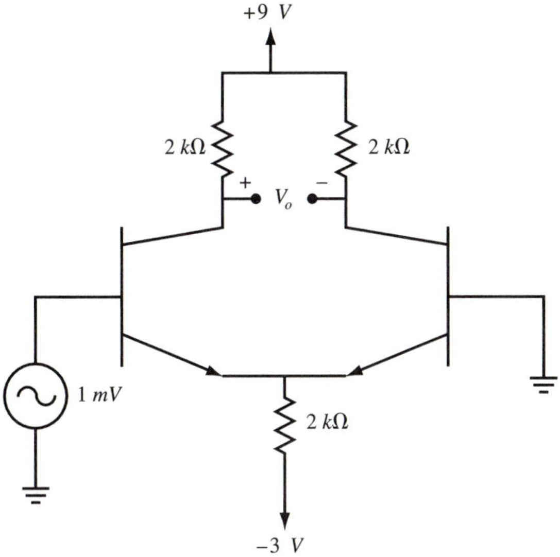


**FIGURE P4.24**

order to get the results of the two simulations to agree, including the time step and RELTOL parameters).

(*b*) Repeat (*a*) with the circuit driven by a voltage source (with a 50-Ω source resistance) driven common-emitter amplifier (using a Q2N2222A with a $\beta$ of 100 and a single 9-V supply). The voltage gain at the resonant frequency should be 10 ($\pm$10 percent).

(*c*) Design a lossless matching network for the tuned circuit so that the amplifier input impedance is 50 Ω resistive at the resonant frequency. Verify the result.

*__4.26__ (*a*) Design an unbalanced parallel-tuned circuit amplifier with a voltage gain of 10 (or greater) at the resonant frequency (using Q2N2222A transistors with a $\beta$ of 100 and a single 9-V supply) which resonates at 1 MHz using a $1 = \mu$H inductor with a $Q$ of 100. Determine the accuracy (or inaccuracy) of Eq. (4.22). Evaluate the circuit's response, using steady-state analysis. The circuit is driven by a voltage source with a 50-Ω source resistance. The circuit should be designed so that the resulting $Q$ is as high as possible.

(*b*) Design a balanced version of the circuit specified in (*a*).

## 4.8

## REFERENCES

1. G. Szentirmai, FILSYN: A General Purpose Filter Synthesis Program, *Proc. IEEE,* **65:** 1443–1458 (1977).
2. G. Valley, Jr., and H. Wallman (eds.), *Vacuum Tube Amplifiers,* McGraw-Hill, New York, 1948.
3. H. J. Blinchikoff and A. I. Zverev, *Filtering in the Time and Frequency Domain,* Wiley, New York, 1976.

■ **4.9**

**ADDITIONAL READING**

Clarke, K. K., and D. T. Hess: *Communication Circuits: Analysis and Design,* Addison-Wesley, Reading, Mass., 1971.

Everitt, W. L.: *Communication Engineering,* 2d ed., McGraw-Hill, New York, 1937.

McIlwain, K., and J. G. Brainerd: *High-Frequency Alternating Currents,* 2d ed., Wiley, New York, 1939.

Szentirmai, G. (ed.): *Computer Aided Filter Design,* IEEE Press, New York, 1973.

Terman, F. E.: *Electronic and Radio Engineering,* 4th ed., McGraw-Hill, New York, 1955.

# High-Frequency Amplifiers and Automatic Gain Control

## INTRODUCTION

Various applications in high-frequency communication circuits require amplifiers. Amplifiers connected to the output of mixers, modulators, and demodulators serve as buffer amplifiers, where they provide a constant-load impedance and gain. Other applications include IF amplifiers, repeaters for cable communications, power amplifiers, power-amplifier drivers, and video amplifiers. This chapter considers the design of amplifiers for applications in which the actual frequency response of the transistors must be considered. Transistors have an upper frequency limit beyond which they do not provide any gain; the frequency response of various transistor amplifier configurations is evaluated in this chapter, and then methods for extending the frequency range of the amplifiers are described. Amplifiers must often automatically adjust gain in response to variations in the input signal amplitude. This ability is called *automatic gain control,* or *AGC.* Several AGC systems are analyzed in Sec. 5.4.

## ■ 5.2

### HIGH-FREQUENCY PERFORMANCE OF BIPOLAR AND FIELD-EFFECT TRANSISTOR AMPLIFIERS

### BJT High-Frequency Model

The bipolar transistor's frequency-dependent properties are accounted for by including the base-emitter capacitance $C_\pi$, the base-collector junction capacitance $C_\mu$, and the collector-to-emitter terminal capacitance $C_o$ in the small-signal equivalent circuit shown in Fig. 5.1. The capacitance $C_\pi$ is actually composed of a diffusion capacitance plus a capacitance that represents the space charge layer present at the base-to-emitter junction. The junction capacitances vary inversely with the voltage

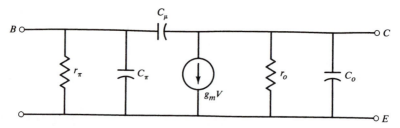

**FIGURE 5.1**
A small-signal, high-frequency equivalent circuit of the bipolar transistor.

across the junction. A detailed description of these capacitances can be found in the literature. For our purposes all parameters in the model can be assumed to be independent of frequency. Although this is not true for all transistors, there are a large number of devices for which the model is accurate at frequencies up to 100 MHz.

At higher frequencies parasitic components, such as lead inductance, become important. How the capacitances limit high-frequency performance can be seen by calculating the short-circuit current gain. To do this, the output is short-circuited as shown in Fig 5.2. Since $r_o$ and $C_o$ are short-circuited, no current flows through them and

$$I_o = -(g_m - j\omega C_\mu)V$$

It is the case that in the frequency region of interest $I_o \approx -g_m V$. The voltage $V$ is given by

$$V = \frac{I_b r_\pi}{j\omega r_\pi (C_\pi + C_\mu) + 1}$$

so the short-circuit current gain is

$$A_i = \frac{I_o}{I_b} = \frac{-g_m r_\pi}{j\omega r_\pi (C_\pi + C_\mu) + 1} = \frac{-\beta}{j\omega r_\pi (C_\pi + C_\mu) + 1} \qquad (5.1)$$

The low-frequency short-circuit current gain is $\beta$. At higher frequencies the gain is reduced because of the two capacitances. The $-3$-dB frequency is referred to as $\omega_\beta$, and

$$\omega_\beta = [r_\pi (C_\pi + C_\mu)]^{-1} = 2\pi f_\beta \qquad (5.2)$$

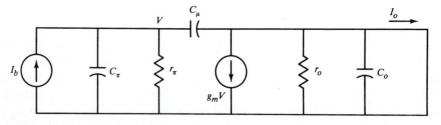

**FIGURE 5.2**
Equivalent circuit for calculating the high-frequency, short-circuit current gain.

At the frequencies near and above $\omega_\beta$ the current gain becomes frequency-dependent, and the base and collector currents are no longer in phase.

An important figure of merit for the transistor is the frequency $f_T$ at which the magnitude of short-circuit current gain becomes unity. That is,

$$|A_i| = 1 = \frac{\beta}{\{1 + [\omega_T r_\pi (C_\pi + C_\mu)]^2\}^{1/2}} \approx \frac{\beta}{\omega_T r_\pi (C_\pi + C_\mu)}$$

or

$$\omega_T = \frac{\beta}{r_\pi (C_\pi + C_\mu)} = \frac{g_m}{C_\pi + C_\mu} \tag{5.3}$$

Note that both $\omega_\beta$ and $\omega_T$ depend upon the transistor's operating point. The transconductance $g_m$ is directly proportional to the collector current: Although $C_\pi$ and $C_\mu$ are not as dependent on the operating point, they do depend on the junction voltages. Increasing the collector bias current will increase $f_T$. Although $C_\pi$ is normally not specified on data sheets, $C_\mu$ is usually given as $C_{cb}$ (collector-to-base capacitance) or $C_{ob}$ (output capacitance, common-base configuration).

**EXAMPLE 5.1.** The 2N3904 is specified to have an $f_T$ of $3 \times 10^8$ Hz, a $\beta$ of 100, and $C_{ob} = 4\,\text{pF}$. Determine the parameters to be used in the hybrid-$\pi$ model. The collector direct current is 10 mA.

*Solution.* Since the collector current is 10 mA,

$$r_\pi = \frac{0.026 \times 100}{0.01} = 260\ \Omega$$

and

$$g_m = 40 \times 10^{-2} = 0.4\ \text{S}$$

Then $C_\pi$ is determined from Eq. (5.3):

$$\omega_T = 2\pi (3 \times 10^8) = \frac{0.4}{C_\pi + C_\mu}$$

or

$$C_\pi + C_\mu = 210\ \text{pF}$$

So

$$C_\pi = 206\ \text{pF}$$

The resistance $r_o$ is normally not given, but a value of 15 k$\Omega$ is a good approximation.

**Current Gain-Bandwidth Product**

The unity gain frequency $f_T$ is also referred to as the *current gain-bandwidth product* of the device, since

$$\omega_T = g_m r_\pi \omega_\beta$$

and $g_m r_\pi$ is the magnitude of the low-frequency short-circuit current gain:

$$A_i = \frac{-g_m r_\pi}{j\omega r_\pi (C_\pi + C_\mu) + 1} \approx \frac{-g_m}{j\omega(C_\pi + C_\mu)} = \frac{-\omega_T}{j\omega} \tag{5.4}$$

So, at any frequency greater than $\omega_\beta$ the magnitude of the short-circuit current gain is readily determined.

**EXAMPLE 5.2.** What would the short-circuit current gain be at 10 MHz for the 2N3904 transistor of the preceding example?

**Solution.** Since the gain-bandwidth product is $3 \times 10^8$ Hz, at 10 MHz the gain will be

$$A_i = \frac{3 \times 10^8}{10^7} = 30$$

The gain-bandwidth product of the device serves as a useful rule of thumb for estimating the gain at any frequency. At frequencies below $\omega_\beta$ the magnitude of the current gain is equal to $\beta$.

## FET High-Frequency Model

The performance of a field-effect transistor is limited at high frequencies by parasitic capacitances, as is that of the bipolar transistor. The high-frequency performance can be described in terms of the equivalent circuit shown in Fig. 5.3. The gate-to-source capacitance $C_{gs}$ and the gate-to-drain capacitance $C_{gd}$ are junction capacitances that are inversely related to the voltage across the junctions. The actual relationship depends upon the particular manufacturing process used, but since the source voltage will be less than the drain voltage, $C_{gs}$ will usually be larger than $C_{gd}$. The drain-to-source capacitance $C_{ds}$ is the stray capacitance associated with the device package. The capacitances of junction and insulted-gate field-effect transistors differ in how they vary with the operating point, but the small-signal amplifier models are similar. We will now discuss the high-frequency behavior of these amplifiers.

The hybrid-$\pi$ models of the junction and field-effect transistors are shown in Fig. 5.4. The only difference between the BJT and FET equivalent circuits is the resistor $r_\pi$ included in the small-signal model for the bipolar transistor. The mid-frequency input impedance of the common-emitter amplifier is $r_\pi$, whereas that of the common-source amplifier is very large and is considered to be infinite. The output circuits of the two devices are the same except that the bipolar transistor is essentially a current-controlled device, while the FET is a voltage-controlled device. Also, the transconductance of bipolar transistors is proportional to the collector direct current, while the transconductance of a field-effect transistor is

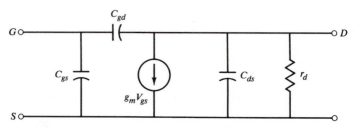

**FIGURE 5.3**
A small-signal, high-frequency equivalent circuit of a field-effect transistor.

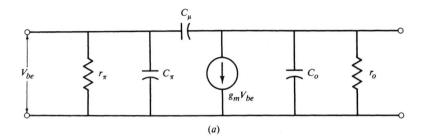

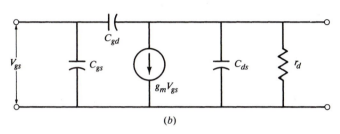

$(b)$

**FIGURE 5.4**
High-frequency equivalent circuits of $(a)$ the BJT and $(b)$ the FET.

proportional to the square root of the drain direct current. Nevertheless, the circuit models are identical, and the same equations can be used for both. Before studying the frequency-dependent properties of these amplifiers, we will discuss a network theorem that is extremely useful for the analysis of transistor amplifiers.

## Miller's Theorem

Consider the network configuration indicated in Fig. 5.5. Nodes 1 and 2 are interconnected by the impedance $Z$. At any frequency the voltage gain between these two nodes can be given as

$$\frac{V_2}{V_1} = K$$

Generally $K$ is frequency-dependent.

We will now show that the current $I_1$ drawn from node 1 through $Z$ can be obtained by disconnecting $Z$ from terminal 1 and by bridging an impedance $Z/(1 - K)$ from node 1 to ground. This follows since

$$I_1 = \frac{V_1 - V_2}{Z} = \frac{V_1 - K V_1}{Z} = \frac{V_1(1 - K)}{Z} = \frac{V_1}{Z_1} \qquad (5.5)$$

where

$$Z_1 = \frac{Z}{1 - K}$$

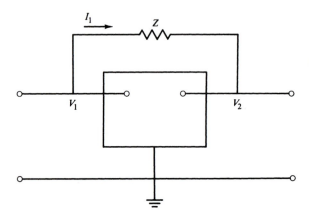

**FIGURE 5.5**
A voltage amplifier with impedance $Z$ connected between the input and output.

Therefore, if $Z$ is replaced at node 1 by an impedance $Z_1$ connected between node 1 and ground, the current leaving node 1 will be the same. The current leaving node 2 is

$$I_2 = \frac{V_2 - V_1}{Z} = \frac{V_2 - V_2/K}{Z} = \frac{V_2}{Z_2} \tag{5.6}$$

where

$$Z_2 = \frac{KZ}{K-1}$$

If $Z_2$ is connected between node 2 and ground, the current leaving node 2 will be the same. Miller's theorem states that if the network illustrated in Fig. 5.5 has a voltage gain $K$ between two nodes, then any impedance connected between those two nodes can be replaced by an impedance at each node connected to ground, as illustrated in Fig. 5.6. The values of the impedances are given by Eqs. (5.5) and (5.6).

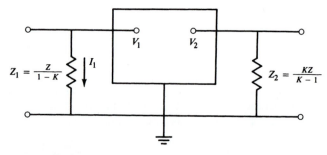

**FIGURE 5.6**
An equivalent circuit of the amplifier shown in Fig. 5.5 but without the feedback resistor.

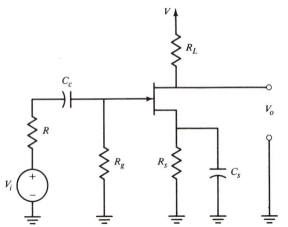

**FIGURE 5.7**
A common-source amplifier.

## High-Frequency Response of FET Amplifiers

### Common-source amplifier

The utility of Miller's theorem can be appreciated by applying it to the analysis of the high-frequency performance of the FET amplifier illustrated in Fig. 5.7. The small-signal, high-frequency equivalent of the amplifier is given in Fig. 5.8. If $R_g \gg R$, then $V_g \approx V_i$, and the midfrequency voltage gain is

$$A_v = \frac{V_2}{V_1} = -g_m R'_L$$

where

$$R'_L = \frac{R_L r_d}{R_L + r_d}$$

We will now use Miller's theorem to replace the gate-to-drain capacitance by two capacitors, one connected from gate to ground and one connected from drain to

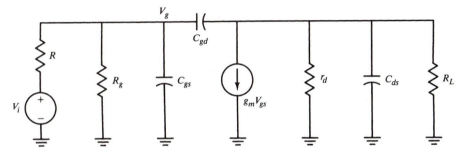

**FIGURE 5.8**
A small-signal, high-frequency equivalent circuit of the common-source amplifier.

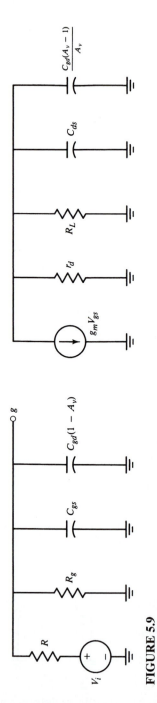

**FIGURE 5.9**

A small-signal, high-frequency equivalent circuit of the common-source amplifier obtained by applying Miller's theorem to the circuit shown in Fig. 5.8.

ground, as illustrated in Fig. 5.9. The impedance from gate to drain is $Z = (j\omega C_{gd})^{-1}$, so the impedance to be added from gate to ground is [Eq. (5.5)]

$$Z_1 = [j\omega C_{gd}(1 - A_v)]^{-1} = [j\omega C_{gd}(1 + g_m R'_L)]^{-1}$$

and the impedance to be added from drain to ground is [Eq. (5.6)]

$$Z_2 = \frac{A_v}{j\omega C_{gd}(A_v - 1)} = \left[j\omega C_{gd}\left(\frac{1 + g_m R'_L}{g_m R'_L}\right)\right]^{-1}$$

The gate-to-drain capacitance is replaced by a capacitor

$$C_1 = C_{gd}(1 + g_m R'_L) \tag{5.7}$$

connected from gate to ground, and a capacitor

$$C_2 = C_{gd}\frac{1 + g_m R'_L}{g_m R'_L} \tag{5.8}$$

connected from drain to ground. If the voltage gain is large, the capacitance shunting the input is large. In a high-gain inverting amplifier any capacitance connected from input to output is equivalent to the same capacitance shunting the input, but with its value amplified by the voltage gain. This is known as the *Miller effect*. The Miller effect limits the high-frequency performance of high-voltage-gain amplifiers.

If we again assume that $R \ll R_g$, the complete frequency-dependent transfer function for the circuit of Fig. 5.9 is

$$\frac{V_o}{V_i} = \frac{-g_m R'_L}{(j\omega RC_T + 1)(j\omega R'_L C_o + 1)} \tag{5.9}$$

where

$$C_T = C_{gs} + C_{gd}(1 + g_m R'_L)$$

and

$$C_o = \frac{C_{gd}(1 + g_m R'_L)}{g_m R'_L} + C_{ds}$$

The amplifier has two high-frequency poles in the left-hand plane:

$$\omega_1 = -(RC_T)^{-1} \tag{5.10}$$

and

$$\omega_2 = -(R'_L C_o)^{-1} \tag{5.11}$$

Unless $R$ is very small, the magnitude of $\omega_1$ will usually be much smaller than the magnitude of $\omega_2$ because $C_T$ is much bigger than $C_o$. Actually, the approximation used in applying Miller's theorem is such that the frequency response is not accurate beyond the lower of the two poles. At higher frequencies the voltage gain $A_v$ is complex, but it was assumed to be real in deriving Eq. (5.9).

**EXAMPLE 5.3.** A 2N5486 FET with a transconductance $g_m$ of $2 \times 10^{-3}$ S and an $r_d$ of 13 k$\Omega$ is used in the amplifier shown in Fig. 5.10. Also $C_{gs} = 5$ pF and $C_{gd} = 1$ pF $= C_{ds}$. Determine the high-frequency response of the amplifier.

**Solution.** The small-signal, high-frequency equivalent circuit is illustrated in Fig. 5.11. Since $R \leq R_g$, the midfrequency voltage gain is given as

$$A_v = -g_m R_L = \frac{(-2 \times 10^{-3})(6 \text{ k}\Omega \times 13 \text{ k}\Omega)}{6 \text{ k}\Omega + 13 \text{ k}\Omega} = -8.2$$

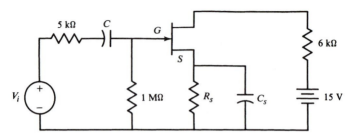

**FIGURE 5.10**
The common-source amplifier discussed in Example 5.3.

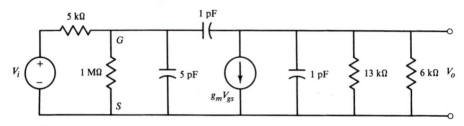

**FIGURE 5.11**
The high-frequency equivalent circuit of the amplifier shown in Fig. 5.10.

Therefore,

$$C_T = 5 + 1(1 + 8.2) = 14.2 \text{ pF}$$

and
$$C_o = 1 + \frac{1 + 8.2}{8.2} = 2.1 \text{ pF}$$

The high-frequency pole due to the input circuit is [Eq. (5.10)]

$$-\omega_1 \approx (RC_T)^{-1} = [(5 \times 10^3)(14.2 \times 10^{-12})]^{-1} = 1.41 \times 10^7 \text{ rad/s}$$

and the high-frequency pole due to the output circuit is [Eq. (5.11)]

$$-\omega_2 = (R_L C_o)^{-1} = [(4.1 \times 10^3)(2.1 \times 10^{-12})]^{-1} = 11.6 \times 10^7 \text{ rad/s}$$

The upper corner frequency is approximately $1.4 \times 10^7$ rad/s and is determined by the input circuit. The estimate for a second high-frequency pole obtained using the approximation is not accurate.

The application of Miller's theorem in the preceding example was somewhat of an approximation. The approximation involved using the midband gain, which is frequency-independent, as an approximation for the voltage gain at all frequencies. To be more precise, the frequency-dependent gain should be used. Since the gain will generally be complex, the capacitance must be replaced by complex impedances in the application of Miller's theorem. However, the approximation is sufficiently accurate if Miller's theorem is used with the midband gain. To illustrate the accuracy, the circuit of Fig. 5.8 can be analyzed directly. If we again assume that $R_g \geq R$, the equilibrium equations can be written as

$$\frac{V_g - V_i}{R} + V_g s C_{gs} + (V_g - V_o) s C_{gd} = 0$$

and

$$-V_g s C_{gd} + V_o s C_{gd} + \frac{V_o}{R'_L} + g_m V_g + V_o s C_{ds} = 0$$

where

$$R'_L = \frac{r_d R_L}{r_d + R_L}$$

If $V_g$ is eliminated, the voltage gain is found to be

$$\frac{V_o}{V_i} = A_v = \frac{-(g_m - s C_{gd}) R'_L}{\begin{array}{c} s^2 R R'_L [C_{gd} C_{gs} + C_{ds} (C_{gd} + C_{gs})] + s[g_m C_{gd} R R'_L \\ + (C_{gd} + C_{ds}) R'_L + (C_{gd} + C_{gs}) R] + 1 \end{array}} \quad (5.12)$$

This transfer function differs from that obtained using the approximation for Miller's theorem. The main difference is that a high-frequency zero has been added in the right half of the $s$ plane. However, the frequency response is close to that obtained previously; the validity of the approximation is illustrated by the following example.

**EXAMPLE 5.4.** The previous example used the approximation to Miller's theorem to determine the high-frequency response. Determine the high-frequency response using Eq. (5.12).

**Solution.** For this amplifier $g_m = 2 \times 10^{-3}$, $C_{gs} = 5 \times 10^{-12}$, $C_{gd} = 10^{-12}$, $R = 5 \times 10^3$, and $R'_L = 4.1 \times 10^3$. If these values are substituted into Eq. (5.12), the transfer function zero is found to be located at

$$\omega_z = 2 \times 10^9 \text{ rad/s}$$

and the two poles are located at

$$\omega_{p1} = -1.3 \times 10^7 \quad \text{rad/s}$$

and

$$\omega_{p2} = -33.7 \times 10^7 \quad \text{rad/s}$$

The actual frequency response obtained from a computer simulation of the amplifier is plotted in Fig. 5.12. The $-3$-dB corner frequency is located at $1.3 \times 10^7$ rad/s ($2.1 \times 10^6$ Hz), whereas a corner frequency of $1.4 \times 10^7$ rad/s was obtained with the approximation. The zero is located at a much higher frequency and has a negligible effect on the transfer function within the passband of the amplifier, but it does affect the rate of attenuation and phase shift at high frequencies.

In general, the midfrequency gain can be used with Miller's theorem to obtain a sufficiently accurate high-frequency amplifier model. This approach greatly simplifies the analysis since it isolates the input and output stages of the amplifier. At frequencies above the $-3$-dB corner frequency the approximation error increases. At sufficiently high frequencies the amplifier rolls off at $-6$ dB per octave, not at $-12$ dB per octave as predicted by Eq. (5.9).

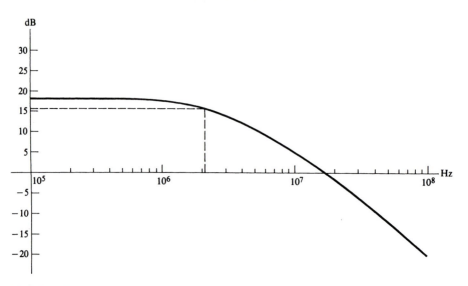

**FIGURE 5.12**
Frequency response of the voltage gain (magnitude) of the amplifier circuit shown in Fig. 5.11.

Since the common-gate and source-follower small-signal, high-frequency equivalent circuits are simplified versions of the common-base and emitter-follower small-signal circuits, they will not be analyzed here. The results derived in the following section for the high-frequency response of BJT amplifiers are directly applicable to FET amplifiers.

## High-Frequency Response of BJT Amplifiers

### Common-emitter amplifier

The small-signal, high-frequency model for the common-emitter amplifier is shown in Fig. 5.13. If $R_s$ and $r_\pi$ are combined into the single resistor

$$R = \frac{R_s r_\pi}{R_s + r_\pi}$$

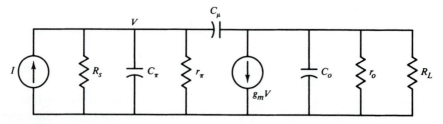

**FIGURE. 5.13**
A small-signal, high-frequency equivalent circuit of a common-emitter amplifier.

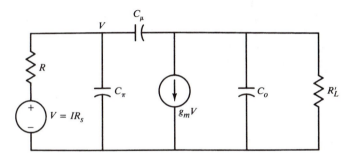

**FIGURE. 5.14**
A simplified equivalent circuit of a common-emitter amplifier.

and if $r_o$ and $R_L$ are also combined

$$R'_L = \frac{r_o R_L}{r_o + R_L} \tag{5.13}$$

the equivalent circuit is as shown in Fig. 5.14. This is identical to the high-frequency circuit for the common-source amplifier; hence the results derived for the frequency response of the common-source amplifier apply equally well to the common-emitter amplifier.

EXAMPLE 5.5.  The amplifier shown in Fig. 5.15 uses a 2N3904 BJT with an $f_T$ of $3 \times 10^8$ Hz, a $\beta$ of 100, and a $C_{ob}$ of 4 pF. Here $C_{ob}$ is the output capacitance of a common-base amplifier. It is usually included on a company's transistor specification sheet because it is easy to measure. The capacitance $C_\mu$ of the hybrid-$\pi$ model is appreciably less[1] than $C_{ob}$, but 4 pF has been assumed for $C_\mu$ in this example. This means that the actual high-frequency bandwidth will be greater than predicted. Calculate the upper 3-dB frequency of the amplifier. The collector direct current is 10 mA. Here $C_o$ is assumed to have a negligible effect on the frequency response.

*Solution.*  Example 5.1 showed that a 2N3904 transistor with a collector direct current of 10 mA, had $r_\pi = 260\ \Omega$, $g_m = 0.4$ S, and $C_\pi = 206$ pF.

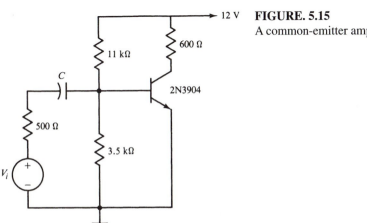

12 V   **FIGURE. 5.15**
A common-emitter amplifier.

The small-signal, high-frequency equivalent circuit is shown in Fig. 5.16. The parallel combination of the two base-biasing resistors is shown as a 2.66-k$\Omega$ resistor. The parallel combination of this resistor and the 260-$\Omega$ emitter resistance is 237 $\Omega$. It has also been assumed that the transistor output impedance is much greater than the 600-$\Omega$ load impedance. The midfrequency output voltage is

$$V_o = \frac{-g_m R_L \times 237}{237 + 500} V_i$$

$$= \frac{-0.4 \times 600 \times 237}{737} V_i = -77.2 V_i$$

The midfrequency base-to-emitter voltage gain is $-g_m R_L = 240$, so the approximate high-frequency, small-signal circuit is as shown in Fig. 5.17. The equivalent input Miller capacitance is 960 pF. If the input circuit is replaced by its Thévenin equivalent, the circuit is as shown in Fig. 5.18. This circuit topology is the same as that given in Fig. 5.9 for the FET amplifier. Hence Eqs. (5.10) and (5.11) can be used to find the two high-frequency poles

$$-\omega_1 = [161(1166 \times 10^{-12})]^{-1} = 5.3 \times 10^6 \text{ rad/s}$$

and
$$-\omega_2 = [600(4 \times 10^{-12})]^{-1} = 33.2 \times 10^7 \text{ rad/s}$$

Since $|\omega_1| \ll |\omega_2|$, the high-frequency performance is determined by the input circuit. The upper $-3$-dB frequency of this amplifier is equal to $|\omega_1|$.

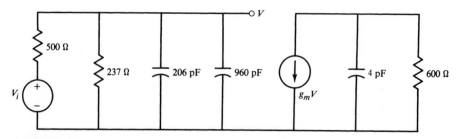

**FIGURE. 5.16**
A high-frequency equivalent circuit of the amplifier shown in Fig. 5.15.

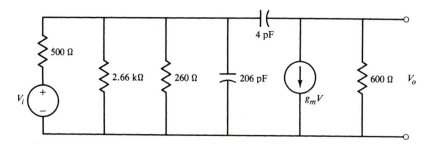

**FIGURE 5.17**
An approximate equivalent circuit of the amplifier shown in Fig. 5.15.

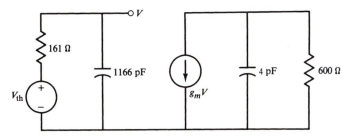

**FIGURE 5.18**
A simplified equivalent of the circuit shown in Fig. 5.17.

The high-frequency common-emitter circuit of Fig. 5.14 is identical in topology to the small-signal, high-frequency equivalent circuit given in Fig. 5.8 for the common-source amplifier. Therefore, Eq. (5.12) is also correct for the common-emitter circuit. The corresponding equation for the BJT is found by changing the notation to correspond to that of the BJT circuit:

$$\frac{V_o}{V_{th}} = A_v = \frac{-(g_m - sC_\mu)R'_L}{s^2RR'_L[C_\pi C_\mu + C_o(C_\pi + C_\mu)] + s[R(C_\pi + C_\mu)]}$$
$$+ R'_L(C_\mu + C_o) + g_m C_\mu RR'_L] + 1 \quad (5.14)$$

Now $R'_L$ is given by Eq. (5.13), and $R$ is equal to the parallel combination of $R_s, r_\pi$, and any bias resistors connected between the base terminal and ground. Also $V_{th}$ is the Thévenin equivalent voltage of the input circuit consisting of these resistors.

**EXAMPLE 5.6.** Calculate the exact frequency response of the amplifier shown in Fig. 5.15, using Eq. (5.14).

*Solution.* If the values from Example 5.5 are substituted into Eq. (5.14), the transfer function is

$$\frac{V_o}{V_{th}} = \frac{-(0.4 - s \times 4 \times 10^{-12})(966 \times 10^3)}{s^2(79.6 \times 10^{-18}) + s(190.2 \times 10^{-9}) + 1}$$

The transfer function has a zero

$$\omega_z = 10^{11} \text{ rad/s}$$

and two left half-plane poles

$$\omega_1 = -5.5 \times 10^6 \text{ rad/s}$$

and

$$\omega_2 = -2.41 \times 10^9 \text{ rad/s}$$

In this example the bandwidth approximation obtained using the Miller theorem is almost the same as that obtained from the more complex formula. This approximation does provide an easy method for estimating the frequency response. Precise estimates require a great deal more calculation or, preferably, some method of simulation.

## Common-Base Amplifiers

Common-base amplifiers are used in many high-frequency circuits because they have a wider bandwidth than does the equivalent common-emitter amplifier. A typical common-base amplifier circuit is shown in Fig. 5.19. The small-signal, high-frequency model for this amplifier is shown in Fig. 5.20. The reasons for the wider bandwidth of the common-base amplifier are that (1) the midfrequency input resistance, given by Eq. (2.17)

$$Z_i \approx \frac{r_\pi}{\beta} = (g_m)^{-1}$$

is $1/\beta$ times that of the common-emitter amplifier and (2) the equivalent Miller capacitance connected across the input is

$$C_M = C_o(1 - A_v)$$

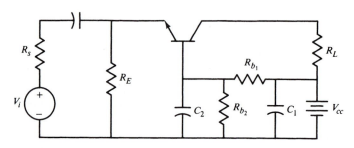

**FIGURE 5.19**
A common-base amplifier.

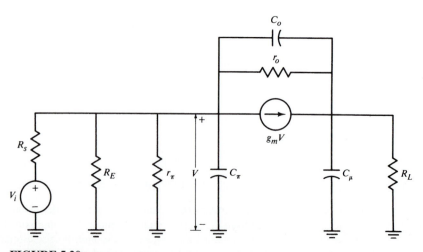

**FIGURE 5.20**
A small-signal, high-frequency equivalent circuit of the common-base amplifier.

as for the common-emitter amplifier, but in the common-base amplifier $A_v$ is positive. If the voltage gain is greater than 1, the Miller capacitance will be negative and will reduce the total capacitance from emitter to base. The total input capacitance can, in fact, become negative, which means that the input impedance is inductive. Because of the small input impedance the transistor collector-to-emitter resistance $r_o$ can be ignored with no sacrifice in accuracy.

The equivalent circuit can then be drawn as shown in Fig. 5.21. In this figure the voltage source has been replaced by an equivalent current source $I = V_i/R_s$, and the resistor $R$ is the parallel combination of the three resistors

$$R = R_s \| R_E \| r_\pi / \beta$$

The emitter voltage $V$ is then

$$V = (I - g_m V)Z$$

or

$$V = \frac{IZ}{1 + g_m Z}$$

where

$$Z = \frac{R}{1 + j\omega R C_T}$$

and

$$C_T = C_\pi + C_o(1 - A_v)$$

Therefore

$$\frac{V}{I} = \frac{R/(1 + g_m R)}{1 + j\omega C_T R/(1 + g_m R)} \tag{5.15}$$

The corner frequency of the input circuit is then

$$\omega_1 = \frac{1 + g_m R}{|C_T| R} \tag{5.16}$$

and the corner frequency of the output circuit is

$$\omega_2 = [R_L (C_\mu + C'_M)]^{-1} \tag{5.17}$$

where

$$C'_M = C_o \frac{A_v - 1}{A_v}$$

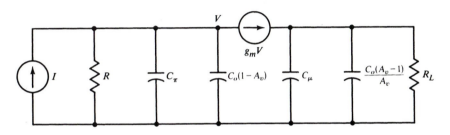

**FIGURE 5.21**
An equivalent circuit of the common-base amplifier, obtained using Miller's theorem.

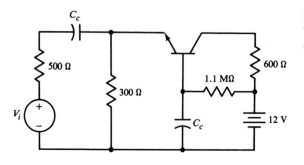

**FIGURE 5.22**
The common-base amplifier discussed in Example 5.7.

For the common-base amplifier the output-circuit corner frequency can be lower than the corner frequency of the input circuit. The particular characteristics of each amplifier must be evaluated.

> **EXAMPLE 5.7.** Calculate the gain and frequency response of the common-base amplifier illustrated in Fig. 5.22. The same transistor (and collector direct current) is used as in Example 5.5 (where the frequency response of the common-emitter amplifier was calculated). The collector-to-emitter capacitance is assumed to be $C_o = 1$ pF.

**Solution.** The high-frequency equivalent circuit is given in Fig. 5.23. The midband emitter-to-collector voltage gain is

$$\frac{V_o}{V} = g_m R_L = 240$$

so the Miller capacitance is

$$C_M = 1 - 240 = -239 \text{ pF}$$

and the collector-to-ground Miller capacitance is

$$C'_M = \frac{239}{240} \approx 1 \text{ pF}$$

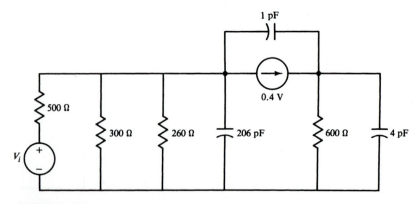

**FIGURE 5.23**
A high-frequency equivalent circuit of the amplifier shown in Fig. 5.22.

Since $R_i = (g_m)^{-1} = 2.5\ \Omega$, the midfrequency voltage gain is

$$\frac{V_o}{V_i} = A_v = 240\left(\frac{2.5}{500}\right) = 1.2$$

The voltage gain is much less than that of the common-emitter amplifier because of the low input resistance.

The high-frequency pole due to the output circuit is [Eq. (5.17)]

$$-\omega_2 = (600 \times 5\text{ pF})^{-1} = 3.33 \times 10^8\text{ rad/s}$$

The high-frequency pole due to the input circuit can be determined using Eq. (5.16). The equivalent resistance is

$$R = 500 \parallel 300 \parallel 260 = 109\ \Omega$$

and the total capacitance

$$C_T = 206 - 239 = -33\text{ pF}$$

is negative. The sign of the capacitance affects the phase shift but not the magnitude of the frequency response or the corner frequency. The high-frequency pole of the input circuit [Eq. (5.16)]

$$-\omega_1 = (RC_T)^{-1} = 2.78 \times 10^8\ \text{ rad/s}$$

is slightly smaller than the pole due to the output circuit, so the high-frequency response is determined by the input circuit. A plot of the input susceptance, as determined by computer simulation, is shown in Fig. 5.24. At low frequencies the susceptance is negative (inductive), as predicted using Miller's theorem. It does, however, increase in magnitude with increasing frequency. At higher frequencies, the voltage gain decreases and the susceptance becomes capacitive (about $1.5 \times 10^7$ Hz).

## Emitter-Followers

The emitter-follower has a positive voltage gain less than unity, so one would predict from Miller's theorem that this circuit has a wider bandwidth than the common-emitter amplifier. The emitter-follower does, in fact, usually have the widest bandwidth of the three amplifier configurations. The high-frequency response can be determined from the emitter-follower equivalent circuit shown in Fig 5.25. For purposes of analysis we will redraw the circuit as shown in Fig. 5.26. In this circuit $I = V_i/R_s$ and

$$Z_i = \frac{R_i}{j\omega C_\mu R_i + 1} \tag{5.18}$$

where

$$R_i = \frac{R_s R_b}{R_s + R_b}$$

Also,

$$Z_\pi = \frac{r_\pi}{j\omega r_\pi C_\pi + 1} \tag{5.19}$$

and

$$Z_L = \frac{R_o}{j\omega R_o C_o + 1} \tag{5.20}$$

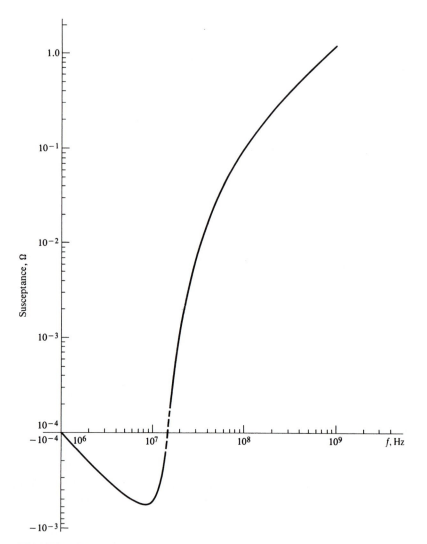

**FIGURE 5.24**

Input susceptance of the circuit shown in Fig. 5.23.

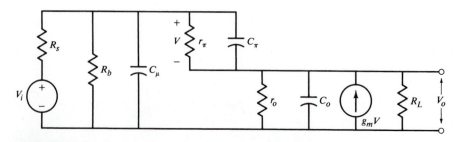

**FIGURE 5.25**

A small-signal, high-frequency equivalent circuit of the emitter-follower.

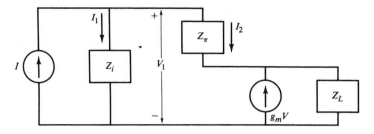

**FIGURE 5.26**
An emitter-follower equivalent circuit.

where
$$R_o = \frac{r_o R_L}{r_o + R_L}$$

The input current divides into the two currents $I_1$ and $I_2$. The amplifier response is derived from the three equations:

$$I = I_1 + I_2$$

$$V_1 = I_1 Z_i = I_2 Z_\pi + V_o$$

and
$$V_o = (1 + g_m Z_\pi) I_2 Z_L$$

If $I_1$ and $I_2$ are eliminated from these three equations, the transimpedance is found to be

$$\frac{V_o}{I} = Z_T = Z_i \frac{Z_L(1 + g_m Z_\pi)}{Z_i + Z_\pi + (1 + g_m Z_\pi)Z_L} \tag{5.21}$$

This complicated equation can be simplified for most practical applications. It will usually be the case that $Z_L \approx R_L$. If $Z_i \gg Z_\pi$, which will be the situation if the amplifier is driven by a current source, the transimpedance is then

$$Z_T \approx R_L(1 + g_m Z_\pi) = R_L \frac{1 + g_m r_\pi + j\omega r_\pi C_\pi}{j\omega r_\pi C_\pi + 1} \tag{5.22}$$

The amplifier will have a pole at

$$-\omega_p = (r_\pi C_\pi)^{-1} \tag{5.23}$$

which is approximately equal in magnitude to the current gain-bandwidth product of the device divided by $\beta$. The amplifier also has a zero at

$$\omega_z = -\frac{(1 + g_m r_\pi)}{r_\pi C_\pi} \tag{5.24}$$

If the amplifier is driven by a voltage source with a source resistance $R_s \ll R_b$, then $Z_i \approx R_s$, and the voltage transfer function simplifies to

$$\frac{V_o}{V_i} = \frac{R_L(1 + g_m Z_\pi)}{R_s + Z_\pi + R_L(1 + g_m Z_\pi)} \tag{5.25}$$

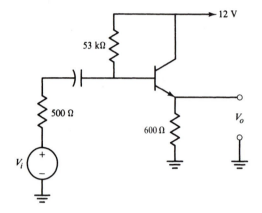

**FIGURE 5.27**
The emitter-follower discussed in Example 5.9.

which has a zero at

$$-\omega_Z = \frac{1 + g_m r_\pi}{r_\pi C_\pi} \tag{5.26}$$

and a pole at the lower frequency

$$-\omega_p = \frac{R_s + r_\pi + (1 + g_m r_\pi) R_L}{r_\pi C_\pi (R_s + R_L)} \tag{5.27}$$

These equations contain several approximations, but they do provide good estimates for the design and analysis of high-frequency emitter-follower amplifiers. The following example illustrates their application.

**EXAMPLE 5.8.** Estimate the frequency response of the amplifier illustrated in Fig. 5.27. The same values for load and source resistance are used as in Example 5.7 with the common-base amplifier. The transistor is biased so that $g_m$ is again 0.4 S.

**Solution.** For this amplifier $R_b \gg R_s$ and $R_L \ll r_o$, so the amplifier bandwidth can be estimated using Eq. (5.27):

$$-\omega_p = \frac{500 + 260 + 105(600)}{260(206 \times 10^{-12})(500 + 600)} = 1.112 \times 10^9 \text{ rad/s}$$

The zero frequency is found from Eq. (5.26) to be

$$-\omega_Z = 2.01 \times 10^9 \text{ rad/s}$$

This model predicts that the frequency response does not decrease above $\omega_z$, but as with the Miller approximation, these equations are also an approximation and the high-frequency gain will decrease at $-6$ dB per octave.

## Differential Amplifiers

### Differential mode frequency response

A BJT differential amplifier with matching transistors is shown in Fig. 5.28. For differential input signals $(V_1 = -V_2)$ the two base currents are equal and

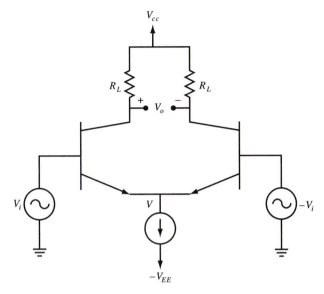

**FIGURE 5.28**

opposite, and the emitter node is at virtual ground. Thus the circuit, for the small-signal response to the differential input, can be drawn as shown in Fig. 5.29. The two halves of the circuit are the same, with responses equal in magnitude and opposite in sign to the two input signals. The frequency response of the differential amplifier can thus be determined from an analysis of the simpler circuit shown in Fig. 5.30. A small-signal circuit equivalent to that of Fig. 5.30 is shown in Fig. 5.31.

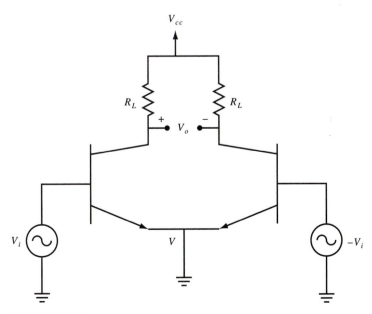

**FIGURE 5.29**

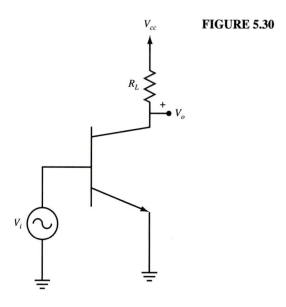

**FIGURE 5.30**

This small-signal circuit is equivalent to the circuit shown in Fig. 5.8 (for the common-source amplifier). Thus

$$\frac{V_o/2}{V_i} = \frac{-g_m R_L r_\pi}{(j\omega R C_T + 1)(j\omega R'_L C_{oT} + 1)(r_\pi + R_s)}$$

where

$$R'_L = r_o R_L/(r_o + R_L)$$

and

$$R = \frac{R_s r\pi}{R_s + \pi r}$$

$$C = C + C_\mu(1 + g_m R'_L)$$

and

$$C_{oT} = C_\mu \frac{(1 + g_m R'_L)}{g_m R'_L}$$

This differential amplifier has the same Miller capacitances as does the single-transistor common-emitter amplifier. Since

$$A_d = \frac{V_o}{V_d} = \frac{V_o}{2V_i}$$

The frequency-dependent small-signal gain of the differential amplifier is

$$A_d = \frac{-g_m R'_L r_\pi}{(r_\pi + R_s)(j\omega R C_T + 1)(j\omega R'_L C_{oT} + 1)}$$

This circuit has two high-frequency poles

$$\omega_{p1} = \frac{1}{R C_T}$$

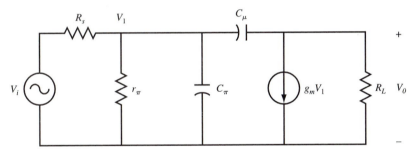

**FIGURE 5.31**

and
$$\omega_{p2} = \frac{1}{R_L' C_{oT}}$$

Because of the approximations involved in applying Miller's theorem, only the smaller of these two poles is an accurate representation of the high-frequency response. This smaller pole is approximately the $-3$-dB bandwidth of the differential amplifier.

### Common-mode frequency response

The equivalent circuit that is used to determine the frequency response of the common-mode gain is shown in Fig. 5.32.

As long as all components are matched, the differential amplifier's common-mode gain (balanced output) is zero. Component mismatches do create frequency-dependent changes in the common-mode gain, but these are usually second-order effects. Even with ideal components, the single-ended (unbalanced) common-mode

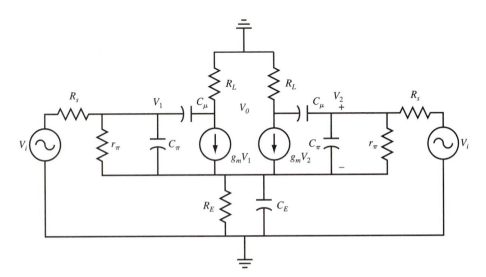

**FIGURE 5.32**

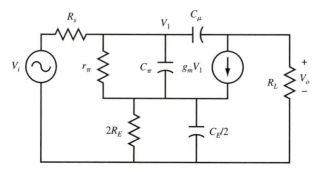

**FIGURE 5.33**

gain is not zero. Because of the circuit symmetry, the common-mode (unbalanced output) gain can be determined from the circuit shown in Fig. 5.33.

A detailed analysis of this circuit is quite involved, but usually the emitter resistance is a very large impedance, since it is desired to have a small common-mode gain, and the higher the value of the emitter resistance $R_E$, the lower the common-mode gain. For the case in which the emitter time constant dominates (the most common case), the single-ended common-mode gain is

$$A_{cm} \approx \frac{-R_L(1 + j\omega C_E R_E)}{2R_E}$$

This equation shows that the common-mode gain starts increasing with frequency at

$$\omega_Z = \frac{1}{C_E R_E}$$

This increase in common-mode gain with frequency represents a degradation in the performance of the differential amplifier. At frequencies higher than $\omega_z$, additional parasitic effects also come into play, and the common-mode gain begins to decrease with increasing frequency. The complete high-frequency analysis is well suited for computer simulation.

**EXAMPLE 5.9.** A BJT differential amplifier with current-source biasing is shown in Fig. 5.34. The emitter current in $Q_1$ and $Q_2$ is approximately 1 mA per transistor. The equivalent load resistance $R'_L$ is approximately 10 times as large as the equivalent source resistance $R$, but the input Miller capacitance is more than 80 times larger than the capacitance shunting the load, so one would expect the upper corner frequency to be determined by the input pole

$$\omega_{p1} = \frac{1}{RC_T}$$

The frequency response of the differential gain is plotted in Fig. 5.35 and the common-mode gain frequency response is plotted in Fig. 5.36.

The differential gain begins to decrease before the common-mode gain has increased significantly, and within the region of constant differential mode gain, the common-mode frequency response is not a factor in amplifier performance.

FIGURE 5.34

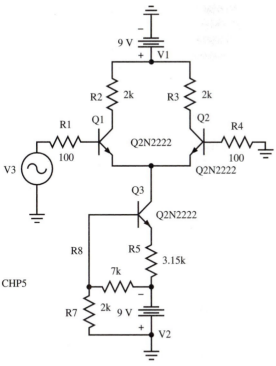

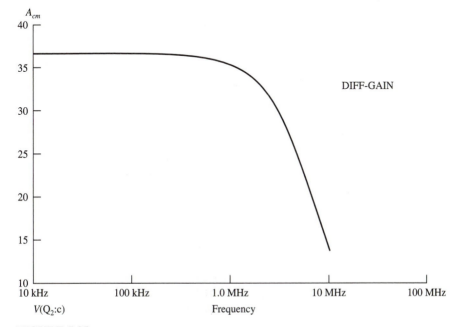

FIGURE 5.35

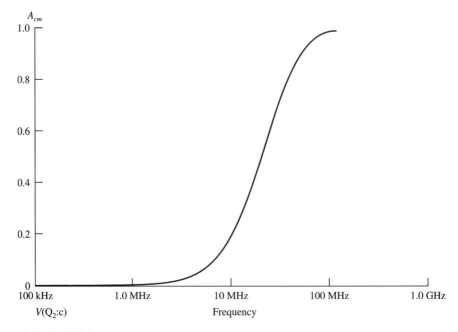

**FIGURE 5.36**

## High-Speed Operational Amplifiers

Operational amplifiers are finding more and more high-frequency applications as their frequency of operation extends to several hundred megahertz. In addition to the extremely versatile voltage feedback operational amplifiers, current feedback amplifiers with high bandwidths are now available. Voltage feedback op amps are now available with gain-bandwidth products in excess of $2 \times 10^9$ Hz (2 GHz) and current feedback operational amplifiers with gain-bandwidth products in excess of 8 GHz. Both types of operational amplifiers find application in high-frequency circuits.

### Voltage feedback operational amplifiers

The commonly used operational amplifier is a voltage feedback (VFB) device which develops a signal proportional to the error voltage $(V_+ - V_-)$. It was shown in Chap. 2 that the gain-bandwidth product of this device (also called the *unity gain frequency* $\omega_T$) is

$$\omega_T = A_o \omega_o$$

where $\omega_o$ is the $-3$-dB frequency (rad/s) of the internally compensated operational amplifier and $A_o$ is the dc gain. When the operational amplifier is used to realize an amplifier, the bandwidth, due to negative feedback, will be much wider than the op-amp bandwidth, but the op-amp bandwidth will be the main factor in determining the closed-loop bandwidth.

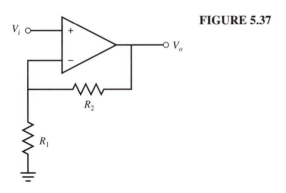

**FIGURE 5.37**

Consider the noninverting amplifier shown in Fig. 5.37. For this circuit the closed-loop transfer function is

$$\frac{V_o}{V_i} = \frac{A}{1 + A/a}$$

where $A$ is the ideal closed-loop gain

$$A = \frac{1 + R_2}{R_1}$$

and $a$ is the transfer function of the operational amplifier

$$a = \frac{A_o}{1 + j\omega/\omega_o}$$

The closed-loop transfer function can be written as

$$\frac{V_o}{V_i} = \frac{A_o/(1 + A_o/A)}{1 + j\omega/[\omega_o(1 + A_o/A)]}$$

The open-loop gain $A_o$ will be much larger than the closed-loop gain $A$, so that the transfer function simplifies to

$$\frac{V_o}{V_i} = \frac{A}{1 + j\omega A/\omega_T}$$

The closed-loop bandwidth is the op amp's gain-bandwidth product (unity gain frequency) divided by the closed-loop gain. As the voltage gain of the amplifier increases, the bandwidth of the amplifier decreases. It is readily shown that the bandwidth of the inverting amplifier also decreases in proportion to the closed-loop gain.

**Slew rate**

A simplified equivalent circuit for a VFB op amp is shown in Fig. 5.38. The capacitor $C_c$ across the high-voltage-gain stage is a compensating capacitor added so that the closed-loop transfer function will be stable. This capacitor causes the

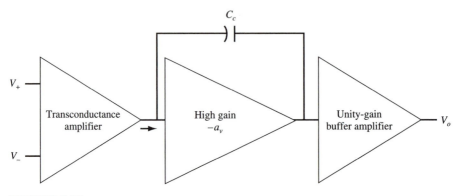

**FIGURE 5.38**

open-loop gain of the circuit to roll off with increasing frequency at a rate of −6-dB per octave, and its value is selected so that additional frequency roll-off caused by the capacitors internal to the op amp does not occur until approximately the op amp's unity gain frequency. This compensating capacitor also limits the maximum rate of change of output voltage. The amplifier's *slew rate* SR is defined as

$$SR \equiv \left( \frac{dV_o(t)}{dt} \right)_{max}$$

For a step input ($e_d = V_+ - V_-$) to the circuit shown in Fig. 5.38,

$$SR = \frac{(g_m e_d)_{max}}{C} = \frac{I_{max}}{C_c}$$

The slew rate is limited by the maximum current out of the transconductance amplifier. Slew rate limiting is a nonlinear effect which plays an important role in VFB op amps in high-speed applications. The SR will be increased by decreasing the transconductance of the first stage. It appears that a lowering of the transconductance of the input stage will lower the slew rate, but the compensating capacitor required for good closed-loop stability is given by

$$C_c = \frac{g_m}{\omega_T}$$

so the slew rate

$$SR = \frac{I_{max} \omega_T}{g_m}$$

is actually inversely proportional to the transconductance of the first stage. Since FETs have a lower transconductance than BJTs for typical input stage operating currents, they can be expected to provide a higher slew rate.

Voltage feedback op amps are very versatile and cost-effective and are available with a wide variety of bandwidth, noise, power, and precision specifications.

**Current feedback operational-amplifier circuits**

The conventional operational amplifier is a voltage-controlled device with high impedance levels. The upper frequency response of these devices is limited by the Miller capacitances and the transistor cutoff frequencies. Transistors can switch currents much more rapidly than they can switch voltages. Devices which feed back current rather than voltage can be expected to have a higher bandwidth. Any amplifier which generates an error signal in the form of current is referred to as a *current feedback amplifier*.

The circuit model of a current feedback operational amplifier is shown in Fig. 5.39. The input stage consists of a unity-gain voltage amplifier with high input impedance and low output impedance. This input buffer amplifier causes the inverting node voltage to follow the non-inverting node voltage. Amplification is provided by a transimpedance amplifier with transimpedance gain $Z$ which creates an output voltage that is $Z$ times that of the current flowing out of the input buffer amplifier. The current feedback amplifier is usually realized by using the current-mirror topology illustrated in Fig. 5.40. The current mirrors function so that the two currents out of the current mirror are equal.

Since the input buffer amplifier has a very high input impedance, it has negligible current flow and thus

$$I_n = I_1 - I_2$$

and because of the two current mirrors, the current through capacitor $C$ is

$$I_c = I_1 - I_2 = I_n$$

Thus the output voltage is

$$V_o(j\omega) = \frac{I_n}{j\omega C}$$

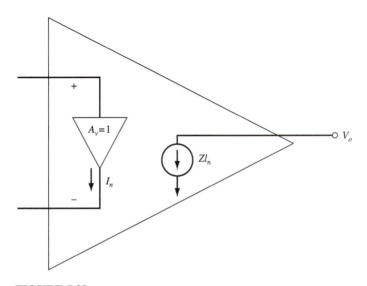

**FIGURE 5.39**

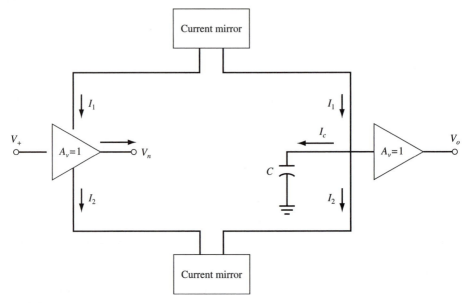

**FIGURE 5.40**

The output current of the op amp's input voltage buffer serves as the error signal in the current feedback (CFB) differential amplifier, just as the error voltage does in the voltage feedback operational amplifier. The performance of the current feedback op amp can be further evaluated by analyzing its performance in realizing the noninverting voltage amplifier shown in Fig. 5.41. In the current feedback op amp, the current out of the inverting node will be $I_n$. Summing currents at the inverting node gives

$$I_n = \frac{V_n}{R_1} + \frac{V_n - V_o}{R_2}$$

Because of the input buffer amplifier

$$V_n = V_i$$

**FIGURE 5.41**

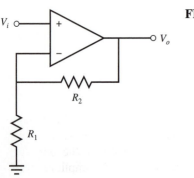

and thus
$$I_n = V_i \left( 1 + \frac{R_2}{R_1} \right) - \frac{V_o}{R_2}$$

Also, in the CFB op amp
$$V_o = I_n Z(j\omega)$$

so the closed-loop transfer function is
$$\frac{V_o}{V_i} = \left( 1 + \frac{R_2}{R_1} \right) \frac{1}{1 + R_2/Z(j\omega)}$$

If the impedance $Z$ is sufficiently large,
$$\frac{V_o}{V_i} = 1 + \frac{R_2}{R_1}$$

The gain of the noninverting voltage amplifier realized with the ideal current feedback op amp is the same as that realized with the more familiar ideal voltage feedback op amp. The impedance $Z(j\omega)$ should be as large as possible to closely realize the ideal transfer function. The impedance $Z(j\omega)$ is conveniently realized with a capacitor $C$ in parallel with a resistor $R_a$, giving the current feedback op amp very high transimpedance at low frequencies provided that $R_a$ is sufficiently large. The frequency-dependent gain properties of the circuit are primarily determined by $Z(j\omega)$.

For the noninverting amplifier with
$$Z(j\omega) = \frac{R_a}{1 + R_a(j\omega C)}$$

the closed-loop transfer for the noninverting amplifier transfer function is
$$\frac{V_o}{V_i} = \left( 1 + \frac{R_2}{R_1} \right) \frac{1}{1 + R_2(j\omega C) + R_2/R_a}$$

Normally,                  $R_2 \ll R_a$

and the closed-loop gain expression simplifies to
$$\frac{V_o}{V_i} = \left( 1 + \frac{R_2}{R_1} \right) \frac{1}{1 + R_2(j\omega C)}$$

An important feature of the current feedback op amp is that the closed-loop bandwidth
$$B = \frac{1}{R_2 C}$$

is independent of the closed-loop gain. The closed-loop gain can be increased by decreasing $R_1$ without affecting the closed-loop bandwidth. And $R_2$ is constrained to be much less than the op amp's $R_a$. When the nonideal model for the amplifier is considered, particularly the output impedance of the input buffer amplifier, current feedback op amps do have some reduction in bandwidth with increasing gain, but bandwidth reduction with increasing gain is much less than that of voltage feedback op amps.

### Cascade of Identical Stages

If a large voltage gain is required, it is often convenient to cascade several identical stages. If each stage has a voltage gain $A$ and a single pole at the upper corner frequency $\omega_p$, the voltage transfer function of each stage is

$$A_v = \frac{A}{j\omega/\omega_p + 1}$$

If $n$ of these stages are cascaded, the overall transfer function will be

$$A_T = \frac{A^n}{(j\omega/\omega_p + 1)^n} \tag{5.28}$$

The overall low-frequency gain is the $n$th power of the voltage gain of a single stage. The overall bandwidth $\omega_1$ is that frequency at which the gain magnitude is 0.707 times its low-frequency value. That is,

$$\frac{A^n}{|(j\omega/\omega_p + 1)|^n} = 0.707 A^n = \frac{A^n}{2^{1/2}}$$

or

$$\left[ \left( \frac{\omega_1}{\omega_p} \right)^2 + 1 \right]^{n/2} = 2^{1/2}$$

The bandwidth is

$$\omega_1 = \omega_p (2^{1/n} - 1)^{1/2} \tag{5.29}$$

The bandwidth $\omega_1$ obtained by cascading $n$ identical stages each of bandwidth $\omega_p$ is reduced by the factor $(2^{1/n} - 1)^{1/2}$ of the single-stage bandwidth. This factor is referred to as the *bandwidth reduction factor*.

**EXAMPLE 5.10.** Two identical stages, each with a voltage gain of 20 dB and a bandwidth of 10 MHz, are cascaded. What are the overall gain and bandwidth?

*Solution.* Since each stage has a gain of 20 dB, the overall low-frequency gain is 40 dB. The overall bandwidth is

$$f_1 = 10^7 (2^{1/2} - 1)^{1/2} = 6.43 \text{ MHz}$$

### ◼ 5.3
### BROADBANDING TECHNIQUES

The preceding analysis emphasized that the most troublesome factor in the design of high-frequency, high-gain amplifiers is the Miller capacitance connected between the input and output. As the gain is increased, the Miller capacitance

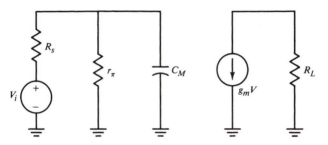

**FIGURE 5.42**
A simplified equivalent circuit of a common-emitter amplifier.

increases; it is this capacitance that invariably limits the upper frequency response. Applications can require narrowband amplifiers, which include some type of tuned circuit, or broadband amplifiers, such as video amplifiers, in which the gain must be kept relatively constant over several decades of frequency. The following section describes methods for improving amplifier frequency response by the modification of the basic amplifier circuit.

Consider the simplified small-signal equivalent circuit of the common-emitter voltage amplifier shown in Fig. 5.42. Here the output capacitance has been neglected, and the equivalent Miller capacitance is

$$C_M = C_\pi + C_\mu(1 + g_m R_L)$$

For this amplifier

$$\frac{V_o}{V_i}(j\omega) = \frac{-g_m r_\pi R_L}{r_\pi + R_s}\left(\frac{j\omega C_M r_\pi R_s}{r_\pi + R_s} + 1\right)^{-1}$$

The midband voltage gain is

$$A_{\text{mid}} = \frac{-g_m r_\pi R_L}{r_\pi + R_s} \tag{5.30}$$

and the bandwidth is

$$B = \frac{r_\pi + R_s}{C_M r_\pi R_s} \quad \text{rad/s} \tag{5.31}$$

This expression shows that the bandwidth can be increased by decreasing $R_s$. There is a limit below which the bandwidth can be reduced by reducing $R_s$, since the analysis ignored the base-spreading resistance $r_b'$, which is in series with $R_s$ (for bipolar transistors). For low values, $R_s$ should be replaced by $R_s + r_b'$ in Eq. (5.31). The analysis does imply that if the source resistance is very large, the bandwidth can be increased by using a source- or emitter-follower for the first stage. There is also a limit beyond which the approximate circuit is no longer valid and the bandwidth is determined by the output circuit capacitance. However, the expression does show that the source impedance should be as small as possible in order to maximize the voltage gain and bandwidth. The bandwidth can often be extended by modifying the amplifier input circuit. Input compensation is a method of extending the circuit bandwidth for applications in which the source resistance cannot be reduced.

## Input Compensation

The bandwidth of a single-stage amplifier can be extended by adding frequency-sensitive components to the circuit. The method is often referred to as *broadbanding*. Figure 5.43 illustrates a method of broadbanding using input compensation. For this circuit the voltage gain is

$$A_v = \frac{-g_m R_L Z_\pi}{Z_\pi + Z_s}$$

where

$$Z_\pi = \frac{r_\pi}{j\omega r_\pi C_M + 1}$$

and

$$Z_s = \frac{R_s}{j\omega R_s C_s + 1}$$

If $C_s$ is selected so that

$$R_s C_s = r_\pi C_M \tag{5.32}$$

then the voltage gain

$$A_v = \frac{-g_m R_L r_\pi}{r_\pi + R_s}$$

is independent of frequency. Input compensation can be used to cancel the effect of the Miller capacitance on the input with no reduction in voltage gain. In this case the bandwidth will be determined by the output capacitance; it is

$$B = [R_L(C_o + C'_M)]^{-1} \tag{5.33}$$

where $C'_M$ is the equivalent Miller capacitance reflected across the output and $C_o$ is the output capacitance (collector-to-emitter capacitance plus any external shunt capacitance).

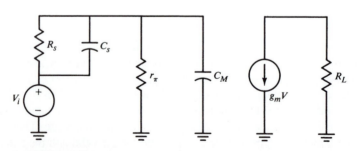

**FIGURE 5.43**
A broadband version of Fig. 5.42 obtained by adding $C_s$.

**EXAMPLE 5.11.** Consider the voltage amplifier shown in Fig. 5.44 which uses a 2N3904 transistor. The biasing is such that the transistor parameters are the same as in Example 5.5. Calculate the voltage gain and bandwidth. Then determine the value of $C_s$ required to cancel the effects of the Miller capacitance on the input side of the circuit, and determine the resulting gain and bandwidth.

**Solution.** The results of Example 5.5 showed that the equivalent circuit for this particular amplifier can be simplified to that shown in Fig. 5.45. The midfrequency voltage gain is $-77.2$, and the calculated bandwidth of the input circuit $\omega_1 = 5.3 \times 10^6$ rad/s. If $C_s$ is selected so that

$$500C_s = 237 \times 1166 \text{ pF} \qquad \text{or} \qquad C_s = 553 \text{ pF}$$

then the bandwidth will be determined by the pole of the output circuit

$$B = (600 \times 4 \times 10^{-12})^{-1} = 4.17 \times 10^8 \text{ rad/s}$$

Input compensation has significantly widened the bandwidth with no reduction in midfrequency gain. A knowledge of $r_b'$ is required for a more accurate prediction of the increase in bandwidth.

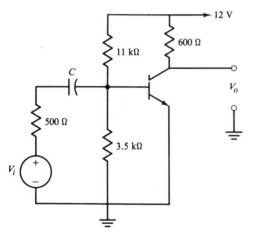

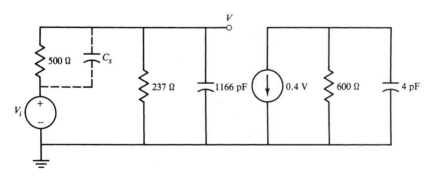

**FIGURE 5.44**
The common-emitter amplifier of Example 5.11.

**FIGURE 5.45**
The equivalent circuit of Fig. 5.44 with the input compensation capacitor $C_s$ added.

The preceding analysis has ignored the base-spreading analysis $r_b'$ of bipolar transistors; $r_b'$ cannot be shunted with a capacitor, since one terminal of the resistor is internal to the device and so limits the minimum value of input impedance. This base-spreading resistance limits the improvement that can be realized with input compensation of BJT amplifiers.

### Feedback

Input compensation does increase the system bandwidth, but it requires that the transistor parameters be known. If this is not possible, or is undesirable, as in designs to be mass-produced, feedback is a good alternative. Negative feedback widens the bandwidth at the cost of a proportional reduction in loop gain. Feedback also has the advantage that it makes the system response much less sensitive to parameters over which the designer has little or no control. The block diagram representation of a negative-feedback amplifier is shown in Fig. 5.46. Here $G$ is the transfer function of the system without the feedback factor $H$, and the transfer function with feedback is

$$\frac{V_o}{V_i} = \frac{G}{1 + GH} \tag{5.34}$$

The forward loop gain $G$ and/or the feedback ratio $H$ can be frequency-dependent. And $GH$ is known as the open-loop gain, and $-GH$ is referred to as the *loop transmission*. If the magnitude of the open-loop gain $|GH| \gg 1$, then the closed-loop gain is

$$\frac{V_o}{V_i} \approx H^{-1} \tag{5.35}$$

The loop response can be designed to depend essentially on the feedback factor $H$, which is under the designer's control, and to be independent of the forward gain $G$, which depends on the transistor parameters. The feedback formulation is usually difficult to apply directly to transistor amplifiers because of the circuit complexity; but it is an accurate approximation that if the amplifier bandwidth without feedback is $B$, then the bandwidth $B_f$ of the system with negative feedback is

$$B_f \approx B|1 + GH| \approx B|GH| \tag{5.36}$$

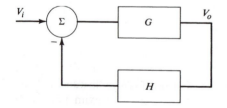

**FIGURE 5.46**
Block diagram representation of a negative-feedback amplifier.

Consider the amplifier with gain $A_o$ and bandwidth $\omega_L$ which can be represented by the transfer function

$$G(j\omega) = \frac{A_o}{1 + j\omega/\omega_L}$$

The gain-bandwidth product of this amplifier is $GB = A_o\omega_L$. If a frequency-independent negative-feedback factor $H$ is included, the closed-loop transfer function becomes

$$\frac{V_o}{V_i} = \frac{G(j\omega)}{1 + G(j\omega)H} = \frac{A_o}{1 + A_oH}\left[1 + \frac{j\omega}{\omega_L(1 + A_oH)}\right]^{-1}$$

The bandwidth has been increased from $\omega_L$ to $B_f = \omega_L(1 + A_oH)$, but the gain has been reduced. The gain-bandwidth product is still

$$G = \frac{A_o}{1 + A_oH}\omega_L(1 + A_oH) = A_o\omega_L$$

**EXAMPLE 5.12.** An amplifier has an open-loop gain of 40 dB and a bandwidth of $10^7$ rad/s. If 20 dB of feedback is added to the amplifier, what are the closed-loop gain and bandwidth?

**Solution.** The closed-loop gain has been reduced by 20 dB (a factor of 10), so the closed-loop gain is $40 - 20 = 20$ dB, and the closed-loop bandwidth has increased proportionally:

$$B_f = 10 \times 10^7 = 10^8 \text{ rad/s}$$

It is generally true that feedback will not increase the gain-bandwidth product. The increased bandwidth is obtained with a proportional reduction in gain. Feedback does have the advantage of also reducing the system's sensitivity to the amplifier components that cannot readily be controlled by the designer.

In transistor amplifiers there are four types of feedback. A current can be fed back that is proportional to the output voltage or current, or a voltage can be fed back which is proportional to the output voltage or current. The input and output impedances are increased or decreased depending upon the particular type of feedback employed. The following discussion concerns the two most frequently used methods of single-stage feedback.

**Current-to-Voltage Feedback**

The amplifier shown in Fig. 5.47 illustrates a frequently used method of single-stage negative feedback. Here a voltage proportional to the load current is fed back to the input via the emitter resistor $R_E$. It is readily shown that this feedback increases the amplifier input impedance and decreases its output impedance. The simplified small-signal, high-frequency equivalent circuit is shown in Fig. 5.48. Although this circuit can be analyzed directly, the resulting frequency-dependent transfer function is too complicated to be of much general utility. A good approximation to the

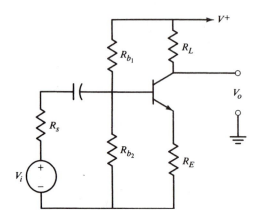

**FIGURE 5.47**
A common-emitter amplifier with current-to-voltage feedback.

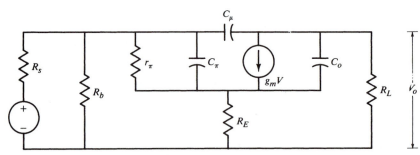

**FIGURE 5.48**
A simplified high-frequency equivalent circuit of the amplifier shown in Fig. 5.47.

circuit's frequency-dependent behavior can be obtained by calculating the gain and bandwidth without feedback and the midfrequency gain with feedback, then determining the feedback ratio $H$ from this. The upper frequency bandwidth can be estimated using Eq. (5.34), provided the upper corner frequency is known for the case in which there is no feedback.

The midfrequency gain without feedback [Eq. (2.7)] is

$$A_{\text{mid}} = \frac{V_o}{V_i'} = \frac{-g_m R_L r_\pi}{r_\pi + R}$$

where

$$R = R_s \| R_b$$

$$V_i' = \frac{V_i R_b}{R_b + R_s}$$

and

$$R_b = \frac{R_{b_1} R_{b_2}}{R_{b_1} + R_{b_2}}$$

It is assumed that $R_L \ll r_o$. The voltage gain with the feedback resistor $R_E$ added is

$$\frac{V_o}{V_i'} = A_{\text{mid}} = \frac{-g_m R_L r_\pi/(r_\pi + R)}{1 + (1 + \beta)R_E/(R + r_\pi)} \tag{5.37}$$

The open-loop gain is

$$GH = \frac{(1+\beta)R_E}{r_\pi + R} = \frac{g_m R_L r_\pi}{r_\pi + R}\frac{(1+\beta)R_E}{g_m R_L r_\pi} \qquad (5.38)$$

Since the gain $G$ without feedback is known, the feedback factor is

$$H = \frac{-(1+\beta)R_E}{g_m R_L r_\pi} \qquad (5.39)$$

If

$$(1+\beta)R_E \gg R_s + r_\pi$$

then

$$(A_{\text{mid}})_f \approx \frac{-g_m r_\pi R_L}{(1+\beta)R_E} \approx H^{-1} \approx \frac{-R_L}{R_E}$$

This equation states that the midfrequency gain with feedback can be made independent of the transistor parameters, which vary from transistor to transistor. The use of negative feedback to make the circuit insensitive to parameters not under the designer's direct control is a most important function. Another important feature is that negative feedback increases the circuit bandwidth. For this circuit the upper $-3$-dB frequency without feedback is

$$\omega_L = \left\{ \frac{r_\pi R}{R + r_\pi}[C_\pi + C_\mu(1 + g_m R_L)] \right\}^{-1} \qquad (5.40)$$

so the upper $-3$-dB frequency with feedback is estimated to be

$$\omega_L' = \omega_L \left[ 1 + \frac{(1+\beta)R_E}{R + r_\pi} \right] \qquad (5.41)$$

This is an approximation since $GH$ is actually a frequency-dependent transfer function. The method and accuracy of the approximation are illustrated by the following example.

**EXAMPLE 5.13.** Figure 5.49 illustrates the same amplifier as used in Example 5.5, only emitter feedback has been added. The small-signal equivalent circuit is as shown in Fig. 5.50. Here $R_b$ is 2.655 k$\Omega$, and $R$ is 421 $\Omega$.

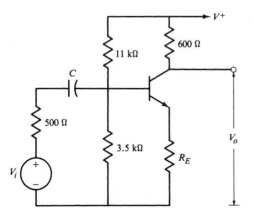

**FIGURE 5.49**
The common-emitter amplifier discussed in Example 5.13.

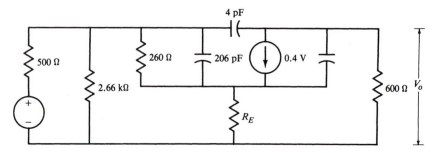

**FIGURE 5.50**
A small-signal equivalent circuit of the amplifier shown in Fig. 5.49.

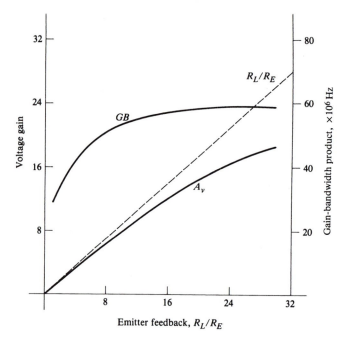

**FIGURE 5.51**
Voltage gain and gain-bandwidth product as a function of the
emitter feedback.

The magnitude of the gain without feedback was found to be $A_v = 77.2$, and the
corresponding bandwidth (obtained with computer simulation) is $B = 0.84 \times 10^6$ Hz.
The gain-bandwidth product is

$$GB = 77.2(0.84 \times 10^6) = 65.15 \times 10^6 \text{ Hz}$$

The gain-bandwidth products for selected values of $R_E$ were obtained by computer
simulation, and the results are plotted in Fig. 5.51. The transconductance $g_m$ was kept
constant and independent of $R_E$. As $R_E$ approaches $R_L$, the gain $V_o/V_i'$ approaches the
asymptotic value of $R_L/R_E$. The gain-bandwidth product decreases as $R_E$ increases.

For $R_L = R_E$ the bandwidth is 36 MHz, and $GB$ is $29.7 \times 10^6$ Hz. The bandwidth can be markedly improved with the addition of emitter feedback.

For the common-emitter amplifier with an emitter resistor, the amplifier mid-frequency input resistance (from base to ground) is

$$R_i = r_\pi + (1 + \beta)R_E$$

Emitter feedback also increases the input impedance. If a small input impedance is needed for impedance matching, then voltage-to-current feedback can be used.

### Voltage-to-Current Feedback

The common-emitter amplifier shown in Fig. 5.52 generates a feedback current through the resistor $R_F$ which is proportional to the output voltage. That this circuit reduces the input impedance can be determined by analyzing the small-signal, mid-frequency equivalent circuit shown in Fig. 5.53. The input current is

$$I_i = \frac{V}{r_\pi} + \frac{V - V_o}{R_F}$$

and the output voltage is

$$V_o = -g_m V R_L + \left( I_i - \frac{V}{r_\pi} \right) R_L$$

If $V_o$ is eliminated from these two equations, the input impedance is found to be

$$Z_i = \frac{V}{I_i} = r_\pi \| \frac{R_F + R_L}{1 + g_m R_L} \tag{5.42}$$

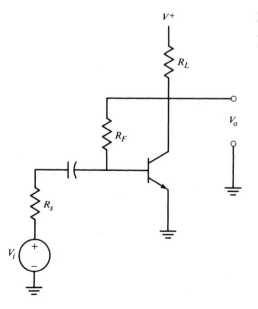

**FIGURE 5.52**
A common-emitter amplifier with voltage-to-current feedback.

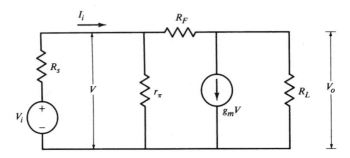

**FIGURE 5.53**
A small-signal equivalent circuit of the amplifier shown in
Fig. 5.52.

This same expression holds true for FET amplifiers with $r_\pi \approx \infty$. The feedback resistor $R_F$ reduces the input impedance; in addition, this form of feedback can reduce the circuit sensitivity to variations in the transistor parameters and can widen the amplifier bandwidth. The midfrequency voltage gain of the circuit shown in Fig. 5.53 can be determined from the two equations

$$\frac{V_i - V}{R_s} = \frac{V - V_o}{R_F} + \frac{V}{r_\pi}$$

and

$$g_m V + \frac{V_o}{R_L} + \frac{V_o - V}{R_F} = 0$$

The voltage gain is

$$A_v = \frac{V_o}{V_i} = R_s^{-1} \frac{g_m - g_F}{(g_L + g_F)(g_s + g_F + g_m) + (-g_m g_F + g_F^2)} \qquad (5.43)$$

Normally $g_m \gg g_F$, $g_m R_L \gg 1$, and $g_s \gg g_\pi$. Under these conditions the voltage gain expression simplifies to

$$A_v = \frac{-g_m R_L}{1 + g_F R_L + g_m R_L (R_s / R_F)}$$

which in turn simplifies to

$$A_v \approx \frac{-R_F}{R_s}$$

provided

$$g_m R_L \frac{R_s}{R_F} \gg 1$$

The gain can be made independent of the transistor parameters by using feedback. The gain is a maximum with $R_F = \infty$ (no feedback); the feedback reduces the gain, so the feedback is negative. The frequency response of this circuit can also be estimated by calculating the loop gain $GH$. If $r_o$ is neglected, the gain without feedback (assuming $R_s \ll r_\pi$) is $A_v = -g_m R_L$. The gain with feedback is

$$A_v = \frac{-g_m R_L}{1 + g_F R_L + g_m R_L (R_s/R_F)}$$

So the open-loop gain

$$GH = g_F R_L + g_m R_L \frac{R_s}{R_F} \tag{5.44}$$

can be used for estimating the closed-loop bandwidth. The following example illustrates the accuracy of the approximations.

EXAMPLE 5.14. The preceding example evaluated the effect of emitter feedback on the gain and bandwidth of the common-emitter amplifier. Here the effects of voltage-to-current feedback are considered by analyzing the same amplifier, except that now voltage-to-current feedback is used. The amplifier is shown in Fig. 5.54. The amplifier is again biased so that $g_m = 0.4$ S, and it is assumed to remain constant. The voltage gain and gain-bandwidth product as a function of $R_F/R_s$ are plotted in Fig. 5.55. In the ratio $R_F/R_s$, note the accuracy of the approximation even through $r_\pi$ is not as large as $R_s$.) The gain-bandwidth product varies from a low of $59.5 \times 10^6$ Hz for a gain of unity to $65.1 \times 10^6$ Hz for a closed-loop gain of 17.9. The constancy of the gain-bandwidth product indicates that the effect of the negative feedback on bandwidth can be easily estimated using the approximations given in this section.

For this amplifier the voltage-to-current feedback results in a more constant gain-bandwidth product than did the current-to-voltage feedback. This is particularly true for small values of loop gain. Each amplifier must be evaluated on an individual basis; the type of feedback to be used will most often be determined by the desired input impedance level. Another feedback technique for controlling the gain and impedance levels uses transformers as the feedback element.

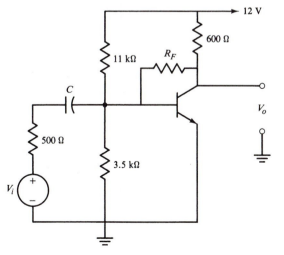

**FIGURE 5.54**
Feedback amplifier analyzed in Example 5.14.

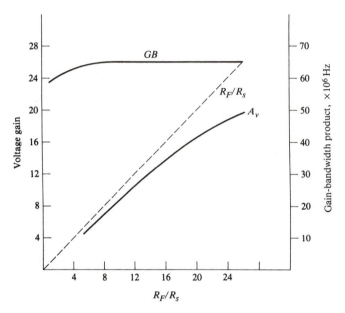

**FIGURE 5.55**
Voltage gain and gain-bandwidth product, as a function of shunt
feedback of the amplifier shown in Fig. 5.54.

## Lossless Feedback Amplifiers[2]

The two methods already described for increasing the bandwidth with negative
feedback used a resistor or resistors to create the feedback signal. Since any resistor
adds noise to the circuit, resistive feedback also decreases the amplifier's noise
performance (that is, it increases the noise figure). Also, the transistor output is
delivered to the feedback network instead of to the load, in proportion to the feedback ratio. Figure 5.56 illustrates a method of lossless feedback that does not have
these two limitations. The feedback is realized with a three-winding transformer
connected to the common-base transistor amplifier so that it can simultaneously
provide gain and impedance-matching. The transformer is wound with the polari-
ties as shown, and the two turns ratios with respect to the primary are $m$: 1 and $n$: 1.
The following analysis is simplified by making the close approximation that the
common-base current gain is unity. If the transformer is lossless, the input power is
equal to the power out, or

$$I_1 V + I_1 n V + I_2 m V = 0 \qquad \text{or} \qquad I_2 = \frac{-(1+n)I_1}{m}$$

Therefore, the current to the load is

$$I_o = I_1 - I_2 = \frac{1+n+m}{m} I_1$$

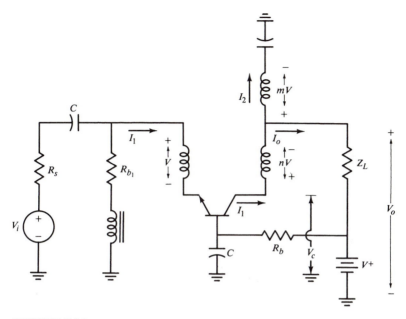

**FIGURE 5.56**
A common-base amplifier with lossless feedback.

The collector voltage is

$$V_c = (n + m)V = \frac{n + m}{m} V_o$$

Therefore

$$\frac{V_c}{I_1} = \frac{V_c}{I_o} \frac{1 + n + m}{m}$$

$$= \frac{V_c}{I_o} \frac{(1 + n + m)(n + m)}{m^2}$$

Since $V_o/I_o = Z_L$,

$$\frac{V_c}{I_1} = \frac{Z_L(1 + n + m)(n + m)}{m^2}$$

If the transistor base-to-emitter voltage $V_{be} \ll V$, then the input impedance is

$$Z_i = \frac{V}{I_1} = \frac{1 + n + m}{m^2} Z_L \tag{5.45}$$

The voltage gain is then

$$A_v = \frac{V_o}{V_i} = \frac{I_o Z_L}{I_1(Z_s + Z_i)} = \frac{1 + n + m}{m} \frac{Z_L}{Z_s + Z_i} \tag{5.46}$$

Another advantage of this circuit is that the transformer turns ratios can be selected for impedance matching. For resistive loads

$$Z_i = \frac{1 + n + m}{m^2} R_L$$

If

$$1 + m + n = m^2 \qquad\qquad (5.47)$$

then $Z_i = R_L$.

Also, the amplifier output impedance can be determined using the equivalent circuit shown in Fig. 5.57 (where it is again assumed that $V_{be} \ll V$).

The output voltage $V_o = mV$, and the output current $I_o$ equals $-I_1 + I_2$. If the transformer is considered ideal, then

$$I_2 = \frac{-(1 + n)I_1}{m}$$

and

$$I_o = \frac{-(m + 1 + n)}{m} I_1$$

so

$$Z_o = \frac{V_o}{I_o} = \frac{m^2}{m + 1 + n} \frac{V}{I_1} = \frac{m^2}{m + 1 + n} R_s$$

Therefore, if $R_s = R_L = R$ and $m^2 = m + 1 + n$, the transistor input and output impedances are also equal: $Z_i = Z_o = R$. The corresponding voltage gain is

$$A_v = \frac{m^2}{(m + n + 1)2} = \frac{m}{2}$$

Since the common-base amplifier has a current gain close to unity, the power gain $A_p \approx A_v$.

For a common-base amplifier without feedback the voltage gain is approximately

$$A_v \approx \frac{R_L}{R_s}$$

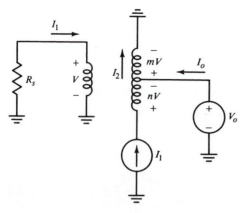

**FIGURE 5.57**
A small-signal equivalent circuit of the amplifier shown in Example 5.14.

If the load and source impedances are equal, the maximum voltage gain is unity. Since the voltage gain with feedback is $A_v = m/2$, the feedback can increase the voltage gain. This is positive feedback, which implies that the feedback will reduce the bandwidth by approximately the same factor that the gain is increased. This feedback technique employs the common-base configuration, which has a much wider bandwidth than the common-emitter configuration, and then sacrifices some of the greater bandwidth for an increase in voltage gain. If $R_L$ is much larger than $R_s$, the feedback can reduce the voltage gain. The analysis always depends on the particular component values. The technique has the additional advantage of being able to match the amplifier input and output impedances. It is well suited as an element in a cascade of identical amplifiers.

> **EXAMPLE 5.15.** The same transistor with the same collector current ($g_m = 0.4$) as used in the two preceding examples is used to implement the common-base feedback amplifier shown in Fig. 5.58. Table 5.1 presents the results obtained from a computer simulation of the circuit.
>
> The *GB* product decreases as the gain is increased. For voltage gains of 6 or more, *GB* is less than that obtained with the common-emitter amplifier using the same transistor and collector direct current. For this amplifier the input and output impedances are 500 Ω.

For the design of high-frequency amplifiers, the most important step is to select a transistor with a sufficiently high gain-bandwidth product. The techniques of high-frequency design become important only when other transistor specifications (such as power level and noise figure) limit the selection to transistors with an

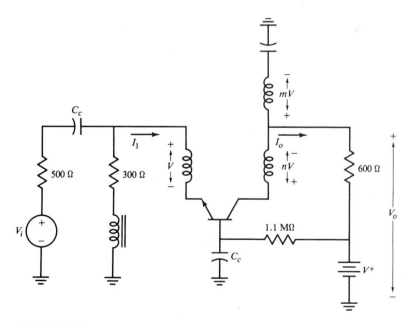

**FIGURE 5.58**
Lossless feedback amplifier discussed in Example 5.15.

**TABLE 5.1**
**Gain-bandwidth product of a common-base amplifier with lossless feedback**

| Turns ratio | Voltage gain | Bandwidth, Hz | Gain-bandwidth product, Hz |
|---|---|---|---|
| 0 | 0.976 | $79.4 \times 10^6$ | $77.5 \times 10^6$ |
| 4 | 1.984 | $11.22 \times 10^6$ | $22.2 \times 10^6$ |
| 6 | 2.975 | $4.67 \times 10^6$ | $13.9 \times 10^6$ |
| 10 | 4.95 | $1.6 \times 10^6$ | $7.92 \times 10^6$ |
| 20 | 9.615 | $4.22 \times 10^5$ | $4.06 \times 10^6$ |

insufficient gain-bandwidth product. The previous discussion has shown how the bandwidth of single-stage amplifiers can be extended.

**Neutralization**

A technique known as *neutralization* was first used in radio-receiver amplifier design to extend the frequency of operation by neutralizing the parasitic capacitance $C_\mu$, which appears between the input and output terminals. It consists of canceling the feedback current through $C_\mu$ with an equal and opposite current. The most frequently used method of neutralization employs the circuit shown in Fig. 5.59. Here the transformer inverts the phase of the output voltage; if the neutralizing impedance equals the impedance of the parasitic capacitance, the neutralizing current will cancel the current through $C_\mu$. If the circuit were accurate and the parasitic capacitances were known precisely, neutralization would be a wideband technique; but in practice it results in a relatively narrowband amplifier. Just as the development of the pentode vacuum tube with much lower feedback capacitance than that of earlier triodes eliminated the need for neutralization in commercial radio receivers, the

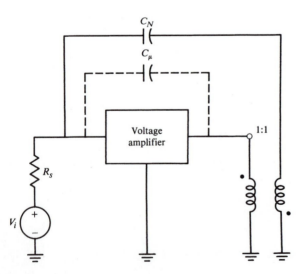

**FIGURE 5.59**
A voltage amplifier with neutralization.

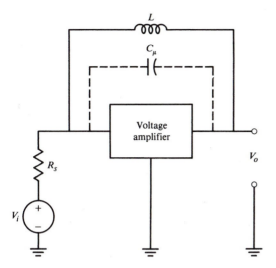

**FIGURE 5.60**
Another method of neutralization.

improvement in high-frequency performance of transistors reduced the need for neutralization in transistor amplifiers. Another method of neutralization that can be used for narrowband amplifiers is given in Fig. 5.60, where the value of inductance is chosen so that it forms a parallel resonant circuit with the capacitance $C_\mu$ at the frequency of interest. Neutralization is particularly useful in tuned-input, tuned-output amplifiers where the feedback capacitance can cause oscillation (see Prob. 7.16). A frequently used neutralization circuit is shown in Fig. 5.61 where the phase inversion is obtained with a center-tapped output transformer.

## Cascode Amplifiers

The analysis of the single-stage amplifiers shows that the upper frequency limit of the amplifier is usually determined by the pole of the input circuit. The bandwidth can be increased either by decreasing the source resistance or by decreasing the

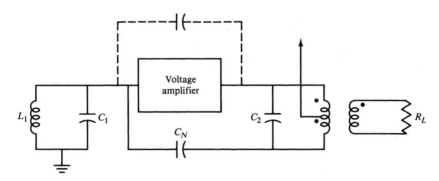

**FIGURE 5.61**
A neutralization circuit employing a center-tapped transformer.

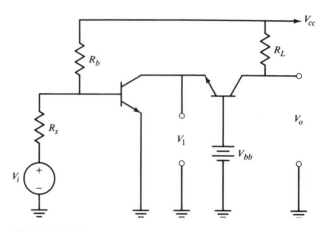

**FIGURE 5.62**
A BJT cascode amplifier.

voltage gain (which reduces the Miller capacitance). If neither of these reductions is possible, it may still be possible to meet the gain and bandwidth specifications by using a two-stage amplifier consisting of a common-emitter (or common source) amplifier followed by a common-base (or common-gate) amplifier. This configuration, illustrated in Fig. 5.62, is known as the *cascode amplifier.* Although the circuit can readily be analyzed by a brute-force approach, we will use an intuitive approach as it provides more insight into the design process. The common-base amplifier has a low input resistance of approximately $r_\pi/\beta$ and a unity current gain. Since the input resistance of the common-base stage, which is low, serves as the load resistance for the first stage, the voltage gain of the first stage will be low. This means that the Miller capacitance for the first stage will be low, and the first-stage bandwidth will be increased. The first-stage input impedance is $r_\pi$, so the voltage gain of the first stage will be

$$A_{v_1} \approx \frac{-r_{\pi_1}}{r_{\pi_1} + R_s} g_{m_1} \frac{r_{\pi_2}}{\beta}$$

Since the collector direct current is the same for both stages and the transistor current gains are assumed equal, $g_{m_1} = g_{m_2}$ and $r_{\pi_1} = r_{\pi_2}$; therefore

$$A_{v_1} = \frac{-r_\pi}{r_\pi + R_s}$$

The voltage gain of the first stage is less than unity. The base alternating current of the first transistor is

$$I_b = \frac{V_i}{R_s + r_\pi}$$

Since the current gain of the first stage is $\beta$ and that of the second stage is approximately 1, the output current is

$$I_o \approx -\beta I_b = \frac{-\beta V_i}{R_s + r_\pi}$$

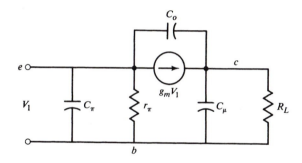

**FIGURE 5.63**
A small-signal model of the cascode amplifier output stage.

and the output voltage is

$$V_o = I_o R_L = \frac{-\beta R_L V_i}{R_s + r_\pi}$$

So the voltage gain of the cascode circuit

$$A_v = \frac{V_o}{V_i} = \frac{-\beta R_L}{R_s + r_\pi} \tag{5.48}$$

is about the same as can be realized with a single-stage amplifier, but since the first stage no longer has a gain greater than unity, its bandwidth is increased even if a large source resistance is used. The second stage, which does have a voltage gain greater than 1, has an input impedance that is small $(r_\pi/\beta)$. The bandwidth for the second stage can be determined from the small-signal model shown in Fig. 5.63. For this stage the midfrequency voltage gain is

$$A_{v_2} = \frac{V_o}{V} = g_m R_L$$

If the voltage gain is large, the small-signal equivalent circuit can be replaced by that of Fig. 5.64. Since the stage is noninverting, the equivalent Miller input capacitance is negative, reducing the input capacitance, and since the equivalent input impedance is $r_\pi/\beta$, the bandwidth of this stage will usually be determined by the capacitance across the output. The magnitude of the output pole is given by

$$\omega_p \approx [R_L(C_o + C_\mu)]^{-1} \tag{5.49}$$

This two-stage cascode amplifier has a voltage gain equal to what can be obtained from a single-stage common-emitter amplifier, but it has an appreciably

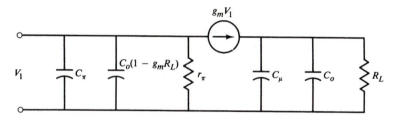

**FIGURE 5.64**
An equivalent small-signal model of the cascode amplifier output stage.

wider bandwidth. The cascode amplifier finds extensive application in high-frequency amplifiers and as the input stage of operational amplifiers.

## ■ 5.4

### AUTOMATIC GAIN CONTROL

Received signals are often subject to fading (slow variations in the amplitude of the received carrier). Fading required the first radio set operators to continually adjust the receiver gain in order to maintain a relatively constant output volume. This led to the design of the automatic volume-control circuit, which detects the amplitude of the received signal and automatically adjusts the gain in order to maintain a constant output signal level.[3] Automatic volume control, now generalized and known as *automatic gain control (AGC)*, is one of the most useful circuits in modern communications receivers. In addition to maintaining constant output signal levels, it is used in a variety of applications, including signal encoding and decoding and oscillator amplitude stabilization.

The basic elements of an AGC system are illustrated in Fig. 5.65. The input signal is amplified by a variable gain amplifier (VGA) whose gain depends on a control signal $V_c$. The amplified signal may be further amplified to produce the output signal $V_o$. Some parameter of the output signal, such as carrier amplitude, sideband power, or depth of modulation, is detected and compared with a reference signal $V_R$. The difference between these two signals is then filtered and used to control the gain of the variable gain amplifier, the gain-controlling signal being either a voltage or current. If the input signal is amplitude-modulated, the AGC circuit must not respond to changes in amplitude modulation or else the AGC loop will distort the modulated signal. This distortion is prevented by restricting the AGC circuit bandwidth so that it does not respond to the modulating frequencies; but AGC can still compensate for signal fading, which is relatively slow compared to the lowest modulating frequency.

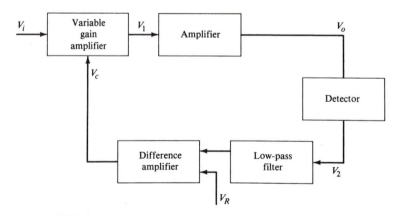

**FIGURE 5.65**
An automatic gain control system.

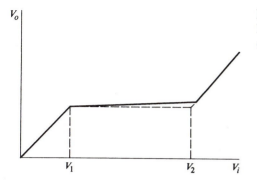

**FIGURE 5.66**
Idealized voltage-transfer characteristics
of an AGC system.

The characteristics of an ideal AGC circuit are shown in Fig. 5.66. For low input signal levels the AGC is inoperative, and the output signal is linearly related to the input signal. When the input signal level is greater than $V_1$, the ideal AGC circuit maintains the output level constant until the input level reaches $V_2$. For larger signal levels the output is again a linear function of the input signal level. The elimination of the AGC action at very high gain levels is often done to prevent system instability problems.

The automatic gain control system described here is a negative-feedback system. The control signal represents an error signal between the amplitude of the output and the reference signal. As the input signal amplitude varies, the control signal varies in a manner to minimize the error signal. The requirements of an AGC loop differ depending upon the particular type of modulation of the input signal. The AGC requirements for AM receivers are usually more stringent than those for FM and pulse-modulation receivers.

The desired signal range at several points in the receiver is an essential design parameter. It is necessary to keep the signal levels sufficiently large that receiver noise does not degrade the performance; but the signal levels must not exceed the linear range of the devices, or else distortion will occur. Signal range requirements are the most critical in the input stage, where differences in signal strength are the greatest.

**Theory of Automatic Gain Control**

An AGC system is an inherently nonlinear system, and there seldom are general solutions to the nonlinear equations that describe the system dynamics. There are, however, certain systems for which a closed-form solution can be found, and for most systems an approximate solution can be derived in terms of a small-signal model.

Figure 5.67 illustrates an AGC system that can be solved analytically. In this system the variable gain amplifier $P$ obeys the control law

$$P = K_1 e^{+aV_c} \tag{5.50}$$

so
$$V_o = V_i K_1 e^{+aV_c}$$

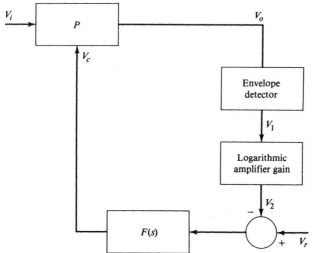

**FIGURE 5.67**
A linear (in decibels) AGC system.

where $V_i$ and $V_o$ represent the envelope amplitudes of the input and output, respectively, and the logarithmic amplifier gain is

$$V_2 = a \ln V_1 = \ln K_2 V_o$$

where $K_2$ is the gain of the envelope detector. The envelope detector output is always positive, so the output of the logarithmic amplifier is a real number (positive or negative). The control voltage is then

$$V_c = F(s)(V_r - V_2) = F(s)(V_r - \ln K_2 V_o)$$

where $F(s)$ is the filter transfer function.

Since the variable gain amplifier obeys an exponential law

$$\ln V_o = a V_c + \ln V_i K_1$$

the control voltage is

$$a V_c = \ln V_o - \ln V_i K_1$$

That is,

$$\ln V_o [1 + a F(s)] = \ln V_i + a F(s) V_r + \ln K_1 - a F(s) \ln K_2$$

The response to the input signal is

$$\ln V_o [1 + a F(s)] = \ln V_i + a F(s) V_r$$

Since $\ln V_o = 2.3 \log V_o$,

$$\ln V_o = \frac{2.3}{20} e_o = 0.115 e_o \qquad \text{dB}$$

Let $e_o$ and $e_i$ denote the output and input, respectively, in decibels. Then

$$e_o = \frac{e_i}{1 + a F(s)} - \frac{8.7 a F(s) V_r}{1 + a F(s)} \tag{5.51}$$

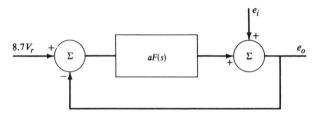

**FIGURE 5.68**
An equivalent block
diagram representation of
the AGC system shown in
Fig. 5.67.

This particular AGC circuit is described by a linear differential equation, provided the input and output quantities are expressed in decibels. The system can then be represented by the linear negative-feedback system shown in Fig. 5.68. The loop dynamics are described by the filter $F(s)$ and the factor $a$ of the variable gain amplifier. Normally $F(s)$ will be a low-pass filter, since the loop bandwidth must be limited so that it does not respond to any amplitude modulation present on the input signal. Loop stability will depend on the order of the filter and the loop gain. The steady-state change in the output, in response to a change in the input, is

$$\frac{\Delta e_o}{\Delta e_i} = [1 + aF(0)]^{-1} \qquad (5.52)$$

where $F(0)$ is the dc gain of the filter. It is desired to keep the change $\Delta e_o$ as small as possible in response to changes in the input amplitude. This is accomplished by making the dc loop gain as large as possible.

If $F(s)$ is a first-order filter described by

$$F(s) = \frac{K}{s/B + 1}$$

where $K$ is the dc gain of the filter and $B$ is the filter bandwidth, then the dc characteristic is

$$\Delta e_o = \frac{\Delta e_i}{1 + aK} \qquad (5.53)$$

The total dc output of the system shown in Fig. 5.68 is

$$e_o = \frac{e_i}{1 + aK} + \frac{8.655 V_r a K}{1 + aK} \qquad (5.54)$$

Since the loop transmission $aK$ is normally much greater than 1, the output $e_o$ equals $8.655 V_r$. The output amplitude in decibels is proportional to the reference voltage $V_r$. AGC loops containing a reference voltage are referred to as *delayed AGC*. This term does not imply that the gain control is delayed because of bandwidth limitation, but rather that the AGC loop contains a reference signal. Simple AGC loops, which do not contain a reference voltage, are common in inexpensive radio receivers.

The closed-loop transfer function for the loop with the first-order low-pass filter is

$$\Delta e_o = \frac{\Delta e_i}{1 + aK} \frac{s/B + 1}{s/[B(1 + aK)] + 1} \qquad (5.55)$$

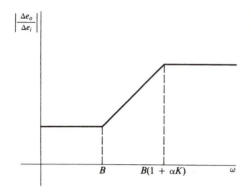

**FIGURE 5.69**
Frequency response of the AGC system
shown in Fig. 5.67.

This system is inherently stable for all $aK > 0$, since the closed-loop pole will be in the left half-plane for all $aK > 0$. The magnitude of the frequency response of the closed-loop system is plotted in Fig. 5.69. AGC loops have a high-pass filter characteristic in response to changes in the amplitude of the input signal. That is, at high frequencies there is little AGC. For amplitude-modulated signals the corner frequency $\omega_L$ should be lower than the lowest modulating frequency $\omega_M$:

$$\omega_L = B(1 + aK) < \omega_M$$

This implies that the filter bandwidth should be much less than the lowest modulating frequency, since the negative feedback increases the closed-loop bandwidth.

It was stated that it is desirable to keep the dc loop gain as large as possible in order to maintain a constant output level. One method is to use an integrator as the filter $F(s)$. That is, $F(s) = C/s$. An ideal integrator has infinite gain at direct current, so the steady-state output amplitude will not change in response to slowly varying changes in the input amplitude. The output for this particular filter is

$$e_o(s) = \frac{e_i(s)s}{s + aC} + \frac{8.6 V_r a}{s + aC}$$

For constant inputs the steady-state output is again proportional to the reference voltage:

$$\lim_{t \to \infty} e_o(t) = \frac{8.6 V_r}{C}$$

**EXAMPLE 5.16.** Determine the time response of the AGC loop illustrated in Fig. 5.67 [with $F(s) = C/s$] to a step change in the input amplitude.

**Solution.** Since the system is described by linear differential equations, superposition can be used to calculate the change in the output . That is,

$$\Delta e_o(s) = \frac{s \, \Delta e_i(s)}{s + aC}$$

For a unit (decibel) step change in the input

$$\Delta e_i(s) = s^{-1}$$

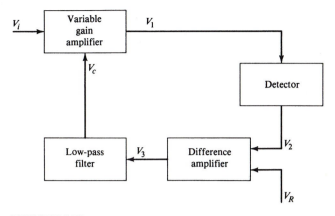

**FIGURE 5.70**
An AGC system with two nonlinear components.

and

$$\Delta e_o(s) = (s + aC)^{-1}$$

so,

$$\Delta e_o(t) = e^{-aCt}$$

The output (in decibels) exponentially decays toward zero in response to a step change in the input amplitude. The time constant of the decay is equal to the $-3$-dB frequency of the AGC loop ($\omega_n = aC$).

## Another AGC Model

Many AGC loops do not contain the logarithmic amplifier which results in a linear model when used with an exponential-type variable-gain amplifier. For some AGC systems it is still possible to derive linear small-signal models. The small-signal limitation implies that the system analysis is valid for small changes from a particular operating point. Figure 5.70 illustrates a block diagram model for an AGC system in which the variable-gain amplifier and the detector are the only nonlinear components in the loop. To simplify the notation without loss of generality, the gains of the detector, the difference amplifier, and the amplifier following the variable-gain amplifier will be assumed to be unity. Then the system can be represented by the simplified block diagram shown in Fig. 5.71 where $V_o$ and $V_i$ now

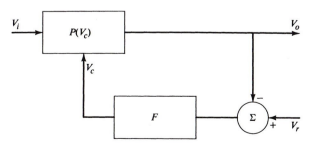

**FIGURE 5.71**
A simplified model of the AGC system shown in Fig. 5.70.

refer to envelope values and $F$ is the frequency-dependent transfer function of the low-pass filter and amplifier combined. The output voltage $V_o$ equals $PV_i$, where the gain $P$ of the variable-gain amplifier is a function of the control voltage $V_c$. The control voltage is

$$V_c = (V_r - V_o)F \tag{5.56}$$

The differential change in output voltage is

$$\frac{dV_o}{dV_i} = \frac{d}{dV_i}(PV_i) = P + V_i\frac{dP}{dV_i} \tag{5.57}$$

Since
$$\frac{dP}{dV_i} = \frac{dP}{dV_c}\frac{dV_c}{dV_i} = \frac{dP}{dV_c}\frac{dV_c}{dV_o}\frac{dV_o}{dV_i} = \frac{dP}{dV_c}(-F)\frac{dV_o}{dV_i}$$

Eq. (5.57) can be written as

$$\frac{dV_o}{dV_i}\left(1 + FV_i\frac{dP}{dV_c}\right) = P$$

or
$$\frac{dV_o/V_o}{dV_i/V_i} = \left(1 + FV_i\frac{dP}{dV_c}\right)^{-1} \tag{5.58}$$

This is the small-signal differential equation for the AGC loop illustrated in Fig. 5.76. It is valid for incremental changes about a particular control voltage. The loop transmission

$$L = -F(s)V_i\frac{dP}{dV_c} \tag{5.59}$$

is a function of the input signal, so the system is, in general, nonlinear. The transient behavior of the system shown in Fig. 5.69 in response to changes in the input amplitude is, in general, difficult to obtain because of the nonlinear nature of the system. Since the loop transmission depends on the input amplitude, so does the pole of the closed-loop system, and hence, the speed of the transient response. This feature of system behavior is illustrated by the following example.

EXAMPLE 5.17. The AGC loop illustrated in Fig. 5.70 has a variable-gain amplifier with the linear characteristic $P(V_c) = V_c$ and includes an integrator as the low-pass filter, $F(s) = K/s$. Determine the transient response to small step changes in the input amplitude.

Solution. For this system the loop transmission [Eq. (5.59)] is

$$L = -\frac{K}{s}V_i$$

and for a small step change in the input signal

$$\frac{\Delta V_i}{V_i}(s) = \frac{\Delta_i}{s}$$

So the normalized change in output voltage is

$$\frac{\Delta V_o}{V_o}(s) = \frac{\Delta_i}{s + K V_i}$$

and

$$\frac{\Delta V_o}{V_o}(t) = \Delta_i e^{-K V_i t}$$

This last example illustrates how the loop dynamics can depend on the amplitude of the input signal. For AGC systems in which control of the transient response is critical, more complex loops are usually required.

If the variable-gain characteristic $P(V_c)$ is known, it is possible to numerically evaluate the dc characteristics of the loop by choosing a value of control voltage as the starting point.

**EXAMPLE 5.18.** The AGC loop illustrated in Fig. 5.71 employs a variable-gain amplifier with a square-law characteristic

$$P(V_c) = V_c^2$$

Determine the dc voltage as a function of input voltage for a reference voltage of 1 V.

*Solution.* The simplest solution is to calculate the input and output voltages for a specified control voltage. For example, if $V_c = 0.5$ V, then $V_o = V_r - V_c = 0.5$ V, and

$$V_i = \frac{V_o}{P(V_c)} = \frac{0.5}{0.5^2} = 2 \text{ V}$$

A plot of the static input-output characteristic is given in Fig. 5.72. The output voltage varies 20 dB as the input voltage increases from 0.1 V to an arbitrarily large voltage. Over this same range of input voltage the control voltage decreases from 1 to 0 V.

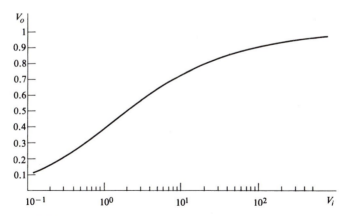

**FIGURE 5.72**
Voltage-transfer characteristic of the AGC system of
Example 5.18.

The AGC systems discussed here all provide a continuous monitoring of the output amplitude and a continuous adjustment of the variable-gain amplifier. There are many systems in which the output load is monitored intermittently and the gain adjusted during these intervals. For the remainder of the time the control loop is an open circuit, and the gain is held constant during this interval. Television receivers are an example of a gated AGC system. If the gated signal that is used for the AGC action does not contain any modulation such as the TV synchronization pulses, then the AGC system bandwidth can be made quite wide to give very fast response, without suppressing the modulation between pulses. Pulse-type AGC systems have been analyzed using sampled-data techniques.[4] This approach has utility when a linear model can be obtained for the system.

## AGC System Components

The designer of AGC systems has several variable-gain-amplifier (VGA) control laws from which to choose. The selection criteria include frequency response, available range of control voltage, and desired range of the variable-gain amplifier. A VGA whose gain is an exponential function of the control voltage can have a wider variation in gain than can, for example, a variable-gain amplifier with a linear control function. Multipliers have a linear control law by definition. Dual-gate MOSFETs and pin-diode attenuators are two of the many circuits that exhibit an exponential control law. Bipolar differential amplifiers, commonly used in integrated circuits, have a voltage gain that is proportional to the collector bias current, so the gain can be varied by adjusting the collector direct current. Figure 5.73 illustrates a simplified differential amplifier with the transistor $Q_3$ serving as a constant-current source. The collector current of $I_3$ is

$$I_C = I_s e^{V_R/V_T}$$

so the gain of the amplifier (proportional to $I_C$) is an exponential function of the control voltage $V_R$. The circuit (with the bias network) shown in Fig. 5.74 can serve

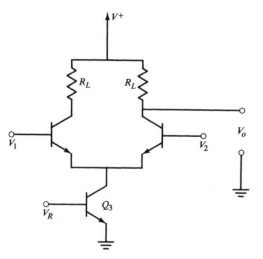

**FIGURE 5.73**
An exponential amplifier for AGC applications.

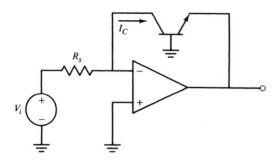

**FIGURE 5.74**
A logarithmic amplifier for AGC loops.

as a logarithmic amplifier for AGC loops that use one. Since the negative terminal of the operational amplifier is at ground potential,

$$I_1 = \frac{V_i}{R_s} = I_C = I_s e^{V_o/V_T}$$

and the output voltage is

$$V_o = V_T \ln \frac{V_i}{R_s I_s}$$

which is a logarithmic function of the input voltage.

## ■ 5.5

## PROBLEMS

**5.1** Add a capacitor to the amplifier shown in Fig. P5.1 which will make the upper 3-dB frequency 20 kHz. Indicate where the capacitor is to be connected. What is the smallest value of capacitance that can be used?

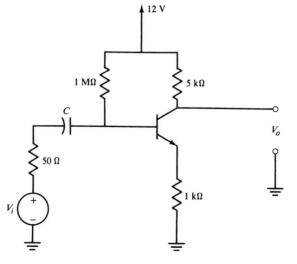

**FIGURE P5.1**
A common-emitter amplifier with voltage-to-current feedback.

**5.2** Add a capacitor to the amplifier shown in Fig. P5.2 which will make the upper 3-dB frequency 20 kHz. Indicate where the capacitor is to be connected. What is the smallest value of capacitance that can be used? (The transistor is biased so that the transconductance $g_m = 10^{-3}$ S.)

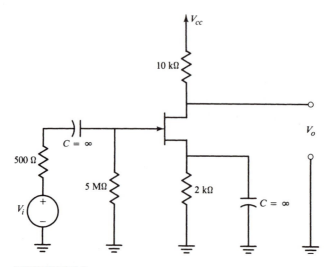

**FIGURE P5.2**
A common-source amplifier.

**5.3** The transistor used in the amplifier shown in Fig. P5.3 has a $\beta$ of 100, a $C_\pi$ of 4 pF, and $C_\mu = 5$ pF. What is the current gain-bandwidth product of the transistor? What is the upper 3-dB frequency of the amplifier?

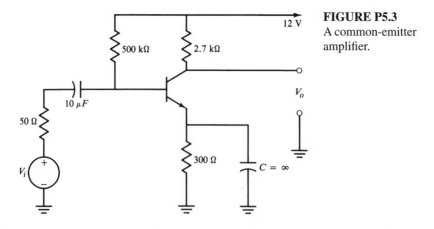

**FIGURE P5.3**
A common-emitter amplifier.

**5.4** What is the upper 3-dB frequency of the amplifier shown in Fig. P5.4? The transistor $\beta = 100$.

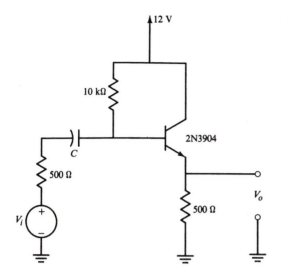

**5.5** Design an amplifier (using discrete Q2V2222A transistors) with a minimum voltage gain of 40 dB. The source impedance is 500 $\Omega$, and the load impedance is 50 $\Omega$. The upper 3-dB frequency is to be 30 kHz and the lower -3dB frequency is 5 Hz.

**5.6** What are the midfrequency voltage gain and upper 3-dB corner frequency of the two-stage amplifier shown in Fig. P5.6? The transistors have identical characteristics: $g_m = 10^{-3}$ S, $r_d = 10^5$ $\Omega$, $C_{ds} = 3$ pF, $C_{gd} = 1$ pF, and $C_{gs} = 3$ pF.

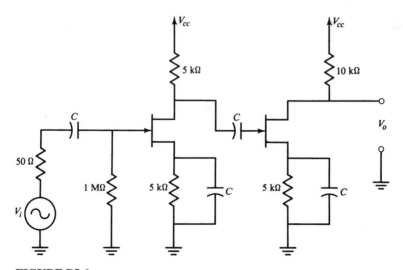

**FIGURE P5.6**
A two-stage FET amplifier.

**5.7** Calculate the gain and frequency response of the common-gate amplifier illustrated in Fig. P5.7. See App. 2 for transistor specifications. The transistor is biased so that the transconductance $g_m = 12 \times 10^{-3}$ S.

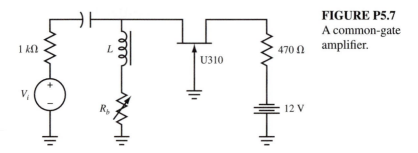

**FIGURE P5.7**
A common-gate amplifier.

**5.8** Calculate the gain and high-frequency performance of the amplifier shown in Fig. P5.8. The U310 transistor is used. See App. 2 for transistor specifications. The transistor is biased for $g_m = 12 \times 10^{-3}$ S.

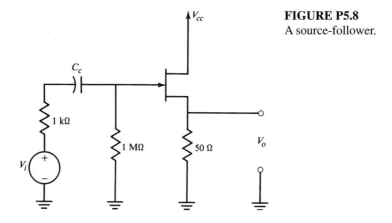

**FIGURE P5.8**
A source-follower.

**5.9** It is desired to build an amplifier with a gain of 40 dB using amplifiers with a gain-bandwidth product of $10^7$ Hz. Will a wider bandwidth be obtained using a single-stage amplifier or cascading two amplifiers, each with a gain of 20 dB?

**5.10** Derive an approximate expression for the high-frequency performance of a Darlington amplifier.

**5.11** Design a lossless feedback amplifier using a 2N3904 transistor that has a voltage gain of 6 and an input impedance of 50 $\Omega$. The source resistance is 50 $\Omega$, and the load impedance is 300 $\Omega$.

**5.12** What will be the gain and bandwidth of the amplifier shown in Fig. P5.3 if the bypass capacitor is removed from the emitter resistor? (*Hint:* Estimate the bandwidth, assuming the gain-bandwidth product remains constant.)

**5.13** What will be the gain of the amplifier shown in Fig. P5.2 if the bypass capacitor is removed from the source resistance? Estimate the bandwidth, assuming the gain-bandwidth product remains constant.

**5.14** Derive an expression for the high-frequency response of an FET cascode amplifier.

**5.15** Design a cascode amplifier using 2N3904 transistors. The load impedance is $1\,\mathrm{k\Omega}$, and the source is $500\,\Omega$. Bias the amplifier so that the voltage gain is approximately 25. What is the predicted upper frequency limit of the amplifier?

**5.16** Show that the control law $K/V_c$ results in the AGC system shown in Fig. 5.70 having constant loop transmission.

**5.17** Repeat Example 5.18 for the case in which the reference voltage is 2 V and the control voltage can vary from 0 to 2 V.

**5.18** Plot the static input-output transfer characteristic of an AGC system in which the variable-gain amplifier obeys the control law

$$P(V_c) = e^{V_c} - 1$$

**5.19** Add a finite output resistance to the input buffer amplifier of the current feedback op-amp model shown in Fig. 5.39, and derive an expression for the bandwidth of a noninverting amplifier realized using a current feedback op amp.

**5.20** A voltage amplifier has a gain-bandwidth product of 1000. Compare the bandwidth obtained by realizing an amplifier with a gain of 1000 by using one of these devices to use three of these devices, each operated with a voltage gain of 10.

**5.21** Determine the output voltage of the differential amplifier in Fig. P5.21. Both transistors have $\beta$ of 100 and a unity gain frequency $f_T = 300$ MHz at a collector current of 2 mA. Also $C_\pi$ and $C_u$ are equal. Determine the upper $-3$-dB frequency of the differential gain.

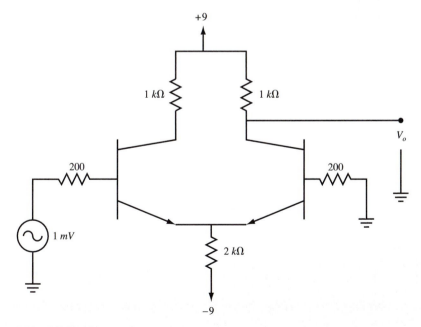

**FIGURE P5.21**

**\*5.22** Design a voltage amplifier with a midband gain of 20 dB, a center frequency of 10 MHz, and a bandwidth of 1 MHz, using capacitor, inductors (all inductors have a $Q$ of 100), chokes, resistors, and transistors. Transistor specifications are as follows: Both NPN and PNP are available with $\beta = 100$ and a unity gain frequency of 300 MHz at a collector current of 2 mA. And $C\pi$ and $C_u$ are equal. The source resistance is 500 ohms and the load resistance is 50 ohms.

**\*5.23** Design a high-frequency video amplifier with a voltage gain of 50 (+10%). The source impedance is 500 $\Omega$, and the load impedance is 50 $\Omega$. The output voltage is to be 2 V p-p ($\pm$10%). The frequency response should be from 50 Hz to as high as possible. Verify your design (including linearity) using computer simulation. Assume both PNP and NPN transistors are available and they can be modeled by

```
MODEL MOD1 NPN BF = 100, VAF = 50, IS = 1.E-12, TF = .6NS, TR = 60NS, CJE
= 5pF, CJC = 5pF, MJC = .5, MJE = .3333, VJC = .75, VJE = .75
```

Also, dual 9-V supplies are available. Limit the total supply current to about 100 mA. State the amount of supply current used. (The source resistance cannot be shunted). Be sure to include a transient response in your report. Use a pulse input for the transient response. Use standard-value resistors and limit the size of the capacitors to 100 $\mu$F. There should be at most 10% overshoot to a step input.

## ▥ 5.6

### REFERENCES

1. C. L. Searle, et al., *Elementary Circuit Properties of Transistors,* Wiley, New York, 1964, p. 103.
2. D. E. Norton, High Dynamic Range Transistor Amplifiers Using Lossless Feedback, *Proc. IEEE Int. Symp. on Circuits and Systems,* 438–440 (1975).
3. H. A. Wheeler, Automatic Volume Control for Radio Receiving Sets, *Proc. IRE,* **16:**30–39 (1928).
4. D. V. Mercy, A Review of Automatic Gain Control Theory, *Radio and Electronic Engineer,* 579–590 (1981).

## ▥ 5.7

### ADDITIONAL READING

Gilbert, B.: A New Wide-Band Amplifier Technique, *IEEE J. Solid-State Circuits,* **3:** 353–365 (1968).

Gray, P. R., and R. G. Meyer: *Analysis and Design of Analog Integrated Circuits,* 3d ed., Wiley, New York, 1993.

Horowitz, P., and W. Hill: *The Art of Electronics,* 2d ed., Cambridge University Press, New York, 1989.

Maclean, D. J. H.: *Broadband Feedback Amplifiers,* Research Studies Press, New York, 1982.

Oliver, B. M.: Automatic Volume Control as a Feedback Problem, *Proc. IRE,* **36:** 466–473 (1948).

Ricker, D. W.: Non-linear Feedback System for the Normalization of Active Sonar Returns, *J. Acoust. Soc. Am.,* **59:** 389–396 (1976).

Vladimirescu, A.: *The Spice Book,* Wiley, New York, 1994.

# Hybrid and Transmission-Line Transformers

## INTRODUCTION

In addition to the transformers described in Chap. 4, there are two other kinds of transformers that are frequently used in communication circuits. The three-winding and transmission-line transformers can both be configured for adding and/or subtracting multiple inputs. Another particularly useful transformer, called the *hybrid transformer*, consists of a four-port circuit that provides isolation between selected signals (ports) and at the same time provides for maximum transfer between the other two ports. In the following sections of this chapter the three-winding and transmission-line transformers will be analyzed, and we will demonstrate how they can both be used to create the hybrid transformer. The idealized behavior of these transformers, most useful for design and analysis, will be presented, along with the description of more detailed models that are best analyzed by computer methods.

## 6.2

### THREE-WINDING TRANSFORMERS

It is often required that voltages in certain sections of an electric circuit induce voltages in selected branches and at the same time isolate other branches. The three-winding transformer is the most widely used means to accomplish this end. To illustrate the technique, the ideal three-winding transformer will first be evaluated.

Consider the ideal transformer of Fig. 6.1 with an $N{:}1$ turns ratio between each half of the primary and the secondary winding. Since the transformer is ideal,

$$V_1 = V_2 = N V_3 \tag{6.1}$$

Also, no power is dissipated in an ideal transformer, so the power output is equal to the power supplied:

$$I_1 V_1 + I_2 V_2 = -I_3 V_3 \tag{6.2}$$

213

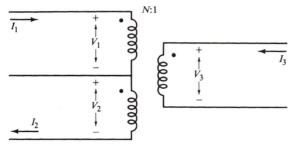

**FIGURE 6.1**
A center-tapped transformer.

or
$$(I_1 + I_2)NV_3 = -I_3V_3$$

Therefore, the current relation in this ideal transformer is

$$-I_3 = N(I_1 + I_2) \tag{6.3}$$

Consider now the ideal transformer used to couple the four voltage sources and source resistances as shown in Fig. 6.2. The three current-loop equations are

$$E_1 - E_4 = I_1(Z_1 + Z_4) - I_2Z_4 + NV_3 \tag{6.4}$$

$$-E_2 + E_4 = -I_1Z_4 + I_2(Z_2 + Z_4) + NV_3 \tag{6.5}$$

$$E_3 = I_3Z_3 + V_3 \tag{6.6}$$

If Eq. (6.6) is used to replace $V_3$, Eqs. (6.3) through (6.5) can be written (in matrix notation) as

$$\begin{bmatrix} E_1 - E_4 - NE_3 \\ -E_2 + E_4 - NE_3 \\ 0 \end{bmatrix} = \mathbf{A} \begin{bmatrix} I_1 \\ I_2 \\ I_3 \end{bmatrix} \tag{6.7}$$

where the matrix

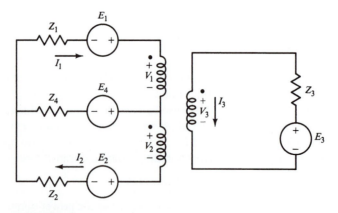

**FIGURE 6.2**
A circuit with four voltage sources and a center-tapped transformer.

$$\mathbf{A} = \begin{bmatrix} Z_1 + Z_4 & -Z_4 & -NZ_3 \\ -Z_4 & Z_2 + Z_4 & -NZ_3 \\ N & N & 1 \end{bmatrix} \tag{6.8}$$

It is possible to select the impedances so that there is isolation between ports. For example, to isolate port 4 from $E_3$, the current in response to $E_3$, through $Z_4(I_1 - I_2)$, should be zero. The current $I_1$ in response to $E_3$ is

$$I_1 = \frac{-NZ_4E_3 - N(Z_2 + Z_4)E_3}{\det \mathbf{A}} \tag{6.9}$$

where $\det \mathbf{A}$ is the determinant of the matrix $\mathbf{A}$. Also, the current $I_2$ in response to $E_3$ is

$$I_2 = -\frac{[N(2Z_4 + Z_1)]E_3}{\det \mathbf{A}} \tag{6.10}$$

so the current through $Z_4$ in response to $E_3$ is

$$I_1 - I_2 = -\frac{N(Z_1 - Z_2)E_3}{\det \mathbf{A}} \tag{6.11}$$

Port 4 will be isolated from port 3 if $Z_1 = Z_2$. That is, if $Z_1 = Z_2$, an applied voltage $E_3$ will not develop a voltage across $Z_4$. This property of the center-tapped transformer is one of the reasons this type of transformer is so often used. An example of its application can be found in the impedance-bridge circuit illustrated in Fig. 6.3. An unknown impedance can be inserted as $Z_2$, and then $Z_1$ is adjusted until the voltage across $Z_4$ is zero. This occurs only when $Z_1$ is equal to $Z_2$. Very fine resolution can be obtained using a null detector to measure and amplify the voltage across $Z_4$, or a current meter can be used in place of a voltmeter. If the voltage across $Z_4$ is zero, the current through $Z_4$ is also zero.

If, in addition, ports 1 and 2 are to be isolated from each other, $I_2$ in response to $E_1$ must be zero. That is,

$$I_2 = \frac{-E_1(N^2Z_3 - Z_4)}{\det \mathbf{A}} \tag{6.12}$$

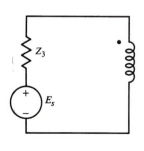
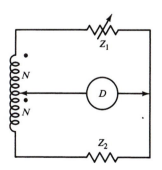

**FIGURE 6.3**
An impedance-bridge circuit that uses a center-tapped transformer.

So $I_2$ will be zero for all $E_1$ if

$$\frac{Z_4}{Z_3} = N^2 \qquad (6.13)$$

Ports that are isolated from each other are said to be *conjugate*. We have seen that ports 3 and 4 will be conjugate if $Z_1 = Z_2$ and that ports 1 and 2 are conjugate if $Z_4/Z_3 = N^2$. If both sets of ports are conjugate, the transformer is said to be *biconjugate*.

EXAMPLE 6.1. If identical loads are placed on the transformer primary so that $I_1 = I_2$ and $V_1 = V_2$, what is the impedance $R'$ seen looking into port 3 of the transformer circuit shown in Fig. 6.4?

*Solution.* Since $R_1 = R_2 = R$, ports 3 and 4 are conjugate; therefore, the current $I_1 - I_2$ is zero, and terminals $a$ and $b$ can be either short- or open-circuited. The secondary resistance is

$$R' = \frac{V_3}{I_3} = \frac{V_1/N}{2NI_1} = \frac{R}{2N^2}$$

If only half of the primary is used ($R_1 = \infty$) so that $I_1 = 0$, then

$$R' = \frac{V_3}{I_3} = \frac{V_2/N}{NI_2} = \frac{R}{N^2}$$

provided that terminals $a$ and $b$ are short-circuited together.

A second property often required of center-tapped transformer circuits is that all four source impedances ($Z_1$, $Z_2$, $Z_3$, and $Z_4$) be matched for maximum power transfer. For the center-tapped transformer of Fig. 6.2, with the port impedances matched for isolation, Eq. (6.8) becomes

$$\begin{bmatrix} E_1 - E_4 - NE_3 \\ -E_2 + E_4 - NE_3 \\ 0 \end{bmatrix} = \begin{bmatrix} Z_1 + Z_4 & -Z_4 & -Z_4/N \\ -Z_4 & Z_1 + Z_4 & -Z_4/N \\ N & N & 1 \end{bmatrix} \begin{bmatrix} I_1 \\ I_2 \\ I_3 \end{bmatrix} \qquad (6.14)$$

The input impedance at port 1 is defined to be

$$Z_{i_1} = \frac{E_1}{I_1} - Z_1 \qquad (6.15)$$

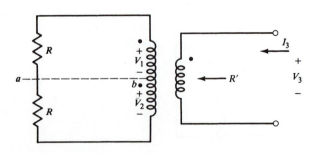

FIGURE 6.4
The transformer circuit discussed in Example 6.1.

since $Z_1$ is taken as the source impedance of $E_1$. If Eq. (6.14) is solved for $I_1$ in response to $E_1$, it is found that

$$\frac{I_1}{E_1} = \frac{Z_1 + 2Z_4}{\det \mathbf{A}} \tag{6.16}$$

where     $\det \mathbf{A} = (Z_1 + Z_4)^2 + 2Z_4(Z_1 + Z_4) + Z_4^2 = (Z_1 + 2Z_4)^2$     (6.17)

Therefore,

$$\frac{E_1}{I_1} = \frac{(Z_1 + 2Z_4)^2}{Z_1 + 2Z_4} = Z_1 + 2Z_4 \tag{6.18}$$

and                                             $Z_{i_1} = 2Z_4$                                             (6.19)

Thus for maximum power transfer from port 1,

$$Z_4 = \frac{Z_1^*}{2} \tag{6.20}$$

where $Z_1^*$ is the complex conjugate of $Z_1$.

The input impedance of port 3 is likewise defined as

$$Z_{i_3} = \frac{V_3}{I_3} = \frac{E_3}{I_3} - Z_3 \tag{6.21}$$

and can be determined by solving Eq. (6.14) for $I_3$ in response to $E_3$. That is,

$$I_3 = \frac{\begin{vmatrix} Z_1 + Z_4 & -Z_4 & -NE_3 \\ -Z_4 & Z_1 + Z_4 & -NE_3 \\ N & N & 0 \end{vmatrix}}{\det \mathbf{A}} \tag{6.22}$$

or     $\dfrac{I_3}{E_3} = \dfrac{N^2 Z_4 + N^2 Z_4 + 2N^2(Z_1 + Z_4)}{(Z_1 + 2Z_4)^2} = \dfrac{2N^2(Z_1 + 2Z_4)}{(Z_1 + 2Z_4)^2}$     (6.23)

Therefore

$$Z_{i_3} = \frac{E_3}{I_3} - Z_3 = \frac{Z_1 + 2Z_4}{2N^2} - Z_3 \tag{6.24}$$

Since ports 1 and 2 are isolated, $Z_4 = N^2 Z_3$ and

$$Z_{i_3} = \frac{Z_1}{2N^2} \tag{6.25}$$

Thus for maximum power transfer from port 3,

$$Z_3 = \frac{Z_1^*}{2N^2} \tag{6.26}$$

which is the same requirement obtained for port 1.

It is easily shown that these relations will also result in maximum power transfer from ports 4 and 2. That is, if the transformer is biconjugate and one port is matched for maximum power transfer, they are all matched for maximum power transfer. To summarize these findings, if in the ideal transformer circuit of Fig. 6.2

$$Z_1 = Z_2 = Z$$

and
$$Z_3 = \frac{Z_4}{N^2} = \frac{Z^*}{2N^2}$$

the transformer will be biconjugate with all ports matched for maximum power transfer. A biconjugate transformer with all four input impedances matched for maximum power transfer is referred to as a *hybrid transformer.*

### Asymmetric Three-Winding Transformers

The ideal transformer we have been considering has equal turns ratios on both sides of the center tap. If the turns ratios are not equal, the impedances can still be matched for port isolation and maximum power transfer between ports. Consider the ideal transformer shown in Fig. 6.5. (Purely resistive impedances will be considered here without loss of generality.) The turns ratio from the secondary to the top section of the primary is $kN$ and from the secondary to the bottom section of the primary is $N$. Since the transformer is ideal,

$$V_1 = kNV_3 \tag{6.27}$$

$$V_2 = NV_3 \tag{6.28}$$

and the power to the transformer is equal to the power output, so

$$V_1 I_1 + V_2 I_2 = -V_3 I_3 \tag{6.29}$$

or
$$kNI_1 + NI_2 = -I_3 \tag{6.30}$$

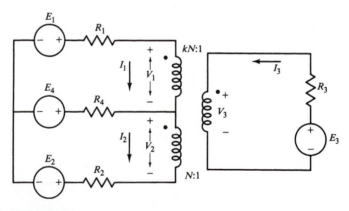

**FIGURE 6.5**
A circuit incorporating an asymmetric three-winding transformer.

The loop equations are

$$-E_4 + E_1 - kNV_3 = I_1(R_1 + R_4) - I_2 R_4 \qquad (6.31)$$

$$E_4 - E_2 - NV_3 = -I_1 R_4 + I_2(R_2 + R_4) \qquad (6.32)$$

$$E_3 = I_3 R_3 + V_3 \qquad (6.33)$$

If Eq. (6.33) is used to eliminate $V_3$, Eqs. (6.31) through (6.33) become

$$\begin{bmatrix} E_1 - E_4 - kNE_3 \\ -E_2 + E_4 - NE_3 \\ 0 \end{bmatrix} = \begin{bmatrix} R_1 + R_4 & -R_4 & -kNR_3 \\ -R_4 & R_2 + R_4 & -NR_3 \\ kN & N & 1 \end{bmatrix} \begin{bmatrix} I_1 \\ I_2 \\ I_3 \end{bmatrix} \qquad (6.34)$$

or

$$E = AI \qquad (6.34a)$$

In order for port 2 to be isolated from port 1, $I_2$ must be zero in response to $E_1$. Since

$$\frac{I_2}{E_1} = \frac{-kN^2 R_3 + R_4}{\det \mathbf{A}} \qquad (6.35)$$

port 2 will be isolated from port 1 (and port 1 from port 2), provided

$$R_4 = N^2 R_3 k \qquad (6.36)$$

In order for port 3 to be isolated from port 4, $I_3$ must be zero in response to $E_4$. Since

$$I_3 = \frac{\begin{vmatrix} R_1 + R_4 & -R_4 & -E_4 \\ -R_4 & R_2 + R_4 & E_4 \\ kN & N & 0 \end{vmatrix}}{\det \mathbf{A}} \qquad (6.37)$$

or

$$\frac{I_2}{E_4} = \frac{kNR_2 - NR_1}{\det \mathbf{A}} \qquad (6.37a)$$

port 3 will be isolated from port 4 if

$$R_1 = kR_2 \qquad (6.38)$$

In the same manner it can be shown that if Eq. (6.38) holds, then port 4 is also isolated from port 3.

Equations (6.36) and (6.38) provide the relations between the resistor values in order that the asymmetric tapped transformer have the first property of the hybrid transformer. We will now determine the resistance required at each port in order for the transformer to also have the second property of maximum power transfer between ports. For maximum power transfer at port 1, it is necessary to determine the port's input resistance

$$R_{i_1} = \frac{E_1}{I_1} - R_1 \qquad (6.39)$$

The response $I_1$ to the voltage $E_1$ can be found by solving Eq. (6.34) [with Eqs. (6.36) and (6.38) substituted for $R_3$ and $R_2$, respectively]

$$I_1 = \frac{\begin{vmatrix} E_1 & -R_4 & -R_4/N \\ 0 & R_1/k + R_4 & -R_4/(Nk) \\ 0 & N & 1 \end{vmatrix}}{\det \mathbf{A}} \tag{6.40}$$

Therefore,

$$\frac{E_1}{I_1} = \frac{\det \mathbf{A}}{R_1/k + R_4 + R_4/k} \tag{6.40a}$$

The value of the determinant $\mathbf{A}$ in the case in which the ports are isolated (biconjugate) is given by

$$\det \mathbf{A} = R_1 \left( \frac{R_1}{k} + R_4 + \frac{R_4}{k} + R_4 \right) + \frac{R_4 R_1}{k} + 2R_4^2 + \frac{R_4^2}{k} + k R_4^2$$

$$= \frac{[R_1 + R_4(1+k)]^2}{k} \tag{6.41}$$

so

$$\frac{E_1}{I_1} = R_1 + R_4(1+k)$$

Therefore, for maximum power transfer from port 1,

$$R_1 = R_4(1+k) \tag{6.42}$$

Likewise, the resistance at port 3 can be calculated using Eqs. (6.34), (6.36), (6.38), and (6.41):

$$I_3 = \frac{\begin{vmatrix} R_1 + R_4 & -R_4 & -kNE_3 \\ -R_4 & R_1/k + R_4 & -NE_3 \\ kN & N & 0 \end{vmatrix}}{\det \mathbf{A}} \tag{6.43}$$

Thus

$$\frac{E_3}{I_3} = \frac{[R_1 + R_4(1+k)]^2}{k} \{N^2(1+k)[R_1 + R_4(1+k)]\}^{-1} \tag{6.44}$$

$$= \frac{R_1 + R_4(1+k)}{N^2 k(1+k)} 1 \tag{6.44a}$$

Since ports 3 and 4 are conjugate [and using Eq. (6.42)],

$$R_{i_3} = \frac{E_3}{I_3} - R_3 = \frac{E_3}{I_3} \frac{R_4}{kN^2} = \frac{R_1}{N^2 k(1+k)} \tag{6.45}$$

That is,

$$R_3 = \frac{R_1}{N^2 k(1+k)} \tag{6.46}$$

for maximum power transfer from port 3. This is also the condition for maximum power transfer from port 1.

It should be observed that once the port isolation properties have been established, the equations can be solved in a much easier manner. For example, if it is desired to calculate the response to an applied voltage $E_3$, the network can be drawn as shown in Fig. 6.6 (since $I_1 - I_2 = 0$). The secondary voltage and current can be found from the two equations

$$-I_3 = kI_1N + I_2N = N(k+1)I_1 \tag{6.47}$$

Since
$$I_1 = \frac{-NkV_3}{Rk} \tag{6.48}$$

we have
$$V_3 = \frac{-I_1 R}{N} = \frac{R}{N}\frac{I_3}{N(k+1)} \tag{6.49}$$

Thus

$$Z_{i_3} = \frac{E_3}{I_3} - R_3 = \frac{V_3}{I_3} = \frac{R}{N^2(k+1)} = \frac{R_4}{N^2 k} \tag{6.50}$$

which agrees with the result previously obtained for the input impedance.

Table 6.1 summarizes the resistance relations required for the asymmetric transformer to have isolated port pairs and maximum power transfer between ports.

**EXAMPLE 6.2.** A transformer circuit is illustrated in Fig. 6.7. Determine the turns ratio $N$ and $R_4$ so that the circuit is a hybrid transformer, and determine the output voltage $E_o$.

**Solution.** For the circuit to be a hybrid transformer. Table 6.1 shows that

$$N^2 = \frac{R_1}{2R_3} = \frac{1}{4}$$

and
$$R_4 = \frac{R}{2} = 25$$

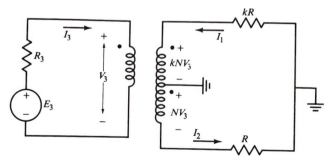

**FIGURE 6.6**
An asymmetric three-winding transformer equivalent circuit in which ports 3 and 4 are conjugate.

■ **TABLE 6.1.**
**Resistance relationships in an**
**asymmetric hybrid transformer**

$R_1 = Rk$

$R_2 = R$

$R_3 = \dfrac{R}{N^2(1+k)}$

$R_4 = \dfrac{R_1}{1+k} = \dfrac{kR}{1+k}$

The circuit will be a hybrid transformer provided $N = \frac{1}{2}$ and $R_4 = 25\,\Omega$. In this case the input impedance of port 1 is $50\,\Omega$, so

$$I_1 = \frac{E_1}{100}$$

The current $I_2$ will be zero, so

$$I_3 = -NI_1 = -\frac{E_1}{200}$$

and

$$E_o = -I_3 R_L = \frac{E_1}{2}$$

## Power Transfer in Hybrid Transformers

If power is applied at port $P_1$, no power is transferred to port $P_2$ because the ports are isolated. Since the transformer is lossless, the power must go to ports 4 and 3.

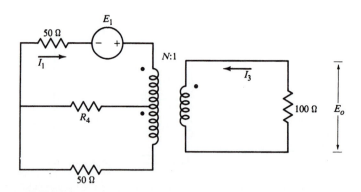

**FIGURE 6.7**
The transformer circuit discussed in Example 6.2.

That is,

$$P_1 = P_3 + P_4 \tag{6.51}$$

Also, since the current through $R_4$ is equal to the current through $R_1$ and

$$R_4 = \frac{R_1}{1+k} \tag{6.52}$$

the power distribution for an applied power $P_1$ is

$$P_4 = \frac{P_1}{1+k} \tag{6.53}$$

and

$$P_3 = P_1 - P_4 = \frac{kP_1}{1+k} \tag{6.54}$$

or

$$\frac{P_3}{P_4} = \frac{N^2 k^2 [R_4/(N^2 k)]}{R_4} = k \tag{6.55}$$

If the power is applied to port 2,

$$P_2 = P_3 + P_4 \tag{6.56}$$

Since $I_4 = I_2$ $(I_1 = 0)$,

$$P_4 = \frac{P_2 k}{1+k} \tag{6.57}$$

$$P_3 = P_2 \left(1 - \frac{k}{1+k}\right) = \frac{P_2}{1+k} \tag{6.58}$$

and

$$\frac{P_3}{P_4} = k^{-1} \tag{6.59}$$

Thus we see that the hybrid transformer can be used to split the power between two ports in a ratio determined by the transformer turns ratio. If the power applied to port 3 is divided between ports 1 and 2,

$$P_3 = P_1 + P_2 \tag{6.60}$$

Since $I_1 = I_2 (P_4 = 0)$, the power is divided so that

$$P_1 = \frac{kP_3}{1+k} \quad \text{and} \quad P_2 = \frac{P_3}{1+k} \quad \text{or} \quad \frac{P_1}{P_2} = k \tag{6.61}$$

If the power is applied to port 4, $I_3 = 0$, so $I_2 = -kI_1$ and

$$\frac{P_1}{P_2} = \frac{R_1}{R_2 k^2} = k^{-1} \tag{6.62}$$

These power distribution results can be summarized as follows: If the power from port $C$ is divided between ports $B$ and $D$ in the ratio $k$, then the power from port $A$, the conjugate port of $C$, will be divided between ports $B$ and $D$ in the ratio $1/k$.

**Phase Distribution in Hybrid Transformers**

One other property of hybrid transformers is concerned with the phase shift. While the phase shift between three of the four ports will be zero, the phase shift between the remaining port will be 180°. Various transformer configurations may be used, but in all of them the transmission path will have a phase shift differing 180° from the other three phase shifts. For example, consider the response $I_3$ to the applied voltages $E_1$ and $E_2$. The current $I_3$ is found using Eq. (6.14):

$$I_3 = \frac{E_1[-Z_4N - N(Z_1 + Z_4)] - E_2[-Z_4N - N(Z_1 + Z_4)]}{\det \mathbf{A}} \qquad (6.63)$$

and
$$E_0 = -I_3 Z_3 = \frac{(E_1 - E_2)(Z_1 + 2Z_4)NZ_3}{\det \mathbf{A}}$$

$$= \frac{(E_1 - E_2)NZ_3}{Z_1 + 2Z_4} \qquad (6.64)$$

The output is proportional to the difference of the input signals, and the device is so used in many communications circuits.

If we calculate $I_4$ in response to $E_1$ and $E_2$, we obtain

$$I_4 = I_1 - I_2 = \frac{E_1[(Z_1 + Z_4) + Z_4]}{\det \mathbf{A}} + \frac{E_2[(Z_1 + Z_4) + Z_4]}{\det \mathbf{A}}$$

$$= \frac{E_1 + E_2}{Z_1 + 2Z_4} \qquad (6.65)$$

If the output is taken across $Z_4$, it is proportional to the sum of the two input signals. The hybrid transformer can therefore also be used to combine or subtract signals and may act as a signal combiner, such as in push-pull circuits. The usefulness of the hybrid transformer arises from its ability to isolate ports and, at the same time, to match impedances between the ports. The device can also be used to split a signal into two components, an operation that is not affected by a failure in the conjugate channel.

**Nonideal Three-Winding Transformer[1]**

The previous discussion of the three-winding transformer assumed that the transformer was ideal, with no mutual coupling between the two sections of the primary. The derivations describe ideal circuit behavior, but they do not describe the limitations encountered in a practical circuit. The actual frequency characteristics of a three-winding transformer will now be described.

The equivalent circuit for a nonideal three-winding transformer with identical primary windings is shown in Fig. 6.8. Here $L_b$ is the inductance of the secondary winding, $L_a$ is the inductance of one primary winding, $M$ is the mutual inductance between the secondary and one-half of the primary, and $M_a$ is the mutual

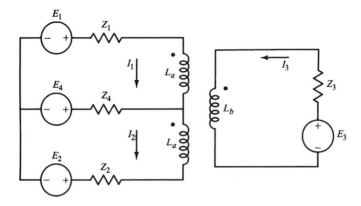

**FIGURE 6.8**
A circuit incorporating a nonideal three-winding transformer.

inductance between the two primary sections. With these definitions, the three loop equations are

$$E_1 - E_4 = (Z_1 + sL_a + Z_4)I_1 + (sM_a - Z_4)I_2 + sMI_3 \qquad (6.66)$$

$$-E_2 + E_4 = (sM_a - Z_4)I_1 + (Z_2 + Z_4 + sL_a)I_2 + sMI_3 \qquad (6.67)$$

$$E_3 = sMI_1 + sMI_2 + (sL_b + Z_3)I_3 \qquad (6.68)$$

or $$E = AI \qquad (6.69)$$

If the current $I_3$ in response to an applied voltage $E_4$ is calculated, it is readily proved that if $Z_1 = Z_2$, ports 4 and 3 are isolated.

The current $I_2$ in response to $E_1$ is

$$I_2 = \frac{E_1[(sM)^2 + (Z_4 - sM_a)(sL_b + Z_3)]}{\det A} \qquad (6.70)$$

where the denominator is the determinant of the $A$ matrix of Eq. (6.69).

In order for port 2 to be isolated from port 1, it is necessary that

$$(sM)^2 + (Z_4 - sM_a)(sL_b + Z_3) = 0 \qquad (6.71)$$

or, in terms of the steady-state response,

$$-\omega^2 M^2 + Z_4 j\omega L_b + \omega^2 M_a L_b + Z_4 Z_3 - j\omega M_a Z_3 = 0 \qquad (6.72)$$

The coefficient of coupling between the two primary sections can be defined as

$$K_a = \frac{M_a}{L_a} \qquad (6.73)$$

and the coefficient of coupling between the primary and secondary can be defined as

$$K = \frac{M}{(L_a L_b)^{1/2}} \qquad (6.74)$$

It follows that

$$M_a L_b = K_a L_a L_b = \frac{K_a M^2}{K^2} \tag{6.75}$$

If
$$K_a = K^2 \tag{6.76}$$

then Eq. (6.72) reduces to

$$Z_4 = \frac{Z_3 j\omega M_a}{Z_3 + j\omega L_b} \tag{6.77}$$

$$= \frac{Z_3}{Z_3/(j\omega K_a L_a) + L_b/M_a} \tag{6.78}$$

which is the required condition for ports 1 and 2 to be conjugate. If, as is the case in most practical situations,

$$K_a \omega L_a \gg Z_3 \tag{6.79}$$

ports 1 and 2 will be isolated, provided that

$$Z_4 \approx \frac{Z_3 M_a}{L_b} = Z_3 N^2 K_a \tag{6.80}$$

where the turns ratio is defined by

$$N^2 = \frac{L_a}{L_b} \tag{6.81}$$

This is identical to the requirement for the ideal transformer $(K_a = 1)$. Hybrid transformers are normally designed with

$$Z_3 = \frac{Z_4}{N^2 K_a} \tag{6.82}$$

This requirement can be met over a broad range of frequencies, but the low-frequency performance of nonideal center-tapped transformers deteriorates when $\omega L_a$ is no longer much greater than $Z_4$; their high-frequency performance is limited by the stray and interwinding capacitances.

## ◼ 6.3

### TRANSMISSION-LINE TRANSFORMERS

Transmission lines can also be used for impedance matching and power transfer. One of the principal functions of transformers in radio communication circuits is to match impedances between two networks. In applications where a narrowband circuit is required, the transformer can also serve as an integral part of the tuned circuit. However, unless special techniques are used, transformers are restricted to relatively narrowband applications, primarily because the interwinding capacitance resonates with the transformer inductance. Transformers built to minimize interwinding capacitances are usually bulky, especially if a great deal of power is to be transferred. Thus, engineers seek other means of matching impedance levels when large bandwidths are desired. A successful approach has been to construct transformers using transmission lines.[2]

The distinguishing feature of the transmission-line transformer is that the coils are so arranged that the interwinding capacitance combines with the inductance to form a transmission line that has no resonant frequencies that would limit the circuit bandwidth. For this reason the windings can be closely spaced while maintaining effective couplings. The net result of this construction of transmission-line transformers is a strong high-frequency response.

Consider the circuit shown in Fig. 6.9. It is assumed that the voltage travels from $V_1$ to $V_2$ via a transmission line. Also $I_1$ is the current at the sending end, and $I_2$ is the current at the receiving end. The loop equations for the circuit can be written as

$$E_i = (I_1 + I_2)R_g + V_1 \tag{6.83}$$

$$= (I_1 + I_2)R_g - V_2 + I_2 R_L \tag{6.84}$$

The two transmission-line equations that relate the voltages and currents on a lossless line are[3]

$$V_1 = V_2 \cos \beta l + j I_2 Z_o \sin \beta l \tag{6.85}$$

and

$$I_1 = I_2 \cos \beta l + \frac{j V_2}{Z_o} \sin \beta l \tag{6.86}$$

where $l$ is the line length, $\beta$ is the line phase constant, and $Z_o$ is the characteristic impedance of the line. Both $\beta$ and $Z_o$ are determined by the line inductance $L$ and capacitance $C$ per unit length:

$$Z_o = \left(\frac{L}{C}\right)^{1/2} \tag{6.87}$$

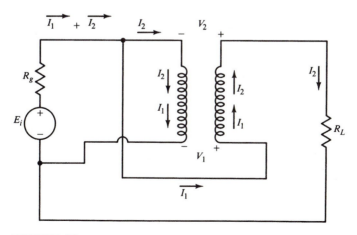

**FIGURE 6.9**
A transmission-line transformer circuit.

and $$\beta = (LC)^{1/2}\omega \tag{6.88}$$

The line wavelength is

$$\lambda = \frac{2\pi}{\beta} \tag{6.89}$$

The line inductance and capacitance depend upon the wire size and distance between conductors. Transmission-line transformers are normally made by twisting the two conductors together. The characteristic impedance decreases as the number of turns per unit length increases.[4] Figure 6.10 illustrates a typical plot of characteristic impedance as a function of the number of turns per centimeter.

If Eq. (6.85) is used to eliminate $V_1$, the equilibrium equations are

$$\begin{aligned} E_i &= \\ E_i &= \\ 0 &= \end{aligned} \begin{bmatrix} R_g & R_g + jZ_o \sin \beta l & \cos \beta l \\ R_g & R_g + R_L & -1 \\ -1 & \cos \beta l & j(\sin \beta l)/Z_o \end{bmatrix} \begin{bmatrix} I_1 \\ I_2 \\ V_2 \end{bmatrix} \tag{6.90}$$

Thus

$$I_2 = \frac{\begin{vmatrix} R_g & E_i & \cos \beta l \\ R_g & E_i & -1 \\ -1 & 0 & (\sin \beta l)/Z_o \end{vmatrix}}{\det \mathbf{A}} \tag{6.91}$$

$$= \frac{E_i(1 + \cos \beta l)}{2R_g(1 + \cos \beta l) + R_L \cos \beta l + j(R_g R_L + Z_o^2)(\sin \beta l)/Z_o}$$

and the output power is

$$P_o = |I_2|^2 R_L$$

$$= \frac{E_i^2(1 + \cos \beta l)^2 R_L}{[2R_g(1 + \cos \beta l) + R_L \cos \beta l]^2 + (R_g R_L + Z_o^2)^2(\sin^2 \beta l)/Z_o^2} \tag{6.92}$$

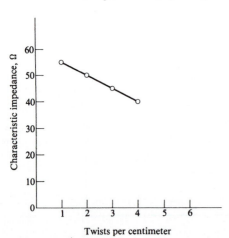

**FIGURE 6.10**
The characteristic impedance of a twisted-pair transmission line as a function of the number of twists.

The value of characteristic impedance $Z_o$ for which the output power is a maximum can be found by setting

$$\frac{dP_o}{dZ_o} = 0$$

The maximizing value of $Z_o$ is found to be

$$Z_o = (R_g R_L)^{1/2} \tag{6.93}$$

independent of the line length $l$.

Now with the value of characteristic impedance $Z_o = (R_g R_L)^{1/2}$,

$$P_o = |I_2^2| R_L = \frac{E_i^2 (1 + \cos \beta l)^2 R_L}{[2R_g(1 + \cos \beta l) + R_L \cos \beta l]^2 + 4R_g R_L \sin^2 \beta l} \tag{6.94}$$

To determine the value of $R_L$ for which $P_o$ is a maximum, the derivative $dP_o/dR_L$ is set to zero. The maximizing value of $R_L$ is

$$R_L = \frac{2R_g(1 + \cos \beta l)}{\cos \beta l} \tag{6.95}$$

which is equal to $4R_g$ for $\beta l \approx 0$. For $Z_o = (R_g R_L)^{1/2}$ and $R_L = 4R_g$, Eq. (6.94) can be written

$$P_o = \frac{E_i(1 + \cos \beta l)^2 R_L}{\{R_L/[2(1 + \cos \beta l)] + R_L \cos \beta l\}^2 + R_L^2 \sin^2 \beta l}$$
$$= \frac{4E_i^2(1 + \cos \beta l)}{R_L(5 \cos^2 \beta l + 6 \cos \beta l + 5)} \tag{6.96}$$

Since the maximum available power from the source $P_A$ is

$$P_A = \frac{E_i^2}{4R_g} = \frac{E_i^2}{R_L} \tag{6.97}$$

the ratio of output power to available power is

$$\frac{P_o}{P_A} = \frac{(1 + \cos \beta l)^2}{\frac{5}{4}(1 + \frac{6}{5} \cos \beta l + \cos^2 \beta l)} \tag{6.98}$$

Equation (6.98) shows that the frequency response of this circuit starts to roll off when $\cos \beta l$ becomes significantly less than 1, and the power gain is zero when

$$\beta l = \pi$$

or

$$l = \frac{\pi}{\omega} = \frac{\lambda}{2} \tag{6.99}$$

The power gain is zero when the line is one-half wavelength long. Transmission-line transformers are designed so that the line length is normally less than one-tenth of a wavelength at the highest frequency of interest. The shorter the physical length of the line, the higher the upper cutoff frequency of the device. Transmission-line transformers can be built with twisted-pair wires to have an upper cutoff frequency above 100 MHz. The upper frequency limit can be extended into the gigahertz region using thin-film fabrication techniques.

Equation (6.98) indicates that the low-frequency response extends down to zero frequency. Actually, the transmission-line model is not valid at the lower frequencies. This is seen when the circuit of Fig. 6.9 is redrawn as shown in Fig. 6.11 where the transformer is modeled as an autotransformer.

The step-down autotransformer can be replaced by the equivalent circuit shown in Fig. 6.12. It is readily shown that for the two circuits to be the same

$$N = \left(\frac{L_2}{L}\right)^{1/2} \tag{6.100}$$

where $L$ is the total inductance shunting $R_L$ and the inductance

$$L_2 = \omega^{-1}\left|\frac{V_2}{I_1}\right|_{I_2=0} \tag{6.101}$$

If the circuit in Fig. 6.12 is closely coupled with $k \approx 1$, the low-frequency behavior is described by

$$V_2 = \frac{NE_i[sL_2R_L'/(R_g + R_L')]}{1 + sL_2(R_g + R_L')/(R_gR_L')} \tag{6.102}$$

Where

$$R_L' = N^2R_L \tag{6.103}$$

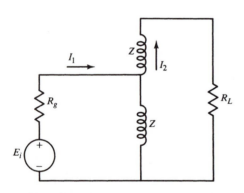

**FIGURE 6.11**
A low-frequency equivalent circuit of the circuit shown in Fig. 6.9.

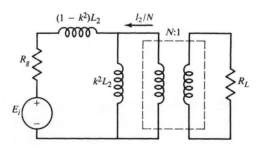

**FIGURE 6.12**
Another low-frequency equivalent circuit of the circuit shown in Fig. 6.9.

the $-3$-dB frequency is

$$\omega_N = \frac{R_g R_L'}{L_2(R_g + R_L')} \tag{6.104}$$

To reduce this cutoff frequency for a given $R_g$, $L$ needs to be increased. This can be done by increasing the line length, but, as previously shown, this reduces the higher-frequency limit of the device. Another method for increasing the inductance $L$ is to wind the wires on a magnetic core of high-permeability material, since the inductance of a conductor is directly proportional to the permeability of the surrounding medium. Since equal and opposite currents flow in the two windings, the core does not influence the internal magnetic fields or the characteristic impedance of the line. The core's sole role is to minimize the shunting current that determine the lower cutoff frequency. The core does not couple energy from input to output (except perhaps at the lowest frequencies); rather, the power is coupled through the dielectric medium of the transmission line. Therefore, small ferrite cores can be used in transmission-line transformers operating at relatively high power levels.

The analysis of circuits containing transmission-line transformers can be simplified by analyzing the transmission-line transformer under ideal conditions (that is, the line is assumed lossless and matched so that $V_1 = V_2$). Consider the transformer in Fig. 6.13. Since the sum of the voltage drops around the loop $V_1 + E_2 - V_2 - E_1$ is 0 if $V_1 = V_2$, then $E_1 = E_2$. Also, in the ideal transmission line, $I_1 = I_2$; thus under ideal conditions the circuit can easily be analyzed.

For the ideal case,

$$I_1 = I_2 \tag{6.105}$$

and $$E_1 = E_2 \tag{6.106}$$

Thus $$V_L = E_2 + E_1 = 2E_1 \tag{6.107}$$

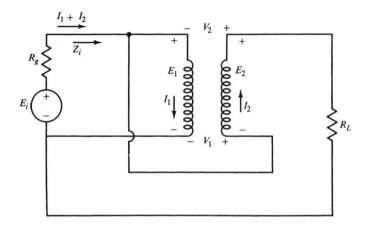

**FIGURE 6.13**
A 4:1 impedance step-down transformer.

and
$$I_L = I_1 \tag{6.108}$$

since
$$R_L = \frac{V_L}{I_L} = \frac{2E_1}{I_1} = 4Z_i \tag{6.109}$$

In this case the transmission line serves as a step-down transformer, thereby reducing the load impedance by a factor of 4. For maximum power transfer, $R_L$ must satisfy the relation

$$R_L = 4R_g \tag{6.110}$$

and the line characteristic impedance must be

$$Z_o = (R_L, R_g)^{1/2} = 2R_g \tag{6.111}$$

## Step-Up Transformer

The transmission-line transformer can also be used as a step-up transformer if used as shown in Fig. 6.14. If the line is ideal,

$$I_1 = I_2 = I \tag{6.112}$$

and
$$E_1 = E_2 = E \tag{6.113}$$

Therefore
$$I_L = 2I \tag{6.114}$$

$$V_L = E \tag{6.115}$$

and
$$R_L = \frac{V_L}{I_L} = \frac{E}{2I} \tag{6.116}$$

The input impedance

$$Z_i = \frac{V_i}{I} = \frac{2E}{I} = 4R_L \tag{6.117}$$

will be 4 times the load impedance.

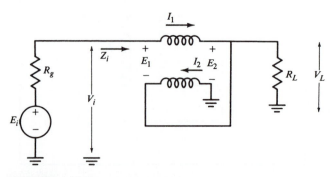

**FIGURE 6.14**
A 4:1 impedance step-up transformer.

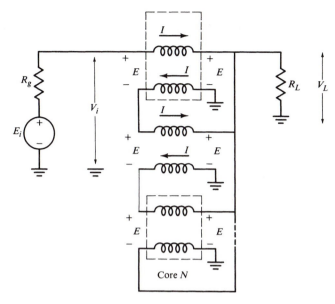

**FIGURE 6.15**
An $(N + 1)^2 : 1$ impedance step-up transformer.

Larger impedance ratios can be obtained using multiple cores, as shown in Fig. 6.15. The current $I$ from the first core flows into the primary of the second core. This same current is also transferred to the secondary of the second core and hence through the third core, etc. It is evident that the total load current is

$$I_L = (N + 1)I \tag{6.118}$$

where $N$ is the number of cores. Also, if each line is matched, the load voltage $V_L = E$ will also be equal to the voltage across the output of each transmission-line transformer. The input voltage $V_i$ must then be

$$V_i = (N + 1)E \tag{6.119}$$

Hence
$$R_L = \frac{E}{(N + 1)I} \tag{6.120}$$

and
$$R_g = \frac{(N + 1)E}{I} \tag{6.121}$$

For maximum power transfer, the load resistance must be

$$R_L = \frac{R_g}{(N + 1)^2} \tag{6.122}$$

A similar method can be used to obtain the step-down transformation

$$R_L = (N + 1)^2 R_g \tag{6.123}$$

## Hybrid Transformer

Transmission-line transformers can also be used to realize the versatile hybrid transformer, as in the circuit illustrated in Fig. 6.16. Three loop equations for the circuit can be written as

$$E_1 - E_4 = I_1(Z_1 + Z_4) - I_2 Z_4 + a_2 \qquad (6.124)$$

$$-E_2 + E_4 = -I_1 Z_4 + I_2(Z_2 + Z_4) + a_1 \qquad (6.125)$$

$$E_3 = I_3 R_3 + a_1 \qquad (6.126)$$

In addition, if the transmission line is ideal,

$$a_1 = a_2 \qquad (6.127)$$

and the following circuit identities hold:

$$I_1 = I \qquad (6.128)$$

$$I_4 = I_1 - I_2 \qquad (6.129)$$

$$I + I_3 + I = I_4 \qquad (6.130)$$

If $I$ and $I_4$ are eliminated from these three equations, it is found that

$$-I_3 = I_1 + I_2 \qquad (6.131)$$

Equations (6.124) through (6.126) and (6.131) are identical to Eqs. (6.3) through (6.6) developed for the center-tapped transformer (with $N = 1$), so the results developed for that circuit apply here also. In particular, the ports of the center-tapped transformer will be biconjugate if

$$Z_1 = Z_2 \qquad (6.132)$$

and $$Z_3 = Z_4 \qquad (6.133)$$

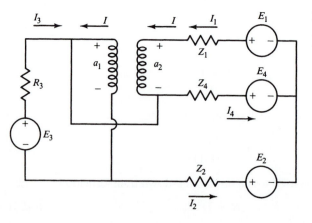

**FIGURE 6.16**
A transmission-line hybrid transformer.

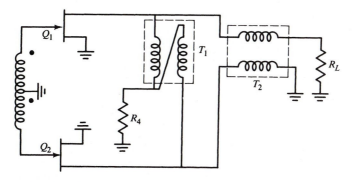

**FIGURE 6.17**
A class B push-pull amplifier incorporating transmission-line transformers in the output stage.

The transformer will be a hybrid transformer provided the additional relation

$$Z_3 = \frac{Z_1^*}{2} \tag{6.134}$$

is satisfied.

## A Power Output Stage

Figure 6.17 illustrates a method for combining the outputs of a class B push-pull amplifier to drive an unbalanced load $R_L$. Two transmission-line transformers are used. The circuit operation can be easily understood by considering the transmission-line transformers one at a time. If the upper transistor $Q_1$ is on and the bottom transistor is off, the circuit can be drawn as shown in Fig. 6.18. (Ignore for the moment that the load is not balanced.) If the current through $R_L$ is $I$, then the current supplied by $E_1$ must be $2I$. If $E_1$ is off (open) and $E_2(Q_2)$ is on, then half the current supplied

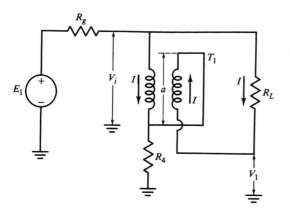

**FIGURE 6.18**
An equivalent circuit for analyzing transformer $T_1$.

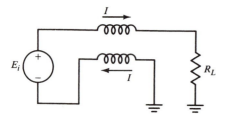

**FIGURE 6.19**

A balanced-to-unbalanced (BALUN) transformer.

by $E_2$ will pass through $R_L$ (but in the direction opposite to that supplied by $E_1$). The transformer $T_1$ serves to combine the two class B push-pull outputs.

The current in both transformer windings is $I$, so the current through $R_L$ is also $I$, and the current supplied by the generator is $2I$. The input impedance of the transformer circuit (Fig. 6.18) is $Z_i = V_i/(2I)$. The voltage at the bottom of $R_L$ is $V_1$, and the voltage drop across $R_4$ is $V_1 + a$, where $a$ is the voltage across the transformer. Then $V_i = V_1 + 2a$ and

$$Z_i = \frac{V_i}{2I} = \frac{V_1 + a}{2I} + \frac{a}{2I} = R_4 + \frac{R_L}{4} \tag{6.135}$$

since $R_L = 2a/I$.

This is the impedance seen by the amplifier with output impedance $R_g$. The remaining problem is that one side of the load in Fig. 6.18 is not grounded. This is solved by using a transmission-line transformer, as shown in Fig. 6.19, to convert from an unbalanced to a balanced load. It is easily shown that the impedance seen by the source $E_i$ is $R_L$. Therefore, the transformer $T_1$ combines the two transistor outputs and serves as an impedance matcher. The transformer $T_2$ converts the balanced output to an unbalanced load.

Transmission-line transformers are relatively broadband and have very effective high-frequency performance (to several hundred megahertz). However, they do not have as good a low-frequency performance as conventional transformers and can only realize the turns ratio of $n + 1$, where $n$ is an integer.

## 6.4

### PROBLEMS

**6.1** Derive the relation between $Z_1$ and $Z_2$ for port 3 of Fig. 6.2 to be isolated from port 4.

**6.2** Show that the equations

$$Z_1 = Z_2 = Z \quad \text{and} \quad Z_3 = \frac{Z_4}{N^2} = \frac{Z^*}{2N^2}$$

will result in maximum power transfer from ports 2 and 4 of the circuit shown in Fig. 6.2.

**6.3** If the applied voltage $E_i = 1$ V rms in the circuit shown in Fig. P6.3, what is the output power?

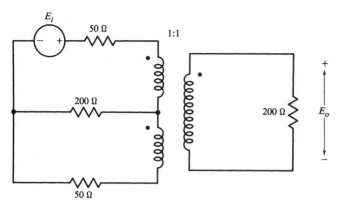

**FIGURE P6.3**
A three-winding transformer circuit.

**6.4** Show that if $R_1 = kR_2$, then port 4 is isolated form port 3 in the transformer illustrated in Fig. 6.5.

**6.5** Derive an equation for the current through the detector $D$ of the bridge circuit in Fig. 6.3 (as a function of $R_1$ and $R_2$) if the detector has zero input impedance.

**6.6** Derive the necessary conditions for ports 3 and 4 of a nonideal transformer to be conjugate.

**6.7** Design a transformer circuit for taking the difference of two signals. The load impedance as well as each source impedance is to be 50 $\Omega$. What is the minimum value of primary inductance which should be used at 20 MHz?

**6.8** An autotransformer with a common ground for all ports is shown in Fig. P6.8. Determine the necessary relationships between the resistances in order for the transformer to be a hybrid transformer. Assume the transformer is ideal with no mutual coupling between turns.

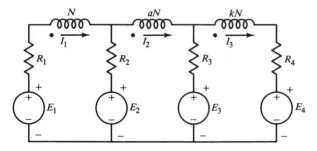

**FIGURE P6.8**
A three-winding autotransformer that can be used as a
hybrid transformer.

**6.9** Derive the relation required for the characteristic line impedance in order to obtain the maximum power transformer relation [Eq. (6.93)].

**6.10** Derive Eq. (6.95).

**6.11** Verify Eq. (6.100).

**6.12** Figure P6.12 illustrates a circuit in which neither side of the load is grounded. Determine the value of $R_L$ for maximum power transfer.

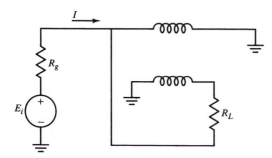

**FIGURE P6.12**
An unbalanced-to-balanced transformer circuit.

**6.13** For the hybrid transformer illustrated in Fig. 6.5 ($k = 1$), determine the responses at ports 1 and 2 to signals applied at ports 3 and 4. Compare the phases of the output signals with those obtained at ports 3 and 4 from signals applied at ports 1 and 2.

**6.14** Figure P6.14 illustrates a transmission-line transformer circuit that can be used with balanced loads. What should $R_L$ be for maximum power transfer?

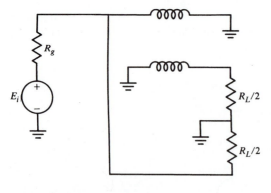

**FIGURE P6.14**
A transformer circuit with a balanced load.

**6.15** Design a transmission-line transformer circuit with an input impedance

$$Z_i = \frac{R_L}{9}$$

**6.16** Derive the relation among $R_1$, $R_2$, $R_3$, and $R_4$ required for the circuit shown in Fig. P6.16 to be a hybrid transformer. Assume the transmission line is ideal.

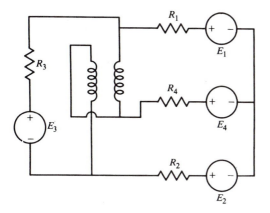

**FIGURE P6.16**
A transmission-line transformer circuit that can realize a hybrid transformer.

**6.17** Design a transmission-line transformer circuit for combining the outputs of two class A amplifiers operating in push-pull. The load is 50 Ω unbalanced.

**6.18** What is the input impedance of the transmission-line transformer circuit shown in Fig. P6.18?

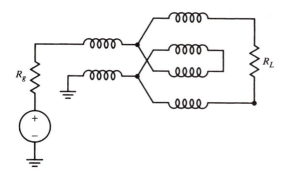

**FIGURE P6.18**
A circuit incorporating three transmission-line transformers.

**6.19** A class B push-pull amplifier requires that the load seen by each transistor be 6 Ω. The load is 50 Ω unbalanced. Design a transmission-line coupling circuit which results in maximum output from the transistors. What should the characteristic impedance of each transmission line be?

**\*6.20** Design a three winding transformer for splitting a signal from a 50 ohm source equally between two 50 ohm loads. Assume the transformer coefficient of coupling is .95. Select the inductors so that the lower cutoff frequency is approximately 100 Hz. Verify your design using computer simulation.

## ▩ 6.5
### REFERENCES

1. K. McIlwain and J. G. Brainerd, *High-Frequency Alternating Currents,* 2d ed., Wiley, New York, 1939.
2. C. L. Ruthroff, Some Broadband Transformers, *Proc. IRE,* **47:** 1337–1342 (1959).
3. H. H. Skilling, *Electric Transmission Lines,* McGraw-Hill, New York, 1951.
4. H. L. Krauss and C. W. Allen, Designing Toroidal Transformers to Optimize Wideband Performance, *Electronics,* Aug. 16, 1973, pp. 113–116.

## ▩ 6.6
### ADDITIONAL READING

Guanella, G.: New Method of Impedance Matching in Radio Frequency Circuits, *Brown-Boveri Rev.,* **31:** 327–329 (1944).

Nagle, J. J.: Use of Wideband Autotransformers in RF Systems, *Electronic Design,* Feb. 2, 1976, pp. 64–70.

Pitzalis, O., and T. Couse: Broadband Transformer Design for Transistor Power Amplifiers, *USAECOM Tech. Rep.* 2989, July 1968 (AD 676–816).

———— and ———— : Practical Design Information for Broadband Transmission Line Transformers, *Proc. IEEE,* **56:** 738–739 (1968).

Satori, E. F.: Hybrid Transformers, *IEEE Trans. Parts, Materials and Packaging,* **4:** 59–66 (1968).

Schlicke, H. M.: *Essentials of Dielectromagnetic Engineering,* Wiley, New York, 1961.

# Oscillators

## 7.1

### INTRODUCTION

An *electronic oscillator* is a circuit with a periodic output signal but with no periodic input signal. It functions by converting dc power to a periodic output signal (ac power). A *harmonic oscillator* is one in which the output signal is approximately sinusoidal. If a crystal resonator is used in the circuit to closely control the oscillating frequency, the oscillator is referred to as a *crystal-controlled oscillator.* Modern communication systems often contain several oscillators, including a crystal-controlled reference oscillator, a voltage-controlled oscillator (VCO), and a voltage-controlled crystal oscillator (VCXO). There exist many integrated circuits that can be used for generating a periodic output signal, but oscillator design is one area in which discrete transistors have a marked advantage. Communication circuits require high-frequency, low-noise oscillators, and integrated-circuit components often cannot meet the noise and frequency specifications.

In this chapter, these various oscillators and three different methods of oscillator analysis are described. First, two mathematical approaches based on linear feedback theory and circuit analysis will be presented. Subsequently, a less rigorous approach will be presented, one which provides additional insight into the design process. After the conditions for oscillations are derived, we will discuss the crystal resonator. It will be shown that the crystal serves in oscillators as a very narrowband circuit and that the design of crystal-controlled oscillators is essentially the same as that of oscillators that do not contain a crystal.

## 7.2

### CONDITIONS FOR OSCILLATION

Electronic oscillator circuits are feedback networks, and the extensive results of linear feedback analysis can be used for oscillator analysis and design. Oscillators are inherently a nonlinear circuit, but linear analysis techniques are the most useful for analysis and design. They provide accurate information for predicting the frequency of oscillation but have limited use for predicting the amplitude of oscillation. In this section we will consider the most widely developed method of linear feedback analysis, basing our discussion on interpretation of the block diagram model.

### Nyquist Stability Criteria

Figure 7.1 shows, in block diagram form, the necessary components of an oscillator. It contains an amplifier with frequency-dependent forward loop gain $G(j\omega)$ and a frequency-dependent feedback network $H(j\omega)$. The output voltage is given by

$$V_o = \frac{V_i G(j\omega)}{1 + G(j\omega)H(j\omega)}$$

For an oscillator, the output $V_o$ is nonzero even if the input signal $V_i$ is zero. This can be possible only if the forward loop gain is infinite (which is not practical) or if the denominator is

$$1 + G(j\omega)H(j\omega) = 0$$

at some frequency $\omega_o$. This leads to the well-known condition for oscillation—the Nyquist criterion: At some frequency $\omega_o$

$$G(j\omega_o)H(j\omega_o) = -1 \tag{7.1}$$

That is, the magnitude of the open-loop transfer function is equal to 1, or

$$|G(j\omega_o)H(j\omega_o)| = 1 \tag{7.2}$$

and the phase shift is $180°$, or

$$\arg G(j\omega_o)H(j\omega_o) = 180° \tag{7.3}$$

This can be expressed more simply as follows: If in a negative-feedback system, the open-loop gain has a total phase shift of $180°$ at some frequency $\omega_o$, the system will oscillate at that frequency, providing the open-loop gain is unity. If the gain is less than unity at the frequency where the phase shift is $180°$, the system will be stable; if the gain is greater than unity, the system will be unstable. If positive feedback is used, the loop phase shift must be $0°$. That is, $\arg G(j\omega_o)H(j\omega_o) = 0°$.

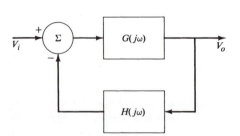

**FIGURE 7.1**
Block diagram representation of a linear negative-feedback system.

The preceding expressions for oscillation are not precise for some complicated systems, but they are correct for those transfer functions normally encountered in oscillator design. The conditions for stability are also known as the *Barkhausen criteria,* which state that if the closed-loop transfer function is

$$\frac{V_o}{V_i} = \frac{\mu}{1 - \mu\beta} \tag{7.4}$$

then the system will oscillate, provided $\mu\beta = 1$. This is equivalent to the Nyquist criterion, the difference being that the transfer function is written for a loop with positive feedback. Both versions state that the total phase shift around the loop must be 360° at the frequency of oscillation, and the magnitude of the open-loop gain must be unity at that frequency.

If a single-stage common-emitter (or common-source) amplifier is used with feedback from collector to base, as illustrated in Fig. 7.2a, then the feedback network must supply 180° phase shift since there is 180° phase shift between the base and collector signals. If a common-base (or common-gate) amplifier (illustrated in Fig. 7.2b) is used, there is no phase shift between the emitter and collector signals; therefore, a necessary condition for oscillation is that there be no phase shift between the input and output of the feedback network. If a small phase shift occurs in the forward loop, this must be compensated for by an equal and opposite phase shift in the feedback network.

An amplifier with feedback but without a frequency-sensitive network can be made to oscillate. However, the frequency of oscillation will be difficult to control. Since the primary purpose of the feedback network is to control the frequency of oscillation, the network is designed so that the Nyquist criteria are satisfied at only a single frequency.

The following analysis of the relatively simple circuit shown in Fig. 7.3 illustrates the method of determining the conditions for oscillation. The linearized (and simplified) equivalent circuit of Fig. 7.3 is given in Fig. 7.4a (with the feedback loop opened), in which the transistor output resistance $r_o$ is ignored, as is the large

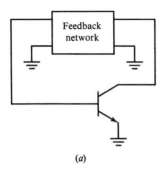
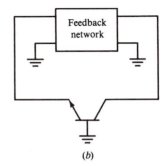

(a)                                          (b)

**FIGURE 7.2**
(a) A system in which the feedback network must provide 180° phase shift in order for oscillations to occur; (b) a system in which the feedback network must provide 0° phase shift for oscillations to occur.

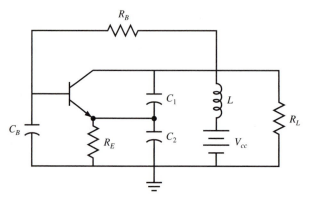

**FIGURE 7.3**
A grounded-base oscillator circuit.

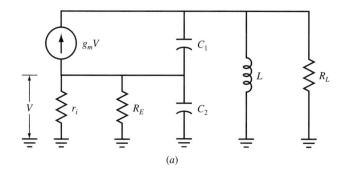

(a)

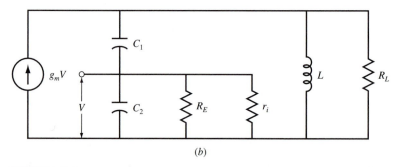

(b)

**FIGURE 7.4**
(a) A small-signal equivalent circuit of the grounded-base oscillator; (b) the circuit of Fig. 7.4a, but with the feedback path opened and terminated in the correct resistance.

biasing resistor $R_B$. Also, the capacitor connected to the base is assumed to be so large that the base is at ground potential for the small-signal analysis. Note that the transistor is connected in the common-base configuration and has no voltage phase inversion (i.e., the feedback is positive). The conditions for oscillation are

$$|G(j\omega_o)||H(j\omega_o)| = 1$$

and because the amplifier is noninverting, the phase shift of the network must also be arg $G(j\omega_o)H(j\omega_o) = 0°$.

The loop gain is calculated by opening the feedback loop, applying a signal, and measuring the return difference. It is necessary that when the loop is opened, the impedance seen at any point be the same as it is with the loop closed. In this case it is convenient to open the loop at the transistor emitter. The impedance shunting the capacitor $C_2$ is the resistor $R_E$ in parallel with the input impedance $r_i$ of the common-base amplifier. In the discussion of the common-base amplifier (Chap. 2), it was shown that the input resistance of the common-base circuit is

$$r_i = \frac{r_\pi}{\beta}$$

so the equivalent circuit of Fig. 7.4a can be redrawn as shown in Fig. 7.4b. The circuit analysis can be greatly simplified by assuming that

$$\{[\omega(C_2 + C_1)]^2\}^{-1} \ll \left(\frac{r_i R_E}{r_i + R_E}\right)^2$$

and that the $Q$ of the load impedance is high. In this case the circuit (see results for capacitive transformers in Chap. 4) reduces to that of Fig. 7.5, where the resistor shunting $C_2$ has been replaced by an equivalent resistance shunting $C_1$ and $C_2$. The circuit shown in Fig. 7.5 is not the small-signal equivalent of Fig. 7.3; rather it is a small-signal equivalent of the open-loop circuit with the loop or feedback network opened at the emitter and the circuit terminated in the correct impedance. The network could be opened at other points, but opening it at the emitter is very convenient because it is relatively easy to determine the terminating impedance at this point.

The feedback voltage is given by

$$V = \frac{V_o C_1}{C_1 + C_2}$$

and the equivalent resistance reflected across the coil is

$$R_{eq} = \frac{r_i R_E}{r_i + R_E}\left(\frac{C_1 + C_2}{C_1}\right)^2 \tag{7.5}$$

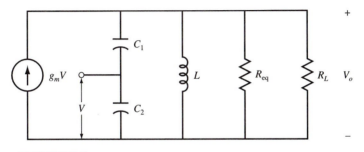

**FIGURE 7.5**
A circuit equivalent to Fig. 7.4b, provided $[\omega(C_1 + C_2)R]^2 \gg 1$.

The forward loop gain is

$$\frac{V_o}{V} = G(j\omega) = g_m Z_L \tag{7.6}$$

where

$$Y_L = Z_L^{-1} = (j\omega L)^{-1} + R_{eq}^{-1} + R_L^{-1} + j\omega C \tag{7.7}$$

and

$$C = \frac{C_1 C_2}{C_1 + C_2}$$

The feedback ratio is

$$\frac{V}{V_o} = H(j\omega) = \frac{C_1}{C_1 + C_2} \tag{7.8}$$

A necessary condition for oscillation is that

$$\arg G(j\omega)H(j\omega) = 0° \tag{7.9}$$

Since $H$ does not depend on frequency in this example, if arg $GH$ is to be zero, the phase shift of the load impedance $Z_L$ must be zero. This occurs only at the resonant frequency of the circuit where

$$\omega_o \left[ \left( L \frac{C_1 C_2}{C_1 + C_2} \right)^{1/2} \right]^{-1} \tag{7.10}$$

At this frequency

$$Z_L = \frac{R_{eq} R_L}{R_{eq} + R_L} \tag{7.11}$$

and

$$G(j\omega)H(j\omega) = g_m \frac{R_{eq} R_L}{R_{eq} + R_L} \frac{C_1}{C_1 + C_2} \tag{7.12}$$

The other condition for oscillation is the magnitude constraint that

$$|G(j\omega)H(j\omega)| = g_m \frac{R_{eq} R_L}{R_{eq} + R_L} \frac{C_1}{C_1 + C_2} = 1 \tag{7.13}$$

**EXAMPLE 7.1.** The preceding results will now be used to design a 20-MHz common-base sinusoidal oscillator using a transistor with a minimum $\beta$ of 100.

**Solution.** Oscillator design consists of a trial-and-error procedure using Eqs. (7.5) through (7.13). As a starting point we will assume a bias current $I_c = 1$ mA; then the common-base input resistance is

$$r_i = g_m^{-1} = \frac{V_T}{I_c} = 26 \ \Omega$$

Since this is so small, we can safely assume that the emitter bias resistor $R_E$ is much larger than $r_i$, so [from Eq. (7.5)]

$$R_{eq} \approx r_i \left( \frac{C_1 + C_2}{C_1} \right)^2$$

Equations (7.5) through (7.13) were based on the assumption that

$$\{[\omega(C_2 + C_1)]^2\}^{-1} \ll r_i^2$$

If $(\omega C_2)^{-1} \approx 8$,

then

$$(\omega C_2)^{-1} < \frac{r_i}{3}$$

and the above inequality is satisfied so the assumption was justified.

Practically, a factor of 10 difference will usually satisfy the $\ll$ inequality. In addition, for oscillations to occur, the loop gain [Eq. (7.12)] must be at least 1. In oscillator design the loop gain is usually selected to be about 3, which allows for some error in the approximation. With a loop gain greater than 1, the system is unstable and the oscillations increase in amplitude until the transistor current begins to saturate. When this occurs, the $\beta$ of the transistor is reduced, and thus $g_m$ is also reduced. This reduces the loop gain and stabilizes the amplitude of oscillation. In this example the loop gain is

$$|GH| \approx g_m r_i \frac{C_1 + C_2}{C_1} = \frac{C_1 + C_2}{C_1} = 3$$

So $C_1 = C_2/2 = 500$ pF. The value of the inductance is found from Eq. (7.10) to be $L = 0.19\ \mu\text{H}$. Also, $R_{eq} = 234\ \Omega$. A load resistor $R_L$ can be shunted across the inductor without affecting the calculations if it is much bigger than $R_{eq}$. In this case a load resistance of 1500 $\Omega$ could be safely added. The complete design would require selecting a supply voltage and bias resistors so that the quiescent collector current is 1 mA. A completed circuit schematic is given in Fig. 7.6.

The preceding oscillator analysis was based on a linear model for the circuit. In practice, the analysis provides a good approximation for the frequency of oscillation and the minimum gain required, but it provides no information about the amplitude of oscillation. The system is initially unstable, and as the output amplitude increases, the transistor begins to saturate, resulting in a reduction of loop gain and a stable oscillating amplitude. Such oscillators are referred to as *self-limiting* to distinguish them from oscillators that use an external means of regulating the oscillation amplitude.

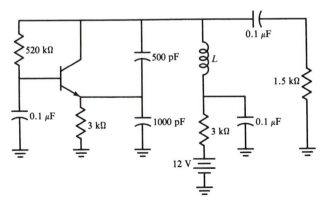

**FIGURE 7.6**
The oscillator circuit designed in Example 7.1.

## Circuit Analysis

Identifying the open-loop gain is particularly useful for oscillator analysis and design. Further illustrations of this technique are given throughout this chapter. There are circuits in which it is difficult to identify the open-loop gain. For this reason two additional methods of oscillator design and analysis will be presented. A direct analysis of the circuit equations is frequently simpler and more informative than the block diagram interpretation (particularly for single-stage amplifiers). Figure 7.7 illustrates a transistor amplifier with three external impedances connected. Terminal 1 represents the base terminal of a bipolar transistor or the gate terminal of a field-effect transistor, and terminal 3 represents the emitter or source terminal. The three external impedances, then, represent feedback connections between the transistor terminals. The small-signal equivalent circuit is given in Fig. 7.8. We will assume that the transistor output impedance is sufficiently large, and it will be neglected in the following analysis. Note that at this point the ground node for the circuit has not been identified. It will be subsequently shown that many different oscillator configurations can be realized for different grounds points.

The loop equations are then

$$V_i = I_1(Z_2 + Z_3) + V + g_m V Z_2 \tag{7.14}$$

and 
$$V = \frac{I_1 Z_1 Z_i}{Z_1 + Z_i} \tag{7.15}$$

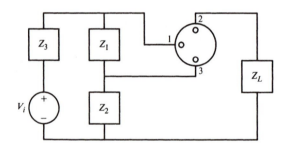

**FIGURE 7.7**
A generalized transistor oscillator circuit.

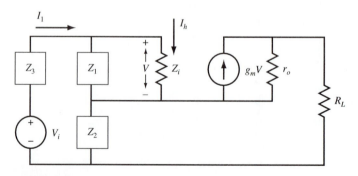

**FIGURE 7.8**
A small-signal equivalent circuit of Fig. 7.7.

For the amplifier to oscillate, currents $I_b$ and $I_1$ must be nonzero, even when $V_i = 0$. This is only possible if the system determinant $\Delta$

$$\Delta = \begin{vmatrix} Z_2 + Z_3 & 1 + g_m Z_2 \\ Z_1 & -\left(1 + \dfrac{Z_1}{Z_i}\right) \end{vmatrix} \tag{7.16}$$

is equal to 0. That is,

$$Z_1 + Z_2 + Z_3 + \frac{Z_1}{Z_i}(Z_2 + Z_3) + g_m Z_1 Z_2 = 0 \tag{7.17}$$

For the bipolar junction transistor (BJT) this reduces to

$$(Z_1 + Z_2 + Z_3)Z_i + Z_1 Z_2 \beta + Z_1(Z_2 + Z_3) = 0 \tag{7.18}$$

Only the case in which the transistor input impedance $Z_i$ is real will be considered here $(Z_i = r_\pi)$. The more complicated case in which $Z_i$ is complex can be analyzed in the same manner. Normally $Z_i$ will have a parallel capacitive component that can be included in $Z_1$. Assume for the moment that $Z_1$, $Z_2$, and $Z_3$ are purely reactive impedances. [It is easily seen that Eq. (7.18) does not have a solution if all three impedances are real and positive.] Since both the real and imaginary parts must be zero, Eq. (7.18) (with $Z_i = r_\pi$) is equivalent to the two equations

$$r_\pi(Z_1 + Z_2 + Z_3) = 0 \tag{7.19}$$

and
$$Z_1[(1 + \beta)(Z_2 + Z_3)] = 0 \tag{7.20}$$

Since $\beta$ is real and positive, $Z_2$ and $Z_3$ must be of opposite signs for Eq. (7.20) to hold. That is,

$$(1 + \beta)Z_2 = -Z_3 \tag{7.21}$$

Therefore, since $r_\pi$ is nonzero, Eq. (7.19) reduces to

$$Z_1 + Z_2 - (1 + \beta)Z_2 = 0 \tag{7.22}$$

or
$$Z_1 = \beta Z_2$$

Since $\beta$ is positive, $Z_1$ and $Z_2$ will be reactances of the same kind. If $Z_1$ and $Z_2$ are capacitors, then $Z_3$ must be an inductor. An example of such a circuit with the ground located at the junction between $Z_2$ and $Z_3$ is shown in Fig. 7.9. This circuit is referred to as a *Colpitts oscillator,* named after the man who first published such a circuit for a vacuum tube oscillator. If the emitter is grounded, the oscillator will be as shown in Fig. 7.10. This circuit has become known as a *Pierce oscillator*. If $Z_1$ and $Z_2$ are inductors and $Z_3$ is a capacitor, the circuit is known as a *Hartley oscillator.* A common-base Hartley oscillator is illustrated is Fig. 7.11.

Equations (7.21) and (7.22) describe the necessary conditions for oscillation when the components external to the transistor are assumed to be ideal. These equations should be used with caution because they apply only to the ideal case. It appears that the conditions for oscillation are independent of the transistor input impedance. This is because the three external impedances are assumed to be purely

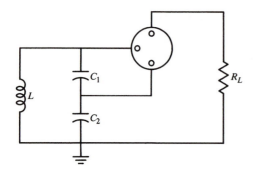

**FIGURE 7.9**
A Colpitts oscillator.

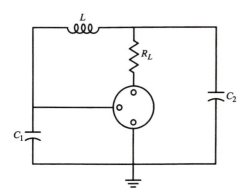

**FIGURE 7.10**
Another Colpitts oscillator, also known as
a Pierce oscillator.

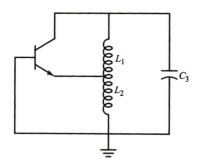

**FIGURE 7.11**
A Hartley oscillator.

reactive. In this case, the circuit will oscillate even without the transistor. The transistor is needed in the practical situation to supply the energy dissipated into a nonideal inductor and capacitor, to provide a means for amplitude limiting, and to furnish the energy that the oscillator must supply to the external circuitry. The preceding analysis will now be applied to the case of the nonideal component.

If $Z_1$, $Z_2$, $Z_3$, or $Z_i$ is complex, the preceding analysis is more complicated, but the conditions for oscillation can still be obtained from Eq. (7.18). For example, if in the Colpitts circuit, there is a resistor $r$ in series with $L(Z_3 = r + j\omega L)$, Eq. (7.18) reduces to the two equations

$$r_\pi[\omega L - (\omega C_1)^{-1} - (\omega C_2)^{-1}] - \frac{r}{\omega C_1} = 0 \qquad (7.23)$$

and
$$r_\pi r - \frac{1+\beta}{\omega^2 C_1 C_2} + (\omega C_1)^{-1} \omega L = 0 \tag{7.24}$$

If we define for notational simplicity

$$C_1' = \frac{C_1}{1 + r/r_\pi} \tag{7.25}$$

then the resonant frequency at which oscillations will occur is found from Eq. (7.23) to be

$$\omega_o = \left\{ \left[ L \left( \frac{C_1' C_2}{C_1' + C_2} \right) \right]^{1/2} \right\}^{-1} \tag{7.26}$$

The resistance in series with the inductor changes the resonant frequency, but since $r$ is normally much less than $r_i$, the change is small. For oscillations to occur, Eq. (7.24) must also be satisfied. That is,

$$rr_\pi = \frac{1+\beta}{\omega_o^2 C_1 C_2} - \frac{L}{C_1} \tag{7.27}$$

It can be shown that if the product $rr_\pi$ is greater than that given by Eq. (7.27), oscillations will die out. If $rr_\pi$ is smaller than that given by Eq. (7.27), the system will be unstable and oscillations will increase in amplitude. Equation (7.27) illustrates the importance of the transistor current gain $\beta$. As $r$ is increased, the current gain must be increased to sustain oscillations. Also, it becomes increasingly difficult to satisfy Eq. (7.27) as the frequency is increased. In general, it is advantageous to have

$$X_{C_1} X_{C_2} = (\omega^2 C_1 C_2)^{-1}$$

as large as possible since then $r$ can be sufficiently large. However, if $C_1$ and $C_2$ are too small (large $X_{C_1}$ and $X_{C_2}$), then the transistor input and output capacitors which shunt $C_1$ and $C_2$, respectively, become important. A good, stable design will always have $C_1$ and $C_2$ much larger than the transistor capacitances they shunt.

**EXAMPLE 7.2.** Design a 5-MHz Colpitts oscillator using a 10-$\mu$H inductor with an unloaded $Q_u$ of 100. The transistor $\beta$ is 100.

*Solution.* From Eq. (7.26) the equivalent capacitance is

$$\frac{C_1' C_2}{C_1' + C_2} = 100 \text{ pF}$$

We will assume that $C_1'$ and $C_1$ are equal; then the maximum $r_\pi$ is found by using Eq. (7.27). The series resistance of the inductor is

$$r = 3.14 \ \Omega$$

One possible solution is to select $C_1 = C_2 = 200$ pF; then from Eq. (7.27),

$$3.14 \, r_\pi = 2.53 \times 10^6 - 5 \times 10^4$$

or
$$r_\pi = 0.81 \times 10^6 \ \Omega$$

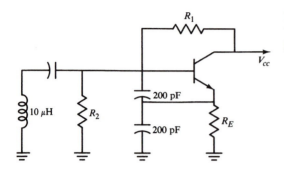

**FIGURE 7.12**
The oscillator circuit designed in
Example 7.2.

Since the transistor transconductance $g_m = 40I$, the transistor must be biased so that

$$I \geq 3.09 \times 10^{-6} \text{ A}$$

A complete circuit is shown in Fig. 7.12. The resistors must be selected so that the bias current exceeds the minimum specified. Note that the assumption that $r$ is much less than $r_\pi$ is justified in this example.

**EXAMPLE 7.3.** In the Colpitts oscillator circuit designed in Example 7.2, what will the frequency of oscillation be if the bias current is increased so that $r_\pi = 1000 \, \Omega$?

*Solution.* From Eq. (7.25)

$$C_1' = 199 \text{ pF}$$

so the increased bias current will have a negligible effect and can be ignored.

### Another Interpretation of the Oscillator Circuit

Although Eqs. (7.19) and (7.20) can be used to determine the exact expressions for oscillation, they are often difficult to use and add little insight to the design process. An alternative interpretation, originally presented by Gouriet[1] for vacuum tube oscillators, will now be presented. It is based on the fact that an ideal tuned circuit (infinite $Q$), once excited, will oscillate indefinitely because there is no resistance element present to dissipate the energy. In the actual case where the inductor $Q$ is finite, the oscillations die out because energy is dissipated in the resistance.

It is the function of the amplifier to maintain oscillations by supplying an amount of energy equal to that dissipated. This source of energy can be interpreted as a negative resistor $r_i$ in series with the tuned circuit, as shown in Fig. 7.13. If the

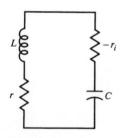

**FIGURE 7.13**
A resonant circuit including a negative resistor.

total resistance is positive, the oscillations will die out, while the oscillation amplitude will increase if the total resistance is negative. To maintain oscillations, the two resistors must be of equal magnitude. To see how a negative resistance is realized, the input impedance of the circuit in Fig. 7.14 will be derived.

If the transistor output impedance is sufficiently large, the equivalent circuit is as shown in Fig. 7.15. The steady-state loop equations are

$$V_i = I_i(X_{C_1} + X_{C_2}) - I_b(X_{C_1} - \beta X_{C_2}) \tag{7.28}$$

$$0 = -I_i(X_{C_1}) + I_b(X_{C_1} + r_\pi) \tag{7.29}$$

After $I_b$ is eliminated from these two equations, $Z_i$ is obtained as

$$Z_i = \frac{V_i}{I_i} = \frac{(1+\beta)X_{C_1}X_{C_2} + r_\pi(X_{C_1} + X_{C_2})}{X_{C_1} + r_\pi} \tag{7.30}$$

If $X_{C_1} \ll r_\pi$, the input impedance is approximately equal to

$$Z_i \approx \frac{1+\beta}{r_\pi}X_{C_1}X_{C_2} + X_{C_1} + X_{C_2} \tag{7.31}$$

$$\approx \frac{-g_m}{\omega^2 C_1 C_2} + \left[ j\omega \left( \frac{C_1 C_2}{C_1 + C_2} \right) \right]^{-1} \tag{7.32}$$

That is, the input impedance of the circuit shown in Fig. 7.14 is a negative resistor

$$r_i = \frac{-g_m}{\omega^2 C_1 C_2} \tag{7.33}$$

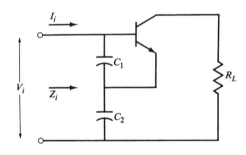

**FIGURE 7.14**
Circuit for generating a negative resistance.

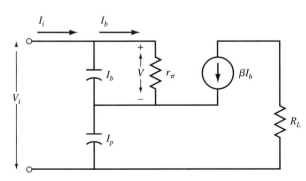

**FIGURE 7.15**
The small-signal equivalent circuit of Fig. 7.14 with the transistor output impedance neglected.

in series with a capacitor

$$C_i = \frac{C_1 C_2}{C_1 + C_2} \tag{7.34}$$

which is the series combination of the two capacitors. With an inductor $L$ (with series resistance $r$) connected across the input, it is clear that the condition for sustained oscillation is

$$r = \frac{g_m}{\omega^2 C_1 C_2} \tag{7.35}$$

and the frequency of oscillation is

$$f_o = \left[ 2\pi \left( L \frac{C_1 C_2}{C_1 + C_2} \right)^{1/2} \right]^{-1} \tag{7.36}$$

This interpretation of the oscillator readily provides several guidelines that can be used in the design. First, $C_1$ should be as large as possible so that $X_{C_1} \ll r_\pi$. Also, $C_1$ and $C_2$ should be much larger than the transistor output capacitances so that the transistor base-to-emitter and collector-to-emitter capacitances have a negligible effect on the circuit's performance. However, Eq. (7.35) limits the maximum value of the capacitances, since

$$r \le \frac{g_m}{\omega^2 C_1 C_2} \le \frac{G}{\omega^2 C_1 C_2} \tag{7.37}$$

where $G$ is the maximum value of $g_m$. For a given series capacitance the product is a minimum when $C_1 = C_2 = C_m$. Then Eq. (7.37) can be written as

$$(\omega C_m)^{-1} > \left( \frac{r}{g_m} \right)^{1/2} \tag{7.38}$$

This equation states that for oscillations to be maintained, the minimum permissible reactance $(\omega C_m)^{-1}$ is a function of the resistance of the inductor and the transistor's mutual conductance $g_m$. If the two capacitors are not equal, then Eq. (7.37) must be satisfied.

The analytical approach of calculating the input impedance seen from the inductor (or capacitor in Hartley oscillators) can be used with all oscillators and is often the easiest way to analyze the circuit. The real part of the input impedance must be negative in order that the active device supply the energy dissipated in the inductor (or capacitor).

An oscillator circuit known as the *Clapp,* or *Clapp-Gouriet, circuit* is shown in Fig. 7.16. This oscillator is equivalent to the one just discussed, but it was the practical advantage of being able to provide another degree of design freedom by making $C_o$ much smaller than $C_1$ and $C_2$. It is possible to use $C_1$ and $C_2$ to satisfy the condition of Eq. (7.37) and then adjust $C_o$ for the desired frequency of oscillation $\omega_o$, which is determined from

$$\omega_o L - (\omega_o C_o)^{-1} - (\omega_o C_1)^{-1} - (\omega_o C_2)^{-1} = 0 \tag{7.39}$$

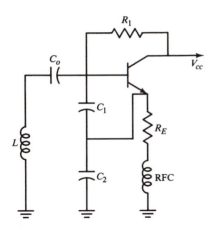

**FIGURE 7.16**
A Clapp-Gouriet oscillator.

The following example illustrates the design procedure.

EXAMPLE 7.4.  Consider the Clapp-Gouriet oscillator shown in Fig. 7.16. The transistor is operated at a $g_m$ of 6 mS. The coil used has an unloaded $Q_u$ of 200 at 1 MHz and a reactive impedance of 800 $\Omega$ ($r = 4\ \Omega$). What are the required conditions for the circuit to oscillate?

*Solution.* In order to satisfy Eq. (7.38), we must have

$$(\omega C_m)^{-1} \geq \left(\frac{r}{g_m}\right)^{1/2} = 25.8\ \Omega$$

Therefore, at 1 MHz

$$C_m = 6200\ \text{pF}$$

And $C_m$ corresponds to the case of maximum series capacitance of the parallel combination of $C_1$ and $C_2$ and occurs for $C_1 = C_2 = C_m$. If both $C_1$ and $C_2$ equal 6200 pF, the reactance of the series combination of $C_1$ and $C_2$ is 51.6 $\Omega$. Then $C_o$ must be selected so that $X_L = X_C$ at 1 MHz. If, for example, $L = 82\ \mu H$, then $X_L = 515\ \Omega$ and $C_o$ should be approximately 343 pF.

The Clapp-Gouriet oscillator is particularly effective at low frequencies. At higher frequencies $C_1$ and $C_2$ must be smaller in order that Eq. (7.37) be satisfied, and little fine-tuning advantage is obtained by adding $C_o$.

The gain requirement for oscillation [Eq. (7.35)] obtained using this model is somewhat different from that obtained with the matrix analysis [Eq. (7.27)]. However, it is readily shown that when

$$\frac{1+\beta}{\omega_o^2 C_2} \gg L$$

then Eq. (7.27) is approximated by

$$r \approx \frac{\beta}{r_\pi \omega_o^2 C_1 C_2} = \frac{g_m}{\omega_o^2 C_1 C_2}$$

Thus the two methods arrive at the same gain requirement for sustained oscillations. An exact expression for the required gain cannot be obtained with either model, since the actual oscillator analysis requires a nonlinear model; nor is an exact expression for the gain necessary. Oscillator design usually consists of realizing an open-loop gain 3 or 4 times larger than what the linear analysis predicts to be necessary for oscillations. This ensures that oscillations will begin. As the oscillation amplitude increases, the $g_m$ of the active device begins to decrease until a stable oscillation amplitude that is self-limiting is reached.

## The Pierce Oscillator

So far in our discussion we have considered two forms of the Colpitts oscillator. In the analysis of the generalized oscillator circuit, no terminal was designated as ground. There are often practical reasons for grounding a particular terminal. Figure 7.3 describes a form of the Colpitts oscillator in which the base is grounded. If the emitter is grounded at the oscillating frequency, the circuit appears as shown in Fig. 7.17 or 7.18. And $C_b$ and $C_E$ serve to short-circuit the dc bias resistors $R_b$ and $R_E$ at the oscillating frequency. This circuit, often referred to as a *Pierce oscillator,* has the beneficial feature that the bias resistors do not shunt the tuned circuit. Resistors connected across the tuned circuit have the effect of reducing the circuit $Q$, which reduces the frequency stability. The Pierce oscillator (grounded emitter) normally has the best frequency stability. (The common-base amplifier is most often used in high-frequency oscillators because the cutoff frequency of the common-base current gain is approximately $\beta$ times greater than that of the common-emitter or common-collector configuration.)

The Pierce oscillator can be analyzed using the small-signal equivalent circuit shown in Fig. 7.19. It is assumed that the transistor output impedance is much larger than $X_{C_1}$. We will analyze the circuit by calculating the loop gain $GH$.

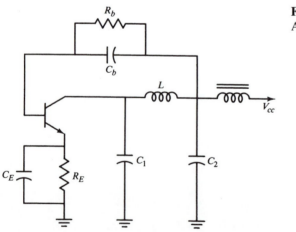

**FIGURE 7.17**
A Pierce oscillator.

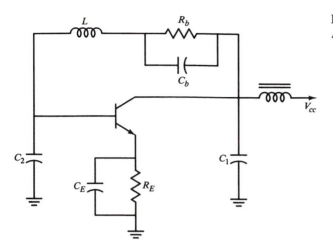

**FIGURE 7.18**
Another Pierce oscillator.

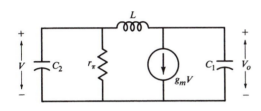

**FIGURE 7.19**
A small-signal equivalent circuit of the
Pierce oscillator.

The voltage

$$V_o = \frac{-g_m V[jX_L - jr_\pi X_{C_2}/(r_\pi - jX_{C_2})] - jX_{C_1}}{jX_L - jr_\pi X_{C_2}/(r_\pi - jX_{C_2}) - jX_{C_1}} \tag{7.40}$$

and the voltage fed back is

$$V = \frac{V_o[-jX_{C_2}r_\pi/(-jX_{C_2} + r_\pi)]}{-jX_{C_2}r_\pi/(-jX_{C_2} + r_\pi) + jX_L} \tag{7.41}$$

If $V$ is eliminated from these two equations, we obtain

$$V_o = \frac{-g_m V_o r_\pi}{(j\omega L j\omega C_1)(j\omega r_\pi C_2 + 1) + j\omega r_\pi (C_2 + C_1) + 1}$$

$$= \frac{-g_m V_o r_\pi}{1 - \omega^2 LC_1 + j\omega r_\pi (C_1 + C_2 - \omega^2 LC_1 C_2)} \tag{7.42}$$

In order for the loop phase shift to be $360°$, the imaginary term must vanish. Therefore, the frequency of oscillation is

$$\omega_o = \left(\frac{C_1 + C_2}{LC_1 C_2}\right)^{1/2} \tag{7.43}$$

and it is also required that $\omega_o^2 LC_1 > 1$ in order for the phase shift to be $360°$. That is [using Eq. (7.42)],

$$\frac{(C_1 + C_2)LC_1}{LC_1C_2} > 1$$

which reduces to

$$\frac{C_1 + C_2}{C_2} > 1 \qquad (7.44)$$

Since this is always the case, Eq. (7.43) determines the frequency of oscillation. For the open-loop gain $GH$ to be greater than 1 at the resonant frequency, the magnitude of Eq. (7.42) must be greater than 1 at the resonant frequency. That is,

$$g_m r_\pi = \beta > \omega^2 LC_1 - 1 = \frac{C_1 + C_2}{C_2} - 1 \qquad (7.45)$$

or

$$\beta C_2 > C_1 \qquad (7.46)$$

This is the same requirement for oscillation as was obtained using the generalized circuit analysis [Eq. (7.22)].

The type of analysis to use depends on the circuit configuration. Sometimes calculating the loop gain is the most convenient. This is often the case with FET oscillators, where the gate-to-source resistance can be neglected. The negative-impedance interpretation is usually the easiest to use and provides the most insight for design.

## ▪ 7.3

### AMPLITUDE STABILITY

Linearized analysis of oscillator circuits is convenient for determining the frequency of oscillation, but not for determining the amplitude of the oscillation. The Nyquist stability criterion states that the frequency of oscillation is the frequency at which the loop phase shift is $360°$, but it says nothing about the oscillation amplitude. If no procedures are taken to control the oscillation amplitude, it is susceptible to appreciable drift.

The two most frequently used methods for controlling the amplitude employ a self-limiting oscillator and an additional circuit or circuit element for amplitude regulation. The self-limiting oscillator is designed to be unstable; i.e., the loop gain is made greater than 1 at the frequency where the phase shift is $180°$. (Usually the loop gain is designed to be 2 or 3 times that needed for oscillation.) As the amplitude increases, the transistor begins to saturate, causing the loop gain to decrease until the amplitude stabilizes—this is a self-limiting oscillator. There are nonlinear analysis techniques for predicting the amplitude of oscillation, but the results are only approximate, except in special idealized cases, and the designer must rely on an empirical approach to establish the oscillation amplitude.

An example of a two-stage emitter-coupled oscillator is shown in Fig. 7.20. In this circuit, amplitude stabilization occurs as a result of current limiting in the

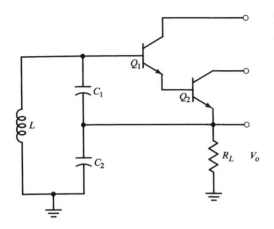

**FIGURE 7.20**
A two-stage emitter-coupled oscillator.

second stage.[2] This circuit has the additional advantage of output terminals that are isolated from the feedback path. The emitter of $Q_2$, which is rich in harmonics, is normally used for the output. Harmonics of the fundamental frequency can be obtained at the emitter of $Q_2$ by using an appropriately tuned circuit. Note that the collector of $Q_2$ is isolated from the feedback path.

## 7.4

### PHASE STABILITY

An oscillator has a frequency or phase stability that can be considered in two separate parts. First there is the long-term stability in which the frequency changes over a period of minutes, hours, days, weeks, or even years. This frequency stability is normally limited by the circuit component's temperature coefficients and aging rates. The other part, short-term frequency stability, is measured in seconds or even much shorter periods. One form of short-term instability is due to changes in phase of the system; here the term *phase stability* is used synonymously with *frequency stability*. It refers to how sensitive the frequency of oscillation is to small changes in phase shift of the open-loop system. It can be intuited that the system with the largest rate of change of phase as a function of frequency $(d\phi/df)$ will be the most stable in terms of frequency stability.

Figure 7.21 contains the phase plots of two open-loop systems used in oscillators. At the system crossover frequency, the phase shift is $-180°$ (with negative feedback). If now some external influence causes a change in phase, say, it adds $10°$ of phase lag, then the frequency will change so that the total phase shift is again $0°$. In this case the frequency will decrease to the point where the open-loop phase shift is $170°$. Figure 7.21 shows that $\Delta f_2$, the change in frequency associated with the $10°$ change in phase of $GH_2$, is greater than the change in frequency $\Delta f_1$, associated with open-loop system $GH_1$, whose phase is changing more rapidly near the open-loop crossover frequency.

This qualitative discussion illustrates that $d\phi/df|_{f=f_o}$ is a measure of an oscillator's phase stability. It provides a good means of quantitatively comparing the

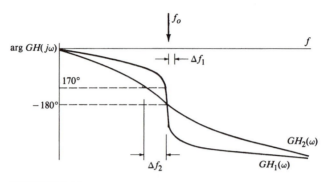

**FIGURE 7.21**
Phase plots of open-loop systems.

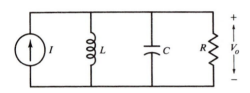

**FIGURE 7.22**
A parallel resonant circuit.

phase stability of two oscillators. Consider the simple parallel tuned circuit shown in Fig. 7.22. For the circuit the transimpedance [see Eq. (4.20)] is

$$\frac{V_o(j\omega)}{I(j\omega)} = \frac{R}{1 + jQ(\omega/\omega_o - \omega_o/\omega)}$$

where              $\omega_o = (LC)^{1/2}$      and      $Q = \dfrac{R}{\omega_o L}$

The circuit phase shift is

$$\arg \frac{V_o}{I} = \phi = -\tan^{-1} Q\left(\frac{\omega}{\omega_o} - \frac{\omega_o}{\omega}\right) \tag{7.47}$$

and the derivative with respect to frequency is

$$\frac{d\phi}{d\omega} = \frac{-Q^{-1}}{(Q^2)^{-1} + [(\omega_o^2 - \omega^2)/(\omega_o\omega)]^2} \frac{\omega^2 + \omega_o^2}{\omega_o\omega^2} \tag{7.48}$$

at the resonant frequency $\omega_o$

$$\left.\frac{d\phi}{d\omega}\right|_{\omega=\omega_o} = \frac{-2Q}{\omega_o} \tag{7.49}$$

The frequency stability factor $S_F$ is defined as the change in phase divided by the normalized change in frequency $\Delta\omega/\omega_o$. That is,

$$S_F = 2Q \tag{7.50}$$

where $S_F$ is a measure of the short-term stability of an oscillator. Equation (7.50) indicates that the higher the circuit $Q$, the higher the stability factor. This is one reason that high-$Q$ circuits are widely used in oscillator circuits. Another reason is that the tuned circuit can be used to filter out undesired harmonics and noise. A piezoelectric or ceramic crystal can function in an electronic circuit as an inductor with a very high $Q$.

## ■ 7.5

### CRYSTAL OSCILLATOR CHARACTERISTICS

The previous discussion has shown that a high-$Q$ circuit is desirable in an oscillator for short-term frequency stability. Piezoelectric (and ceramic) crystals are electromechanical devices that have very small dissipative losses and very high and stable electric circuit $Q$'s. For these reasons they are usually employed when an oscillator with a very stable operating frequency is desired.

Crystals are three-dimensional, mechanically oscillating bodies with many modes of oscillation. These oscillations are excited through the piezoelectric properties and by the arrangement and shape of the electrodes and the crystal itself. Through the fabrication of the crystal, one has the capability of selecting certain oscillating modes and harmonics. At the electrical terminals of the crystal, the observable equivalent circuit contains an infinite number of (lossy) series resonant circuits, all connected in parallel and all in parallel with a capacitance $C_o$, which represents the static capacitance of the electrode arrangement. This static capacitance represents the capacitance between the electrodes plus the capacitance of the leads and crystal holder.

An equivalent electric circuit of a crystal is shown in Fig. 7.23. The circuit contains several series resonant circuits whose frequencies are all approximate (but not exact) odd harmonics of the fundamental frequency $f_1$; the higher resonant frequencies are referred to as *overtones of the fundamental frequency*. In a narrow frequency region around any resonant frequency $f_i$, the circuit can be simplified to that shown in Fig. 7.24. This simplified model can be considered sufficiently accurate for oscillator design, but precautions are often necessary to prevent the circuit from oscillating at an unwanted resonant frequency.

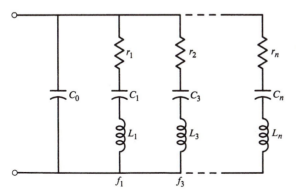

**FIGURE 7.23**
Electric circuit equivalent of a crystal.

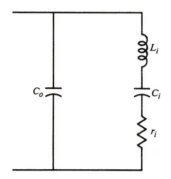

**FIGURE 7.24**
Electric circuit equivalent of a crystal, valid near the crystal's $i$th overtone frequency.

Quartz crystals have a $Q$ ranging from about 10,000 to over 1 million. Ceramic resonators can be fabricated with a $Q$ of several thousand. The inductance $L_1$ and capacitance $C_1$ of the equivalent circuit primarily depend on the mass and compliance of the quartz. The resistance $r_1$, which represents the losses of the circuit, is mainly attributable to damping resulting from the electrodes, crystal mounting structure, internal friction, and lead resistance. The circuit components are referred to as the *crystal motion elements,* since they are the electrical equivalents of the vibratory (mechanical) motion of the crystal. Typical circuit parameters for fundamental, third, and fifth overtone crystals are given in Table 7.1. Notice that at low frequency the series resistance increases exponentially. This is due to the crystal's edge effect, which rapidly decreases the $Q$ and therefore increases the series resistance. The $Q$ decreases with increasing frequency at higher frequencies, corresponding to an increase in series resistance.

Table 7.1 indicates that the dissipation resistance $r_1$ of a crystal is relatively small. Figure 7.25 illustrates how the resistance $r_1$ varies as a function of funda-

**TABLE 7.1**
**Typical crystal data**

| Frequency MHz | Mode | $R_1$, $\Omega$ | $C_1$, fF | $C_0$, pF |
|---|---|---|---|---|
| 1.0 | Fund. | 400 | 8 | 3.2 |
| 2.097 | Fund. | 270 | 10 | 4.3 |
| 5.7 | Fund. | 25 | 21 | 5.1 |
| 7.16 | Fund. | 30 | 29 | 6.4 |
| 8.5 | Fund. | 20 | 27 | 5.9 |
| 9.5 | Fund. | 30 | 27 | 5.5 |
| 20 | Fund. | 20 | 26 | 5.8 |
| 26 | 3 | 40 | 3 | 6.2 |
| 80 | 5 | 60 | 0.5 | 6.1 |
| 100 | 5 | 60 | 0.11 | 2.9 |

Sample crystal data courtesy of JAN Crystals.

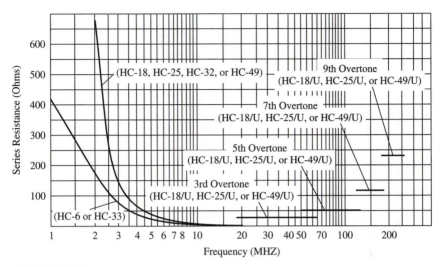

**FIGURE 7.25**

Crystal resistance $r_1$ as a function of fundamental frequency for three holder types. *(Courtesy of Savoy Electronics Inc.)*

mental frequency for several holder types. If $r_1$ is zero, the crystal input impedance is (in a narrow frequency region about $f_o$)

$$Z(j\omega) = \frac{(j\omega C_o)^{-1}[j\omega L_1 + (j\omega C_1)^{-1}]}{j\omega L_1 + (j\omega C_1)^{-1} + (j\omega C_o)^{-1}}$$

$$= \frac{-[j/\omega(C_1 + C_o)](1 - \omega^2 L_1 C_1)}{1 - \omega^2 L_1[C_o C_1/(C_o + C_1)]}$$

(7.51)

The impedance $Z(j\omega)$ will be zero when the inductor $L_1$ and $C_1$ are in resonance $[\omega^2 = (LC)^{-1}]$. The frequency

$$f_s = [2\pi(L_1 C_1)^{1/2}]^{-1}$$

(7.52)

is referred to as the *series resonant frequency* of the crystal. The crystal impedance will be infinite at the frequency

$$f_a = \left\{ 2\pi \left[ L_1 \left( \frac{C_o C_1}{C_o + C_1} \right) \right]^{1/2} \right\}^{-1}$$

(7.53)

where $f_a$ is referred to as the *antiresonant frequency* of the crystal. The ideal crystal $(r = 0)$ behaves as both a series resonant and a parallel resonant circuit with infinite $Q$. The actual crystal also functions as both series and parallel resonant circuits, but with finite $Q$. The effect of a nonzero $r$ on the circuit can be calculated from the equation

$$Z(j\omega) = \frac{(j\omega C_o)^{-1}[j\omega L_1 + r_1 + (j\omega C_1)^{-1}]}{j\omega L_1 + r_1 + (j\omega C_1)^{-1} + (j\omega C_o)^{-1}}$$

(7.54)

but it is usually not necessary, since the solution can be closely approximated with little difficulty. For all practical purposes, the series resonant frequency is unchanged for a nonzero $r_1$ (using typical crystal parameters). The effect of $r_1$ is primarily a reduction of the circuit $Q$. The effect of $r_1$ on the antiresonant behavior is readily evaluated by making a series-to-parallel transformation in the equivalent circuit, as shown in Fig. 7.26. The transformation, of course, is valid only at a single frequency. The equivalent parallel impedances are given by

$$R_p = r_1\left(1 + \frac{X_s^2}{r_1^2}\right) = r_1(1 + Q_s^2) \tag{7.55}$$

and

$$X_p = X_s\left(1 + \frac{r_1^2}{X_s^2}\right) = X_s[1 + (Q_s^2)^{-1}] \tag{7.56}$$

where

$$X_s = \omega L_1 - (\omega C_1)^{-1} \quad \text{and} \quad Q_s = \frac{X_s}{r_1}$$

At antiresonance, the equivalent parallel reactance must be equal to the reactance of the shunt capacitor $X_p = X_{C_o}$, and at $f_a$ the series reactance $X_s$ is large, so $Q_s \gg 1$. Therefore, $X_{C_o} \approx X_s$ and

$$R_p \approx \frac{X_{C_o}^2}{r_1} \tag{7.57}$$

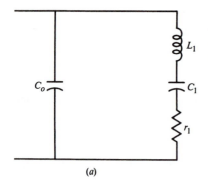

(a)

**FIGURE 7.26**
(a) Electric circuit equivalent, valid near the crystal's fundamental frequency; (b) a parallel equivalent of Fig. 7.26a, valid at a particular frequency.

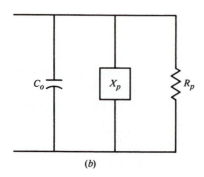

(b)

Circuits containing crystals are frequently designed so that the frequency range of interest is between the series resonant and antiresonant frequencies of the crystal. In this frequency range, the crystal impedance is reactive, as can be seen from an examination of Eq. (7.51). The ratio of the antiresonant to the resonant frequency is [from Eqs. (7.52) and (7.53)]

$$\frac{f_a}{f_s} = \frac{2\pi (L_1 C_1)^{1/2}}{2\pi \{L_1[C_1 C_o/(C_1 + C_o)]\}^{1/2}} = \left(1 + \frac{C_1}{C_o}\right)^{1/2} \tag{7.58}$$

The shunt capacitance $C_o$ is normally much greater than $C_1$ so that the Taylor expansion reduces to

$$\frac{f_a}{f_s} = \left(1 + \frac{C_1}{C_o}\right)^{1/2} \approx 1 + \frac{C_1}{2C_o} = 1 + (2k)^{-1} \tag{7.59}$$

Typical values of $k$ lie between 250 and 300. The crystal's antiresonant frequency is higher than its series resonant frequency.

**EXAMPLE 7.5.** If the antiresonant frequency of the 1-MHz crystal in Table 7.1 is 1 MHz, what is the series resonant frequency $f_s$?

**Solution.** Since $k = C_o/C_1 = 3.2/0.008 = 400$ for this crystal,

$$f_s = 1 \times 10^6 \left(1 + \frac{1}{800}\right)^{-1} = 0.999 \text{ MHz}$$

The antiresonant frequency is 1 kHz higher than the series resonant frequency.

If Eq. (7.51) for the crystal impedance is rewritten as

$$Z(j\omega) = \frac{-(C_o + C_1)^{-1}(j/\omega)[1 - (\omega/\omega_s)^2]}{1 - (\omega/\omega_a)^2}$$

it is seen that the impedance is inductive for $\omega_s \leq \omega \leq \omega_a$, and it is capacitive for other frequency ranges. A plot of $Z(j\omega)$ is given in Fig. 7.27. This plot ignores the overtone circuits illustrated in Fig. 7.24. The actual crystal impedance will have multiple resonant and antiresonant frequencies, with the impedance inductive between each resonant and antiresonant frequency.

**EXAMPLE 7.6.** Consider the 5.7-MHz crystal whose characteristics are given in Table 7.1. At the antiresonant frequency $f_a$ the $Q_s$ is large, so

$$X_s \approx X_{C_o}$$

If the total shunt capacitance is 6 pF, the magnitude of the shunt reactance at 5.7 MHz is approximately 4654 $\Omega$, which must be approximately the same as $X_s$. Therefore, since the series resistance $r = 25 \Omega$, the circuit $Q$ is

$$Q = \frac{4654}{25} = 186$$

(Note that this is not the crystal $Q$, but solely an equivalent $Q$ derived for easily expressing the series-to-parallel transformation.) And $X_s$ is composed of the series combination of inductive and capacitive reactances. That is,

$$X_L = X_s + X_{C_1} \approx X_{C_o} + X_{C_1}$$

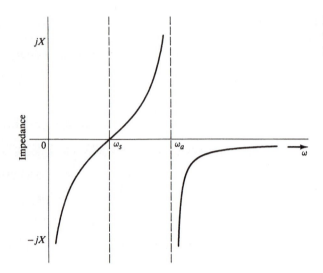

**FIGURE 7.27**
Crystal impedance as a
function of frequency.

Since $C_1 = 21 \times 10^{-3}$ pF,

$$X_{C_1} = [(21 \times 10^{-15})(2\pi)(5.7 \times 10^6)]^{-1} = 1.3 \times 10^6 \ \Omega$$

Now $C_1$ is much smaller than the shunt capacitance, so the inductive reactance is

$$X_L \approx 1.3 \times 10^6 \ \Omega$$

and the crystal $Q$ is

$$Q = \frac{1.3 \times 10^6}{25} = 52{,}000$$

Actually $X_L$ and $X_{C_1}$ differ by a small amount (4654 $\Omega$), but the calculation of $Q$ is sufficiently accurate.

## Parallel-Mode Crystal Oscillators

Crystals often serve as either parallel or series resonant circuits in oscillators. Their high $Q$ provides greater frequency stability than is attainable with discrete inductors and capacitors. If the crystal is used in the antiresonant mode, the circuit is referred to as a *parallel-mode crystal oscillator.* Series-mode oscillators use the crystal as a series resonant circuit, while in parallel-mode oscillators the crystal actually serves as an inductor. The oscillator design is the same as for noncrystal oscillators except that the biasing network can be different since crystals block dc voltages. A parallel-mode crystal is cut to be antiresonant at the desired oscillating frequency when the external capacitance across the crystal is a specified amount. The external capacitance is usually large enough that circuit stray capacitances can be neglected. A typical value for the specified crystal load capacitance is 32 pF.

**EXAMPLE 7.7.** Design 20-MHz Colpitts parallel-mode crystal-controlled oscillator.

*Solution.* The oscillator can be converted to a crystal-controlled oscillator simply by replacing the inductor with a parallel-mode crystal antiresonant at 20 MHz. If the

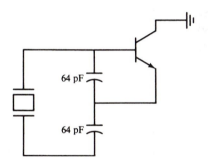

**FIGURE 7.28**
The crystal oscillator circuit designed in Example 7.7.

crystal load capacitance is specified to be 32 pF, then the series combination of $C_1$ and $C_2$ must be 32 pF. This could be satisfied by using 64-pF capacitors for both $C_1$ and $C_2$. In order for oscillations to occur, the loop gain must still be greater than 1, or [from Eq. (7.35)]

$$\frac{g_m X_{C_1} X_{C_2}}{r_1} > 1$$

where $r_1$ is the series resistance of the crystal. The series resistance of crystals in this frequency range is approximately 20 $\Omega$, so the inequality is easily satisfied. The complete circuit (except for biasing) would appear as shown in Fig. 7.28.

A difficulty in designing parallel-mode crystal oscillators is the selection of the bias circuitry so that it does not reduce the circuit $Q$. Any resistance shunting a crystal will reduce the $Q$ if the crystal is being used in the parallel mode. Figure 7.29 contains three parallel-mode oscillator circuits. A grounded-base oscillator is shown in Fig. 7.29a. And $C_b$ serves as a short circuit at the oscillating frequency. The crystal is shunted by a resistance $R_s$, consisting of the common-base input impedance $r_i$ (a low impedance) in parallel with $R_E$ increased by the turns ratio squared. This is,

$$R_s = \frac{r_i R_E}{r_i + R_E} \left(1 + \frac{C_2}{C_1}\right)^2 \tag{7.60}$$

In addition, the impedance of the RF choke must be high so that $R_3$ does not reduce the crystal $Q$. Usually $R_s$ is so small that the crystal $Q$ is markedly degraded; this is not a good crystal oscillator circuit.

Figure 7.29b illustrates a grounded-collector oscillator. The crystal is shunted by the bias resistors $R_1$ and $R_2$. If these resistors are not sufficiently large, they will significantly reduce the $Q$ of the circuit. The bias resistors $R_1$ and $R_2$ do not shunt the crystal in the Pierce oscillator illustrated in Fig. 7.29c. Capacitor $C_E$ shunts the bias resistor $R_E$ at the oscillating frequency. The Pierce circuit configuration is usually the best choice for a parallel-mode crystal oscillator—provided one side of the crystal does not have to be grounded. Since the bias resistors do not shunt the crystal, the Pierce oscillator normally has the highest $Q$ and hence the best frequency stability.

**Capacitor in parallel with the crystal**

In many applications (such as in voltage-controlled oscillators) it is necessary to adjust the frequency of oscillation. Equation (7.53) describes how the parallel-mode (antiresonant) frequency can be varied by adding an external capacitor in

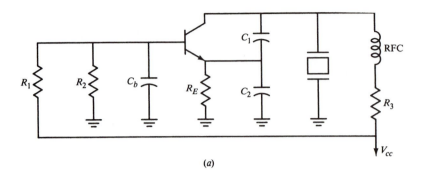

(a)

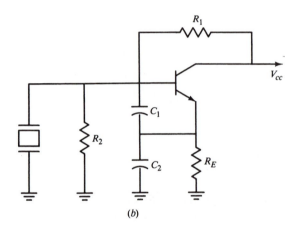

(b)

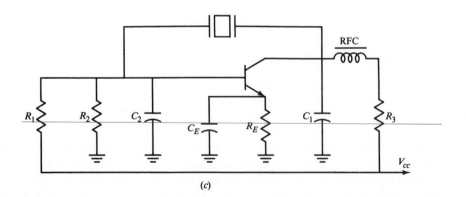

(c)

**FIGURE 7.29**
Three parallel-mode oscillator circuits: (a) a grounded-base circuit,
(b) a grounded-collector circuit, and (c) a Pierce oscillator circuit.

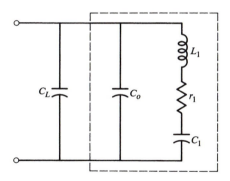

**FIGURE 7.30**
Adding a capacitor $C_L$ in parallel with a crystal will reduce its antiresonant frequency.

parallel with the crystal, as shown in Fig. 7.30. By simplifying the Taylor expansion, the antiresonant frequency becomes

$$f_a = \left\{ 2\pi \left[ L \frac{C_1(C_o + C_L)}{C_1 + C_o + C_L} \right]^{1/2} \right\}^{-1} \approx f_s \left[ 1 + \frac{C_1}{2(C_o + C_L)} \right] \qquad (7.61)$$

By increasing $C_L$, $f_a$ can be decreased until $f_a \approx f_s$. It is left as an exercise to show that adding $C_L$ does not significantly change the series resonant frequency. The frequency range

$$f_a - f_s = \frac{C_1}{2C_o} f_s \qquad (7.62)$$

is referred to as the *pulling range* of the crystal. Figure 7.31 illustrates how $f_a$ varies as a function of the external load capacitance $C_L$. Although Eq. (7.61) indicates that the parallel antiresonant frequency can be "pulled" down to the series resonant frequency, in practice this results in poor performance for parallel-mode oscillators, since the crystal $Q$ is simultaneously reduced.

Equation (7.55) shows that the equivalent parallel resistance at resonance is

$$R_p = r_1 \left( 1 + \frac{X_s^2}{r_1^2} \right) \qquad (7.55)$$

For the antiresonant mode the parallel reactance

$$X_p = [\omega(C_o + C_L)]^{-1} \qquad (7.63)$$

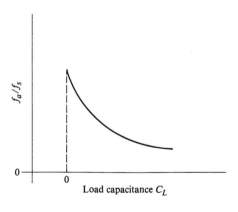

**FIGURE 7.31**
Variation in crystal antiresonant frequency as a function of load capacitance $C_L$.

must equal the series impedance $X_s$ [Eq. (7.56)]. Therefore the equivalent parallel resistance is

$$R_p = r_1\left[1 + \left(\frac{X_p}{r_1}\right)^2\right] \tag{7.64}$$

As $C_o + C_L$ increases, $X_p$ and hence $R_p$ decrease, resulting in a reduction in the loop gain and eventually a cessation of oscillations. A graph of $R_p$ versus $C_L$ is plotted in Fig. 7.32 for a typical crystal. Equation (7.64) indicates that $R_p$ will be small for high-frequency crystals, since $X_p$ will be small. This is one reason why parallel-mode crystal oscillators are not used at frequencies above 20 MHz. The $Q$ of the series resonant mode does not depend significantly upon the shunt capacitance $C_o + C_L$, so series-mode oscillators are used at the higher frequencies.

A practical rule of thumb is that the combination of the crystal capacitance $C_o$ and $C_L$ can be used in an antiresonant circuit as long as

$$[\omega(C_o + C_L)r_1]^{-1} > 4 \tag{7.65}$$

This relation is based upon the fact that for values less than 4, the slope of the phase shift of the crystal near frequency $f_a$ is not sufficiently steep to provide good phase-frequency stability.[3]

EXAMPLE 7.8  Consider again the 5.7-MHz crystal of Table 7.1. Assume that the specified load capacitance is 32 pF. If it is desired to decrease the antiresonant frequency, an additional capacitor $C_L$ must be added in parallel with the 32-pF-load capacitor.

For example, if an additional 22-pF capacitor is added in parallel, the new antiresonant frequency is [from Eq. (7.61)]

$$f_a' = f_s\left[1 + \frac{C_1}{2(C_o + C_L)}\right] = f_a\frac{1 + C_1/[2(C_o + C_L)]}{1 + C_1/(2C_o)}$$

$$f_a' = f_a\frac{1 + 0.021/[2(37.1 + 22)]}{1 + 0.021/[2(37.1)]} = f_a\frac{1.002}{1.003} = 5.6994 \text{ MHz}$$

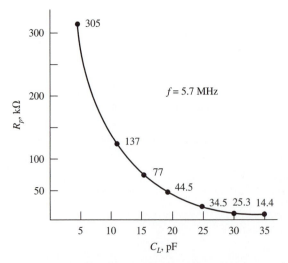

**FIGURE 7.32**
Equivalent parallel load resistance of a typical crystal as a function of additional load capacitance $C_L$.

That is, an additional 22 pF in parallel with $C_o$ will reduce the antiresonant frequency by 600 Hz. The equivalent parallel resistance is found [using Eq. (7.55)] to be

$$R_p = 25\{1 + [(25 \times 2\pi \times 5.6994 \times 10^6 \times 59.1 \times 10^{-12})^{-1}]^2\} = 8.96 \text{ k}\Omega$$

The additional 22 pF reduces the antiresonant frequency by 600 Hz and decreases $R_p$ from 22.7 to 8.96 kΩ.

## Series-Mode Crystal Oscillators

The results of the analysis of the Clapp-Gouriet oscillator can be used to show that, for crystal oscillators operating in the parallel mode, the $g_m$ of the active device must satisfy the relation

$$g_m \geq C_1 C_2 \omega^2 r_1$$

As the frequency is increased, $C_1$ and $C_2$ must be reduced in order for this relation to hold. Once the value of the capacitors approaches the size of the transistor terminal capacitances, the oscillator stability is seriously degraded, since the transistor capacitances cannot be precisely controlled. To overcome this problem, high-frequency oscillators are operated with the crystals in the series resonant mode. (Most oscillators operating above 20 MHz are, in fact, in the series resonant mode.)

Another characteristic of crystals is that their fundamental frequency is inversely proportional to the crystal thickness (for most crystal cuts). High-frequency crystals require thin plates that are very fragile and sensitive to contamination. For this reason, high-frequency crystals usually operate on an overtone of the fundamental frequency, which allows for thicker, less fragile crystals. Overtone crystals are almost always used in the series mode. As a general rule, third-overtone crystals are used from 20 to 60 MHz and fifth-overtone crystals from 60 to 125 MHz.

A crystal operating in the series mode functions as a short circuit at the oscillating frequency and as a large impedance at other frequencies. An example of a series-mode crystal oscillator is shown in Fig. 7.33. At frequencies other than its series resonant frequency, the crystal impedance is large enough to prevent current

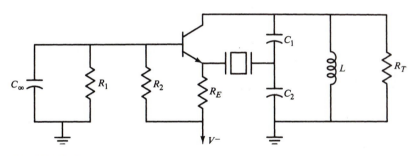

**FIGURE 7.33**
A series-mode crystal oscillator.

from being fed back to the emitter. The tank circuit is designed to be antiresonant at the series resonant frequency of the crystal. It is far from obvious that the circuit $Q$ is determined by the crystal and not the tank circuit, but the following analysis shows that it is. The small-signal equivalent circuit is shown in Fig. 7.34. And $Z_x$ is the crystal impedance, and $r_i$ is the transistor common-base input resistance, which is assumed to be much smaller than the emitter bias resistor $R_E$. The voltage $V$ is

$$V = \frac{V_o C_1}{C_1 + C_2} \frac{r_i}{r_i + Z_x}$$

provided $(\omega C_2)^{-1} \ll |Z_x + r_i|$, which will be the case near the series resonant frequency in a well-designed oscillator. The output voltage is

$$V_o = g_m V Z_L$$

where $Z_L$ is the equivalent load seen by the collector. At the series resonant frequency, $Z_L$ will consist of the crystal resistance $r_x$ and $r_i$, transformed by the capacitance turns ratio squared, in parallel with $R_T$. That is,

$$Z_L(j\omega_o) = R_T \| (r_x + r_i) \left( \frac{C_2 + C_1}{C_1} \right)^2$$

The open-loop gain is

$$A_o \approx \frac{Z_L C_1}{C_1 + C_2} (Z_x + r_i)^{-1} \tag{7.66}$$

It was explained earlier in the chapter that a figure of merit for an oscillator is the rate of change of the phase shift of the open-loop gain, evaluated at the resonant frequency. That is,

$$\frac{S_F}{\omega_o} = \left. \frac{d \arg A_o}{d\omega} \right|_{\omega=\omega_o} = \frac{d \arg Z_L}{d\omega} + \frac{d \arg (Z_x + r_i)^{-1}}{d\omega} \tag{7.67}$$

The impedance $Z_L$ includes the shunting effects of the crystal circuit, but it is easily shown that near the resonant frequency

$$Z_L \approx \frac{R_p}{1 + jQ(\omega/\omega_o - \omega_o/\omega)}$$

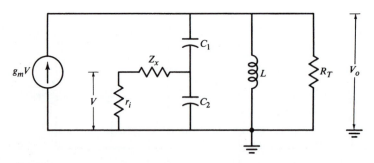

**FIGURE 7.34**
A small-signal equivalent circuit of the oscillator shown in Fig. 7.33, with the feedback path opened and terminated.

where
$$R_p = Z_L(j\omega_o)$$

Likewise,
$$(Z_x + r_i)^{-1} = \left[1 + jQ_x\left(\frac{\omega}{\omega_o} - \frac{\omega_o}{\omega}\right)\right]^{-1}$$

where
$$Q_x = \frac{\omega_o L_i}{r_x + r_i} = \frac{Q_1}{1 + r_i/r_x}$$

and $Q_1$ is the crystal $Q$. Since the phase shift of a resonant circuit at the resonant frequency is [Eq. (7.50)]

$$\frac{d\,\arg Z_L}{d\omega} = \frac{2Q_L}{\omega_o}$$

the stability factor is

$$S_F = \omega_o \frac{d\,\arg A_o}{d\omega} = 2(Q_L - Q_x) \tag{7.68}$$

and $Q_L$ is the $Q$ of the parallel tuned circuit which is much less than that of the crystal $Q_x$; that is, $Q_L \ll Q_x$. Any parallel loading of the tank circuit will reduce $Q_L$, which justifies the assumption of ignoring the effect of crystal loading on the tank circuit. Therefore,

$$S_F \approx -2Q_x$$

The stability factor is approximately that of $2Q_x$. And $Q_x$ is proportional to the crystal $Q$. The transistor common-base input resistance $r_i$ appears in series with the crystal, but since $r_i$ is of the same order of magnitude as the crystal resistance $r_x$, the circuit is controlled by the high $Q$ of the crystal. The crystal could also be inserted in the base circuit, but then the crystal would be in series with the common-emitter input resistance, which is much larger, and the phase stability factor would be reduced proportionally.

Another series-mode oscillator is shown in Fig. 7.35. This circuit is often referred to as the *impedance-inverting Pierce oscillator*. A simple test to determine whether a crystal operates in the series or parallel mode in a circuit is to replace the crystal by a short circuit. If the circuit will not oscillate with the crystal

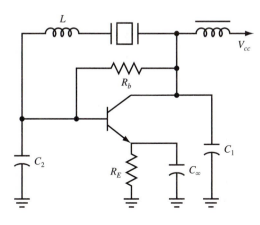

**FIGURE 7.35**
An impedance-inverting Pierce oscillator.

short-circuited, it is a parallel-mode oscillator; otherwise, it usually is a series-mode oscillator.

### Capacitor in series with the crystal

The frequency of series-mode crystal oscillators cannot be adjusted by adding a capacitor in parallel with the crystal, since such a capacitor has a negligible effect on the series resonant frequency $f_s$. The series resonant frequency can be altered by adding a capacitor in series with the crystal, as shown in Fig. 7.36. For this circuit the input impedance is

$$Z(j\omega) = -jX_s + \frac{-jX_o(jX_1 - jX_{C_1} + r_1)}{jX_1 - jX_{C_1} - jX_o + r_i} \tag{7.69}$$

where $\qquad\qquad X_s = (\omega C_s)^{-1} \qquad X_{C_o} = (\omega C_o)^{-1}$

and $\qquad\qquad X_1 = \omega L_1 \qquad\qquad X_{C_1} = (\omega C_1)^{-1}$

The input impedance can be written as

$$\begin{aligned} Z(j\omega) &= \frac{-jX_s[jX_1 - j(X_{C_1} + X_o) + r_1] - jX_o(jX_1 - jX_{C_1} + r_1)}{jX_1 - jX_{C_1} - jX_o + r_1} \\[6pt] &= \frac{(jX_1 + r_1)(-jX_s - jX_o) + j(X_o + X_s)jX_{C_1} + j^2X_oX_s}{jX_1 + r_1 - j(X_o + X_{C_1})} \\[6pt] &= \frac{-j(X_o + X_s)\{jX_1 + r_1 - j[X_{C_1} + X_oX_s/(X_o + X_s)]\}}{jX_1 + r_1 - j(X_o + X_{C_1})} \end{aligned} \tag{7.70}$$

Let $\qquad\qquad X' = \left[\omega\left(\dfrac{C_oC_s}{C_o + C_s}\right)\right]^{-1} = nX_o \tag{7.71}$

where $\qquad\qquad n = \dfrac{C_o + C_s}{C_s} = \dfrac{X_o^{-1} + X_s^{-1}}{X_s^{-1}} = \dfrac{X_o + X_s}{X_o} \tag{7.72}$

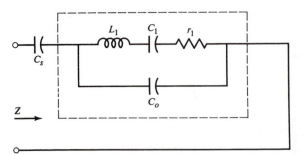

**FIGURE 7.36**
A capacitor placed in series with the crystal will increase the series resonant frequency.

and define

$$X'_{C_1} = X_{C_1} + \frac{X_o X_s}{X_o + X_s} \tag{7.73}$$

Then

$$X_o + X_{C_1} = X'_{C_1} + X_o - \frac{X_o X_s}{X_o + X_s}$$

$$= X'_{C_1} + \frac{X_o^2}{X_o + X_s} \tag{7.73a}$$

and

$$Z(j\omega) = \frac{-j(X_o + X_s)[(jX_1 + r_1) - jX'_{C_1}]}{jX_1 + r_1 - j[X'_{C_1} + X_o^2/(X_o + X_s)]}$$

$$= \frac{-jnX_o(jX_1 + r_1 - jX'_{C_1})}{jX_1 + r_1 - j(X'_{C_1} + X_o/n)} \tag{7.74}$$

$$= \frac{-jnX_o(jn^2 X_1 + n^2 r_1 - jn^2 X'_{C_1})}{jn^2 X_1 + n^2 r_1 - jn^2 X'_{C_1} - jX_o n}$$

That is, the crystal-plus-series capacitor can be represented by the equivalent circuit shown in Fig. 7.37, where

$$(C'_1)^{-1} = C_1^{-1} + (C_o + C_s)^{-1}$$

or

$$C'_1 = \frac{C_1(C_o + C_s)}{C_1 + C_o + C_s} \tag{7.75}$$

This equivalent circuit is of the same form as that of the crystal (minus the series capacitor), so the same expressions can be used as were used in deriving the crystal characteristics. For example, the antiresonant frequency of the circuit with the capacitor in series with the crystal is [Eq. (7.53)]

$$\omega_a = \left\{ \left[ L_1 n^2 \frac{(C'_1/n^2)(C_o/n)}{C'_1/n^2 + C_o/n} \right]^{1/2} \right\}^{-1}$$

This expression is readily simplified to

$$\omega_a = \left[ \left( L \frac{C_1 C_o}{C_1 + C_o} \right)^{1/2} \right]^{-1}$$

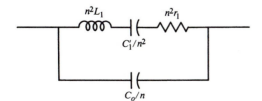

$n^2 L_1$   $n^2 r_1$

$C'_1/n^2$

$C_o/n$

**FIGURE 7.37**
A convenient equivalent circuit of a capacitor in series with a crystal.

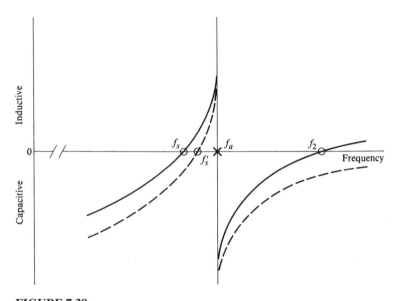

**FIGURE 7.38**
Effect of a series capacitor on crystal impedance (dashed curve).

which is the same as the antiresonant frequency of the crystal without a series capacitance. The series capacitance does not change the antiresonant frequency.

The new series resonant frequency is

$$\omega_s' = \left[ \left( \frac{n^2 L_1 C_1'}{n^2} \right)^{1/2} \right]^{-1} = \left\{ \left[ L_1 \frac{C_1(C_o + C_s)}{C_1 + C_o + C_s} \right]^{1/2} \right\}^{-1} \tag{7.76}$$

The addition of the series capacitor moves the series resonant frequency toward the antiresonant frequency. The crystal-plus-series capacitor reactance plots are illustrated in Fig. 7.38.

Note that without the series capacitance, $C_o$ has a negligible effect on the series resonant frequency, but once $C_s$ is added, $\omega_s$ can be changed by changing $C_o$. Changing $C_o$ will, however, also change $\omega_a$. The ability to change the oscillating frequency by adding a capacitor in series or parallel is used in the design of voltage-controlled crystal oscillators.

## ■ 7.6

### VOLTAGE-CONTROLLED OSCILLATORS AND
### VOLTAGE-CONTROLLED CRYSTAL OSCILLATORS

The preceding sections have shown how the frequency of oscillation can be varied by the addition of a capacitor. Diodes exist (referred to as *varicaps* or *varactors*) which function as voltage-variable capacitors. If the varicap is included in the oscillator circuit and the frequency of oscillation is varied by changing the dc bias volt-

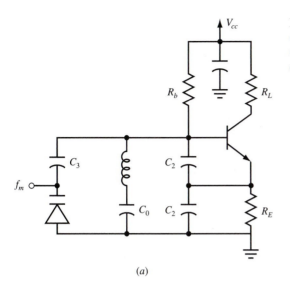

**FIGURE 7.39**
(*a*) A voltage-controlled oscillator;
(*b*) another voltage-controlled
oscillator.

(*a*)

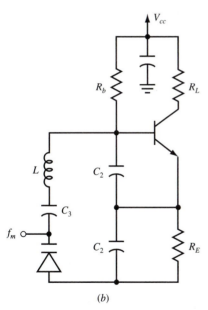

(*b*)

age across the varicap, the oscillator is referred to as a *voltage-controlled oscillator* (*VCO*). If the VCO is crystal-controlled, the oscillator is referred to as a *voltage-controlled crystal oscillator* (*VCXO*). These devices find many applications, such as in frequency modulators, telemetry, Doppler radar, spectrum analyzers, television tuners, phase-locked loops, and frequency synthesizers.

A frequently used VCO is shown in Fig. 7.39*a*. The modulating voltage $f_m$ changes the varicap voltage and thus the capacitance shunting the inductor, thereby changing the frequency of oscillation. If the varicap is added in series with the inductor, as shown in Fig. 7.39*b*, the VCO is Clapp-Gouriet oscillator. In this

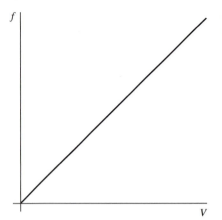

**FIGURE 7.40**
Ideal voltage-frequency transfer characteristics.

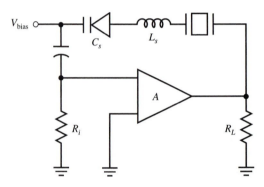

**FIGURE 7.41**
A voltage-controlled crystal oscillator.

configuration a smaller value of capacitance can be used to change the oscillating frequency. A main difficulty in VCO design is to achieve a linear voltage-frequency transfer characteristic. The idealized voltage-frequency transfer characteristic is illustrated in Fig. 7.40.

A simplified model of a VCXO is illustrated in Fig. 7.41. (The biasing of the tuning diode is neglected.) The transistor amplifier, operating near the series resonant frequency of the feedback circuit, is represented by a voltage amplifier with frequency-independent gain $A$. The crystal is operating in the series mode (if the crystal can be replaced by a short circuit without stopping the oscillation, the crystal is being used in the series mode). The frequency is controlled by the bias voltage applied to the voltage-variable capacitance $C_s$. Since the addition of $C_s$ increases the series resonant frequency, the reactance of $L_s$ is chosen equal in magnitude to

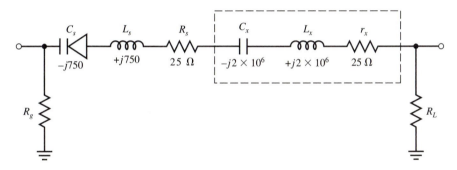

**FIGURE 7.42**
A small-signal equivalent circuit of the voltage-controlled oscillator.

that of $C_s$, making the series resonant frequency again that of the crystal. The resonant frequency can be increased or decreased with the control voltage.

The overall $Q$ will be close to that of the crystal, so the circuit frequency stability is close to that of the fixed-frequency crystal. To verify that the $Q$ remains
high, consider a 20-MHz series-mode crystal-plus-series capacitance and inductor,
as shown in Fig. 7.42. At the resonant frequency, the magnitude of the crystal inductance is 2 M$\Omega$, and the series resistance of the crystal is 25 $\Omega$ ($Q_x = 80,000$).
The magnitude of the series inductances is 750 $\Omega$, and the coil resistance is
25 $\Omega$ ($Q_u = 30$). Therefore, the total series inductance reactance is approximately
2 M$\Omega$, the total series resistance is 50 $\Omega$, and circuit $Q_u = 40,000$. The addition
of the series inductor has reduced the circuit $Q$ by a factor of 2, but it is still a very
high-$Q$ circuit.

Figure 7.43$a$ illustrates the capacitance of a tuning diode as a function of the
reverse-bias voltage across the diode. The capacitance is an approximately inverse
function of voltage for reverse voltages of 1 to 10 V. (The capacitance varies from
16 to 5 pF). If this diode is used as the tuning capacitor in Fig. 7.42 (a capacitive
reactance of 750 $\Omega$ corresponds to 10.6 pF at 20 MHz), the control voltage should
be approximately 2.5 V. If the voltage is reduced to 1 V, the series capacitance will
be 16 pF (497 $\Omega$), and the net series reactance will be equal to 253 (750 − 497) $\Omega$
inductive, which will cause a decrease in the series resonant frequency. If the control voltage is increased to 10 V, the series capacitance will be 5 pF (1591 $\Omega$), and
the net series reactance will be equal to 841 (1591 − 750) $\Omega$ capacitive, resulting
in an increase in the series resonant frequency. If it is desired to use this circuit to
frequency-modulate an audio signal, the diode should be biased approximately in
the middle of its linear region ($V_R = 4.5$ V), and the audio voltage should be
approximately 9 V peak to peak if the entire linear region of the tuning diode is to
be utilized. Figure 7.43$b$ describes the voltage-versus-capacitance characteristics of
another family of tuning diodes.

## SILICON EPICAP DIODES

. . . designed in the popular PLASTIC PACKAGE for high volume requirements of FM Radio and TV tuning and AFC, general frequency control and
tuning applications; providing solid-state reliability in replacement of
mechanical tuning methods.

Also available in Surface Mount Package up to 33 pF.

● High Q
● Controlled and Uniform Tuning Ratio
● Standard Capacitance Tolerance — 10%
● Complete Typical Design Curves

## MMBV2101LT1
## MMBV2103LT1 thru
## MMBV2105LT1
## MMBV2107LT1 thru
## MMBV2109LT1★
## MV2101
## MV2103 thru MV2105
## MV2107 thru MV2109
## MV2111
## MV2113 thru MV2115★

**CASE 318-07, STYLE 8**
**SOT-23 (TO-236AB)**

3 ○──|◀──○ 1
Cathode        Anode

**CASE 182-02, STYLE 1**
**(TO-226AC)**

2 ○──|◀──○ 1
Cathode        Anode

**6.8–100 pF**
**30 VOLTS**
**VOLTAGE-VARIABLE**
**CAPACITANCE DIODES**

★MMBV2101LT1, MMBV2105LT1,
MMBV2109LT1, MV2101, MV2104,
MV2108, MV2109, MV2111, MV2113
and MV2115 are Motorola
designated preferred devices.

### MAXIMUM RATINGS

|  |  | MV21XX | MMBV21XXLT1 |  |
|---|---|---|---|---|
| **Rating** | **Symbol** | **Value** | | **Unit** |
| Reverse Voltage | $V_R$ | 30 | | Volts |
| Forward Current | $I_F$ | 200 | | mA |
| Forward Power Dissipation<br>@ $T_A$ = 25°C<br>Derate above 25°C | $P_D$ | 280<br>2.8 | 200<br>2.0 | mW<br>mW/°C |
| Junction Temperature | $T_J$ | +125 | | °C |
| Storage Temperature Range | $T_{stg}$ | −55 to +150 | | °C |

### DEVICE MARKING

| | | |
|---|---|---|
| MMBV2101LT1 = M4G | MMBV2105LT1 = 4U | MMBV2109LT1 = 4J |
| MMBV2103LT1 = 4H | MMBV2107LT1 = 4W | |
| MMBV2104LT1 = 4Z | MMBV2108LT1 = 4X | |

### ELECTRICAL CHARACTERISTICS ($T_A$ = 25°C unless otherwise noted)

| Characteristic | Symbol | Min | Typ | Max | Unit |
|---|---|---|---|---|---|
| Reverse Breakdown Voltage<br>($I_R$ = 10 μAdc) | $V_{(BR)R}$ | 30 | — | — | Vdc |
| Reverse Voltage Leakage Current<br>($V_R$ = 25 Vdc, $T_A$ = 25°C) | $I_R$ | — | — | 0.1 | μAdc |
| Diode Capacitance Temperature Coefficient<br>($V_R$ = 4.0 Vdc, f = 1.0 MHz) | $TC_C$ | — | 280 | — | ppm/°C |

---

**FIGURE 7.43a**

**MMBV2101LT1 MMBV2103LT1 thru MMBV2105LT1 MMBV2107LT1 thru MMBV2109LT1
MV2101 MV2103 thru MV2105 MV2107 thru MV2109 MV2111 MV2113 thru MV2115**

| Device | $C_T$, Diode Capacitance $V_R$ = 4.0 Vdc, f = 1.0 MHz pF | | | Q, Figure of Merit $V_R$ = 4.0 Vdc, f = 50 MHz | TR, Tuning Ratio $C_2/C_{30}$ f = 1.0 MHz | | |
|---|---|---|---|---|---|---|---|
| | Min | Nom | Max | Typ | Min | Typ | Max |
| MMBV2101LT1/MV2101 | 6.1 | 6.8 | 7.5 | 450 | 2.5 | 2.7 | 3.2 |
| MMBV2103LT1/MV2103 | 9.0 | 10 | 11 | 400 | 2.5 | 2.9 | 3.2 |
| MMBV2104LT1/MV2104 | 10.8 | 12 | 13.2 | 400 | 2.5 | 2.9 | 3.2 |
| MMBV2105LT1/MV2105 | 13.5 | 15 | 16.5 | 400 | 2.5 | 2.9 | 3.2 |
| MMBV2107LT1/MV2107 | 19.8 | 22 | 24.2 | 350 | 2.5 | 2.9 | 3.2 |
| MMBV2108LT1/MV2108 | 24.3 | 27 | 29.7 | 300 | 2.5 | 3.0 | 3.2 |
| MMBV2109LT1/MV2109 | 29.7 | 33 | 36.3 | 200 | 2.5 | 3.0 | 3.2 |
| MV2111 | 42.3 | 47 | 51.7 | 150 | 2.5 | 3.0 | 3.2 |
| MV2113 | 61.2 | 68 | 74.8 | 150 | 2.6 | 3.0 | 3.3 |
| MV2114 | 73.8 | 82 | 90.2 | 100 | 2.6 | 3.0 | 3.3 |
| MV2115 | 90 | 100 | 110 | 100 | 2.6 | 3.0 | 3.3 |

MMBV2101LT1, MMBV2103LT1 thru MMBV2105LT1 and MMBV2107LT1 thru MMBV2109LT1 are also available in bulk. Use the device title and drop the "T1" suffix when ordering any of these devices in bulk.

## PARAMETER TEST METHODS

**1. $C_T$, DIODE CAPACITANCE**

($C_T$ = $C_C$ + $C_J$), $C_T$ is measured at 1.0 MHz using a capacitance bridge (Boonton Electronics Model 75A or equivalent).

**2. TR, TUNING RATIO**

TR is the ratio of $C_T$ measured at 2.0 Vdc divided by $C_T$ measured at 30 Vdc.

**3. Q, FIGURE OF MERIT**

Q is calculated by taking the G and C readings of an admittance bridge at the specified frequency and substituting in the following equations:

$$Q = \frac{2\pi f C}{G}$$

(Boonton Electronics Model 33AS8). Use Lead Length ≈ 1/16"

**4. $TC_C$, DIODE CAPACITANCE TEMPERATURE COEFFICIENT**

$TC_C$ is guaranteed by comparing $C_T$ at $V_R$ = 4.0 Vdc, f = 1.0 MHz, $T_A$ = −65°C with $C_T$ at $V_R$ = 4.0 Vdc, f = 1.0 MHz, $T_A$ = +85°C in the following equation which defines $TC_C$:

$$TC_C = \frac{C_T(+85°C) - C_T(-65°C)}{85 + 65} \cdot \frac{10^6}{C_R(25°C)}$$

Accuracy limited by measurement of $C_T$ to ± 0.1 pF.

**FIGURE 7.43a (continued)**

**MMBV2101LT1 MMBV2103LT1 thru MMBV2105LT1 MMBV2107LT1 thru MMBV2109LT**
**MV2101 MV2103 thru MV2105 MV2107 thru MV2109 MV2111 MV2113 thru MV211**

**TYPICAL DEVICE PERFORMANCE**

FIGURE 1 — DIODE CAPACITANCE versus REVERSE VOLTAGE

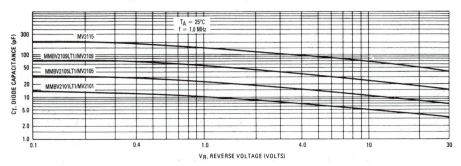

FIGURE 2 — NORMALIZED DIODE CAPACITANCE versus
JUNCTION TEMPERATURE

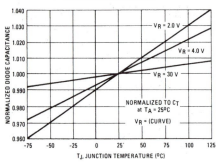

FIGURE 3 — REVERSE CURRENT versus REVERSE
BIAS VOLTAGE

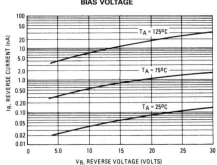

FIGURE 4 — FIGURE OF MERIT versus REVERSE VOLTAGE

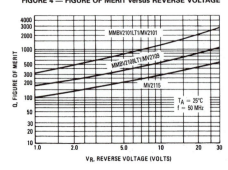

FIGURE 5 — FIGURE OF MERIT versus FREQUENCY

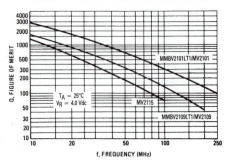

**FIGURE 7.43b**

## ▪ 7.7

### FIELD-EFFECT TRANSISTOR OSCILLATORS

FETs are used extensively in oscillator circuits because FETs have several inherent advantages over bipolar transistors. Their high input impedance permits operation at lower current levels with less power dissipation, and hence the thermal problems introduced by the power dissipation are reduced. Also, when it is operated in the square-law region, the transconductance of the device is not a function of signal level. Figure 7.44 illustrates (with the biasing circuitry removed) three configurations commonly used in FET oscillators. Figure 7.44a contains a Colpitts oscillator, and Fig. 7.44b illustrates a Hartley common-gate oscillator. The oscillator shown in Fig. 7.44c utilizes the phase inversion possible with a transformer to obtain a 360° loop phase shift (180° via the transformer, plus the 180° phase shift present in the common-source configuration). Example 7.9 will illustrate the main points of FET oscillator design.

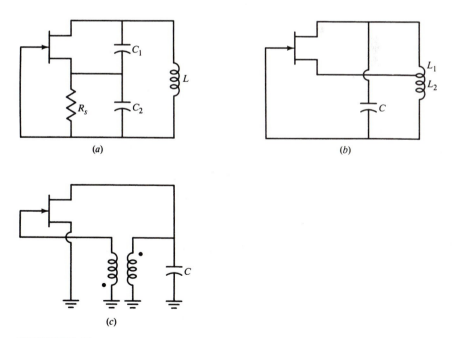

**FIGURE 7.44**
(a) an FET Colpitts oscillator; (b) an FET Hartley oscillator; (c) an FET oscillator with transformer feedback.

## FET Pierce Oscillator

Figure 7.45 illustrates a Pierce oscillator employing an FET. The equivalent circuit is shown in Fig. 7.46. If $r_d$ is neglected, the small-signal loop gain is

$$\frac{g_m V_o (X_{C_1} X_{C_2})}{-jX_{C_1} - jX_{C_2} + jX_L + R_s} = V_o \qquad (7.77)$$

or

$$\frac{g_m (\omega^2 C_1 C_2)^{-1} V_o}{R_s + j(X_L - X_{C_1} - X_{C_2})} = V_o$$

For the loop phase shift to be $360°$, the reactances must cancel so that

$$X_L = X_{C_1} + X_{C_2}$$

If the phase shift is $360°$, the circuit will oscillate, provided

$$\frac{g_m X_{C_1} X_{C_2}}{R_s} \geq 1 \qquad (7.78)$$

In the preceding analysis $r_d$ was assumed large enough to be neglected. In the design $X_{C_1}$ should be selected so that it is much less than $r_d$.

**EXAMPLE 7.9.** Design an FET oscillator with a transistor whose parameters are $r_d = 50 \text{ k}\Omega$ and $g_m = 5 \times 10^{-3}$. The frequency of oscillation is to be 16 MHz.

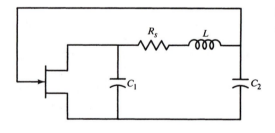

**FIGURE 7.45**
An FET Pierce oscillator.

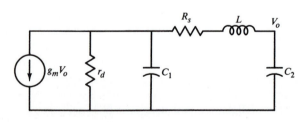

**FIGURE 7.46**
A small-signal equivalent circuit of the FET Pierce oscillator.

*Solution.* The Pierce oscillator shown in Fig. 7.45 can be used. $X_{C_1}$ can be about 1 k$\Omega$; $r_d$ will then have a negligible loading effect. Also for oscillations to occur it is necessary that $g_m X_{C_1} X_{C_2} > R_s$. If $R_s = 15$ and $X_{C_1} = 1$ k$\Omega$, then

$$X_{C_2} > \frac{R_s}{g_m X_{C_1}} = \frac{R_s}{g_m} 10^{-3}$$

or

$$X_{C_2} > 10^{-3} \frac{15}{g_m} = \frac{15}{10^3 \times 5 \times 10^{-3}} = 3$$

The corresponding capacitance values are

$$C_2 = 3300 \text{ pF}$$

and

$$C_1 = 10 \text{ pF}$$

This value of $C_1$ may be so small that the transistor output capacitance has an effect. Therefore it is desirable to increase $C_1$. If $C_1$ is increased by a factor of 10, so that $X_{C_1} = 100$, then $C_2$ must also be increased:

$$X_{C_2} > \frac{R_s}{g_m \times 100} = \frac{15}{0.5} = 30$$

That is, $C_2$ must be less than 330 pF for oscillations to occur. The inductance $L$ is found from $X_L = X_{C_1} + X_{C_2} = 130$ $\Omega$, or $L = 1.3 \times 10^{-6}$ H.

## 7.8

### OSCILLATOR CONTROL USING DELAY LINES

The criteria for oscillation are (1) that the magnitude of the loop gain be unity and (2) that the loop output signal be fed back so that it is in phase with the input at the frequency at which the magnitude of the gain is unity. The oscillators described up to this point have used an $LC$ resonator to obtain the desired phase shift. Another method that can be used is to incorporate a delay line in the feedback path, as illustrated in Fig. 7.47. An ideal delay line has the transfer function

$$H(j\omega) = e^{-j\omega T}$$

The magnitude of this transfer function $|H(j\omega)|$ is 1 at all frequencies, and the phase shift

$$\arg H(j\omega) = -\omega T$$

is a linearly decreasing function of frequency. The phase shift can be described as shown in Fig. 7.48. Recent advances in acoustic surface-wave technology now make the design of delay lines a practical task in the frequency region above 10 MHz. Surface-wave delay lines can be designed to have circuit $Q$'s between those of

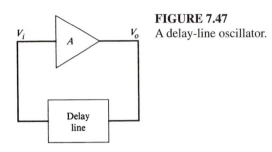

**FIGURE 7.47**
A delay-line oscillator.

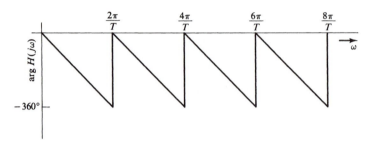

**FIGURE 7.48**
Delay-line phase shift as a function of frequency.

crystals and $LC$ resonators. The delay line will find applications in designs where it is desired to have a greater frequency stability than is provided by an $LC$ oscillator and greater "pullability," or frequency deviation, than can be obtained with a crystal oscillator. With delay-line oscillators, wideband frequency modulation can be achieved without the use of a complex frequency-multiplier chain.

The $Q$ of a delay line is defined by comparing the slope of its phase shift to the slope of an $RLC$ network at its resonant frequency. The slope of the phase shift of a parallel $RLC$ network at its resonant frequency is [Eq. (7.49)]

$$\frac{d\phi}{d\omega} = \frac{-2Q}{\omega_o}$$

The slope of the phase shift of a delay line $d\phi/d\omega = -T$ is equal to the line time delay. The delay line $Q$ is defined as

$$Q = \frac{\omega_o T}{2}$$

The longer the delay of the line, the higher the $Q$ of the line, and the greater will be the frequency stability of the line.

Acoustic surface-wave delay lines provide a good approximation to the ideal delay line.[4] The magnitude and phase characteristics depend upon the fabrication and crystal cut. One of the best fabrication techniques has a time delay of

$$T = \frac{N}{f_o}$$

where $N$ is the number of wavelengths of line length. For this line the $Q$ is

$$Q = \pi N$$

The magnitude of the line frequency response is given by

$$|H(j\omega)| = \left[ \frac{\sin(f - f_o)2\pi/f_o}{(f - f_o)2\pi/f_o} \right]^2$$

## 7.9

### RELAXATION OSCILLATORS

The relation oscillator is another type of oscillator. It consists of active devices plus resistors and at least a capacitor. A simplified version of a relaxation oscillator is shown in Fig. 7.49a. The capacitor is charged by a constant current source (whose direction can be switched) until the capacitor voltage reaches a level at which the Schmitt trigger fires and causes the current source to switch direction, such that the capacitor discharges at a constant rate until the capacitor voltage diminishes to a level which causes the Schmitt trigger to reset and to switch the current source direction so that the capacitor is again charging at a constant rate. The waveform on

**FIGURE 7.49**

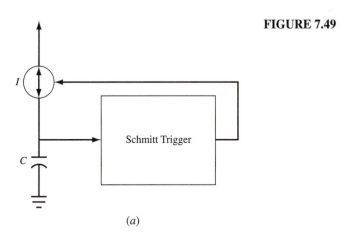

(a)

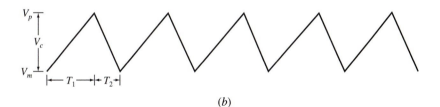

(b)

the capacitor is as shown in Fig. 7.49b, and the waveform of the Schmitt trigger output is a rectangular wave of period $T$, where $T$ is the sum of the charging time $T_1$ plus discharge time $T_2$. Relaxation oscillators have several advantages over $LC$ oscillators. Relaxation oscillators can be linearly controlled over a much broader frequency range than can $LC$ oscillators and have a much wider tuning range. Some relaxation oscillators are tunable over three or four decades of frequency. They can be operated at a very low power level. Relaxation oscillators have always been thought of as low-frequency oscillators, but publications now describe relaxation oscillators that operate above 1 GHz. In general, these relaxation networks can be fully monolithic as all their components (resistors, capacitors, and transistors) are fully integrable. Relaxation oscillators also offer easy ways to generate two signals in phase quadrature relative to each other. Relaxation oscillators can be realized using balanced circuits, and it is possible to increase the circuit's maximum frequency of operation by utilizing the transistor's parasitic capacitance as part of the timing capacitance—in a balanced configuration. An inductor cannot be realized in integrated circuits, and so $LC$ oscillators require an external capacitor. In addition, it is relatively easy to realize a quadrature oscillator (needed for a direct conversion receiver), using relaxation oscillators. The main drawback to the relaxation oscillator is excessive phase noise, and this phase noise is the reason why relaxation oscillators are not more widely used. A very attractive feature of relaxation oscillators is that they can be realized in integrated-circuit form.

## ▮ 7.10

### INTEGRATED-CIRCUIT OSCILLATORS

Oscillators can also be realized using integrated circuits, but their performance is not as good as a well-designed and constructed discrete-component oscillator. Discrete-component oscillators can be built with a better noise performance and are capable of operating at higher frequencies than the IC oscillators. Many IC oscillators that do not use external components are of the relaxation oscillator type shown in Fig.7.50. This circuit contains a free-running multivibrator using transistor $Q_1$ and $Q_2$. Transistors $Q_3$ and $Q_4$ function as constant current sources. The emitter resistors $R_E$ linearize the voltage-current relationship so that the current is proportional to the control voltage $V_c$. The frequency of oscillation is

$$f = \frac{I_o}{4CV}$$

where the voltage $V$ is the voltage required to turn the switching transistor ($Q_1$ or $Q_2$) off (approximately 0.6 $V$). Since

$$I_o \approx \frac{V_c - V_{bc}}{R_E}$$

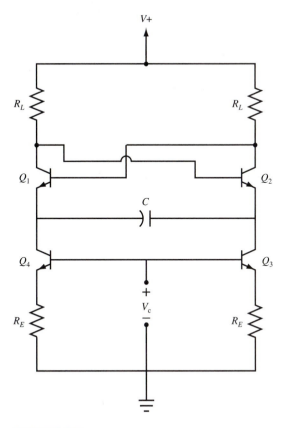

**FIGURE 7.50**
An IC oscillator.

the oscillating frequency is directly proportional to the control voltage. The oscillator output can be either a square or a triangular waveform, but additional wave-shaping circuitry is required to obtain a sinusoidal output waveform.

**EXAMPLE 7.10.** Design a 20-kHz oscillator using the configuration shown in Fig. 7.50. Use SPICE to simulate your design and plot the following waveforms: the capacitor voltage to ground, the capacitor voltage, the $Q_1$ and $Q_2$ collector voltages, and the $Q_1$ and $Q_2$ collector-to-emitter voltages. Assume $V_{be} = 0.6$ V, V+ $= 9$ V, $R_C = 2$ k$\Omega$, $R_E = 1$ k$\Omega$, and $C = 20$ nF. In your SPICE design, you may need to add a transient in the supply voltage in order to trigger the oscillations. For example,

```
VCC   1 0 DC 0 AC .0000001 PWL
+0   0
+.0000001 4.5
+.0000002 12.0
+.001 9.0
```

*Solution.* Given these circuit parameters, the $I_0 = 1.12$ mA and the control voltage needs to be 1.66 V. The schematic and required waveforms are shown in Fig. 7.51*a* and Fig. 7.51*b*.

## NE602

In many communication circuits the frequency of a waveform must be modified. One of the most common ways to modify the frequency of a waveform is to mix it with a local oscillator, as shown in Fig. 7.52*a*. The ideal output of this operation is the sum and difference of the input frequency and local oscillator frequency. Fig. 7.52*b* shows part of a basic superheterodyne receiver. This type of receiver uses

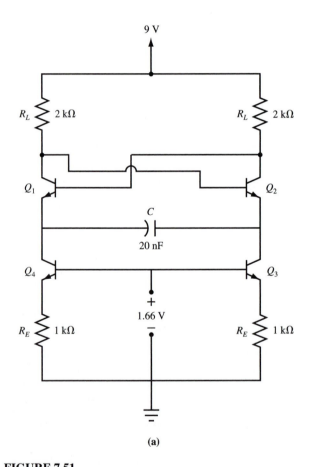

(a)

**FIGURE 7.51**
(*a*) Schematic of 20-kHz relaxation oscillator; (*b*) voltage waveforms at various nodes in the circuit.

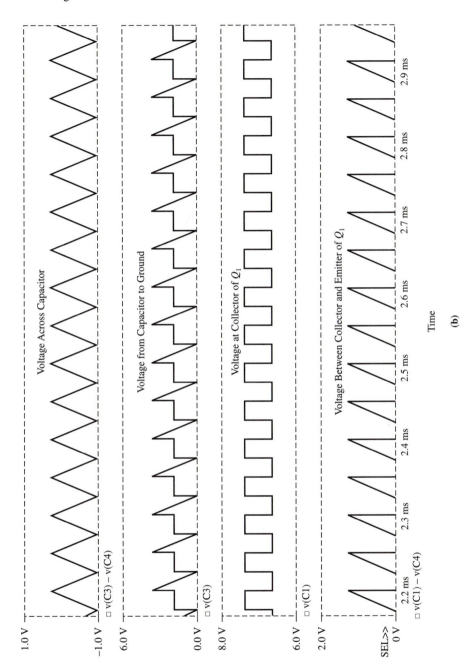

**FIGURE 7.51 (continued)**

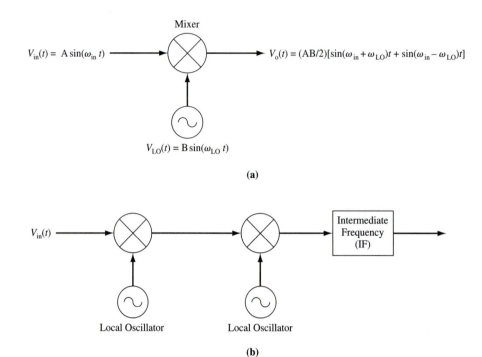

**FIGURE 7.52**
(a) Local oscillator and mixer; (b) part of superheterodyne receiver.

several stages of mixing to down-convert the input signal to an intermediate frequency (IF) band where the information can be demodulated by the appropriate detector. Since frequency conversion is commonplace in communication circuits, special integrated circuits have been developed to include both a mixer and local oscillator. The Phillips (Signetics) NE602 shown in the data sheets of Fig. 7.54 is an example of such an *IC*.

The NE602 includes a Gilbert cell multiplier (see Chapter 12) which acts as a double-balanced mixer. This type of mixer isolates the output sum and difference frequencies from the two input frequencies, thereby eliminating the need to tune out the local oscillator signal. The input and output resistance are $1.5 \text{ k}\Omega$ and the oscillator will operate up to 200 MHz and the mixer up to 500 MHz. Notice from Figure 4 of the NE602 data sheet that the oscillator requires an external tuning network to set the frequency of oscillations.

EXAMPLE 7.11. Design a local oscillator tuned network to drive the NE602 oscillator input in order to convert an input signal with frequency range from 87.5 MHz to 108.0 MHz to a 10.7 MHz intermediate frequency. Ignore the image frequency in your design. Use the varactor data sheets in Fig. 7.43 and if a crystal is required, use the data from Table 7.1.

**Solution.**  In order to convert the input waveform from 87.5 MHz to 10.7 MHz the local oscillator would need to be 76.8 MHz (low-side injection) or 98.2 MHz (high-side injection). To convert the input waveform from 108.0 MHz the local oscillator would need to be 97.3 MHz  (low-side injection) or 118.7 MHz (high-side injection). Using low-side injection, the frequency range of the local oscillator is given by 76.8 to 97.3 MHz. The frequency of oscillation is given by the equation

$$f = \frac{1}{2 \cdot \pi \sqrt{L \cdot C}}$$

Using this equation the inductance-capacitance product can be found using

$$L \cdot C = \left( \frac{1}{2 \cdot \pi \cdot f} \right)^2$$

For the frequency range of 76.8 to 97.3 MHz, this value varies from 4.29E-18 to 2.68E-18. First, choose $L = 0.05 \, \text{uH}$. This gives a value of $C$ that varies between 86 and 53.5 pF. The impedance looking into the NE602 oscillator (pins 6 & 7) is given to be a $1.5 \text{-k}\Omega$ resistor in parallel with 3 pF of capacitance. Therefore, we need a capacitance that varies from 83 pF to 50.5 pF. From the varactor data sheets given in Fig. 7.43, the MV2109 has a capacitance of 30 pF for a 6V reverse voltage and capacitance of 65 pF for a 0.5-V reverse voltage. Place a 20-pF capacitance in parallel with these varactors to offset their capacitance into the tuning range and the $LC$ network is complete. A Colpitts oscillator is chosen to implement the design as shown in Fig. 7.53.

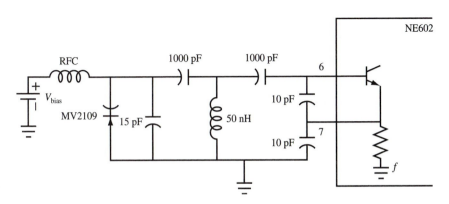

**FIGURE 7.53**
The Colpitts implementation for Example 7.11 using varactor tuning.

Philips Semiconductors RF Communications Products                          Product specification

## Double-balanced mixer and oscillator                                          NE/SA602A

### DESCRIPTION
The NE/SA602A is a low-power VHF monolithic double-balanced mixer with input amplifier, on-board oscillator, and voltage regulator. It is intended for high performance, low power communication systems. The guaranteed parameters of the SA602A make this device particularly well suited for cellular radio applications. The mixer is a "Gilbert cell" multiplier configuration which typically provides 18dB of gain at 45MHz. The oscillator will operate to 200MHz. It can be configured as a crystal oscillator, a tuned tank oscillator, or a buffer for an external LO. For higher frequencies the LO input may be externally driven. The noise figure at 45MHz is typically less than 5dB. The gain, intercept performance, low-power and noise characteristics make the NE/SA602A a superior choice for high-performance battery operated equipment. It is available in an 8-lead dual in-line plastic package and an 8-lead SO (surface-mount miniature package).

### FEATURES
- Low current consumption: 2.4mA typical
- Excellent noise figure: <4.7dB typical at 45MHz
- High operating frequency
- Excellent gain, intercept and sensitivity
- Low external parts count; suitable for crystal/ceramic filters
- SA602A meets cellular radio specifications

### PIN CONFIGURATION

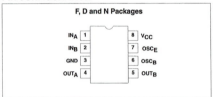

F, D and N Packages

| | | |
|---|---|---|
| IN$_A$ 1 | | 8 V$_{CC}$ |
| IN$_B$ 2 | | 7 OSC$_E$ |
| GND 3 | | 6 OSC$_B$ |
| OUT$_A$ 4 | | 5 OUT$_B$ |

### APPLICATIONS
- Cellular radio mixer/oscillator
- Portable radio
- VHF transceivers
- RF data links
- HF/VHF frequency conversion
- Instrumentation frequency conversion
- Broadband LANs

### ORDERING INFORMATION

| DESCRIPTION | TEMPERATURE RANGE | ORDER CODE | DWG # |
|---|---|---|---|
| 8-Pin Plastic Dual In-Line Plastic (DIP) | 0 to +70°C | NE602AN | 0404B |
| 8-Pin Plastic Small Outline (SO) package (Surface-mount) | 0 to +70°C | NE602AD | 0174C |
| 8-Pin Ceramic Dual In-Line Package (Cerdip) | 0 to +70°C | NE602AFE | 0580A |
| 8-Pin Plastic Dual In-Line Plastic (DIP) | -40 to +85°C | SA602AN | 0404B |
| 8-Pin Plastic Small Outline (SO) package (Surface-mount) | -40 to +85°C | SA602AD | 0174C |
| 8-Pin Ceramic Dual In-Line Package (Cerdip) | -40 to +85°C | SA602AFE | 0580A |

### ABSOLUTE MAXIMUM RATINGS

| SYMBOL | PARAMETER | | RATING | UNITS |
|---|---|---|---|---|
| V$_{CC}$ | Maximum operating voltage | | 9 | V |
| T$_{STG}$ | Storage temperature range | | -65 to +150 | °C |
| T$_A$ | Operating ambient temperature range | NE602A | 0 to +70 | °C |
| | | SA602A | -40 to +85 | °C |
| θ$_{JA}$ | Thermal impedance | D package | 90 | °C/W |
| | | N package | 75 | °C/W |

*(a)*

**FIGURE 7.54**

Philips Semiconductors RF Communications Products — Product specification

## Double-balanced mixer and oscillator — NE/SA602A

### BLOCK DIAGRAM

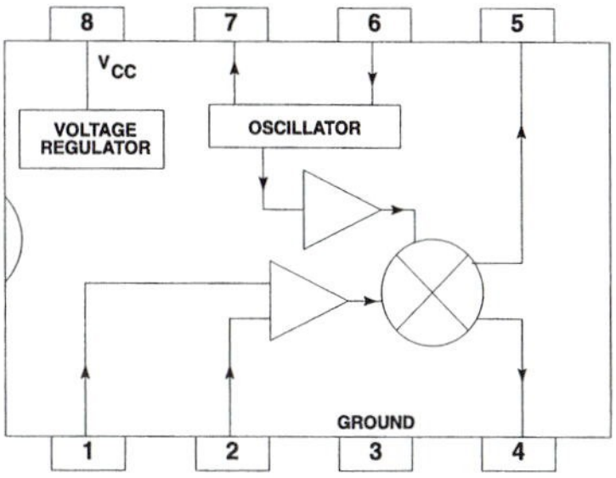


### AC/DC ELECTRICAL CHARACTERISTICS

$V_{CC}$ = +6V, $T_A$ = 25°C; unless otherwise stated.

| SYMBOL | PARAMETER | TEST CONDITIONS | LIMITS NE/SA602A | | | UNITS |
|---|---|---|---|---|---|---|
| | | | MIN | TYP | MAX | |
| $V_{CC}$ | Power supply voltage range | | 4.5 | | 8.0 | V |
| | DC current drain | | | 2.4 | 2.8 | mA |
| $f_{IN}$ | Input signal frequency | | | 500 | | MHz |
| $f_{OSC}$ | Oscillator frequency | | | 200 | | MHz |
| | Noise figure at 45MHz | | | 5.0 | 5.5 | dB |
| | Third-order intercept point | $RF_{IN}$ = -45dBm: $f_1$ = 45.0MHz $f_2$ = 45.06MHz | | -13 | -15 | dBm |
| | Conversion gain at 45MHz | | 14 | 17 | | dB |
| $R_{IN}$ | RF input resistance | | 1.5 | | | kΩ |
| $C_{IN}$ | RF input capacitance | | | 3 | 3.5 | pF |
| | Mixer output resistance | (Pin 4 or 5) | | 1.5 | | kΩ |

### DESCRIPTION OF OPERATION

The NE/SA602A is a Gilbert cell, an oscillator/buffer, and a temperature compensated bias network as shown in the equivalent circuit. The Gilbert cell is a differential amplifier (Pins 1 and 2) which drives a balanced switching cell. The differential input stage provides gain and determines the noise figure and signal handling performance of the system.

The NE/SA602A is designed for optimum low power performance. When used with the SA604 as a 45MHz cellular radio second IF and demodulator, the SA602A is capable of receiving -119dBm signals with a 12dB S/N ratio. Third-order intercept is typically -13dBm (that is approximately +5dBm output intercept because of the RF gain). The system designer must be cognizant of this large signal limitation. When designing LANs or other closed systems where transmission levels are high, and small-signal or signal-to-noise issues are not critical, the input to the NE602A should be appropriately scaled.

Besides excellent low power performance well into VHF, the NE/SA602A is designed to be flexible. The input, RF mixer output and oscillator ports can support a variety of configurations provided the designer understands certain constraints, which will be explained here.

The RF inputs (Pins 1 and 2) are biased internally. They are symmetrical. The equivalent AC input impedance is approximately 1.5k II 3pF through 50MHz. Pins 1 and 2 can be used interchangeably, but they should not be DC biased externally. Figure 3 shows three typical input configurations.

The mixer outputs (Pins 4 and 5) are also internally biased. Each output is connected to the internal positive supply by a 1.5kΩ resistor. This permits direct output termination yet allows for balanced output as well. Figure 4 shows three single ended output configurations and a balanced output.

The oscillator is capable of sustaining oscillation beyond 200MHz in crystal or tuned tank configurations. The upper limit of operation is determined by tank "Q" and required drive levels. The higher the "Q" of the tank or the smaller the required drive, the higher the permissible oscillation frequency. If the required LO is beyond

April 17, 1990

(*b*)

**FIGURE 7.54 (continued)**

Philips Semiconductors RF Communications Products                                                    Product specification

## Double-balanced mixer and oscillator                                                      NE/SA602A

oscillation limits, or the system calls for an external LO, the external signal can be injected at Pin 6 through a DC blocking capacitor. External LO should be at least 200mV$_{P-P}$.

Figure 5 shows several proven oscillator circuits. Figure 5a is appropriate for cellular radio. As shown, an overtone mode of operation is utilized. Capacitor C3 and inductor L1 suppress oscillation at the crystal fundamental frequency. In the fundamental mode, the suppression network is omitted.

Figure 6 shows a Colpitts varactor tuned tank oscillator suitable for synthesizer-controlled applications. It is important to buffer the output of this circuit to assure that switching spikes from the first counter or prescaler do not end up in the oscillator spectrum. The

dual-gate MOSFET provides optimum isolation with low current. The FET offers good isolation, simplicity, and low current, while the bipolar transistors provide the simple solution for non-critical applications. The resistive divider in the emitter-follower circuit should be chosen to provide the minimum input signal which will assure correct system operation.

When operated above 100MHz, the oscillator may not start if the Q of the tank is too low. A 22kΩ resistor from Pin 7 to ground will increase the DC bias current of the oscillator transistor. This improves the AC operating characteristic of the transistor and should help the oscillator to start. A 22kΩ resistor will not upset the other DC biasing internal to the device, but smaller resistance values should be avoided.

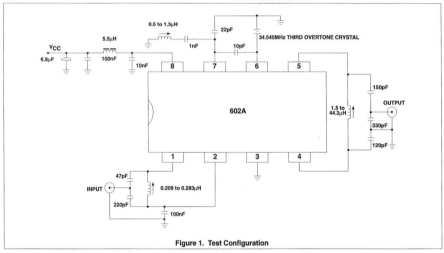

**Figure 1. Test Configuration**

($c$)

April 17, 1990

**FIGURE 7.54 (continued)**

Philips Semiconductors RF Communications Products

Product specification

## Double-balanced mixer and oscillator

## NE/SA602A

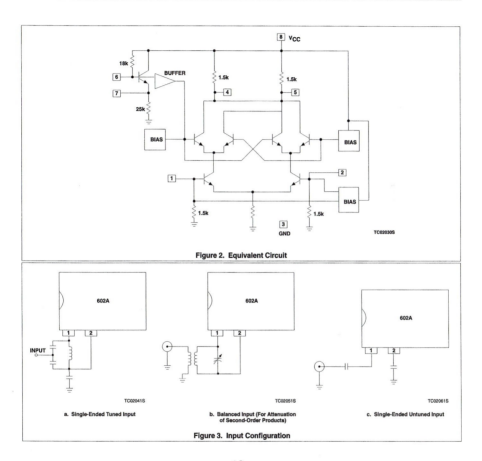

Figure 2. Equivalent Circuit

a. Single-Ended Tuned Input

b. Balanced Input (For Attenuation of Second-Order Products)

c. Single-Ended Untuned Input

Figure 3. Input Configuration

(d)

April 17, 1990

**FIGURE 7.54 (continued)**

Philips Semiconductors RF Communications Products                              Product specification

## Double-balanced mixer and oscillator                                      NE/SA602A

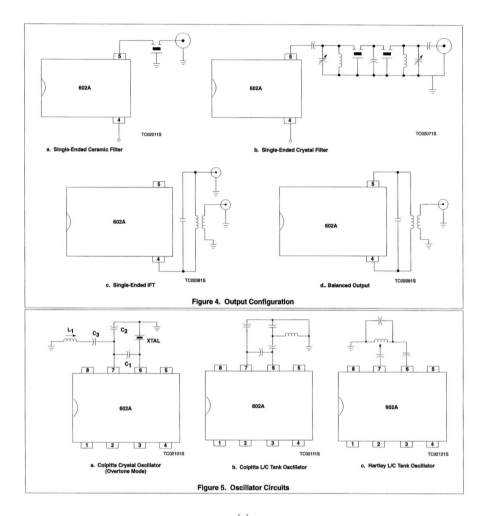

a. Single-Ended Ceramic Filter

b. Single-Ended Crystal Filter

c. Single-Ended IFT

d.. Balanced Output

**Figure 4.  Output Configuration**

a. Colpitts Crystal Oscillator
(Overtone Mode)

b. Colpitts L/C Tank Oscillator

c. Hartley L/C Tank Oscillator

**Figure 5.  Oscillator Circuits**

(*e*)

April 17, 1990

**FIGURE 7.54 (continued)**

Philips Semiconductors RF Communications Products                    Product specification

## Double-balanced mixer and oscillator                    NE/SA602A

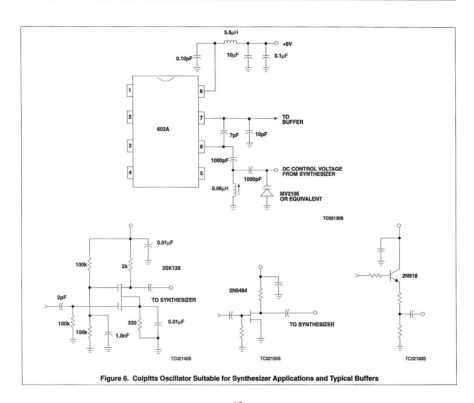

Figure 6.  Colpitts Oscillator Suitable for Synthesizer Applications and Typical Buffers

(f)

April 17, 1990

**FIGURE 7.54 (continued)**

Philips Semiconductors RF Communications Products                                      Product specification

## Double-balanced mixer and oscillator                                   NE/SA602A

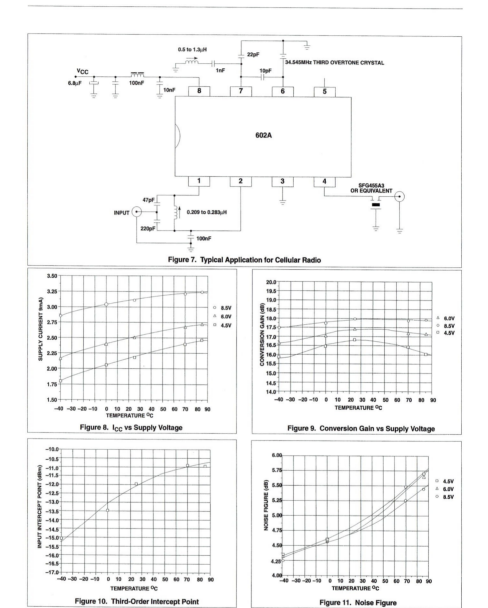

Figure 7. Typical Application for Cellular Radio

Figure 8. $I_{CC}$ vs Supply Voltage

Figure 9. Conversion Gain vs Supply Voltage

Figure 10. Third-Order Intercept Point

Figure 11. Noise Figure

April 17, 1990                                      (g)

**FIGURE 7.54 (continued)**

Philips Semiconductors RF Communications Products                                      Product specification

## Double-balanced mixer and oscillator                                      NE/SA602A

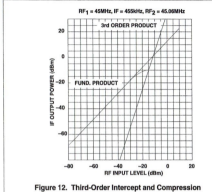

Figure 12.  Third-Order Intercept and Compression

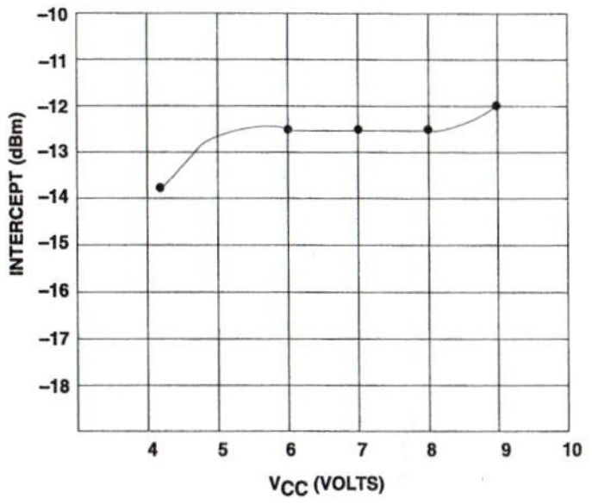

Figure 13.  Input Third-Order Intermod Point vs $V_{CC}$

$(h)$

April 17, 1990

**FIGURE 7.54 (continued)**

### ▨ 7.11

## PROBLEMS

**7.1** The circuit of Fig. P7.1 represents the small-signal equivalent circuit of a two-stage amplifier with feedback from output to input. Determine the value of $L$ required for the circuit to oscillate at 10 MHz. What is the minimum value of $g_m$ required for the circuit to oscillate at this frequency?

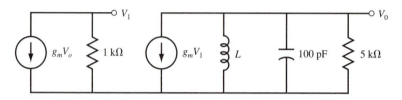

**FIGURE P7.1**
A second-order feedback circuit.

**7.2** Show that if one or more of $Z_1$, $Z_2$, and $Z_3$ of Eq. (7.18) are positive real resistors, then there are no conditions for which the circuit will oscillate.

**7.3** Determine the value of inductance $L$ and the turns ratio $N_1/N_2$ so that the circuit illustrated in Fig. P7.3 will oscillate at 5 MHz. The loop gain should initially be approximately equal to 3. Assume the transistor input impedance is sufficiently large so that it does not load down the autotransformer. (The transistor $\beta = 100$.)

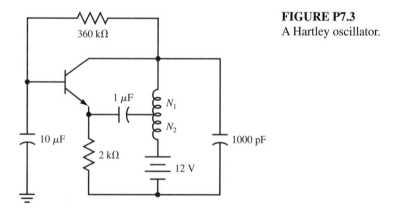

**FIGURE P7.3**
A Hartley oscillator.

**7.4** The FET in the circuit of Fig. P7.4 is biased so that $g_m = 5$ mS. Determine capacitors $C_1$ and $C_2$ so that the circuit will oscillate at 10 MHz. The open-loop gain should be at least 2.5 to ensure that oscillations begin. The unloaded inductor $Q_u = 100$.

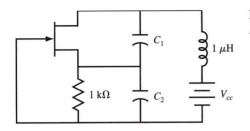

**FIGURE P7.4**
A common-gate Colpitts oscillator.

**7.5**  The FET in the circuit illustrated in Fig. P7.5 is biased so that $g_m = 5$ mS. Determine the inductance $L$ and $N_1/N_2$ so that the circuit will oscillate at 10 MHz.

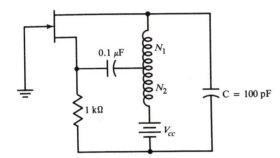

**FIGURE P7.5**
A common-gate Hartley oscillator.

**7.6**  Design a Colpitts 3.5-MHz oscillator using a 1.5-$\mu$H coil with a $Q_u$ of 150. The load resistance is $4\,\text{k}\Omega$, and the transistor has a minimum $\beta$ of 100. The supply voltage is 12 V. Specify the complete circuit, including bias resistors.

**7.7**  Repeat the oscillator design of Prob. 7.6, using an FET with a $g_m$ of $6 \times 10^{-3}$ S.

**7.8**  Design a 20-MHz oscillator using a 2N3904 transistor ($\beta_{\min} = 100$) and a 12-V supply. Specify the complete circuit, including bias resistors. Assume all inductors have $Q_u = 100$.

**7.9**  If a capacitor is added in series with a crystal, does the antiresonant frequency of the composite circuit change?

**7.10**  Design a 15-MHz crystal-controlled oscillator using a crystal that is antiresonant at 15 MHz, provided a 32-pF load is connected across it.

**7.11**  A crystal has $C_o = 3$ pF, $C_1 = 0.01$ pF, $L_1 = 0.1$ H, and $r = 15\ \Omega$. Calculate the series and parallel resonant frequencies of the crystal. How much capacitance must be added to change the antiresonant frequency by 0.01 percent? How much capacitance must be added to change the series resonant frequency by 0.01 percent?

**7.12** Calculate the output impedance of the common-base amplifier shown in Fig. P7.12. Under what conditions will this circuit oscillate when the inductor with a finite $Q_u$ is connected across it.

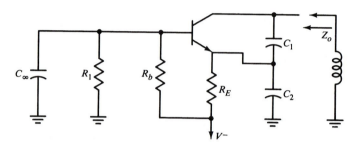

**FIGURE P7.12**
A common-base oscillator.

**7.13** Derive an expression for the loop gain of the grounded-base crystal oscillator illustrated in Fig. P7.13. Show that the loop gain is always less than 1.

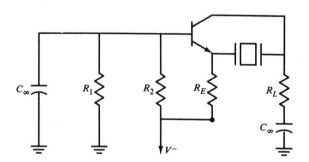

**FIGURE P7.13**
A common-base crystal oscillator.

**7.14** Figure P7.14 illustrates an FET Pierce oscillator. Assume the 10-MHz crystal of Table 7.1 is antiresonant at 10 MHz if the external load capacitance is 32 pF. Select $C_1$ and $C_2$ and determine the minimum $g_m$ for this circuit to oscillate at 10 MHz. The transistor input impedance is 10 M$\Omega$ shunted by 3 pF, and the transistor output impedance is 15 k$\Omega$.

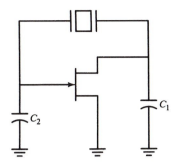

**FIGURE P7.14**
An FET Pierce crystal oscillator.

**7.15** Design a series resonant 40-MHz crystal oscillator. The supply voltage is 20 V. Show all component values, including the bias network. All inductors have a $Q_u = 100$.

**7.16** The circuit shown in Fig. P7.16 is frequently used as a tuned-input, tuned-output amplifier. Show that the collector-to-base capacitance $C_1$ can cause the circuit to oscillate.

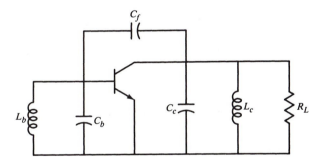

**FIGURE P7.16**
A tuned-input, tuned-output amplifier with capacitive feedback.

**7.17** The 45-MHz third-overtone crystal described in Table 7.1 is used in a series-mode oscillator. An inductor with a $Q$ of 100 is placed across the crystal to resonate out $C_o$ at 45 MHz. Estimate the resulting $Q$ of the combination at 45 MHz.

**7.18** Derive an expression for the change in the series resonant frequency if an inductor is added in series with a crystal.

**7.19** Derive an expression for the change in the antiresonant frequency if an inductor is added in parallel with a crystal.

**7.20** Given a crystal that is antiresonant at 10 MHz with a 32-pF load and that has $r_i = 60\,\Omega$ plus an NPN transistor with a minimum $\beta$ of 100 and a 12-V supply, design a 10-MHz oscillator. Show all circuit details, including the bias network. Any additional components used can be assumed to be ideal.

**7.21** The circuit shown in Fig. P7.21 uses a time-delay network in the feedback path. What value of $\tau$ will be required for the circuit to oscillate at 10 MHz?

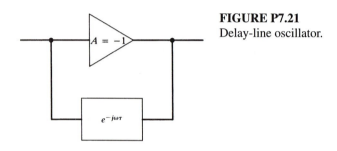

**FIGURE P7.21**
Delay-line oscillator.

**7.22** The FET of the 20-MHz circuit illustrated in Fig. P7.22 is biased so that $g_m = 4$ mS. If $C_1$ and $C_2$ are each 64 pF, what is the maximum value of crystal resistance $r_1$ for which the circuit will oscillate? Show how you would modify the circuit so that you can use a 20-MHz series-mode crystal with one terminal connected to the transistor gate.

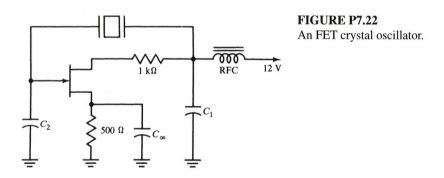

**FIGURE P7.22**
An FET crystal oscillator.

**7.23** Select values for $C_1$ and $C_2$ that will enable the circuit shown in Fig. P7.23 to oscillate at 40 MHz. The inductor $Q_u = 100$. What will be the minimum $g_m$ of the transistor necessary for the circuit to oscillate with the selected capacitance values?

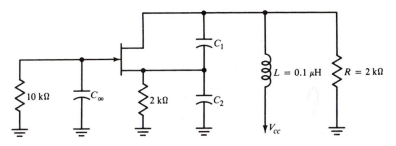

**FIGURE P7.23**
A grounded-gate oscillator.

**7.24** Consider the series-mode oscillator circuit of Fig. P7.24. It uses the 45-MHz crystal described in Table 7.1.

    (*a*) What value of capacitance must be added to increase the oscillating frequency by 0.01 percent?

    (*b*) Where should the capacitor be added?

    (*c*) What value of $L$ is required? Assume the inductor $Q$ is infinite.

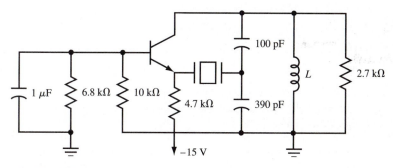

**FIGURE P7.24**
A grounded-base oscillator.

**7.25** Figure P7.25 illustrates an FET VCO with a buffered output stage. Explain the operation of the circuit, including the purpose of each transistor, and estimate the turns ratio $N_1/N_2$ required for an open-loop gain of 3.

**7.26** Design a 9.5-MHz crystal oscillator using a crystal specified in Table 7.1 (the required crystal load is 32 pF). Use an NPN transistor (minimum beta of 100), and show the complete circuit, including biasing. Estimate the $Q$ of the resulting circuit. Describe how and by how much the oscillating frequency could be changed.

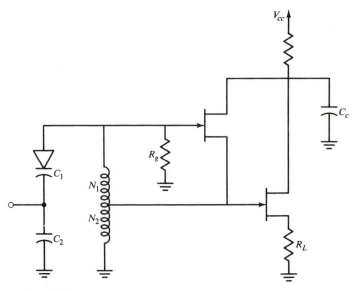

**FIGURE P7.25**
A voltage-controlled oscillator.

**\*7.27**    Design and evaluate a 10-MHz Colpitts oscillator (common-collector). The transistor model is: MODEL MOD1 NPN BF = 100, VAF = 50, IS =
1.E-12, TF =.6 NS, TR = 60 NS, CJE = 5 pF, CJC = 5 pF, MJC = 0.5,
MJE = 0.3333, VJC = 0.75, VJE = 0.75. Use a single 9-V supply; the unloaded
$Q$ of any of the inductors is 100. What is the minimum value of unloaded $Q$ for
which the circuit will oscillate? Does this agree with the simulation results? (The
transconductance should be kept fixed.)

**\*7.28**    (*a*)  Simulate the 5.7-MHz crystal whose parameters are given in Table 7.1. Add a
trimmer capacitance, if necessary, to move the antiresonant frequency as close
to 5.7 MHz as possible.

  (*b*)  Design a 5.7-MHz Pierce oscillator using the crystal of part *a*. Use a single 9-
V supply and a bipolar junction transistor with a $\beta$ of 100.

## ▉ 7.12

### REFERENCES

1. G. G. Gouriet, High Stability Oscillator, *Wireless Engineer,* April, pp. 105–112.
2. M. M. Driscoll, Two-Stage Self-Limiting Series Mode Type Quartz Crystal Oscillator
   Exhibiting Improved Short-Term Frequency Stability, *IEEE, Trans. on Instrumentation
   and Measurement,* **22:** 130–138 (1973).
3. D. Firth, *Quartz Crystal Oscillator Circuits, Design Handbook,* Magnavox Co., Fort
   Wayne, IN, 1965. Available from Nat. Tech. Information Service as AD 460377.
4. H. G. Vollmers and L. T. Claiborne, R. F. Oscillator Control Utilizing Surface Wave
   Delay Lines, *Proc. 28th Ann. Frequency Control Symposium,* 1974, pp. 256–259.

## ▩ 7.13
### ADDITIONAL READING

Adler: A Study of Locking in Oscillators, *Proceedings. IRE,* **3 4**: 351–357, (1946).

Anderson, T. C., and F. G. Merrill: Crystal Controlled Primary Frequency Standards: Latest Advances for Long Term Stability, *IRE Trans. on Instrumentation* **9**: 136–140 (1960).

Banu, M.: MOS Oscillators with Multi-Decade Tuning Range and Gigahertz Maximum Speed, *IEEE J. Solid-State Circuits,* **23**: 1386–1393 (December 1988).

Barnes, J. A., et. al.: Characterization of Frequency Stability, U.S. Dept. of Comm. Natl. Bureau of Standards, *NBS Technical Note 394,* 1970.

Baxandall, P. J.: Transistor Crystal Oscillators and the Design of a 1 Mc/s Oscillator Circuit Capable of Good Frequency Stability, *Radio and Electronic Engineer,* April 1965.

Buchwald, A. W., and K. W. Martin: High-Speed Voltage-Controlled Oscillator with Quadrature Outputs, *Electronics Letters,* **27**: 309–310 (1990).

Campbell, Colin: *Surface Acoustic Wave Devices and Their Signal Processing Applications,* Academic Press, New York, 1989.

Clapp, J. K.: An Inductance Capacitance Oscillator of Unusual Frequency Stability, *Proc. IRE,* **36**: 356–358 (1949).

Cote, A. J., Jr.: Matrix Analysis of Oscillators and Transistor Applications, *IRE Trans. on Circuit Theory,* **5**: 181–189 (1958).

Edson, W. A.: *Vacuum Tube Oscillators,* Wiley, New York, 1953.

Felch, E. O., and J. O. Israel: A Simple Circuit for Frequency Standards Employing Overtone Crystals, *Proc. IRE,* **43**: 596–603 (1955).

Frerking, M. R.: *Crystal Oscillator Design and Temperature Compensation,* Van Nostrand Reinhold, New York, 1978.

Gerber, E. A., and R. A. Sykes: State of the Art Quartz Crystal Units and Oscillators, *Proc. IEEE,* **54**: 103–116 (1966).

Heising, R. A.: *Quartz Crystals for Electrical Circuits,* Van Nostrand, New York, 1946.

Helle, J.: VCXO Theory and Practice, *Proc. 29th Ann. Symp. on Frequency Control,* 1975, pp. 300–307.

Kent, R. L.: The Voltage Controlled Crystal Oscillator (VCXO), Its Capabilities and Limitations, *Proc. 19th Ann. Freq. Control Symposium,* 1965, pp. 642–654.

Kukielka, J.F., and R. G. Meyer: A High-Frequency Temperature-Stable Monolithic VCO, *Circuits,. IEEE J. Solid-State* **16**: 639–647 (1981).

Kushner, L. J.: The Composite DDS—A New Direct Digital Synthesizer Architecture, *Proceedings of the 47th IEEE Annual International Frequency Control Symposium,* 1993, pp. 255–260.

Lane, M.: Transistor Crystal Oscillators to Cover Frequency Range 1 kHz–100 MHz, Australian Post Office Research Laboratories, Report 6513, 1970.

Layden, O. P., W. L. Smith, A. E. Anderson, M. B. Bloch, D. E. Newell, and P. C. Sulzer: Crystal Controlled Oscillator, *IEEE Trans. on Instrumentation and Measurement,* **21**: 277–286 (1972).

Marker, Thomas F.: Crystal Oscillator Design Notes, *Frequency,* **6**: 12–16 (1968).

Mathys, R. J.: *Crystal Oscillation Circuits,* Wiley, New York, 1983.

Mortley, W. S.: Circuit Giving Linear Frequency Modulation of a Quartz Crystal Oscillator, *Wireless World,* **57**: 399–403 (October 1951).

Parker, T. E., and G. K. Montress: Precision Surface-Acoustic-Wave (SAW) Oscillators. *IEEE Trans. on Ultrasonics, Ferroelectrics, and Frequency Control,* **35**: 342–364 (1988).

Parzen, B.: *Design of Crystal and Other Harmonic Oscillators,* Wiley, New York, 1983.

Smith, W. L.: Miniature Transistorized Crystal Controlled Precision Oscillators, *IRE Trans. on Instrumentation,* **9**: 141–148 (1960).

Sneep, J., and C. J. M. Verhoeven: A New Low-Noise 100-MHz Balanced Relaxation Oscillator, *IEEE J. Solid-State Circuits* **25:** 692–698 (1990).

Steyaert, M., and R. Roovers: A 1-GHz Single-Chip Quadrature Modulator, *IEEE J. Solid State Circuits* **27:** 1194–1197 (1992).

Verhoeven, C. J. M.: A High-Frequency Electronically Tunable Quadrature Oscillator, *IEEE J. Solid-State Circuits,* **27:** 1097–1100 (1992).

Warner, A. W.: High Frequency Crystal Units for Primary Frequency Standards, *Proc. IRE,* **40:** 1030–1033 (1952).

Ziegler, R. R.: Know Your Oscillators, *Microwave J.,* June 1976, pp. 44–47.

# Phase-Locked Loops

## ▦ 8.1
### INTRODUCTION

A *phase-locked loop* (*PLL*) is a feedback system in which the feedback signal is used to lock the output frequency and phase to the frequency and phase of an input signal. The input waveform can be of many different types, including sinusoidal or digital. The first known application of the phase-locked technique was in 1932 for the synchronous detection of radio signals.[1,2] These early applications were all concerned with the detection of a transmitted signal.

Starting in the 1960s, the NASA satellite programs used the phase-locked technique to determine the frequency of the signals transmitted by satellites. Although the transmission was designed to take place at 108 MHz, oscillator drift and Doppler shift resulted in an uncertainty of several kilohertz in the received signal. The transmitted signal was of very narrow bandwidth, but because of the frequency drifts it was necessary that the receiver bandwidth be much wider, with a resultant increase in noise power. (It was demonstrated in Chap. 3 that the receiver noise power is propotional to the bandwidth.) However, the satellite communication system was improved by using a phase-locked loop to lock onto the transmitted frequency, and thus permit a much narrower receiver bandwidth with much less output noise power.

The phase-locked loop has been used for filtering, frequency synthesis, motor-speed control, frequency modulation, demodulation, signal detection, and a variety of other applications. The realization of the phase-locked loop as a relatively inexpensive integrated circuit has made it one of the most frequently used communication circuits. Phase-locked loops can be analog or digital, but the majority are composed of both analog and digital components. Some authors apply the term *Digital PPL* to a digital phase-locked loop that contains one or more digital components. But since virtually all PLLs contain digital components, in this book the notation *digital phase-locked loop* (*DPLL*) will be reserved for PLLs in which all the components are digital.

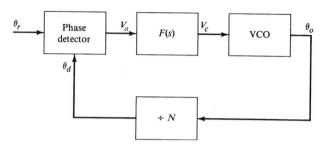

**FIGURE 8.1**
Block diagram of a phase-locked loop.

Figure 8.1 illustrates the basic architecture of the phase-locked loop. The phase detector generates an output signal that is a function of the difference between the phases of the two input signals. The detector output is filtered (and perhaps amplified), and the dc component of the error signal is applied to the voltage-controlled oscillator. The signal fed back to the phase detector is the VCO output frequency divided by $N$. The VCO control voltage $V_c(t)$ forces the VCO to change frequency in the direction that reduces the difference between the input frequency and the divider output frequency. If the two frequencies are sufficiently close, the PLL feedback mechanism forces the two-phase detector input frequencies to be equal, and the VCO is "locked" with the incoming frequency. That is,

$$f_r = f_d$$

and the divider output frequency is

$$f_d = \frac{f_o}{N}$$

The output frequency

$$f_o = Nf_r$$

is an integral multiple of the input frequency. If a divider is not used, $N$ equals 1. Once the loop is in lock, there will be a small phase difference between the two phase detector input signals. This phase difference results in a dc voltage at the phase detector output which is required to shift the VCO from its free-running frequency and keep the loop in lock. (This is not true for type II PLLs, which are described in the next chapter.)

The self-correcting ability of the PLL allows it to track frequency changes in the input signal once it is locked. The range of frequencies over which the PLL can remain locked to an input signal is known as its *lock range*. The *capture range* is the range of frequencies over which the loop can acquire lock, and this range is less than the lock range.

Since the PLL output frequency is an integral multiple of the reference frequency, it can be changed simply by changing the divide ratio $N$. Integrated circuitry has made the digital programmable divider an inexpensive circuit component.

This provides a means of easily generating multiple frequencies from a single input frequency. Frequency synthesis is a major application of PLLs, and an entire chapter is devoted to it in this text. Before examining various applications of the phase-locked loop, we will first develop a mathematical model for the system and examine the characteristics of the various types of phase detectors.

## ■ 8.2
### LINEAR MODEL OF THE PHASE-LOCKED LOOP

Although the PLL is nonlinear because the phase detector is nonlinear, it can be accurately modeled as a linear device when the phase difference between the phase detector input signals is small. For the linear analysis, it is assumed that the phase detector output is a voltage which is a linear function of the difference in phase between its inputs; that is,

$$V_a = K_d(\theta_r - \theta_d) \tag{8.1}$$

where $\theta_r$ and $\theta_d$ are the phases of the input and feedback signals, respectively, and $K_d$ is the *phase detector gain factor* and has units of volts per radian. The characteristics of several types of phase detectors are discussed in detail later in the chapter. It will also be assumed that the VCO can be modeled as a linear device whose output frequency deviates from its free-running frequency by an increment of frequency

$$\Delta\omega = K_o V_c \tag{8.2}$$

where $V_c$ is the VCO input voltage and $K_o$ is the *VCO gain factor,* in units of radians per second per volt. The output frequency is

$$\omega_o = \omega_c + \Delta\omega = \omega_c + K_o V_c$$

where $\omega_c$ is the free-running frequency of the VCO. Since frequency is the time derivative of phase, the VCO operation can be described as

$$\Delta\omega = \frac{d\theta_o}{dt} = K_o V_c \tag{8.3}$$

The output of the frequency divider $f_d$ is the divider input frequency divided by $N$. That is,

$$f_d = \frac{f_o}{N}$$

or, since phase is the time integral of frequency,

$$\theta_d = \frac{\theta_o}{N}$$

For the PLL model, the divide-by-$N$ circuit can be replaced by a frequency-independent scalar equal to $1/N$. With these assumptions, the PLL can be represented by the linear model shown in Fig. 8.2, where $F(s)$ is the transfer function of

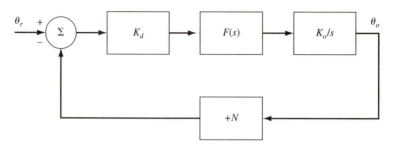

**FIGURE 8.2**
A small-signal linear model of a phase-locked loop.

the low-pass filter. The linear transfer function relating the output phase $\theta_o(s)$ and the input phase $\theta_r(s)$ is

$$\frac{\theta_o(s)}{\theta_r(s)} = \frac{K_d K_o F(s)/s}{1 + K_d K_o F(s)/(Ns)} = \frac{G(s)}{1 + G(s)/N} \tag{8.4}$$

The same transfer function relates the input and output frequencies $f_r(s)$ and $f_o(s)$.
    If no low-pass filter is used, the transfer function is

$$\frac{\theta_o}{\theta_r} = \frac{K_d K_o}{s + K_d K_o/N} = \frac{NK_v}{s + K_v}$$

which is equivalent to the transfer function of a simple low-pass filter with a dc gain of $N$ and a bandwidth equal to $K_v$, where

$$K_v = \frac{K_d K_o}{N}$$

is defined to simplify the notation.
    This PLL is referred to as a *first-order loop* since it can be described by a first-order differential equation and is also of type I.
    With the mathematical model in use here, the phase-locked loop appears to be a low-pass filter, but the output phase and frequency represent deviations from the free-running frequency $\omega_c$. The PLL is actually a bandpass filter centered at the frequency of the input waveform. The phase detector output is a low-frequency signal that is filtered by a low-pass filter. It is much easier to build narrow-bandwidth, low-pass filters than the high-$Q$ filters that would otherwise be required. This is one of the principal advantages of the PLL.

    **EXAMPLE 8.1.** A frequency synthesizer uses a PLL to synthesize a 1-MHz signal from a 25-kHz reference frequency. To realize an output frequency of 1 MHz, a division of

$$N = \frac{10^6}{25 \times 10^3} = 40$$

must be included in the feedback path. If no filtering is included, the closed-loop transfer function will be

$$\frac{\theta_o}{\theta_r} = \frac{K_d K_o/s}{1 + K_d K_o/(sN)} = \frac{K_d K_o}{s + K_d K_o/N}$$

A typical value for $K_d$ is 2 V/rad, and a typical value for the VCO gain factor $K_o$ (for a 1-MHz VCO) is 100 Hz/V. With these values the closed-loop transfer function is

$$\frac{\theta_o}{\theta_r} = \frac{(2 \times 100)2\pi}{s + (2 \times 100 \times 2\pi)/40}$$

The synthesizer bandwidth will be $(2 \times 100)/40$, or 5 Hz.

Normally the loop will also contain a filter to filter out undesirable components from the phase detector and to provide further control over the loop's frequency response. If $F(s)$ is a simple low-pass filter, then

$$F(s) = \left(\frac{s}{\omega_L} + 1\right)^{-1}$$

and the closed-loop transfer function is

$$\frac{\theta_o(s)}{\theta_r(s)} = \frac{NK_v}{s(s/\omega_L + 1) + K_v} = \frac{N}{(s^2/\omega_n^2) + (2\zeta/\omega_n)s + 1} \tag{8.5}$$

where

$$K_v = \frac{K_d K_o}{N} \tag{8.6}$$

$$\omega_n^2 = K_v \omega_L \tag{8.7}$$

$$2\zeta = \frac{\omega_n}{K_v} = \left(\frac{\omega_L}{K_v}\right)^{1/2} \tag{8.8}$$

Equation (8.5) is the general form of the second-order low-pass transfer function. It occurs so frequently in PLL analysis that its characteristics are described in detail here. The magnitude of the steady-state frequency response is

$$\left|\frac{\theta_o}{\theta_r}(j\omega)\right| = \frac{N}{[(1 - \omega^2/\omega_n^2)^2 + (2\zeta\omega/\omega_n)^2]^{1/2}} \tag{8.9}$$

and the phase shift is

$$\arg\frac{\theta_o}{\theta_r}(j\omega) = -\tan^{-1}\frac{2\zeta\omega}{\omega_n(1 - \omega^2/\omega_n^2)} \tag{8.10}$$

The magnitude of the frequency response [Eq. (8.9)] of this second-order transfer function is plotted in Fig. 8.3 for selected values of $\zeta$. For $\zeta = 0.707$, the transfer function becomes the second-order "maximally flat" Butterworth response. For values of $\zeta < 0.707$, the gain exhibits peaking in the frequency domain. The maximum value of the frequency response $M_p$ as a function of the damping ratio can be found by setting the derivative of Eq. (8.9)—with respect to frequency—equal to zero. Then $M_p$ is found to be

$$M_p = \frac{N}{2\zeta(1 - \zeta^2)^{1/2}} \tag{8.11}$$

and the frequency $\omega_p$ at which the maximum occurs is

$$\omega_p = \omega_n(1 - 2\zeta^2)^{1/2} \tag{8.12}$$

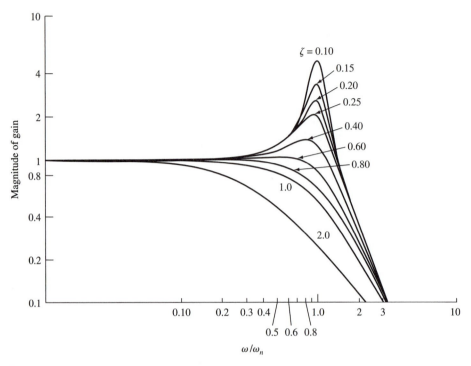

**FIGURE 8.3**
Magnitude of a second-order PLL as a function of frequency for selected damping ratios.

The 3-dB bandwidth $\omega_h$ can be derived by solving for the frequency $\omega_n$ at which the value of Eq. (8.9) is equal to 0.707 for the dc gain $(0.707N)$. And $\omega_h$ is found to be (provided $\zeta < 1$)

$$\omega_h = \omega_n[1 - 2\zeta^2 + (2 - 4\zeta^2 + 4\zeta^4)^{1/2}]^{1/2} \tag{8.13}$$

The time it takes for the output to rise from 10 to 90 percent of its final value is called the *rise time* $t_r$. Rise time is approximately related to the system bandwidth by the relation

$$t_r = \frac{2.2}{\omega_h} \tag{8.14}$$

which is exact for first-order systems.

Normally the designer would like to have the bandwidth narrow for maximum filtering and have the rise time as short as possible so that the loop can follow changes in the input waveform. Equation (8.14) shows that this is not possible; rather, the designer must make a tradeoff between the system speed of response and system bandwidth.

**EXAMPLE 8.2.** Example 8.1 described a frequency synthesizer with $K_v = 10\pi$ rad/s. The closed-loop bandwidth is $10\pi$ rad/s. What value of low-pass filter should be used so that the closed-loop system approximates a second-order Butterworth filter?

***Solution.*** For a Butterworth filter the damping ratio $\zeta = 0.707$. From Eq. (8.8)

$$2\zeta = 1.414 = \left(\frac{\omega_L}{K_v}\right)^{1/2} = \left(\frac{\omega_L}{10\pi}\right)^{1/2}$$

so the required low-pass filter bandwidth is

$$\omega_L = 20\pi \text{ rad/s}$$

The bandwidth of the closed-loop system is [from Eq. (8.13)]

$$\omega_h = \omega_n \qquad (\zeta = 0.707)$$
$$= (K_v\omega_L)^{1/2} = 14.14\pi \text{ rad/s}$$

The corresponding system rise time is estimated to be [Eq. (8.14)]

$$t_r = \frac{2.2}{\omega_n} = 49.4 \times 10^{-3} \text{ s}$$

The system characteristics can be changed by changing the loop gain or the filter bandwidth or by adding a higher-order filter. A detailed analysis of phase-locked loops is provided in the next chapter.

## 8.3

### PHASE DETECTORS (PDs)

PLL performance characteristics vary depending upon the type of phase detector used. The three most frequently used forms of phase detector are the digital detector in which the output signal is restricted to two or three possible levels, the analog mixer or multiplier, and the sampling phase detector. These three types of phase detectors will now be described.

### Digital Phase Detectors

Logic circuits now serve as the most frequently used phase detectors because they are readily available as small, inexpensive integrated circuits. The output of logic circuit PDs is a constant-amplitude pulse whose width is proportional to the phase difference between the two input signals (which can be either analog or digital).

### Exclusive-OR Phase Detectors

The exclusive-OR circuit shown in Fig. 8.4 often serves as one of the simplest types of PDs. The output of the exclusive-OR circuit is high if, and only if, one of the two input signals is high. In digital PDs, *phase error* is defined as

$$\phi_\epsilon = \frac{\tau}{T}2\pi \tag{8.15}$$

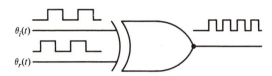

**FIGURE 8.4**
An exclusive-OR phase detector.

where $T$ is the period of the input signals and $\tau$ is the time difference between the leading edges of the two signals. (If the two inputs are not of the same frequency, phase error is ambiguous.) The average value of the exclusive-OR gate output as a function of phase error is plotted in Fig. 8.5. It is assumed that both input signals have a 50 percent duty cycle. The output is a maximum (the gate output is high at all times) when the two signals are $180°$ out of phase. There are two values of phase error for each value of output voltage, but one value will correspond to a negative loop gain and the other value to a positive loop gain. For a positive value of loop gain, the closed-loop system is unstable, and the error will adjust itself to the phase error corresponding to a negative-feedback loop. One disadvantage of the exclusive-OR phase detector is that the output depends on the duty cycle of the input waveforms.

## Flip-Flop Detectors

The simple set-reset flip-flop illustrated in Fig. 8.6 can also be used as a phase detector. The signals, $f_A$ and $f_B$ consisting of narrow pulses, are connected to the set and reset inputs. The average value of the $Q$ output will be proportional to the phase difference between the two signals. The average-voltage-versus-phase transfer characteristic will be as shown in Fig. 8.7. This flip-flop phase detector has an advantage over the exclusive-OR circuit in that the detector has twice the phase range (0 to $2\pi$). That is, the output is $V$ V only when the phase error reaches $2\pi$ rad. A disadvantage of this phase detector is that the output requires more filtering than the

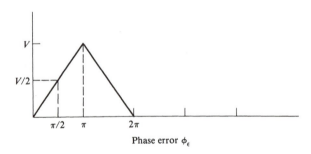

**FIGURE 8.5**
Average voltage output as a function of phase error for the exclusive-OR phase detector.

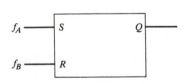

**FIGURE 8.6**
An *RS* flip-flop used as a phase detector.

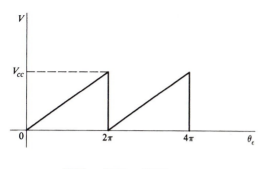

**FIGURE 8.7**
Average voltage output as a function of phase error for the *RS* flip-flop phase detector.

**FIGURE 8.8**
Exclusive-OR and *RS* flip-flop phase detector outputs in response to the detector inputs $f_A$ and $f_B$.

exclusive-OR phase detector output. Consider the timing diagram of Fig. 8.8. (It is assumed here that some means was used to convert the input signals $A$ and $B$ to digital pulses.) The exclusive-OR circuit output is at twice the frequency of the input signals, whereas the flip-flop output frequency is the same as the input frequency. This implies that the first ac component of the exclusive-OR output is twice as fast as that of the flip-flop output, and therefore the low-pass filter requirements will be less stringent if an exclusive-OR phase detector is used. The *RS* flip-flop works best with low duty-cycle input waveforms. The output will have a flat spot of width corresponding to the width of the input waveform, which will have a negative effect on PLL performance.

**EXAMPLE 8.3.** A flip-flop with a 0-V output is used as a phase detector, and the reference frequency is $f_r$. What will be the amplitude and frequency of noise components generated in the phase detector when the loop is in frequency lock?

*Solution.* When the loop is in lock, the phase detector output $\theta_\epsilon(t)$ will be a rectangular pulse train, as shown in Fig. 8.9. The error signal is

$$\theta_\epsilon(t) = \sum_{n=0}^{\infty} p(t - nT)$$

where
$$p(t) = V \qquad 0 \leq t \leq \tau$$
$$= 0 \qquad T \geq t > \tau$$

and
$$T = f_r^{-1}$$

and $\tau$ is the time delay between the reference pulse and the divider output. If the time origin is shifted by $\tau/2$ (which does not alter the amplitude of the harmonics), $\theta_\epsilon(t)$ can

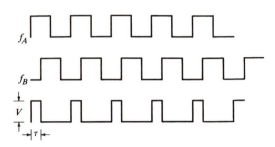

**FIGURE 8.9**
Output of an *RS* flip-flop phase detector to input signals $f_A$ and $f_B$.

be expanded in a Fourier series as

$$\theta_\epsilon(t) = \sum_{n=0}^{\infty} C_n \cos n\omega_r t$$

where

$$C_o = \frac{V}{T}\tau$$

and

$$C_n = \frac{2V}{T}\frac{\sin(n\omega_r\tau/2)}{n\omega_r}$$

Each $C_n$ is a maximum when $\sin(n\omega_r\tau/2) = 1$. That is,

$$\frac{n\omega_r\tau}{2} = \frac{\pi}{2}$$

which can be written as

$$\tau = \frac{2\pi}{2n\omega_r} = \frac{T}{2n}$$

The amplitude of the component at the reference frequency ($n = 1$) will have a maximum value of

$$C_1 = \frac{2V}{T\omega_r} = \frac{V}{\pi}$$

when the two input signals are 180° out of phase. The maximum amplitudes of the other harmonics will occur at different time delays between the two input signals; the maximum amplitude of the $n$th harmonic is

$$(C_n)_{\text{max}} = \frac{V}{n\pi} = \frac{C_1}{n}$$

## Dual-D Flip-Flops

The phase-voltage characteristics of the preceding set-reset flip-flop are sensitive to the width of the input signals. If they are of finite width, nonlinearities will occur in the characteristics. The dual-D flip-flop shown in Fig 8.10 is less sensitive to the duty cycle of the waveforms. The D flip-flops go high on the leading edge of the

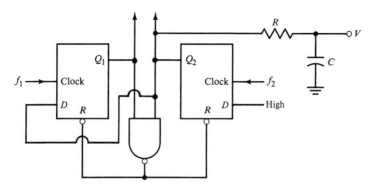

**FIGURE 8.10**
A dual-D flip-flop phase detector, including a low-pass filter.

input waveform and remain high until they are reset. The reset signal occurs when both inputs are high. When both signals are in phase and of the same frequency, both outputs will remain low and no pump signals will be applied to the low-pass filter. When the two signal frequencies are the same, but not necessarily in phase, the dc output voltage transfer characteristic will be the same as shown in Fig. 8.7 for the *RS* flip-flop. If the two signal frequencies are not the same, the output voltage will depend on both the relative frequency and phase differences. The timing diagram of Fig. 8.11 illustrates the case in which $f_2 = 2f_1$. In Fig. 8.11$a$, the leading edge of $f_1$ occurs just after that of $f_2$, so $Q_2$ (which goes high when $f_2$ goes high and then resets when $f_1$ goes high) is high 50 percent of the time, and the average value of the PD output is 0.5 V. In Fig. 8.11$b$, the leading edge of $f_1$ occurs just

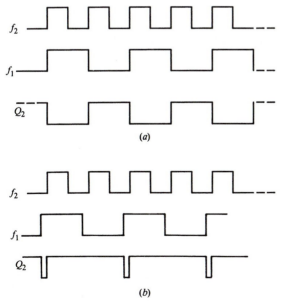

**FIGURE 8.11**
(*a*) Dual-D flip-flop phase-detector output ($Q_2$) when $f_1$ lags $f_2$; (*b*) dual-D flip-flop phase detector output ($Q_2$) when $f_2$ lags $f_1$.

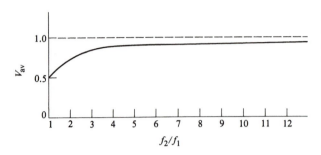

**FIGURE 8.12**
Average dual-D flip-flop output as a function of input frequency difference.

before that of $f_2$, so $Q_2$ is high almost all the time, and the average output voltage is approximately $V$. The output voltage averaged over all the phase differences is then $0.75V$ for $f_2 = 2f_1$. In general, it can be deduced that the average output (averaged over all phase differences) is given by

$$V_{av} = \left(1 - \frac{f_1}{2f_2}\right) V$$

provided $f_2$ is greater than $f_1$. This expression is plotted in Fig. 8.12.

### Phase-Frequency Detectors

The exclusive-OR and flip-flop phase detectors, although simple, have several limitations. One limitation is that the output requires substantial filtering to extract the dc value. Also, the loop can be slow to respond if the two input signals are not of the same frequency. The *phase-frequency,* or *three-state,* phase detector is designed to reduce these limitations. The phase-frequency detector acts as a phase detector during lock and provides a frequency-sensitive signal to aid acquisition when the loop is out of lock. Phase-frequency detectors are available in integrated-circuit form and usually contain a charge pump as an integral part of the device. The essential idea of a charge pump is illustrated in Fig. 8.13. The charge pump consists of a voltage-controlled current source that outputs a current of plus or minus I depending on the value of control voltage. For yet other values of control voltage, the current is zero (i.e., open-circuited). If the capacitor is part of an integrator, another pole at the origin is added to the transfer function, and the loop becomes a type II loop. If $C$ is shunted by a resistor, the loop remains a type I loop.

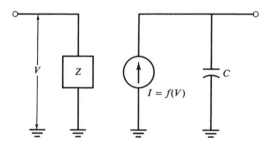

**FIGURE 8.13**
A charge pump circuit.

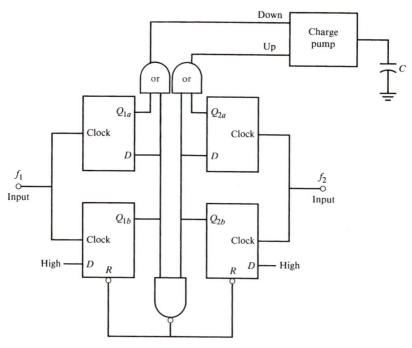

**FIGURE 8.14**
A quad-D phase-frequency detector.

Many manufacturers now produce a quad-D phase-frequency detector (as shown in Fig. 8.14). If the two input frequencies are the same, flip-flops $Q_{1a}$ and $Q_{2a}$ are never set, and the circuit functions as the dual-D flip-flop. If the frequencies are not equal, then either $Q_{1a}$ or $Q_{2a}$ will be set. These two flip-flops then serve as an out-of-frequency detector. For example, if $f_1$ is at least twice as fast as $f_2$, then $Q_{1a}$ or $Q_{1b}$ will be high all the time.

The average-voltage versus relative-frequency characteristic plotted in Fig. 8.15 applies for $f_2 \geq f_1$. Note that if the frequencies are not the same, the average output voltage is greater than that of the dual-D flip-flop. Therefore, a larger voltage is applied to the VCO, and the loop is quicker to respond. Once the loop reaches frequency lock, then the phase error can be obtained from $Q_{1b}$ and $Q_{2b}$ just as in the

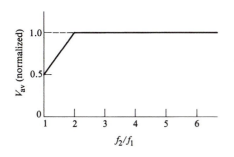

**FIGURE 8.15**
Average output voltage of a quad-D phase-frequency detector as a function of input frequency differences.

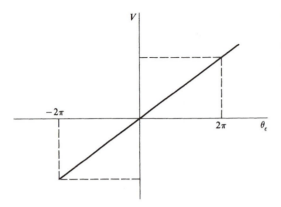

**FIGURE 8.16**
Average output voltage as a function of phase error for the quad-D phase-frequency detector.

dual-D flip-flop. If the loop is in lock, the average-voltage-versus-phase characteristic is as shown in Fig. 8.16. The phase range for the phase-frequency detector is 720°.

**Mixers**

Mixers (and multipliers) are often used as phase detectors in analog PLLs. If the input signal is $\theta_i = A_i \sin \omega_o t$ and the reference signal is $\theta_r = A_r \sin (\omega_o t + \phi)$, where $\phi$ is the phase difference between the two signals, then the output signal $\theta_\epsilon$ is

$$\theta_\epsilon = \theta_i \theta_r = \frac{A_i A_r}{2} K \cos \phi - \frac{A_i A_r}{2} K \cos (2\omega_o t + \phi) \qquad (8.16)$$

where $K$ is the mixer gain. One of the primary functions of the loop's low-pass filter is to eliminate the second harmonic term before it reaches the VCO. The second harmonic will be assumed to be filtered out, and only the first term will be considered. Therefore,

$$\theta_\epsilon = \frac{A_i A_r}{2} K \cos \phi \qquad (8.17)$$

When the error signal is zero, $\phi = \pi/2$. The error signal is proportional to phase differences about 90°. For small changes in phase $\Delta\phi$,

$$\phi \cong \frac{\pi}{2} + \Delta\phi$$

$$\theta_\epsilon = \frac{A_i A_r}{2} K \cos \left( \frac{\pi}{2} + \Delta\phi \right) = \frac{A_i A_r}{2} K \sin \Delta\phi$$

For a small phase perturbation $\Delta\phi$

$$\theta_\epsilon \simeq \frac{A_i A_r K}{2} \Delta\phi \qquad (8.18)$$

since the phase detector output was assumed to be $\theta_\epsilon = K_d(\theta_i - \theta_o)$. The phase-detector scale factor $K_d$ is given by

$$K_d = \frac{A_i A_r K}{2} \tag{8.19}$$

The phase-detector scale factor $K_d$ depends on the input signal amplitudes; the device can only be considered linear for constant-amplitude input signals and for small deviations in phase. For larger deviations in phase,

$$\theta_\epsilon = K_d \sin \Delta\phi \tag{8.20}$$

which describes the nonlinear relation between $\theta_\epsilon$ and $\phi$.

**Sampling Detectors**

Phase detection can also be accomplished with a linear time-varying switch that is closed periodically. Mathematically, the switch can be described as a pulse modulator, as shown in Fig. 8.17. If the operation of the sampling switch is time-periodic, that is, if the sampler closes for a short interval $P$ at instants $T = 0, T, 2T, \ldots, nT$, then the sampling is uniform. The waveshapes of the input and output signals of a uniform-rate-sampling device are shown in Fig. 8.18. The output can be considered to be

$$\theta_\epsilon(t) = \theta_i(t)\theta_r(t) \tag{8.21}$$

where $\theta_r(t)$ can be assumed to be a periodic train of constant-amplitude pulses of amplitude $A_r$, width $P$, and period $T$. Since $\theta_r(t)$, illustrated in Fig. 8.19, is periodic, it can be expanded in a Fourier series as

$$\theta_r(t) = \sum_{n=-\infty}^{\infty} C_n e^{jn\omega_o t} \tag{8.22}$$

where
$$C_n = T^{-1} \int_0^P A_r e^{-jn\omega_o t} \, dt \tag{8.23}$$

$$= \frac{A_r}{n\pi} \sin \frac{n\omega_o P}{2} e\left(\frac{-jn\omega_o P}{2}\right) \qquad n \neq 0 \tag{8.24}$$

$$= \frac{A_r}{T} P \qquad n = 0 \tag{8.25}$$

**FIGURE 8.17**
A switch modeled as a phase modulator.

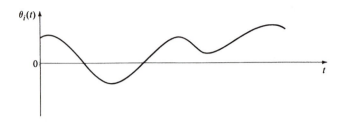

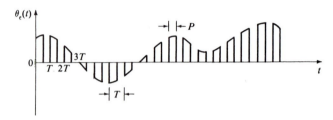

**FIGURE 8.18**
An example of the input and output waveforms of a uniform-rate sampling device.

Thus,
$$\theta_r(t) = \frac{A_r}{T}P + \sum_{n=1}^{\infty}\frac{2A_r}{n\pi}\sin\frac{n\omega_o P}{2}\cos\left[n\omega_o\left(t - \frac{P}{2}\right)\right] \tag{8.26}$$

If the input signal is a sine wave
$$\theta_i(t) = A_i\sin(\omega_i t + \phi)$$

then
$$\theta_\epsilon(t) = \theta_r(t)\theta_i(t) = A_i A_r\left(\frac{P}{T}\sin(\omega_i t + \phi) + \frac{1}{\pi}\sum_{n=1}^{\infty}\frac{\sin}{n}\frac{n\omega_o P}{2}\right.$$

$$\times\left\{\sin\left[(n\omega_o + \omega_i)t + \phi - \frac{n\omega_o P}{2}\right] + \sin\left[(\omega_i - n\omega_o)t + \phi + \frac{n\omega_o P}{2}\right]\right\}\right) \tag{8.27}$$

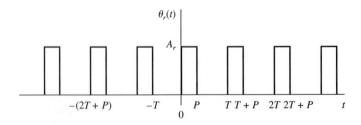

**FIGURE 8.19**
Pulse modulation is used to model the uniform-rate sampler.

when the loop is in lock ($\omega_i = \omega_o$). The dc term is

$$\theta_e(t)_{dc} = \frac{A_i A_r}{T} \sin \frac{\omega_o P/2}{\omega_o} \sin \left( \phi + \frac{\omega_o P}{2} \right) \tag{8.28}$$

The term $\omega_o P/2$ occurs because the pulse of $\theta_r(t)$ is assumed to start at $t = 0$. If the time origin is shifted to the middle of the pulse, this term does not appear. For small phase differences $\phi$, the error signal is proportional to the phase difference. Therefore, the linear time-varying switch is able to serve as a phase detector. It differs from the mixer in that the dc output is zero when $\phi = -\omega_o P/2$. That is, the output is zero when the oscillator and reference signal are in phase. This differs from the mixer type of phase detector, which is nulled when the two signals are in phase quadrature. As with the mixer, the sampling-phase-detector gain constant $K_d$ is proportional to the amplitude of the applied signals. The necessary conditions for both types of loops to exhibit linear characteristics are that the input signal amplitudes be constant and the phase error be sufficiently small that

$$\sin \left( \phi + \frac{\omega_o P}{2} \right) \approx \phi + \frac{\omega_o P}{2}$$

When the loop is in lock ($\omega_i = \omega_o$), the mixer output contains a dc term and the second harmonic, whereas the sampled output contains a dc term plus all harmonics of the input frequency. Therefore, the low-pass filter requirements for the sampling type of phase detector are more stringent than those for the sinusoidal mixer. Fortunately, there are filters that can easily be implemented for the sampling PD. The most commonly used is the *zero-order data hold (ZODH)* or *boxcar generator.* The zero-order data hold is a device that converts the pulses of width $P$ to constant-amplitude pulses of width $T$, as shown in Fig. 8.20. The output of the zero-order data hold $\theta_o(t)$ between the sampling instants $t_i$ and $t_{i+1}$ is

$$\theta_o(t) = \theta_\epsilon(t_i)[u(t) - u(t_i)] \tag{8.29}$$

where $\theta_\epsilon(t_i)$ is the value of $\theta_\epsilon(t)$ at the sampling time $t_i$. Although the exact analysis of the finite-pulse-width sampler and ZODH combination is complex, the frequency response can be closely approximated if the sampling process is replaced by

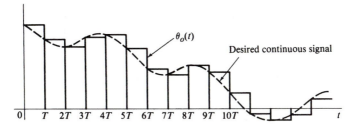

**FIGURE 8.20**
Output of a zero-order data hold compared with the ideal output
(dashed line).

an "ideal sampler" whose output is a train of impulses. That is, the sampled signal $\theta^*(t)$ is a train of amplitude-modulated impulses

$$\theta^*(t) = \theta_i(t)\delta_T(t) \tag{8.30}$$

where $\delta_T(t)$ is a unit-impulse train of period $T$,

$$\delta_T(t) = \sum_{n=-\infty}^{\infty} \delta(t - nT) \tag{8.31}$$

and $\delta(t - nT)$ represents an impulse of unit area occurring at time $t = nT$. Since $\delta_T(t)$ is periodic, it can be expressed by the Fourier series

$$\delta_T(t) = \sum_{n=-\infty}^{\infty} C_n e^{-jn\omega_o t} \tag{8.32}$$

where $\omega_o = 2\pi/T$. The constants $C_n$ are determined from

$$C_n = T^{-1} \int_{-T/2}^{T/2} \delta_T(t) e^{-jn\omega_o t} \, dt = T^{-1} \tag{8.33}$$

Now $\delta_T(t)$ can be expanded in a Fourier series

$$\delta_T(t) = T^{-1} \sum_{n=-\infty}^{\infty} e^{jn\omega_o t} \qquad \omega_o = \frac{2\pi}{T} \tag{8.34}$$

and since $e^{jn\omega_o t} + e^{-jn\omega_o t} = 2\cos n\omega_o t$,

$$\delta_T(t) = T^{-1} + \frac{2}{T} \sum_{n=1}^{\infty} \cos n\omega_o t \tag{8.35}$$

That is, the frequency spectrum of an impulse train of period $T$ contains a dc term plus the fundamental frequency and all harmonics with an amplitude of $1/T$. Therefore, Eq. (8.30) can be written

$$\theta^*(t) = \theta_i(t)\left( T^{-1} + \frac{2}{T} \sum_{n=1}^{\infty} \cos n\omega_o t \right) \tag{8.36}$$

If the input $\theta_i(t)$ is a sine wave $\theta_i(t) = A_i \sin(\omega_i t + \phi)$,

$$\theta^*(t) = \frac{A_i}{T}\left[ \sin(\omega_i t + \phi) + 2\sum_{n=1}^{\infty} \cos n\omega_o t \, \sin(\omega_i t + \phi) \right] \tag{8.37}$$

This equation is similar to the result obtained [Eq. (8.27)] for the more realistic finite-pulse-width model of the sampler. The difference is that for the finite-pulse-width model the harmonics are attenuated by the factor

$$\frac{\sin(n\omega_o P/2)}{n\omega_o}$$

With the impulse sampler, all harmonics are attenuated by $2/T$.

The impulse response of the ZODH is a pulse $T$ s wide, or

$$\theta_o(t) = u(t) - u(t - T)$$

so its frequency-dependent transfer function is

$$G_z(s) = \frac{1 - e^{-sT}}{s}$$

(8.38)

and the ZODH frequency response is

$$G_z(j\omega) = \frac{1 - e^{-j\omega T}}{j\omega} = \frac{T}{2} e^{-j\omega T/2} \frac{e^{j\omega T/2} - e^{-\omega T/2}}{j\omega T/2}$$

$$= Te^{-j\omega T/2} \frac{\sin{(\omega T/2)}}{\omega T/2}$$

(8.39)

which is a low-pass filter with a linear phase shift, as illustrated in Fig. 8.21. An important feature of this filter is that it has zero gain at the sampling frequency and at all harmonics of the sampling frequency. As Eq. (8.37)—or Eq. (8.27)—shows, when the input and sampling frequencies are equal, the output of the sampler contains a dc term and all harmonics of the sampling frequency. Since the ZODH has zero gain at these nonzero frequencies, the unwanted harmonics are completely removed by the filter. This is one of the primary reasons for the widespread application of samplers in phase-locked loops. The ZODH also has a phase lag that increases linearly with frequency. The effect of this negative phase shift on loop stability is discussed in the next chapter.

Although a PLL containing a sampling detector is often analyzed by using Z-transform techniques, an equally accurate approximation to loop performance can be obtained by using continuous techniques. The inaccuracy inherent in the Z-transform analysis of PLLs is further considered in the section on large-signal behavior in the following chapter. When the input and feedback frequencies are equal, Eq. (8.37) can be written

$$\theta^*(t) = A_i \frac{\sin\phi}{T} + \text{high-frequency terms}$$

(8.40)

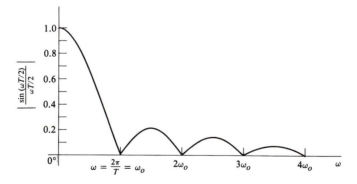

**FIGURE 8.21**
Frequency response of a zero-order data hold.

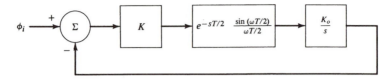

**FIGURE 8.22**
A simplified model of a phase-locked loop containing a sample-and-hold module.

Since the high-frequency terms are filtered out by the zero-order data hold,

$$\theta^*(t) \approx A_i \frac{\phi}{T} \qquad (8.41)$$

for small $\phi$. The frequency response of the loop can then be estimated by using the model shown in Fig. 8.22, provided the VCO is the only other frequency-dependent component in the loop. In this model, $K$ is composed of the phase detector factor $A_i$ times any additional gain in the loop. This model is used to analyze loop stability and frequency-response characteristics in the next chapter.

### Phase Detector Comparisons

Which type of phase comparator to select for a particular application depends on many factors, including cost, size, speed, and noise performance. The double-balanced mixer has the best noise performance of all the PDs, but it is only capable of producing approximately 0.5-V output. Since most VCOs require 2- to 10-V input, a preamplifier will be required with this type of PD, but the additional noise contributed may be so large that it is no longer the best choice. The double-balanced mixer finds application primarily in loops where little VCO pulling range is necessary, such as when VCXOs are used. The sample-and-hold discriminator works well from 20 to 100 kHz, but above this range there is too much harmonic leakage with existing systems for the sample-and-hold PD to be the best choice. For high-speed performance, digital phase detectors using emitter-coupled logic (ECL) are usually preferred.

## ■ 8.4

### VOLTAGE-CONTROLLED OSCILLATORS

VCOs are described in detail in Chap. 7, and the effect of their noise on the performance of PLLs used as frequency synthesizers is discussed in Chap. 10. Here we will be concerned with only the main points of their importance when used in PLLs. The main properties of a VCO used in a PLL are discussed below.

1. *Frequency deviation.* The maximum PLL capture range is equal to the open-loop gain, provided the VCO frequency deviation capability is at least this great. If it

is less, then the PLL capture range is limited by the maximum VCO frequency deviation capability.

2. *Frequency stability.* If high-frequency stability is required, VCXOs are normally employed. Frequency stability is of the utmost importance in frequency synthesizers. However, as mentioned previously, the VCXO has a small frequency deviation and is not able to follow signals with a large frequency deviation.

3. *Modulation sensitivity.* The modulation sensitivity $K_o$ should be high. A small change in dc voltage should produce a relatively large change in VCO frequency.

4. *Response.* The VCO should respond quickly enough that it does not affect the loop stability characteristics. Normally, the VCO poles should lie outside the dominant poles of the system.

5. *Frequency-voltage characteristics.* The VCO frequency-voltage characteristics must be linear. The tolerance on linearity depends on the particular application. PLLs that include a microprocessor in the loop can use the microprocessor plus a digital-to-analog converter to compensate for VCO nonlinearities.

6. *Spectral purity.* In some applications, such as analog frequency synthesizers, the VCO output should be as pure a sine wave as possible. In other applications, the VCO output can be a rectangular wave train.

## 8.5
### LOOP FILTERS

The loop filter is a low-pass filter, usually of the first order, but higher-order filters are used when additional suppression of the ac components of the phase detector output is desired. In some instances a notch network is included in the filter for suppression of a particular frequency. The network configuration depends on whether the phase detector output can be modeled as a voltage source (low output impedance) or a current source (high output impedance). Figure 8.23*a* illustrates

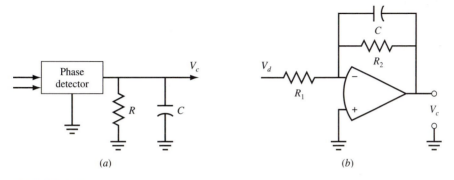

(*a*)

(*b*)

**FIGURE 8.23**
A first-order filter realized with a charge pump output phase detector; (*b*) a first-order filter realized with a voltage output phase detector.

a first-order filter that can be used with a charge pump output. The voltage is given by

$$V_c = \frac{I(V)R}{sRC + 1}$$

where $I(V)$ is the amplitude of the phase detector output. Figure 8.23$b$ illustrates a first-order active filter, which can be used with a phase detector with a low output impedance. The transfer function

$$\frac{V_c}{V_d} = -\frac{R_2}{R_1}(sR_2C + 1)^{-1}$$

is a first-order filter with a dc gain of

$$A_o = -\frac{R_2}{R_1}$$

which can be adjusted to modify loop performance. The selection of loop filters for loop stability, transient performance, and noise suppression is described in greater detail in the following chapter.

## ◼ 8.6

### LINEAR AND NONLINEAR PLL SIMULATIONS USING SPICE

When one is simulating a complex circuit using SPICE, sometimes it is difficult to find the correct models for some of the components, or models may not even exist. There are also times when an exact simulation is not required or the designer wants to get a quick feel for how the circuit will operate under certain constraints. For example, a complete PLL with a varactor-tuned VCO, a double-balanced mixer as the phase detector, and an $RC$ network for the low-pass filter (LPF) can be constructed in SPICE and can result in accurate simulations. However, this level of simulation would not be a good starting place for a preliminary PLL design. It is a good design practice to begin with top-level block diagram models that can represent basic system performance.

Earlier in this chapter we saw that designing PLLs involved tradeoffs. Equation (8.14) demonstrated that system speed of response and system bandwidth complicate the PLL design because they work against each other. In addition, we will see in Chap. 9 that the behavior of PLLs can become quite complex when they are designed around a desired transient response and loop stability is taken into account. These complications indicate that a top-level design should represent system components only at an abstract level so that the PLL's performance can be established. SPICE's Analog Behavior Modeling (ABM) feature provides a quick, accurate, and easily changeable model. In this method an input-output relationship is developed that models the black-box behavior of the circuit. In the following sections we will use this technique to model phase detectors, VCOs, and complete PLLs. In addition, we can use the techniques developed in this section to model more complicated PLL designs in Chap. 9.

## A Sinusoidal Phase Detector Model

Suppose that a sum/difference phase detector is to be modeled and that the reference and feedback frequencies are equal (frequency lock). Figure 8.24 shows the response of this model to nonlinear inputs. For small phase differences, the series expansion of the sine function gives $\sin(\theta_r - \theta_d) \cong \theta_r - \theta_d$ and the phase error reduces to the linear difference in phases. Using the ABM *E VALUE* primitive in SPICE, the phase detector can be modeled by the following equation: OUT $=$ $K_d^*(V(\text{reference}) - V(\text{feedback}))$. The schematic for a phase detector modeled in this way is given in Fig. 8.25. Note that although the series expansion reduces to a linear function, the output of the phase detector is inherently nonlinear.

## Voltage-controlled Oscillator Simulations

The VCO was also modeled as a linear device in Sec. 8.2. In this model we found that the output frequency followed the relation

$$\omega_o = \omega_c + \Delta\omega$$

and that

$$\Delta\omega = \frac{d\vartheta_o(t)}{dt} = K_o V_c(t)$$

where $K_o$ is the VCO gain constant [rad/(s·V)] and $V_c$ is the input control voltage. From this equation the VCO's output waveform is given by

$$V_o(t) = K \sin\left[\omega_c t + K_o \int V_c(t)\, dt\right]$$

and that for a linear control voltage is

$$V_o(t) = K \sin\left[(2\pi f_c + K_o V_c)t\right]$$

EXAMPLE 8.4. Using SPICE and ABM, simulate a VCO with a center frequency of 10 kHz and a VCO gain factor $K_o$ of $2\pi \times 5$ kHz/(s·V). Also, simulate the same circuit with a 45° (0.7854-rad) VCO phase offset. Plot the output frequencies for control values of $-1$ and 1 V.

*Solution.* To model this circuit, we need an ABM block using the SPICE s dt (integral) function to produce the phase integral

$$\text{Phase} = K_o \int V_c(t)\, dt$$

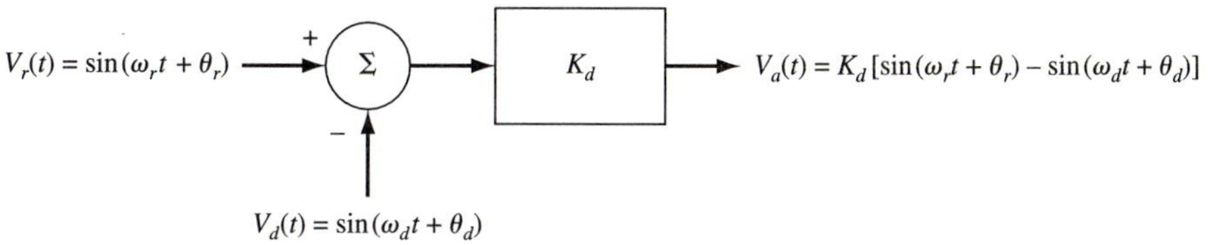

$V_r(t) = \sin(\omega_r t + \theta_r)$ ⟶ $+$ $\Sigma$ ⟶ $K_d$ ⟶ $V_d(t) = K_d[\sin(\omega_r t + \theta_r) - \sin(\omega_d t + \theta_d)]$

$V_d(t) = \sin(\omega_d t + \theta_d)$

**FIGURE 8.24**
Nonlinear phase detector output using summing node with gain stage.

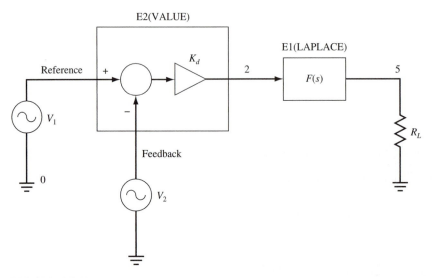

**FIGURE 8.25**
Phase detector circuit model used in SPICE simulations.

and an ABM block to produce the output sinusoid

$$V_o(t) = \sin\left[2\pi(f_c t) + \text{phase} + \text{offset}\right]$$

Note that the offset could have been a node voltage instead of a constant. By using a node voltage, an offset step or ramp change can be simulated. The schematic, SPICE circuit description,[3] and output plots are given in Fig. 8.26a to c.

### Linear and Nonlinear PLL Simulations

Now that we have developed simple models for the phase detector and VCO, we can complete the feedback path and handle PLL simulations. Remember that the models that we have developed are only approximations of PLL behavior, and they were designed for a top-level look at how the PLL behaves.

For a linear simulation, we would like to assume that the loop has achieved frequency lock and subtract the small (approximately linear) phases. Unfortunately, SPICE has a limited number of trigonometric primitives that can be used to extract the phase from the sine functions. Since the arcsine function is not available, the arctangent function must be used. One method to obtain the phase is to use the relationship $\cos^2 x = 1 - \sin^2 x$. An ABM block is used to convert the sinusoidal input to the argument of the sine by using the equation

$$\text{OUT} = \text{ARCTAN}\left(\frac{\text{IN}}{\sqrt{1 - \text{IN}^2}}\right)$$

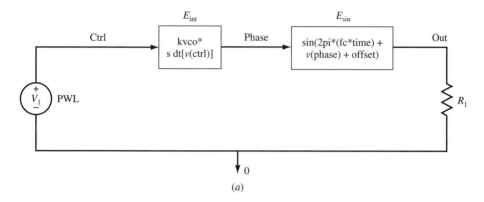

(a)

```
* VCO simulation with center freq = 10 kHz and gain factor = 2*pi*5000 rad/sec/V
.PARAM        twopi=6.283 fc=10k kvco=5000*twopi
*.PARAM       offset=0.7854
.PARAM        offset=0.0
Esine     out   0      VALUE { sin(twopi*(fc*time)+v(phase)+offset) }
R₁        0     out  1G
Eint      phase 0      VALUE { kvco*sdt(v(ctrl)) }
V₁        ctrl  0      PWL 0 -1 v 0.5m -1 v 0.501 ms 1 v
.tran 1u 1m 0  50 m
.OP
.probe
.END
```

(b)

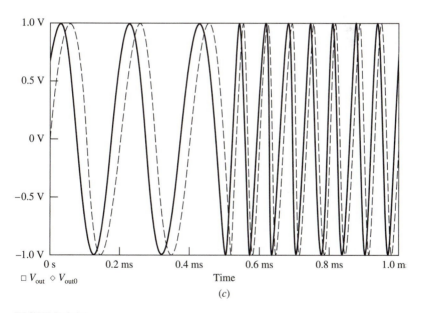

(c)

**FIGURE 8.26**
VCO simulation using analog behavior modeling. (*a*) VCO schematic; (*b*) circuit file; and (*c*) VCO output.

The phase detector takes the difference between the reference's sine argument and the feedback's sine argument. The frequency arguments subtract out, and the difference between the phases remains.

For nonlinear simulations, we just connect the VCO output to the nonlinear phase detector. There are also a number of nonlinearities that can be added to the VCO response by using the ABM technique. For the present simulations, the phase error will be kept small, so that the nonlinear results can be compared to the linear results. The large-signal PLL behavior is presented in Chap. 9.

At this point, when one is simulating the PLLs, it is important to carefully choose $K_d$ and $K_o$ in relation to the VCO center frequency. In Chap. 9 their relation to closed-loop bandwidth, settling time, transient performance, and stability is presented. For now, a rule of thumb for the SPICE simulations is to choose them such that $K_v$ (where $K_o$ is in hertz per volt and $K_d$ is in volts per radian) is at least one-half the VCO center frequency. In the next example, a linear PLL is simulated. A close look at the control voltage into the VCO shows a triangular waveform that forces the VCO frequency forward and backward until phase lock is achieved. If $K_v$ is too small, then there will not be enough gain in the loop to ramp the control voltage high enough in magnitude. In that case, the VCO cannot obtain phase lock.

EXAMPLE 8.5. Design and simulate a linear and nonlinear PLL with the following parameters:

$$\text{Center frequency} = 10 \text{ kHz}$$

$$\text{VCO gain factor} = 5000 \text{ Hz/V}$$

$$\text{Phase detector gain} = 1 \text{ V/rad}$$

$$\text{Reference phase offset} = 5° (0.08727 \text{ rad})$$

Use a VCO for the reference waveform. Compare the VCO input voltages with the following three LPF transfer functions:

$$\frac{1}{1 + (1.2\text{E}^{-4})s} \qquad \frac{1}{1 + (1.2\text{E}^{-5})s} \qquad \frac{1}{1 + (1.2\text{E}^{-6})s}$$

Also, plot the output and reference sinusoids for the second LPF loop.

*Solution.* Figure 8.27a shows a schematic for the nonlinear PLL. By using ABM techniques, the PLL is implemented with four components. An ABM VCO is used to generate a reference frequency of 10 kHz. A phase offset is added to the reference at 0.5 ms, using a piecewise linear voltage source to generate the offset. The resultant signal is connected to a phase detector that takes the difference between the reference and VCO signals. The low-pass filter is implemented by using the ABM Laplace function. Finally, a VCO with a center frequency of 10 kHz integrates the output of the LPF for the control voltage, generates the corresponding sinusoid, and feeds its output back to the phase detector. A linear PLL schematic is shown in Fig. 8.27d. It performs the same

**FIGURE 8.27**
PLL simulation results: (*a*) nonlinear PLL schematic; (*b*) nonlinear SPICE circuit file; (*c*) control voltages, reference, and output; (*d*) linear PLL schematic; (*e*) linear SPICE circuit file; (*f*) control voltages, reference, and output.

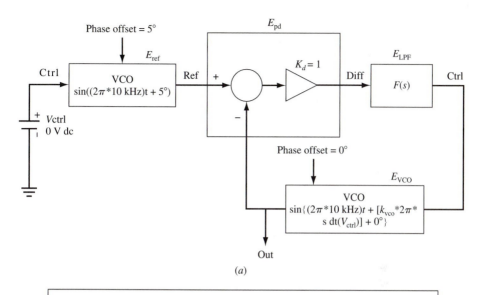

$(a)$

```
*Nonlinear PLL simulation circuit file
*
.PARAM          twopi=6.2831853 fc=10k kvco=5000
.PARAM          Kd=1 offset=0 Va=1
*
*
* This is the reference VCO's phase step offset
*
V1         offsetr 0 PWL 00 0 .5m 0 .500000001m .08727 1 0 .08727
R5         offsetr 0 1G
*
* PLL #1
Eref       ref 0 VALUE {
+ Va*sin(twopi*(fc*time)+(kvco*twopi*sdt(V(ctrlr)))+V(offsetr)) }
Vctrlr     ctrlr   0 0V
Epd        diff    0 VALUE { Kd*(V(ref)-V(out)) }
ELPF       ctrl    0 LAPLACE { V(diff, 0) } = { 1/(1+1.2E-4*s) }
EVCO       out     0 VALUE {
+ Va*sin(twopi* (fc*time)+(kvco*twopi*sdt(V(ctrl)))+offset) }
*
* PLL #2
Eref2      ref2    0 VALUE {
+ Va*sin(twopi* (fc*time)+(kvco*twopi*sdt(V(ctrlr2)))+V(offsetr)) }
Vctrlr2    ctrlr2  0 0V
Epd2       diff2   0 VALUE { Kd*(V(ref2)-V(out2)) }
ELPF2      ctrl2   0 LAPLACE { V(diff2, 0) } = { 1/(1+1.2E-5*s) }
EVCO2      out2    0 VALUE {
+  Va*sin(twopi*(fc*time)+(kvco*twopi*sdt(V(ctrl2)))+offset) }
*
* PLL #3
Eref3      ref3    0 VALUE {
+ Va*sin(twopi*(fc*time)+(kvco*twopi*sdt(V(ctrlr3)))+V(offsetr)) }
Vctrlr3    ctrlr3  0 0V
Epd3       diff3   0 VALUE { Kd* (V(ref3)-V(out3)) }
ELPF3      ctrl3   0 LAPLACE { V(diff3, 0) } = { 1/(1+1.2E-6*s) }
EVCO3      out3    0 VALUE {
+  Va*sin(twopi*(fc*time)+(kvco*twopi*sdt(V(ctrl3)))+offset) }
*
.tran 1u 3m 0 10u
.OP
.probe
.END
```

$(b)$

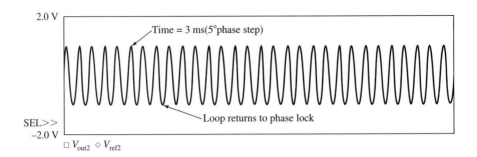

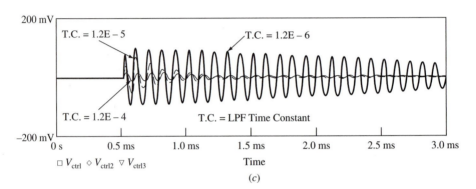

$\square\ V_{\text{out2}}\ \lozenge\ V_{\text{ref2}}$

$\square\ V_{\text{ctrl}}\ \lozenge\ V_{\text{ctrl2}}\ \triangledown\ V_{\text{ctrl3}}$

(c)

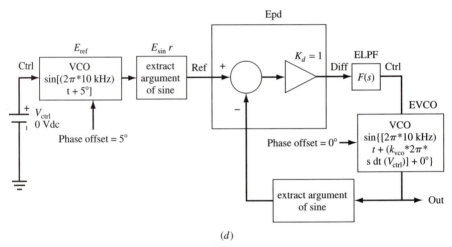

(d)

```
* Linear PLL simulation circuit file
.PARAM          twopi=6.2831853 fc=10k kvco=5000
.PARAM          Kd=1 offset=0 Va=1.0
*
*  This is the reference VCO's phase step offset
*
V_V7      offsetr   0  PWL 00 0 .5m 0 .500000001m .08727 1 0.08727
R_R5      offsetr   0  1G
*
*PLL #1
E_EPD     diff      0  VALUE { Kd*(V(ref)-V(outf)) }
E_Esinf   outf      0  VALUE { ARCTAN((V(out)/SQRT(Va*Va-
(V(out))*(V(out))))))
+ }
E_Esinr   ref       0  VALUE { ARCTAN((V(r)/(SQRT(Va*Va-(V(r))*(V(r)))))) }
}
```

338

```
V_V6        Ctrlr    0 0V
E_ELPF      ctrl     0 LAPLACE { V(diff, 0) } = { 1/(1+1.2E-4*s) }
E_Eref      r 0 VALUE {
+ Va*sin(twopi* (fc*time)+(kvco*twopi*sdt(V(ctrlr)))+V(offsetr)) }
E_Evco      out      0 VALUE {
+ Va*sin(twopi* (fc*time)+(sdt(kvco*twopi*V(ctrl)))+offset) }
*
* PLL #2
E_EPD2      diff2    0 VALUE { Kd*(V(ref2)-V(outf2)) }
E_Esinf2    outf2    0 VALUE {
+ ARCTAN((V(out2)/(SQRT(Va*Va-(V(out2))*(V(out2)))))) }
E_Esinr2    ref2     0 VALUE { ARCTAN((V(r2)/(SQRT(Va*Va-
(V(r2))*(V(r2)))))) }
V_V10       ctrlr2 0 0V
E_ELPF2     ctrl2    0 LAPLACE { V(diff2, 0} } = { 1/(1+1.2E-5*s) }
E_Eref2     r2       0 VALUE {
+ Va*sin(twopi* (fc*time)+(kvco*twopi*sdt(V(ctrlr2)))+V(offsetr)) }
E_Evco2     out2     0 VALUE {
+ Va*sin(twopi*(fc*time)+(sdt(kvco*twopi*V(ctrl2)))+offset) }
*
* PLL #3
E_EPD3      diff3    0 VALUE { Kd*(V(ref3)-V(outf3)) }
E_Esinf3    outf3    0 VALUE {
+ ARCTAN((out3)/(SQRT(Va*Va-(V(out3))*(V(out3)))))) }
E_Esinr3    ref3     0 VALUE { ARCTAN((V(r3)/(SQRT(Va*Va-
(V(r3))*(V(r3)))))) }
V_V12       ctrlr3 0 0V
E_ELPF3     ctrl3    0 LAPLACE { V(diff3, 0} } = { 1/(1+1.2E-6*s) }
E_Eref3     r3       0 VALUE {
+ Va*sin(twopi*(fc*time)+(kvco*twopi*sdt(V(ctrlr3)))+V(offsetr)) }
E_Evco3     out3     0 VALUE {
+ Va*sin(twopi*(fc*time)+(sdt(kvco*twopi*V(ctrl3)))+offset) }
.tran 1u 3m 0 10u
.OP
.probe
.END
```

(*e*)

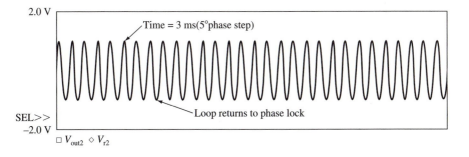

□ $V_{out2}$  ◇ $V_{r2}$

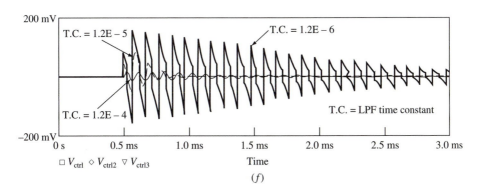

□ $V_{ctrl}$  ◇ $V_{ctrl2}$  ▽ $V_{ctrl3}$     Time

(*f*)

function as the nonlinear PLL, but also includes the two ABM blocks that extract the argument of the sine. The nonlinear PLL SPICE circuit file is given in Fig. 8.27*b,* and the control voltage waveforms are shown in Fig. 8.27*c.* The linear PLL SPICE circuit file is given in Fig. 8.27*e,* and the control voltage waveforms are shown in Fig. 8.27*f.* Notice that the value of the LPF times the constant affects the settling time of the loop. It is left as an exercise to also see that the values of $K_d$ and $K_v$ affect the loop's performance.

## ■ 8.7

### PHASE-LOCKED LOOP APPLICATIONS

The PLL is an exceptionally versatile device suitable for a variety of frequency-selective modulation, demodulation detection, tracking, and synthesis applications. A few of the basic applications are described here. The use of the PLL for frequency synthesis is described in detail in Chap. 10, and additional applications of the PLL for modulation and detection are given in Chap. 12.

### Tracking Filters

A phase-locked loop can filter the noise present on the input (reference) signal. The PLL will track the frequency of the input as long as it is not changing rapidly. The PLL transfer function is then a low-pass filter centered about the VCO frequency, which is the same as the input frequency. It is not a linear filter, since any amplitude information is lost, but it will reduce fluctuations in the input frequency. When the loop is tracking, the transfer function is given by Eq. (8.4) (normally $N$ will be equal to 1 in the tracking filter). The loop functions as a bandpass filter whose center frequency is that of the input reference frequency.

The bandwidth of the second-order loop is given by Eq. (8.13). From this equation and Eq. (8.7) it is seen that the bandwidth $\omega_n \propto (K_v\omega_L)^{1/2}$. That is, the closed-loop bandwidth (from the center frequency to one 3-dB point) is proportional to the square root of the loop filter bandwidth times the loop gain constant. It is possible for the loop to lock on to harmonics or subharmonics of the loop, depending on what types of phase detector and VCO are used. For example, with a mixing-type phase detector, the PLL can lock onto subharmonics of the input if the VCO output is a square wave.

The ability of the PLL to automatically tune its center frequency to that of the input signal makes it an attractive solution to modulation and demodulation problems.

### Angle Modulation

A phase-locked loop provides a ready means of phase modulation and indirect frequency modulation. Figure 8.28, illustrates a phase-locked loop with the modulat-

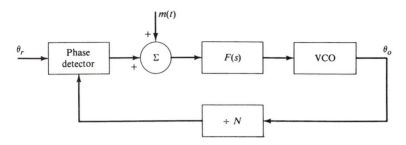

**FIGURE 8.28**
A phase-locked loop with a phase-modulating signal $m(t)$.

ing signal added before the low-pass filter. If a linear model is assumed for the loop, then superposition can be used to find the output phase:

$$\theta_o(s) = \frac{\theta_r(s)[K_o K_d F(s)/s]}{1 + K_o K_d F(s)/(sN)} + \frac{M(s)[K_o F(s)/s]}{1 + K_o K_d F(s)/(sN)} \tag{8.42}$$

where $M(s)$ is the Laplace transform of the modulating signal $m(t)$. At low frequencies

$$\left| \frac{K_o K_d F(j\omega)}{j\omega N} \right| \gg 1 \tag{8.43}$$

and

$$\theta_o(j\omega) = N\left[ \theta_r(j\omega) + \frac{M(j\omega)}{K_d} \right]$$

The phase-locked loop is a low-pass filter, and the expression [Eq. (8.43)] is satisfied at frequencies within the loop bandwidth. In this frequency region, the output phase is modulated by $m(t)$, and

$$f_o(t) = \frac{d\theta_o(t)}{dt} = N\left[ \frac{d\theta_r(t)}{dt} + \frac{1}{K_d}\frac{dm(t)}{dt} \right]$$

If the input phase is constant, the output frequency is proportional to the derivative of the modulating signal. This is referred to as *indirect frequency modulation* to distinguish it from direct FM, in which the frequency is directly proportional to the modulating signal.

## Frequency Demodulation

If the PLL is locked on an input frequency, the VCO control voltage is proportional to the VCO's frequency shift from its free-running frequency. If the input frequency shifts, the control voltage shifts accordingly (provided the frequency shift is within the loop's tracking range). If the input signal is frequency-modulated, then the VCO control voltage will be the demodulated output. The PLL can be used for detecting either narrowband or wideband (high-deviation) FM signals with a higher degree of linearity than can be obtained by other FM detection methods. If the maximum phase detector output voltage is $V$ V, then the maximum control voltage applied to the VCO is $KV$ V, where $K$ is the dc gain of the low-pass filter. The maximum frequency deviation of the VCO is then

$$(\Delta\omega)_{max} = K_oKV \qquad \text{rad/s}$$

This, of course, assumes that the VCO is designed to be linear over this frequency range. If the phase detector output can deviate between $\pm V$ V, the tracking range (TR) will be

$$TR = 2(\Delta\omega)_{max} = 2K_oKV$$

This tracking range must be greater than the frequency deviation of the input signal. FM demodulation can then be obtained by setting the free-running frequency of the VCO equal to the center frequency of the input signal. This detection method assumes that the envelope of the input waveform has a constant amplitude. In many applications, an amplifier and amplitude limiter are added before the phase-locked loop to ensure that this is the case.

A particular application of *frequency-shift keying (FSK)* demodulation is the detection of one of two transmitted frequencies. This detector is frequently referred to as a *touch-tone decoder*. There now exist multitone encoder and decoder integrated circuits. The decoders are able to detect a coded sequence of tones, allowing, for example, three different tones to represent eight transmitters.

Digital data are often transmitted using FSK, which consists of shifting the carrier frequency between two predetermined frequencies. The phase-locked loop can be used to demodulate the FSK signal. The voltage at the output of the loop filter shifts between two discrete voltage levels, reproducing the digital waveform. A realization of FSK demodulation is described in the section on digital PLLs.

## Amplitude Demodulation

The amplitude-modulated signal

$$S(t) = V[1 + m(t)] \sin \omega_o t$$

can be demodulated by multiplying the signal by a local oscillator signal of the same carrier frequency. The method is illustrated in Fig. 8.29. The multiplier output is

$$V(t) = V[1 + m(t)] \sin \omega_o t\, A \sin (\omega_o t + \theta)$$

$$= V[1 + m(t)]\frac{\cos \theta - \cos (2\omega_o t + \theta)}{2}$$

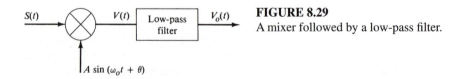

**FIGURE 8.29**
A mixer followed by a low-pass filter.

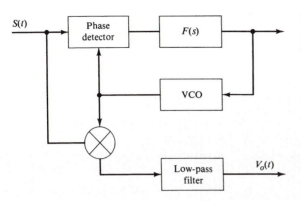

**FIGURE 8.30**
A coherent amplitude demodulator.

The multiplier output consists of the low-frequency modulating signal

$$V[1 + m(t)]\cos\theta$$

and the modulating signal centered about twice the carrier frequency

$$V[1 + m(t)]\cos(2\omega_o t + \theta)$$

The high-frequency term can be removed by the low-pass filter, so the filter output
will be

$$V_o(t) = V[1 + m(t)]\cos\theta$$

This relatively simple direct-conversion receiver has several advantages over
the more frequently used superheterodyne receiver. It eliminates the intermediate-
frequency (IF) filter and the need for additional oscillators. The elimination of the
IF filter also eliminates the need for image rejection filters. It does, however, have
problems, one of them being that if the input signal phase angle $\theta$ is not known,
the output voltage can be small. To ensure maximum output voltage, the phase
angle should be zero; that is, the local oscillator signal should be phase-locked to
the carrier signal. In this case, the demodulation is one of coherent detection, which
performs better than incoherent detection methods when the input signal-to-noise
ratio is low.[4] The local oscillator signal, phase-locked on the input carrier, can be
generated in a phase-locked loop. A complete amplitude demodulator is illustrated
in Fig. 8.30. This circuit assumes that the VCO is in phase with the input. If a phase
detector that causes the loop to lock with the VCO 90° out of phase with the input
is used, then the VCO output must be shifted by 90° before it is mixed with incoming signals. Several phase-locked-loop integrated circuits include an additional
multiplier on the chip so that amplitude demodulation can be easily realized.

## Phase Shifters

Many methods of modulation and demodulation require that the local oscillator signal be phase-shifted by 90°. The circuit shown in Fig. 8.31 presents one method of
obtaining a signal together with the signal shifted in phase by 90°. The phase-locked

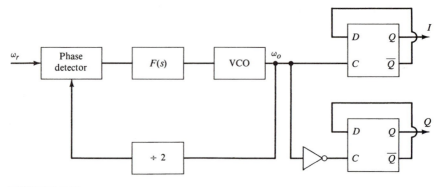

**FIGURE 8.31**
A phase-locked-loop circuit used for generating in-phase and quadrature signals.

loop doubles the reference frequency. The flip-flop output frequency is the same as the reference frequency, and as can be seen from the timing diagram illustrated in Fig. 8.32, the two output signals differ by 90° in phase ($\pi/4 = 90°$).

### Signal Synchronizers and Carrier Recovery

The communications methods that have the lowest error rate are coherent; they require that the carrier frequency and phase be precisely known for demodulation. At the same time, it is an inefficient use of energy to transmit the carrier. Many encoding schemes transmit a low-level pilot carrier. FM stereo and some commercial television encoding schemes use this technique. The color television signal includes a short sinusoidal burst transmitted at the rate of the horizontal synchronization pulses for synchronizing the color signals. These carrier signals can readily be recovered with a phase-locked loop.

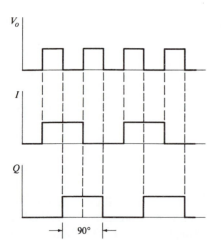

**FIGURE 8.32**
A timing diagram showing the outputs of the circuit in Fig. 8.31.

There are other communications systems in which the carrier is suppressed, and accurate reception requires the generation of signals based on the phase information about the carrier. The receiver circuits that generate the carrier and clock signals are known as *signal synchronizers;* phase-locked loops are often used in these synchronizers. (The phase-locked loop applications that have been discussed so far all require an input signal at the frequency to be tracked.) Signal synchronizers employ

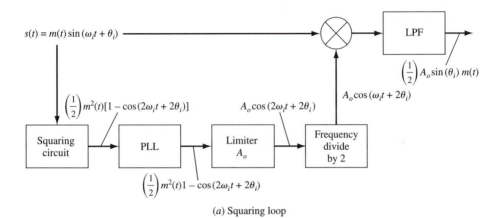

(a) Squaring loop

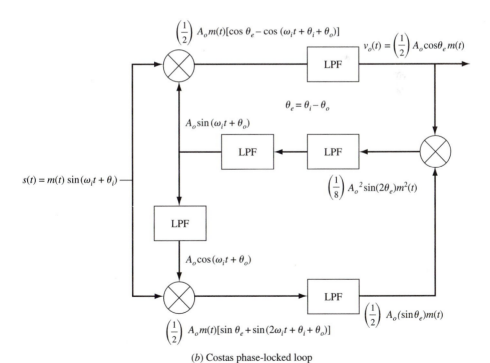

(b) Costas phase-locked loop

**FIGURE 8.33**
Two circuits for demodulating a suppressed carrier signal.

a nonlinear circuit to regenerate a carrier or clock signal together with a phase-locked loop to track the signal. A frequent application is the detection of *binary phase-shift keying (BPSK)*. If the data bit is a 1, then the signal is transmitted with a phase of $+90°$; if the data bit is a 0, then the signal phase is $-90°$. Each pulse has a $T$-s duration, and the pulses are $T$ s apart. If an equal number of 1s and 0s are transmitted, the carrier is completely suppressed. The transmitted signal are can be represented as

$$s(t) = m(t)\sin(\omega_i t + \theta_i)$$

and the average value of the modulating signal $m(t)$ is zero. The Fourier transform of $s(t)$ reveals that the spectrum is

$$S(f) = \frac{1}{2}[M(f - f_i) + M(f + f_i)]$$

This is the same spectrum as that of a *double-sideband suppressed carrier (DSB-SC)* signal. To coherently demodulate the received signal, the carrier must be regenerated. Figure 8.33 illustrates two methods of demodulating the signal.

For constant $\theta_i$ the signal at the output of the squaring loop circuit reduces to the message signal $m(t)$, multiplied by a constant amplitude. Likewise, for constant $\theta_i$ and $\theta_o$, the signal at the output of the Costas phase-locked loop (see Fig. 8.33b) is proportional to the message signal $m(t)$.

Notice that for both circuits there exists a $180°$ phase ambiguity. For example, suppose the transmitted signal is given by

$$s(t) = m(t)\sin(\omega_i t + \theta_i + \pi) = -m(t)\sin(\omega_i t + \theta_i)$$

such that the outputs will be proportional to $-m(t)$. This ambiguity can only be resolved by the special encoding (for example, differential encoding) of the transmitted message. The application of signal synchronization for *quadrature phase-shift keying (QPSK)* detection is discussed in Chap. 12.

## ▓ 8.8

### DIGITAL PHASE-LOCKED LOOPS

In many applications, including digital communications systems, the input signal is digital and is best treated in the digital domain. Digital versions of the phase-locked loop have some advantages over their analog counterparts. A simplified block diagram of a digital PLL (DPLL) is shown in Fig. 8.34. In addition to the phase detector, the loop consists of an accumulator and a digitally controlled oscillator. The digital phase detector is the same as those already described. The phase-detector output is a constant amplitude pulse whose width is proportional to the phase error. The digital phase-locked loop generates a digital representation of this error signal by sampling the pulse-width modulated signal at a rate that is much faster than that of the reference frequency. The digital accumulator is the digital equivalent of the analog filter. It is usually realized with an adder, multipliers, and delay units. The digi-

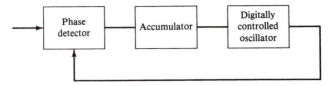

**FIGURE 8.34**
Block diagram of a digital phase-locked loop.

tally controlled oscillator outputs a pulse at a rate that is readily controlled by an external signal.

Figure 8.35 illustrates a practical realization of a first-order DPLL. All components except for the divide-by-$N$ counter are realized in a single integrated circuit. The data sheets for the Texas Instruments 74L5297 digital phase-locked loop and the 74L5292/294 programmable frequency divider are given in Appendix 6. Digital phase-locked loops do not contain a voltage-controlled oscillator, which is sensitive to voltage and temperature changes. Digital integrated-circuit phase-locked loops operate at higher frequencies than do their analog IC counterparts (32 MHz compared with 10 MHz), and it is far easier to generate a linear voltage-versus-frequency characteristic in a digitally controlled oscillator. Therefore, digital phase-locked loops offer easier microprocessor control.

The first function of the $K$ counter is to convert the phase detector output to a counter value that is proportional to the phase error. It contains an up counter and a down counter with carry and borrow outputs, respectively. The two input signals are the high-frequency clock and the phase-detector output that controls whether the clock is applied to the up counter or down counter. When the phase detector output is low, it is applied to the up counter; otherwise, it is presented to the down counter.

The I/D counter is clocked at the rate $2Nf_c$, where $N$ is the modulus of the divide-by-$N$ counter, and outputs a signal at the rate $Nf_c$. When a carry pulse from

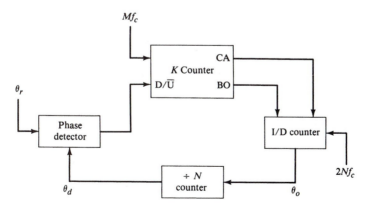

**FIGURE 8.35**
A first-order DPLL.

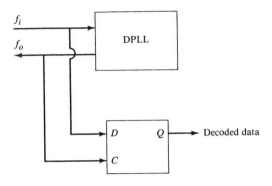

**FIGURE 8.36**
A DPLL used for frequency-shift keying (FSK) decoding.

the $K$ counter occurs, one-half of a cycle is added to the I/D output, and a half-cycle is removed on the occurrence of a $K$ counter borrow pulse. The frequency of the divide-by-$N$ counter output is then $f_c$ when there is no phase error. The output of the $K$ counter can be interpreted as proportional to the phase error

$$K_o = \frac{K_d \phi_\epsilon M f_c}{K}$$

where $M f_c$ is the frequency at which the phase error is sampled, $K_d$ in the phase detector scale factor, and $K$ is the modulus of the $K$ counter. This method of implementation creates both count-up and borrow pulses, even when the phase error is zero, but the numbers of count-up and borrow pulses will be equal. The higher the divide-ratio $K$, the less often the borrow and carry pulses will occur. The length of the $K$ counter can be controlled externally and can be varied from divide by $2^3$ to divide by $2^{17}$. A larger divide number corresponds to a narrower loop filter and thus increases the time to acquire lock, but it reduces any ripple or jitter on the output.

Applications for this DPLL include motor-speed control, noise filtering, tone recognition, frequency synthesis, frequency demodulation, and phase demodulation. A method of FSK decoding is illustrated in Fig. 8.36. Here the input $f_i$ alternates between the two frequencies $f_1$ and $f_2$, and the loop center frequency is selected so that $f_1 < f_c < f_2$. The D flip-flop serves as a simple frequency detector. It is set if the D input is high at the moment the clock pulse goes low. The output will be low if the D input is low when the leading edge of the clock pulse occurs. If the input frequency is $f_1$, a negative phase error is required to reduce the output frequency from $f_c$ to $f_1$. A negative phase error means that $\theta_r$ lags $\theta_d$ so that the D input of the flip-flop is not set. If $f_2$ occurs, the phase error will be positive ($f_i$ leads $f_o$), and the D flip-flop will be set.

## 8.9

### INTEGRATED-CIRCUIT PHASE-LOCKED LOOPS

In many applications it is desirable to have all the system components on a single integrated circuit (IC). In most cases, integration of electrical components on silicon can lead to reductions in the system's size and power consumption and sometimes speed increases. Currently, *application-specific integrated circuits (ASICs)* are becoming commonplace due to these advantages and the increase in reliability that are inherent in IC processes. However, many times it is impractical, and sometimes impossible, to integrate all electrical components on a chip. For example, to increase the flexibility of an IC, the manufacturer may require external components so that the user can adjust the IC's basic functionality for the specific application. Also, since high-$Q$ inductors are currently not available on silicon, it is difficult to build a stable, high-frequency oscillator and some of the more advanced filters without requiring external components.

Due to the many applications of phase-locked loops, the demand for IC versions has increased. Progress in building high-frequency PLL ICs has been hindered by the current inability to integrate a high-$Q$ inductor for the VCO. In addition, since the loop filter controls much of the PLL's performance, it is desirable to let the user implement this part of the loop based on the application's bandwidth and settling-time requirements.

## A CMOS PLL

Figure 8.37 is the block diagram of the Motorola MC 14046B, a low-power, CMOS-based PLL IC (this IC is also manufactured by Harris, National Semiconductor, Phillips, and others). The features and operating parameters for this IC are given in the data sheet in Appendix 7.

**EXAMPLE 8.6.** Using the MC 14046B (see Appendix 7), implement a frequency synthesizer that synthesizes a 100-kHz signal from a 2.5-kHz reference frequency. Use a type I low-pass filter so that the closed-loop system approximates a second-order Butterworth filter (as in Example 8.2). Design the loop under the assumption that the duty cycle of the reference frequency is 50 percent and that $V_{DD}$ is a 9-V regulated power supply.

**Solution.** Since the reference frequency is at 50 percent duty cycle, we should use phase comparator I for the 2.5-kHz reference frequency input. From Appendix 7 the VCO minimum and maximum frequencies are given by

$$f_{\min} = \frac{1}{R_2(C + 32 \text{ pF})} \qquad f_{\max} = \frac{1}{R_1(C + 32 \text{ pF})} + f_{\min}$$

The VCO is designed to oscillate at 100 kHz. Suitable values for $R_1$, $R_2$, and $C$ are 10.8 k$\Omega$, 10.8 k$\Omega$, and 1 nF, respectively. With these values, $K_{\text{VCO}}$ is given by the equation in Appendix 7 as

$$K_{\text{VCO}} = \frac{2\pi \, \Delta f_{\text{VCO}}}{V_{DD} - 2V} \cong 1.61 \times 10^5 \text{ rad/V}$$

and $K_\varphi$ is given by $V_{DD}/\pi = 2.9$ V/rad. These values give

$$K_V = \frac{K_{\text{VCO}} K_\phi}{N} = 11.5 \times 10^3$$

To implement a Butterworth filter, $\xi = 0.707$. From Eq. (8.8), we find that $\omega_n = 16.3 \times 10^3$ (2.6 kHz) and $\omega_L = 23 \times 10^3$ (3.67 kHz). Now, by using filter A in Appendix 7

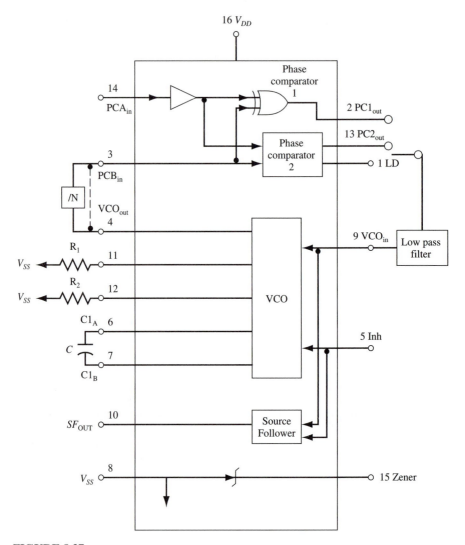

**FIGURE 8.37**
Block diagram of the Motorola MC14046B including external connections and components.

and the equation $\omega_n = \sqrt{K_v/(R_3 C_2)}$, suitable values for $R_3$ and $C_2$ are 43.3 kHz and 1 nF, respectively. The complete circuit is shown in Fig. 8.38. The closed-loop transfer function [Eq. (8.5)] is given by

$$\frac{\theta_o(s)}{\theta_r(s)} = \frac{10.6 \times 10^9}{s^2 + 2.3 \times 10^4 s + 2.66 \times 10^8}$$

and the rise time [Eq. (8.14)] is approximately 0.13 ms. Note that the loop stability, transient performance, and noise suppression are not evaluated for this example. These topics are introduced in the next chapter.

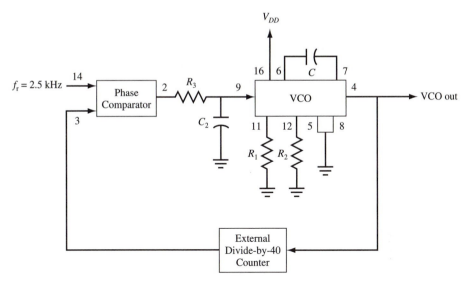

**FIGURE 8.38**
Block diagram of Example 8.6 solution.

## ▉ 8.10

### PROBLEMS

**8.1** A first-order PLL is to be used to synthesize a 1-MHz signal from a 50-kHz reference frequency. The phase-detector gain is 2 V/rad, the VCO sensitivity is 100 Hz/V, and the free-running frrequency of the VCO is 1 MHz. Estimate the rise time of the system. What will the rise time be if it is desired to realize an output frequency of 1.2 MHz? How many different output frequencies can be realized if the VCO has a maximum frequency deviation of ±200 kHz?

**8.2** (*a*) If the PLL of Prob. 8.1 is to synthesize an output frequency of 1 MHz, what would be the required bandwidth of a first-order low-pass filter in order for the closed-loop system transfer function to approximate that of a second-order Butterworth filter?
(*b*) What is the answer to part *a* if the output frequency is to be 1.2 MHz?

**8.3** Design the filters determined in Prob. 8.2, assuming the phase detector has a charge pump output.

**8.4** A PLL has a reference frequency of 25 MHz. Determine the harmonic with the maximum amplitude at the phase detector output for

(*a*) An exclusive-OR phase detector. The input waveforms can be assumed to have a 50 percent duty cycle.
(*b*) A dual-D flip-flop phase detector.

**8.5** Determine the condition between the phases of the two input signals for zero phase error in the output of

(*a*) A mixing-type phase detector

*(b)* A sampling phase detector
*(c)* An exclusive-OR phase detector
*(d)* An *RS* flip-flop phase detector
*(e)* A dual-D phase detector

**8.6** Describe how to construct a motor-speed controller using a phase detector and oscillator. [*Hint:* A constant-field dc motor can be described by the transfer function

$$\frac{\theta_o}{V_c} = \frac{K}{s(\tau_M s + 1)}$$

where $V_c$ is the motor control voltage and $\theta_o$ is the motor output angle in radians.]

**8.7** Use a 74LS297 integrated circuit (App. II) to realize a digital phase-locked loop with a free-running frequency of 1 MHz. Determine the clock frequency (plus $M$ and $K$) as well as the counter contents in order for the DPLL to discriminate between the two input frequencies of 0.9 and 1.1 MHz.

**8.8** For the system described in Prob. 8.7, show how the system response time will vary as a function of the contents of the $K$ register. From this result describe the number of input signals required for detection.

**8.9** Calculate the maximum output phase change possible in one reference period for a second-order type I system that uses a digital phase detector with maximum output voltage $V$.

**8.10** Show that in second-order type I systems, which use a charge pump output phase-frequency detector, the maximum frequency change (tracking range) is given by

$$(\Delta\omega)_{\text{max}} = K_o V$$

**8.11** Derive the voltage transfer characteristic for the quad-D flip-flop for the case in which $f_1 f_2$.

**8.12** Determine the loop gain $K_d K_o$ of the MC14046B phase-locked loop.

**8.13** Design a low-pass filter for the MC14046B so that the filter bandwidth is 1 kHz using phase comparator *I* and no frequency offset ($R_2 = \infty$).

**8.14** Design a frequency demodulator using an MC14046B integrated-circuit phase-locked loop. The carrier frequency is 100 kHz, and the frequency deviation is 2 kHz. Use phase comparator *I*.

**\*8.15** Simulate a nonlinear PLL with a center frequency of 100 kHz, a VCO gain factor of 25 kHz/V, and a phase detector gain of 2 V/rad. Offset the phase detector by 3° and the VCO by $-2°$ at 50 $\mu$s. Use an LPF with the following transfer function:

$$\frac{1}{1 + 1.23E^{-6}s}$$

**\*8.16** Simulate a nonlinear and linear PLL with a center frequency of 10 kHz, a VCO gain factor of 5000 Hz/V, and a phase detector gain of 2 V/rad. Offset the phase detector by

30° and 60°, and compare the VCO input voltages. Use an LPF with the following transfer function:

$$\frac{1}{1 + 1.2E^{-5}s}$$

## 8.11
### REFERENCES

1. H. de Bellescize, La Reception Synchrone, *Onde Electr.,* **11:** 230–240 (1932).
2. S. Barab and A. McBride, Uniform Sampling Analysis of a Hybrid Phase-Locked Loop with a Sample-and-Hold Phase Detector, *IEEE Trans., AES,* **11:** 210–216 (1975).
3. Voltage-Controlled Oscillators, *The Design Center Source Newsletter Minosim Corporation,* July 1990, pp. 312–321.
4. F. M. Gardner, *Phaselock Techniques,* 2d ed., Wiley, New York, 1979.
5. A. Viterbi, *Principles of Coherent Communications,* McGraw-Hill, New York, 1966.
6. J. P. Costas, Synchronous Communications, *Proc. IRE,* **44:** 1713–1718 (1956).
7. W. T. Green, Jr., and B. Kean, Digital Phase-Locked Loops Move into Analog Territory, *Electronic Design,* March 31, 1982.
8. K. Feher, *Digital Communications Satellite/Earth Station Engineering,* Prentice-Hall, Englewood Cliffs, NJ, 1983.
9. Donald G. Troha, "Digital Phase-Locked Loop Design using SN54/74LS297," *Application Note,* Texas Instruments, 1994.

## 8.12
### ADDITIONAL READING

Blanchard, A.: *Phase-Locked Loops,* Wiley, New York, 1976.
Gupta, S. C.: Phase-Locked Loops, *Proc. IEEE,* **63:** 291–306 (1975).
Lindsey, W. C., and C. M. Chie: A Survey of Digital Phase-Locked Loops, *Proc. IEEE,* **69:** 410–431 (1981).
——— and M. K. Simon (eds.): *Phase-Locked Loops and Their Application,* IEEE Press, New York, 1978.
Rohde, U. L.: *Digital PLL Frequency Synthesizers,* Prentice-Hall, Englewood Cliffs, NJ, 1983.

# 9

# Phase-Locked Loop Analysis

## 9.1

**INTRODUCTION**

In Chap. 8 it was demonstrated that the phase-locked loop is an exceptionally versatile circuit with many applications. The system designer first selects the phase detector and voltage-controlled oscillator and then determines the loop gain and loop filter frequency response. These parameters determine both the loop's transient performance and the system's noise performance. As mentioned in the last chapter, there is an inverse relationship between rise time and bandwidth. The designer also faces tradeoffs in establishing the system's speed of response and its noise performance. Finally, because it is a feedback loop, its loop stability must be ensured.

Loop performance is first analyzed in this chapter assuming a linear PLL model, and then the transient analysis of some nonlinear PLL models containing a digital phase detector will be determined. This latter analysis will allow an evaluation of the results obtained from the less accurate linearized model.

## 9.2

**STEADY-STATE ERROR ANALYSIS**

Steady-state error analysis determines the final error in response to inputs, which can be expressed as a time polynomial:

$$\theta_r(t) = \sum_{n=0}^{K} a_n t^n$$

For phase-locked loops the two inputs of greatest interest are the step input in phase

$$\theta_r(t) = \theta_o$$

and the step input in frequency

$$f_r(t) = f_o$$

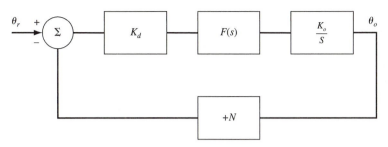

**FIGURE 9.1**
A linear PLL model.

or
$$\theta_r(t) = f_o t$$

It is important to know what the steady-state error will be in response to these inputs. This can be readily determined for the PLL linear model illustrated in Fig. 9.1 and derived in Chap. 8.

The error signal $\theta_\epsilon$, defined as $\theta_r - \theta_o/N$, can be expressed as

$$\theta_\epsilon(s) = \frac{\theta_r(s)}{1 + K_v F(s)/(Ns)} \tag{9.1}$$

where $K_v = K_o K_d$. If the systems are stable, the steady-state error for polynomial inputs $\theta_r(t) = t^n$ can be obtained from the final-value theorem:

$$\lim_{t \to \infty} \theta_\epsilon(t) = \lim_{s \to 0} s\theta_\epsilon(s) \tag{9.2}$$

$$= \lim_{s \to 0} \frac{s^2 \theta_r(s) N}{K_v F(s)} \tag{9.3}$$

If $\theta_r(t)$ is a step function representing a sudden increase in phase of $\phi°$, $\theta_r(s) = \phi/s$, and

$$\lim_{t \to \infty} \theta_\epsilon(t) = \lim_{s \to 0} \frac{s\phi}{K_v F(s)} \tag{9.4}$$

then $F(s)$ is either a constant or a low-pass filter that may include poles at the origin. That is,

$$\lim_{s \to 0} F(s) = \frac{K}{s^n} \neq 0 \tag{9.5}$$

Therefore, for a step increase in phase, the steady-state error can be written as

$$\lim_{t \to \infty} \theta_\epsilon(t) = \lim_{s \to 0} \frac{s^{n+1} \theta_o}{K_v K} = 0 \tag{9.6}$$

This equation shows that a stable phase-locked loop will track step changes in phase with a zero steady-state error. If there is a constant amplitude change in the input frequency (ramp) of $A$ rad/s, $\theta_r(s) = f_o/s^2$, Eq. (9.4) becomes

$$\lim_{t \to \infty} \theta_\epsilon(t) = \lim_{s \to 0} \frac{f_o}{K_v F(s)} = \frac{f_o}{K_v F(0)} \tag{9.7}$$

If $F(0) = 1$, the steady-state phase error will be inversely proportional to the loop gain $K_v$. Recall that the larger $K_v$ is, the larger the closed-loop bandwidth is, and thus the faster the loop response. To increase the response speed and reduce the tracking error, the loop gain should be as large as possible. If $F(0)$ is finite, there will be a finite steady-state phase error. The frequency error $f_\epsilon(t) = d/dt\, \theta_\epsilon(t)$ will be zero in the steady state. That is, the input and VCO frequencies will be proportional $(\omega_r N = \omega_o)$. (These conclusions are not correct for PLLs containing digital phase detectors. The steady-state behavior of these systems will be discussed in the section on large-signal behavior.)

If it is necessary to have zero phase error in response to step changes in the input frequency, then $\lim_{s \to 0} F(s)$ must be infinite. That is, the dc gain of the low-pass filter must be infinite. This can be realized by including in $F(s)$ a pole at the origin. In this case $F(s)$ will be of the form

$$F(s) = s^{-1}\frac{s/\omega_z + 1}{s/\omega_p + 1} \tag{9.8}$$

and the system will be type II, since the open-loop system now has two poles at the origin. However, the addition of the pole at the origin creates difficulties with the loop stability, and in fact the system will be unstable unless a lead network $(\omega_z)$ is also included in $F(s)$. Loop stability will now be examined in detail in order to determine how to design the stabilizing networks for adequate loop stability.

## 9.3

### STABILITY ANALYSIS

Feedback systems whose open-loop transfer function has one pole at the origin are known as *type I systems*. If the open-loop system has $N$ poles at the origin, it is a type $N$ system. Most phase-locked loops are either type I or type II systems. The PLL has an open-loop pole at the origin because of the VCO, so the system is at least type I. A second pole at the origin is often added to the filter to reduce steady-state errors and increase the noise suppression; the system is then type II. The stability characteristics of type I and type II systems will now be analyzed. The methods can be extended to higher-type systems.

A linear model block diagram for the PLL is given in Fig. 9.1. The open-loop gain (negative of the loop transmission) will be denoted as

$$G(s) = K_v \frac{F(s)}{s}$$

where

$$K_v = \frac{K K_o K_d}{N}$$

where $K_d$ is the phase detector gain, $K_o$ is the VCO sensitivity, $K$ is any additional loop amplification, and $N$ is the divide ratio. Also, $K_v$ is known as the *velocity constant*. Stability requires that the closed-loop poles all be located in the left half of the $s$ plane (the real part of the poles is less than zero). Stability analysis and system design are best carried out by deducing the closed-loop characteristics from the open-loop transfer function and loop gain. The application of Nyquist's criterion (see Chap. 7) to a polar plot of the open-loop frequency response is the most general method. All the PLL systems of practical significance are minimum-phase (no open-loop poles or zeros in the right half-plane), so Nyquist's stability criterion can be simplified to studying the open-loop gain and phase characteristics near the unit-gain (crossover) frequency.

The easiest way to analyze the loop stability is to plot the magnitude and phase of the open-loop transfer function versus frequency. The *phase margin* is defined as

$$\phi_M = 180° + \arg G(j\omega_c) \tag{9.9}$$

where the open-loop crossover frequency $\omega_c$, is the frequency at which the open-loop gain is unity. That is, the phase margin is equal to $180°$ plus the phase shift of the open-loop transfer function (a negative number) at the open-loop crossover frequency $\omega_c$. The greater the phase margin, the more stable the system and the more phase lag from parasitic effects can be tolerated. Additional phase lag invariably arises from neglected poles or from the time delay arising in the phase detector.

**EXAMPLE 9.1.** Consider a phase-locked loop which has $K_v = 50$ rad/s and which contains a low-pass filter with a corner frequency of 100 rad/s. What is the phase margin?

*Solution.* The magnitude and phase of the open-loop transfer function are plotted in Fig. 9.2. The system crossover frequency is approximately 50 rad/s. At this frequency the phase shift of the open-loop transfer function is $-112.5°$, so the phase margin is $67.5°$.

In this example the complete phase plot was presented, but once one is familiar with phase plots, they no longer need to be included. Since these are minimum-phase systems, the phase is uniquely determined by the magnitude characteristics. One can simply calculate the phase shift after determining the open-loop crossover frequency from the magnitude plot.

**EXAMPLE 9.2.** In Example 9.1, if the filter corner frequency had been 10 rad/s rather than 100 rad/s, what would have been the system phase margin?

*Solution.* To determine the phase margin, first plot the magnitude of the open-loop gain and determine the crossover frequency. The straight-line approximation of the magnitude is plotted in Fig. 9.3. And $\omega_c$ is found to be approximately 22 rad/s. Thus, the system phase margin is $180° - (90° + \tan^{-1} 2.2) = 23.40°$, which is too small for satisfactory loop stability. This is in agreement with Bode's rule, which states that if the magnitude of the open-loop response crosses the 0-dB line with a slope of $-12$ dB per octave, the system is unstable.[1] In this example, the straight-line approximation for the gain decreases at $-12$ dB per octave, but the actual response crosses the 0-dB line with a slope slightly more positive than $-12$ dB per octave, hence the small phase margin.

Although the most important design parameters concerning system frequency response are the closed-loop bandwidth $\omega_h$ and the peak value $M_P$ of the closed-loop frequency response, no design techniques exist that allow one to easily specify $\omega_h$ and $M_P$ for higher-order systems. However, it is relatively easy to design for

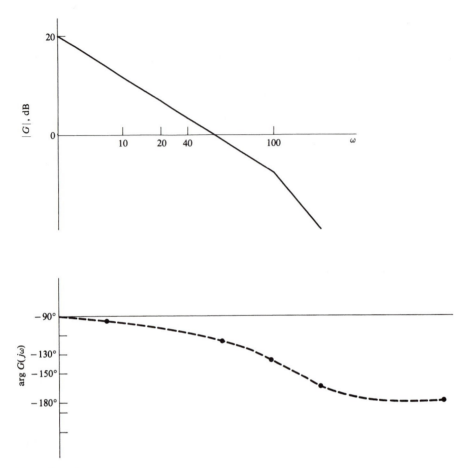

**FIGURE 9.2**
Magnitude and phase response of the open-loop system discussed in Example 9.1.

specified open-loop parameters $\omega_c$ and $\phi_M$. There are approximations which relate $\omega_c$ and $\phi_M$ to $\omega_n$, $M_P$, $\zeta$, and thus to the system rise time and overshoot. Fortunately, the conditions under which these approximations are valid are satisfied by most PLLs.

Since the interpretation is different for type I and type II systems, they will be discussed individually.

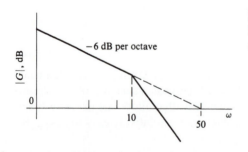

**FIGURE 9.3**
Magnitude response of the system discussed in Example 9.2.

**Type I Systems**

Consider first the case in which $F(s)$ is a first-order low-pass filter. For this case the open-loop transfer function is

$$G(j\omega) = \frac{K_v}{j\omega(j\omega/\omega_L + 1)} \tag{9.10}$$

where $\omega_L$ is the filter bandwidth and the loop velocity constant $K_v$ is the product of the phase detector gain $K_d$, the VCO gain $K_o$, any additional gain $K$, and $1/N$; that is,

$$K_v = \frac{K_d K_o K}{N}$$

If $N$ is unity, the closed-loop transfer function is

$$\frac{\theta_o}{\theta_r} = \left(\frac{s^2}{\omega_n^2} + \frac{2\zeta}{\omega_n}s + 1\right)^{-1} \tag{9.11}$$

where

$$\omega_n^2 = K_v \omega_L \tag{9.12}$$

and

$$\zeta = \frac{1}{2}\left(\frac{\omega_L}{K_v}\right)^{1/2} \tag{9.13}$$

By setting the magnitude of the open-loop transfer function (Eq. 9.10) to unity, the open-loop unity-gain frequency is easily shown to be

$$\omega_c = \omega_L\left\{\frac{[1 + 4(K_v/\omega_L)^2]^{1/2} - 1}{2}\right\}^{1/2} \tag{9.14}$$

Once $\omega_c$ is known, the phase margin

$$\phi_M = 90° - \tan^{-1}\frac{\omega_c}{\omega_L} = 90° - \tan^{-1}\left(\frac{\{1 + [(2\zeta^2)^2]^{-1}\}^{1/2} - 1}{2}\right)^{1/2} \tag{9.15}$$

can be calculated. This equation is plotted in Fig. 9.4. The parameters of the closed-loop system that are most important are adequate stability (which is related to phase margin), system bandwidth (which determines the speed of the transient response), and system transient response (which is described by the rise time and overshoot). The system bandwidth for a low-pass transfer function is defined as the frequency at which the gain is equal to 0.707 times its dc value. The bandwidth of the system represented by Eq. (9.11) is

$$\omega_h = \omega_n[1 - 2\zeta^2 + (2 - 4\zeta^2 + 4\zeta^4)^{1/2}]^{1/2} \tag{9.16}$$

For the underdamped second-order system given by Eq. (9.10) ($\zeta < 1$), the peak value of the time response to a unit step input can be shown to be

$$P_o = 1 + \exp\left[\frac{-\pi\zeta}{(1 - \zeta^2)^{1/2}}\right] \tag{9.17}$$

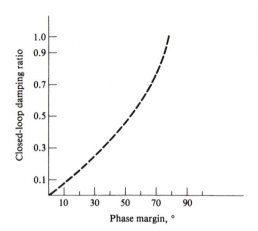

**FIGURE 9.4**
Phase margin as a function of damping ratio for a type I system.

and the percent overshoot

$$P_o = \left( \frac{P_o - 1}{1} \right) 100 = (P_o - 1)100$$

The overshoot is determined solely by $\zeta$. And $P_o$ as a function of $\zeta$ is plotted in Fig. 9.5. For zero damping the overshoot is 100 percent, and it decreases to zero for a unity damping ratio.

For high-order systems, the overshoot and bandwidth are not as readily related to the open-loop system parameters, but a good first approximation is that Eqs. (9.16) and (9.17) hold true for higher-order type I systems. The damping ratio is defined in terms of the phase margin by Eq. (9.15). It is relatively easy to design a system to have a given phase margin, and the design can then be evaluated by using computer simulation. If the simulation indicates that the overshoot is too high (or too low), then the phase margin can be increased (or reduced), but the relations between phase margin, damping, and overshoot are amazingly accurate for higher-order type I systems.

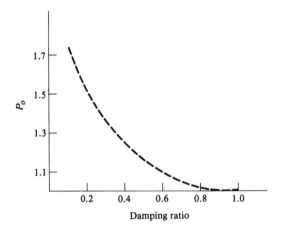

**FIGURE 9.5**
Step response peak overshoot as a function of damping ratio for a type I second-order system.

This implies that the response of many feedback systems can be described by a second-order model. Also, the closed-loop bandwidth can be related to the open-loop crossover frequency $\omega_c$. (It is usually about 50 percent greater).

If a design is desired for a specified peak transient overshoot, Eq. (9.17) can be used to determine the required damping, and then Eq. (9.15) can be used to determine the required phase margin.

EXAMPLE 9.3.  Consider a type I PLL with the open-loop transfer function

$$G(s) = \frac{100}{s(s/\omega_L + 1)^2}$$

which corresponds to a PLL with a second-order filter for increased suppression of the high-frequency components. What should $\omega_L$ be so that the system has approximately 20 percent overshoot to a step input?

*Solution.*  Although the relationships between overshoot, damping ratio, and phase margin have been derived for a second-order system, we will assume that they apply to the third-order system. From Fig. 9.5 it is seen that 20 percent overshoot to step input corresponds to the damping ratio $\zeta = 0.5$, and Fig. 9.4 shows that the damping ratio of 0.5 corresponds to a phase margin of 50°. For a phase margin of 50° the low-pass filter must have phase shift of $-40°$ at the crossover frequency $\omega_c$. Therefore, each filter pole can contribute $-20°$ at the crossover frequency. That is, $\tan^{-1}(\omega_c/\omega_L) = 20°$. Therefore, $\omega_L$ must be significantly larger than $\omega_c$. Using Eq. (9.14) with $K_v = 100$ rad/s and $\omega_L = 258$ rad/s, the crossover frequency is approximately 94 rad/s. The open-loop gain and phase plots are given in Fig. 9.6. It is seen that the crossover frequency is 15 Hz, and the phase margin is 50°. When the closed-loop step response of the system is simulated, the peak overshoot is found to be 14 percent. The use of the second-order approximations gives a fairly good estimate of the third-order characteristics. Also, we know that the overshoot can be increased by decreasing the phase margin. In fact, in this case selecting $\omega_L = 233$ rad/s corresponding to phase margin of 43.5° gives an overshoot of 20 percent. The step response is plotted in Fig. 9.7. The rise time (10 to 90 percent) is approximately 13 ms.

## Pole-Zero Filter

A zero can be added to the loop filter, resulting in the transfer function

$$F(s) = \frac{1 + s/\omega_z}{1 + s/\omega_p} \tag{9.18}$$

The addition of the zero increases the ability to shape the loop's frequency response, but it decreases the amount of reference-frequency filtering. If $\omega_z > \omega_p$, then the open-loop frequency is as shown in Fig. 9.8. In this case $F(s)$ is used to reduce $\omega_c$ and the closed-loop bandwidth. For good loop stability, the frequency of the zero filter $\omega_z$ should be less than the open-loop crossover frequency.

If $\omega_p > \omega_z$, then the open-loop frequency response is as shown in Fig. 9.9. In this case the addition of the filter zero is used to increase the loop bandwidth. It also increases any high-frequency noise which may be present. However, since

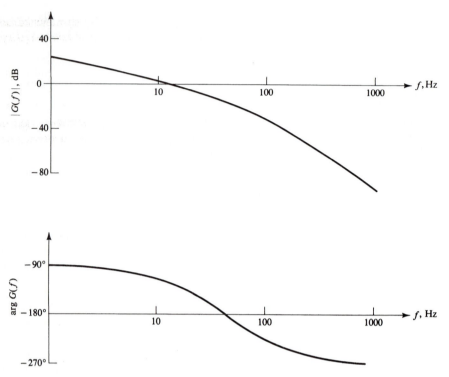

**FIGURE 9.6**
Open-loop magnitude and phase response of the system discussed in Example 9.3.

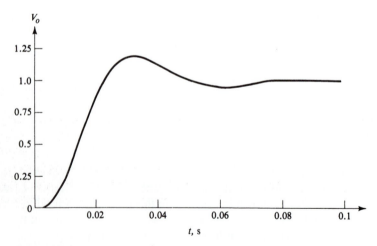

**FIGURE 9.7**
Step response of the system discussed in Example 9.3.

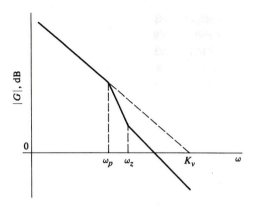

**FIGURE 9.8**
Magnitude response of a type I system
with a pole-zero filter with $\omega_z > \omega_p$.

$F(s)$ is assumed to have a dc gain of unity, it can only be realized using active components.

In the preceding example, the loop gain $K_v$ was fixed, and it was necessary to select the loop filter bandwidth to achieve the desired transient performance. Often the filter bandwidth selection is based on the required suppression of the phase detector *spurious* (the reference frequency and its harmonics). The design procedure is the same in this case, only now the loop gain is chosen to give the desired phase margin.

## Control of Loop Bandwidth

In some instances it is also necessary to specify the loop bandwidth. To control both the loop damping and bandwidth, an amplifier can be added in series with the low-pass filter. If the filter is implemented using active components, the additional gain can be obtained without any additional components.

> **EXAMPLE 9.4.** Consider again Example 9.3, with the additional specification that the rise time in response to a unit step input be less than 1.3 ms.

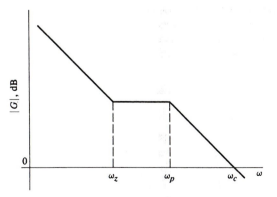

**FIGURE 9.9**
The pole of the pole-zero filter is
larger than the zero frequency.

*Solution.* Since the overshoot is to be 20 percent, the phase margin must still be approximately $45°$. To design for the rise time specification, it often suffices to use the approximation

$$t_r = \frac{2.2}{\omega_h} \tag{9.19}$$

which is exact only for first-order systems, but provides a good design guideline for higher-order systems. Thus, $\omega_h$ should be greater than $2.2/t_r = 1.69 \times 10^3$ rad/s.

There are several approximations for estimating the closed-loop bandwidth from the open-loop frequency response. The simplest approximation is to assume the closed-loop bandwidth is the same as the open-loop crossover frequency. This is exact for first-order type I systems. A more accurate approximation for higher-order systems is usually obtained by using Eq. (9.16). In this example, assuming that $\zeta = 0.45$,

$$\omega_n = 1.27 \times 10^3 = (K_v \omega_L)^{1/2}$$

Since the phase margin is to be the same as in the previous example, it is reasonable to assume that the ratio $K_v/\omega_L$ remains constant. The filter frequency must also be increased to maintain the same phase margin. In this case, $K_v = 0.83 \times 10^3$ rad/s.

For this value of $K_v$, the closed-loop bandwidth is 1250 rad/s and the rise time is 1.6 ms. For this example, increasing $K_v$ to 1000 and $\omega_L$ to 2330 rad/s meets the specifications.

A plot of the step response is shown in Fig. 9.10. The peak overshoot is 20 percent, and the rise time is 1.3 ms. The two specifications are now met. In general, two adjustable parameters, such as loop gain and filter bandwidth, are needed to independently specify overshoot and rise time.

A plot of the magnitude of the closed-loop frequency response is given in Fig. 9.11. The actual closed-loop bandwidth is 1500 rad/s. This example illustrates that the approximations can work well for higher-order systems.

The use of the phase margin specification provides an easy method for meeting the system design specifications. Third-order and higher systems normally

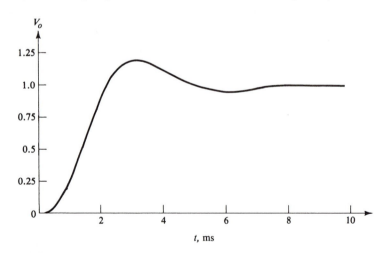

**FIGURE 9.10**
Step response of the system discussed in Example 9.4.

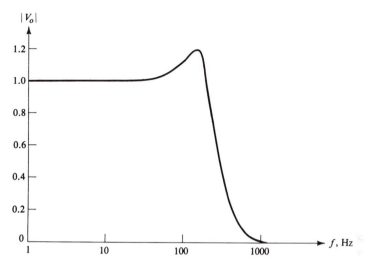

**FIGURE 9.11**
Closed-loop frequency response of the system discussed in
Example 9.4.

cannot be solved analytically. For these systems, designing to meet a specified
phase margin allows one to make a very good first approximation. Closer specifications, if necessary, are then achieved with computer simulation. A similar
approach can be used for type II systems, but a different set of approximations is
needed.

## Type II Systems

The previous steady-state error analysis showed that for zero steady-state phase
error in response to step changes in the input frequency, the low-pass filter $F(s)$
must contain a pole at the origin. The open-loop system will then have two poles at
the origin. This is referred to as a *type II feedback system.* Type II systems are inherently unstable unless a phase lead network is added inside the loop. If $F(s)$ is simply $1/s$, then the open-loop transfer function is

$$G(s) = \frac{K_a}{s^2} \tag{9.20}$$

which, as shown in Fig. 9.12, crosses the 0-dB axis with a slope of $-12$ dB per
octave, and the phase margin is $0°$. The 0-dB point is the frequency at which

$$\frac{K_a}{\omega^2} = 1$$

or

$$\omega_c = K_a^{1/2} \tag{9.21}$$

The loop gain $K_a$ has dimensions expressed in radians per second squared.

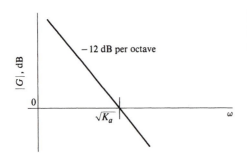

$-12$ dB per octave

$|G|$, dB

0

$\sqrt{K_a}$

$\omega$

**FIGURE 9.12**
Magnitude response of a type II
second-order system.

Bode developed a rule of thumb for interpreting the system stability from a plot of the magnitude of the open-loop frequency response.[1] Specifically, if the open-loop frequency response crosses the 0-dB line with a slope of $-6$ dB per octave, the system is stable. If the slope is $-12$ dB per octave or even more negative, the system is unstable.

In order to stabilize this system, a network must be added that alters the gain so that it crosses the 0-dB line with a slope of $-6$ dB per octave. Such a network is referred to as a *lead network,* since it has a positive or leading phase shift. Figure 9.13 illustrates the gain modified with the simplest lead compensation that can be used. Here a zero at $\omega_z$ has been added to the open-loop transfer function. The composite low-pass filter transfer function is

$$F(s) = \frac{s/\omega_z + 1}{s} \tag{9.22}$$

The zero location $\omega_z$ is selected to give the desired phase margin. The smaller $\omega_z$ is, the greater will be the phase margin and the greater will be the crossover frequency $\omega_c$. Note that this system is also a second-order control system.

Figure 9.14 illustrates an easy method for realizing the filter pole and zero for the case in which the phase detector output is a charge pump. The VCO control voltage is

$$V_c(s) = I(s)\frac{RsC + 1}{sC} \tag{9.23}$$

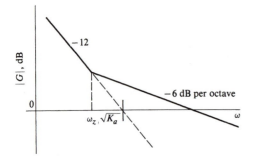

**FIGURE 9.13**
Magnitude response of a type II second-order system including a lead network.

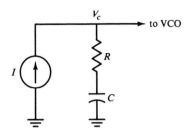

**FIGURE 9.14**
A lead network used with a charge pump.

The capacitor realizes the additional pole at the origin, and the addition of the resistor $R$ realizes the zero at

$$\omega_z = (RC)^{-1} \tag{9.24}$$

Note that without the resistor the system would be unstable.

If the phase detector output can be modeled as a voltage source (low output impedance), then the circuit illustrated in Fig. 9.15 can be used to realize the desired transfer function. In this case,

$$V_c(s) = V_D(s)\frac{R_2 sC + 1}{(R_1 + R_s)sC} \tag{9.25}$$

The open-loop transfer function is

$$G = \frac{K_a(s/\omega_z + 1)}{s^2}$$

and the closed-loop transfer function is

$$\frac{G(s)}{1 + G(s)} = \frac{s/\omega_z + 1}{s^2/K_a + s/\omega_z + 1} \tag{9.26}$$

The new transfer function is not as easily analyzed as is Eq. (9.11) because of the presence of the zero at $\omega_z$. The design of type II systems is most easily carried out

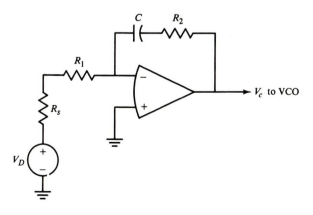

**FIGURE 9.15**
An active lead network to be used with a voltage source.

using Bode design techniques; note that if $\omega_z = \omega_c$, which requires that $\omega_c = K_a^{1/2}$, the calculated phase margin (using the straight-line approximation) is 45°. The phase margin will be greater than this since the actual crossover frequency will be increased by the addition of the zero at $\omega_z$. The following example illustrates the application of Bode design techniques for the design of type II systems.

**EXAMPLE 9.5.** For a PLL with a $K_a$ of 1000 (rad/s)$^2$, design a low-pass filter such that the system will have zero steady-state phase error in response to constant-frequency inputs, less than the 20 percent overshoot to step changes in the input phase, and a rise time of less than 1 ms.

*Solution.* The solution calls for a type II system; the simplest open-loop transfer will be of the form

$$G(s) = \frac{K_a(s/\omega_z + 1)}{s^2}$$

Since all stable type II systems meet the steady-state error specification, the only problem is to select $\omega_z$ to meet the overshoot and rise-time specifications. If $\omega_z$ is made much less than $K_a$, the phase margin will approach 90° and the open-loop crossover frequency will be that of a first-order system. The open-loop crossover frequency will then be approximately $K_a/\omega_z$, and the system will behave as a first-order system meeting the overshoot specification, since there will be no overshoot. However, there is much less reference filtering.

If the characteristics of a second-order system are to be maintained, it will be necessary that $\omega_c \cong (K_d K_o)^{1/2} = K_a^{1/2}$. Note, however, that $\omega_c \simeq K_a^{1/2} = 1000^{1/2}$ in this case, so the crossover frequency is too low to meet the rise-time specification $(t_r \approx 2.2/1000^{1/2}\text{ s})$. Therefore, additional gain will be required in the loop. For the first estimate the open-loop gain $K_a$ will be selected so that the open-loop crossover frequency $\omega_c = 2.2 \times 10^3$. This crossover frequency should result in the system's meeting

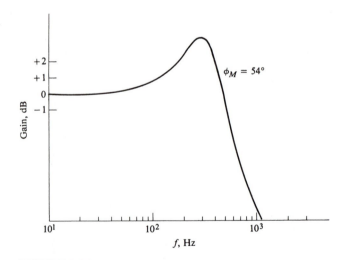

**FIGURE 9.16**
Closed-loop frequency response of a type II system with 54° phase margin (Example 9.5).

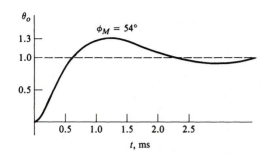

**FIGURE 9.17**

Step response of a type II system with 54° phase margin (Example 9.5).

the rise-time specification. And $K_a$ must equal $(2.2 \times 10^3)^2$. The filter zero $\omega_z$ will also be set equal to $\omega_c$, since this will give a phase margin greater than 45° and will actually increase the crossover frequency above that estimated using the straight-line approximation. The actual crossover frequency turns out to be close to $3 \times 10^3$ rad/s, and the phase margin is 54°. The magnitude of the closed-loop frequency response, plotted in Fig. 9.16, has a peak value of $M_p = 3.5$ dB. The system step response, plotted in Fig. 9.17, has a rise time of approximately $0.5 \times 10^{-3}$ s and a 30 percent overshoot.

Because the overshoot is far greater than would have been predicted from the phase margin, this indicates that the required phase margin predicted from the analysis of type I systems is not always a good predictor of step response in type II systems. Nevertheless, system performance can readily be improved by increasing the phase margin. In this system the phase margin can be increased by increasing $K_a$ or decreasing $\omega_z$, both of which will increase the crossover frequency $\omega_c$. For this example $\omega_z$ was first reduced to $1.665 \times 10^3$ rad/s, and the overshoot was found to be (using computer simulation) 24 percent. For the next trial $\omega_z$ was reduced to $1.25 \times 10^3$ rad/s (the open-loop crossover frequency increased to $3.8 \times 10^3$ rad/s); the phase margin was approximately 70°. The step response, plotted in Fig. 9.18, has a peak overshoot of 17 percent, and the frequency response, plotted in Fig. 9.19, has an $M_p$ of 1.6 dB. The rise time is approximately $0.45 \times 10^{-3}$ s. Since this is less than one-half of the specified rise time, the loop bandwidth could be reduced by 50 percent and still meet the transient response specifications. This can easily be accomplished by reducing both $\omega_z$ and the square root of $K_a$ by 50 percent. The closed-loop frequency response and the step response are plotted in Figs. 9.20 and 9.21 for the case in which $\omega_z = 0.625 \times 10^3$ and $K_a = 1.21 \times 10^6$. The closed-loop bandwidth has been reduced to 350 Hz, the step response rise time is $0.85 \times 10^{-3}$ s, and the overshoot is 17 percent.

The preceding example illustrates that the relationships derived for type I systems between phase margin, damping ratio, and overshoot are not accurate for the

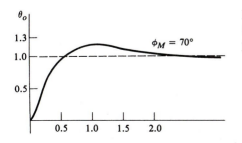

**FIGURE 9.18**

Step response of a type II system with 70° phase margin.

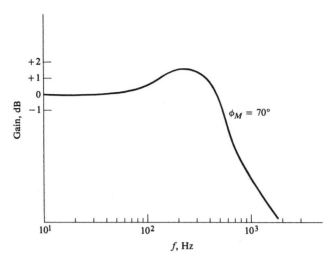

**FIGURE 9.19**
Closed-loop frequency response of a type II system with
70° phase margin.

type II systems under discussion. Therefore, type II systems will now be analyzed
and appropriate design approximations derived. For the open-loop system

$$G(s) = \frac{K_a(s/\omega_z + 1)}{s^2}$$

The error signal transfer function is

$$\theta_\epsilon(s) = \frac{\theta_r(s)}{1 + K_a G(s)} = \frac{\theta_r(s)s^2}{s^2 + K_a(s/\omega_z + 1)} \tag{9.27}$$

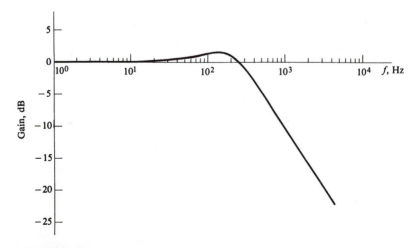

**FIGURE 9.20**
Closed-loop frequency response of a type II system that has a smaller
bandwidth than the system of Fig. 9.19.

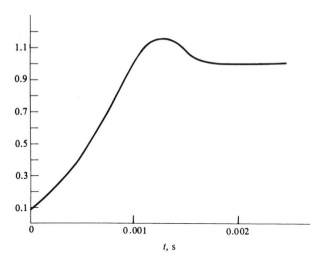

**FIGURE 9.21**
Frequency step response of
system discussed in
Example 9.5 (smallest-
bandwidth case).

and the error in response to a unit step input is

$$\theta_\epsilon(s) = \frac{s}{s^2 + K_a(s/\omega_z + 1)} \tag{9.28}$$

The corresponding time domain response is

$$\theta_\epsilon(t) = e^{-\zeta \omega_n t}\left[\cos \omega_n(1 - \zeta^2)^{1/2}t - \frac{\zeta}{1 - \zeta^2}\sin \omega_n(1 - \zeta^2)^{1/2}t\right] \tag{9.29}$$

where

$$\omega_n^2 = K_a \tag{9.30}$$

and

$$2\zeta \omega_n = \frac{K_a}{\omega_z}$$

or

$$\zeta = \frac{K_a^{1/2}}{2\omega_z} \tag{9.31}$$

The peak overshoot can be determined by calculating the minimum error signal since for a step input

$$\theta_o(t) = 1 - \theta_\epsilon(t)$$

The peak output signal is found by first finding the time at which the error signal is at a minimum; this is determined by setting the first derivative equal to zero. That is,

$$\frac{d\theta_\epsilon}{dt} = -\zeta \omega_n e^{-\zeta \omega_n t}\left[\cos \omega_n(1 - \zeta^2)^{1/2}t + \frac{-\zeta}{1 - \zeta^2}\sin \omega_n(1 - \zeta^2)^{1/2}t\right]$$

$$- \omega_n(1 - \zeta^2)^{1/2}e^{-\zeta \omega_n t}\sin \omega_n(1 - \zeta^2)^{1/2}t \tag{9.32}$$

$$- \frac{\zeta}{(1 - \zeta^2)^{1/2}}\omega_n e^{-\zeta \omega_n t}\cos \omega_n(1 - \zeta^2)^{1/2}t = 0$$

The time $t_p$ at which the minimum error occurs is found to be

$$t_p = \omega_n^{-1}[(1 - \zeta^2)^{1/2}]^{-1}\tan^{-1}\frac{2\zeta(1 - \zeta^2)^{1/2}}{2\zeta^2 - 1} \tag{9.33}$$

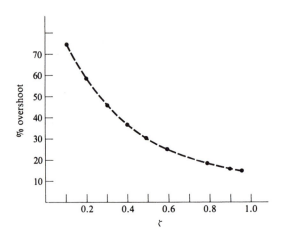

**FIGURE 9.22**
Percent overshoot of the step
response of a type II second-order
system.

If we define

$$t_p' = \omega_n(1-\zeta^2)^{1/2}t_p = \tan^{-1}\frac{2\zeta(1-\zeta^2)^{1/2}}{2\zeta^2-1}$$

then the corresponding peak value of the output is

$$P_o = 1 + \exp\frac{-\zeta}{(1-\zeta^2)^{1/2}}t_p'\left[\cos t_p' - \frac{\zeta}{(1-\zeta^2)^{1/2}}\sin t_p'\right] \tag{9.34}$$

The peak overshoot $P_o$ as a function of $\zeta(K_a^{1/2}/\omega_z = 2\zeta)$ is plotted in Fig. 9.22.
The open-loop crossover frequency of this type II system can be found by solving

$$\left|\frac{K_a(j\omega/\omega_z + 1)}{\omega^2}\right| = 1$$

It is readily shown that

$$\omega_c = K_a^{1/2}[2\zeta^2 + (4\zeta^4 + 1)^{1/2}]^{1/2} \tag{9.35}$$

The normalized open-loop crossover frequency $\omega_c/K_a^{1/2}$ as a function of $\zeta$ is plotted
in Fig. 9.23. Figure 9.23 shows that increasing $\zeta$ (by either increasing $K_a$ or decreasing $\omega_z$) increases the open-loop crossover frequency, and hence the closed-loop
bandwidth $B$. The closed-loop bandwidth can be determined analytically by solving

$$\frac{|K_a(j\omega/\omega_z + 1)/\omega^2|^2}{|1 + K_a(j\omega/\omega_z + 1)/(j\omega)^2|^2} = \frac{1}{2}$$

It is found that the closed-loop bandwidth is

$$B = K_a^{1/2}[1 + 2\zeta^2 + (2 + 4\zeta^2 + 4\zeta^4)^{1/2}]^{1/2} \qquad \text{rad/s} \tag{9.36}$$

The normalized closed-loop bandwidth $B/K_a^{1/2}$ as a function of $\zeta$ is plotted in Fig.
9.24.

The type II systems under discussion have two variables, the open-loop gain constant and $\omega_z$, the filter zero location. These two variables can be used to determine

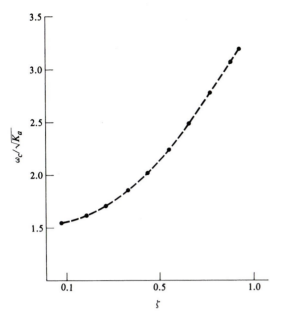

**FIGURE 9.23**
Normalized crossover frequency $\omega_c/K_a^{1/2}$ of a type II second-order system.

the system bandwidth and transient response to a step input. The procedure is illustrated by the following design example.

**EXAMPLE 9.6.** A type II system is to be designed to have 20 percent overshoot to a unit step input of phase and a closed-loop bandwidth of $10^3$ rad/s.

**Solution.** From Fig. 9.22 we see that 20 percent overshoot ($P_o = 1.2$) corresponds to $\zeta = 0.8$. For this value of $\zeta$, Fig. 9.24 indicates that the closed-loop bandwidth is

$$B = K_a^{1/2} \ (2.18)$$

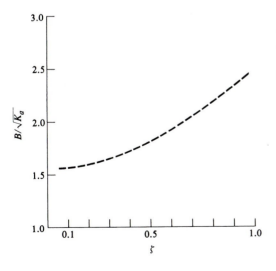

**FIGURE 9.24**
Normalized closed-loop bandwidth $B/K_a^{1/2}$ of a type II second-order system as a function of the closed-loop damping ratio.

so
$$K_a = \left(\frac{10^3}{2.18}\right)^2 = 0.21 \times 10^6 \text{ (rad/s)}^2$$

Therefore,

$$\omega_n = 459 \text{ rad/s}$$

and
$$\omega_z = \frac{\omega_n}{2\zeta} = 287 \text{ rad/s}$$

If this system is to be realized using the phase-locked loop filter illustrated in Fig. 9.14, the loop gain $I/(CK_o) = K_a$ must be $0.21 \times 10^6$; or if the gain is less, additional amplification will be required. The resistor $R$ is selected so that

$$(RC)^{-1} = 287 \text{ rad/s}$$

## Type II Third-Order Systems

Type II third-order systems cannot be solved analytically as easily as the type II second-order systems, but the results obtained for second-order systems can be used as approximations in the design of third-order systems. The simplest open-loop transfer function of a type II third-order system is of the form

$$G(s) = \frac{K_a(s/\omega_z + 1)}{s^2(s/\omega_p + 1)}$$

The additional pole at $\omega_p$ may be inherent in one of the transfer functions, or it may be added to the loop filter to increase the high-frequency filtering. A third-order type II system will have a high-frequency attenuation rate of 12 dB per octave, while a second-order type II system will roll off at $-6$ dB per octave. The third-order type II system is most easily analyzed by studying the phase characteristics of

$$G_F(j\omega) = \frac{j\omega/\omega_z + 1}{j\omega/\omega_p + 1} \tag{9.37}$$

Since
$$\arg G(j\omega) = -\pi + \arg G_F(j\omega)$$

The system will be unstable if $\omega_p \le \omega_z$. Therefore, this case does not need to be considered. If $\omega_z < \omega_p$, then the phase margin is

$$\phi_M = \arg G_F(j\omega_c) = \tan^{-1}\frac{\omega_c}{\omega_z} - \tan^{-1}\frac{\omega_c}{\omega_p} \tag{9.38}$$

A plot of $\arg G_F(j\omega)$ for various values of $\alpha = \omega_p/\omega_z$ is contained in Fig. 9.25. The frequency at which the peak value of $\arg G_F(j\omega)$ occurs can be found by setting the derivative $d/d\omega \arg G_F(j\omega)$ equal to 0. It is readily shown that the frequency $\omega$ at which the $\arg G_F(j\omega)$ is a maximum is

$$\omega_m = (\omega_z\omega_p)^{1/2} \tag{9.39}$$

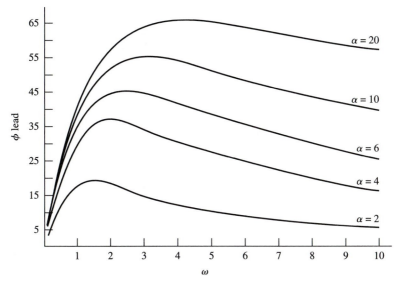

**FIGURE 9.25**
Lead network phase shift as a function of frequency for selected lead ratios $\alpha$.

If this form of transfer function is to be used as the stabilizing filter, then selecting the crossover frequency equal to

$$\omega_c = (\omega_z \omega_p)^{1/2} \tag{9.40}$$

will result in the maximum phase margin. If $\omega_c \neq (\omega_z \omega_p)^{1/2}$, then a larger lead ratio $\alpha$ will be required to obtain the same phase margin. A plot of the maximum phase lead available from $G_F(j\omega)$ as a function of $\alpha$ is plotted in Fig. 9.26. Once the designer knows the necessary phase margin, the necessary lead ratio can be

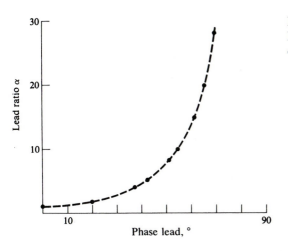

**FIGURE 9.26**
Phase lead available as a function of lead ratio $\alpha$.

obtained from that figure. The required phase margin can be estimated by again considering the characteristics of the type II second-order system. Equation (9.35) gives the open-loop crossover frequency of this system as a function of $K_a$ and $\zeta$ (or $\omega_z$). For the second-order type II system, the phase margin is

$$\phi_M = \tan^{-1} \frac{\omega_c}{\omega_z}$$

Substituting $\omega_c$ from Eq. (9.35), we obtain

$$\phi_M = \tan^{-1}\left\{ \frac{K_a^{1/2}}{\omega_z}[2\zeta^2 + (4\zeta^4 + 1)^{1/2}]^{1/2} \right\}$$
$$= \tan^{-1}\{2\zeta[2\zeta^2 + (4\zeta^4 + 1)^{1/2}]^{1/2}\}$$
(9.41)

The corresponding phase margin as a function of $\zeta$ is plotted in Fig. 9.27. The percent overshoot as a function of phase margin is plotted in Fig. 9.28. The use of these curves in the design of a type II third-order system is illustrated in the following example.

**EXAMPLE 9.7.** Consider again the type II system of the previous example.

$$K_a = 0.21 \times 10^6 (\text{rad/s})^2$$

Determine a lead-lag network so that the closed-loop system has approximately the same overshoot to step changes in phase.

**Solution.** Figure 9.28 indicates that for a type II second-order system, the phase margin must be at least 70° for the overshoot to be 20 percent or less. If we assume that the same phase margin will be required for the type II third-order system, then we can determine the minimum lead ratio required from Fig. 9.26. This figure indicates that a lead ratio of 28 is required for a phase margin of 70°. The remaining design problem is to determine $\omega_z$, which is to be placed so that Eq. (9.39) is satisfied. At $\omega_c$ the magnitude

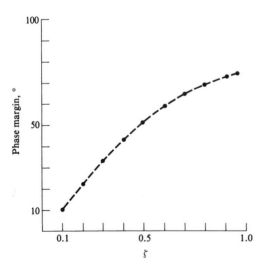

**FIGURE 9.27**
Phase margin of a type II second-order system as a function of damping ratio.

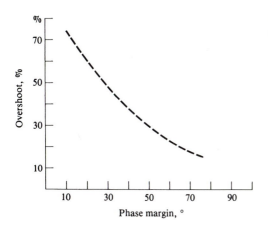

**FIGURE 9.28**
Percent overshoot in response to a step input as a function of phase margin for a type II second-order system.

of the lead network response is (in decibels)

$$20 \log G_F(j\omega_c) = 20 \log \frac{|j\omega_c/\omega_z + 1|}{|j\omega_c/(\alpha\omega_z) + 1|} = 10 \log \frac{\omega_c^2/\omega_z^2 + 1}{\omega_c^2/(\alpha^2\omega_z^2) + 1} \quad (9.42)$$

and since $\omega_c^2 = \alpha\omega_z^2$,

$$20 \log G_F(j\omega) = 10 \log \frac{\alpha + 1}{\alpha^{-1} + 1} \approx 20 \log \alpha^{1/2} \quad (9.43)$$

That is, the magnitude of the gain of the lead-lag network is equal to $\alpha^{1/2}$ at the crossover frequency, provided Eq. (9.40) is satisfied. In this example, with $\alpha$ equal to 28, $G_F(j\omega_c)$ increases the gain by

$$20 \log 28^{1/2} = 14.5 \, \text{dB}$$

at $\omega_c$, so we must determine where the uncompensated frequency response is $-14.5$ dB. That is,

$$20 \log \frac{K_a}{\omega^2} = -14.5 \, \text{dB}$$

which occurs at

$$\omega_c = 28^{1/4} K_a^{1/2} = 1.06 \times 10^3 \, \text{rad/s}$$

The 28:1 lead ratio will increase the crossover frequency by a factor of 2.3. The zero is placed at

$$\omega_z = \frac{\omega_c}{28^{1/2}} = 200 \, \text{rad/s}$$

and the pole at

$$\omega_p = 28 \, \omega_z = 5600 \, \text{rad/s}$$

The frequency response of the systems described in the preceding two examples is compared in Fig. 9.29. The third-order system has a wider bandwidth, and for frequencies above $10^3$ Hz it also has greater attenuation. This attenuation could be desirable for the additional filtering of input noise that is created in the phase detector. The bandwidth of the third-order loop could be reduced by reducing $K_a$ (plus $\omega_z$ and $\omega_p$). The

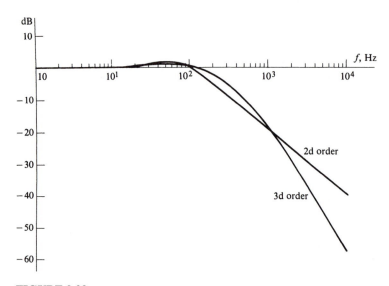

**FIGURE 9.29**
Frequency response of second- and third-order type II systems.

transient responses of the two systems are shown in Fig. 9.30. The third-order system has a shorter rise time, as expected, since it has the greater bandwidth. Note that its overshoot is 13 percent, significantly below the specified 20 percent maximum. The overshoot could be increased, if desired, by reducing the phase margin either by reducing the lead ratio or simply by increasing $\omega_p$.

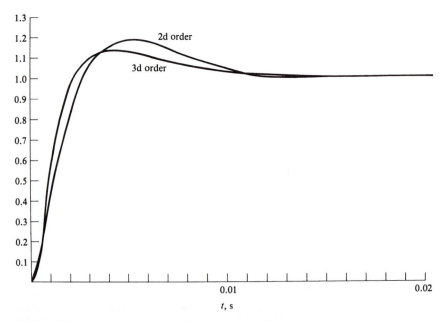

**FIGURE 9.30**
Step response of second- and third-order type II systems.

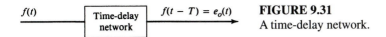

FIGURE 9.31
A time-delay network.

The analysis of type II systems selected the filter bandwidth after the loop gain has been determined. Often the filter frequency response is first determined based on other considerations, particularly suppression of the high-frequency components of the phase detector output. If the filter is specified, then the loop gain is selected so that the desired phase margin is obtained.

## Loops Including a Time Delay

Many phase-locked loops include a time delay, which can seriously degrade loop stability. Time delays, for example, can occur in sampling phase detectors and digital dividers. Figure 9.31 illustrates a time-delay network in block diagram form. For an input $f(t)$ to a time-delay network, the output is

$$e_o(t) = f(t - T) \tag{9.44}$$

Therefore,

$$E_o(s) = \int_0^\infty f(t - T)e^{-sT}\, dt$$
$$= F(s)e^{-sT}$$

That is, the transfer function of a time-delay network is

$$\frac{E_o(s)}{F(s)} = e^{-sT} \tag{9.45}$$

The delay network is shown is block diagram form in Fig. 9.32.

The magnitude of the frequency response of a time-delay network is $|e^{-j\omega T}| = 1$, so the time delay does not affect the magnitude of the frequency response. However, the phase shift

$$\arg e^{-j\omega T} = -\omega T$$

is a linearly decreasing function of frequency. The effect that a time delay can have on stability is illustrated by the following examples.

EXAMPLE 9.8. A simple linear model of a PLL consisting of a VCO and phase detector (modeled as a time delay) is illustrated in Fig. 9.33. The problem is to determine for what value of delay $T$ the loop will become unstable.

*Solution.* Since the open-loop transfer function is

$$G(s) = \frac{e^{-Ts} K_o}{s}$$

FIGURE 9.32
Block diagram of a time-delay network.

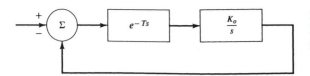

**FIGURE 9.33**
Linear model of a type I PLL including a time delay.

the crossover frequency is found from

$$\left| \frac{e^{-j\omega T} K_o}{j\omega} \right| = 1$$

or

$$K_o = \omega_c$$

The phase shift at the crossover frequency is

$$\phi = -\frac{\pi}{2} - \omega_c T$$

and the system will be marginally stable if

$$\omega_c T = \frac{\pi}{2} \qquad \text{or} \qquad T = \frac{\pi}{2K_o}$$

For smaller values of $T$, the system will be stable, and for larger values of $T$ it will be unstable.

In the preceding example the system would have been stable for all values of gain if it had not been for the phase lag due to the time delay. In more complex systems, it is usually not possible to arrive analytically at a relationship between loop gain and time delay $T$, but the effect of the time delay on phase margin is readily determined. This is illustrated by the following example.

**EXAMPLE 9.9.** Consider the phase-locked loop system with the open-loop transfer function

$$G(s) = \frac{1000}{s(s/1192 + 1)}$$

This system is found to have a phase margin of 50° and a crossover frequency of approximately 100 rad/s. What time delay can the phase detector introduce and still have a phase margin of 40°?

*Solution.* Since the phase margin without time delay is 50°, a 10° phase lag can be introduced by the time delay at the crossover frequency. That is,

$$\phi_T = -\omega_c T \le -0.174 \text{ rad } (10°)$$

So

$$T \le \frac{0.174}{\omega_c} = 0.174 \times 10^{-3}$$

In a well-designed system with adequate phase margin, an additional phase lag of 0.1 rad can usually be tolerated. In such a system, any time delay $T$ introduced will not be a significant factor, provided $T \le 0.1/\omega_c$.

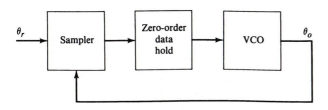

**FIGURE 9.34**
A PLL including a sampling phase detector.

## Loops Containing a Sample-and-Hold Phase Detector

Consider the phase-locked loop shown in Fig. 9.34, which contains a sampling phase detector and zero-order data hold. It was shown in Chap. 8 that this PLL can be replaced by the linear model shown in Fig. 8.22.

The open-loop transfer function is

$$G(j\omega) = \frac{K_v e^{-j\omega T/2}}{j\omega} \frac{\sin(\omega T/2)}{\omega T/2} \tag{9.46}$$

This system appears to be a type II system since there are two poles at the origin, but since

$$\lim_{\omega \to 0} \frac{\sin(\omega T/2)}{\omega T/2} \to 1$$

there is also a zero at the origin. Hence the system is type I. At the crossover frequency $\omega_c$, the open-loop phase shift is

$$\phi = -\frac{\pi}{2} - \frac{\omega_c T}{2}$$

and the phase margin is

$$\phi_M = \pi + \phi = \frac{\pi}{2} - \frac{\omega_c T}{2}$$

Since the magnitude of the open-loop gain at the crossover frequency is unity,

$$\frac{T}{2} K_v \frac{\sin(\omega_c T/2)}{(\omega_c T/2)^2} = 1$$

or

$$\frac{T}{2} K_v = \frac{(\omega_c T/2)^2}{\sin(\omega_c T/2)}$$

$$= \frac{(\pi/2 - \phi_M)^2}{\sin(\pi/2 - \phi_M)} \tag{9.47}$$

This equation describes the relation between phase margin $\phi_M$, sampling period $T$, and loop gain $K_v$ for a PLL composed of an ideal sampling phase detector, zero-order data hold, and VCO. The plot of $(T/2)K_v$ as a function of $\phi_M$ given in Fig. 9.35 shows that for each value of $\phi_M$ there is a single value of $(T/2)K_v$. For a $(T/2)K_v$ of $(\pi/2)^2$, the phase margin is $0°$. As $K_v$ is decreased, the phase margin increases and reaches $90°$ for $K_v = 0$.

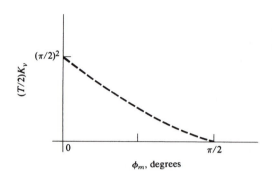

**FIGURE 9.35**

Phase margin as a function of sampling rate for a second-order PLL using a sampling phase detector.

The effect on loop performance of changes in the other system variable (the sampling rate $T$) can be determined in the same manner. Since at the crossover frequency $\omega_c$ the magnitude of the open-loop gain is unity,

$$\frac{T}{2} K_v \frac{\sin(\omega_c T/2)}{(\omega_c T/2)^2} = 1$$

If $K_v T$ remains constant and the sampling rate $T$ is changed, $\omega_c$ must change such that $\omega_c T$ is constant. That is, if the sampling rate is decreased ($T$ increases), and $K_v$ decreases so that $TK_v$ remains constant, the crossover frequency must decrease so that $\omega_c T$ remains constant.

**EXAMPLE 9.10.** Calculate the value of $K_v$ required for a 45° phase margin in a PLL whose open-loop transfer function is given by Eq. (9.46).

**Solution.** In order to have a 45° phase margin, the phase lag of the sample-and-hold detector must be 45° at the crossover frequency. Therefore, $\omega_c T/2 = \pi/4$ and the crossover frequency must be $\omega_c = \pi/(2T)$. Since the magnitude of the open-loop gain is unity at the crossover frequency, $K_v$ is determined from

$$\frac{T}{2} K_v \frac{\sin(\pi/4)}{(\pi/4)^2} = 1$$

or
$$K_v = \frac{(\pi/4)^2 2^{3/2}}{T}$$

An important characteristic of the sample-and-hold phase detector is the time delay, which is equal to one-half of the reference period. The effect on the magnitude of the frequency response can usually be ignored, but the time delay can significantly affect the loop stability.

## 9.4

### PLL TRANSIENT PERFORMANCE

Since the PLL is a nonlinear device over much of its operating range, the complete performance of the loop depends especially on the type of phase comparator used in the system. Nevertheless, an approximate analysis of the system performance provides insight for formulating generalized design guidelines. Before we consider the dynamic performance of the loop, it is instructive to consider the range of input frequencies over which the loop can remain locked. First consider a type I loop with a sinusoidal phase detector. If the PLL is in lock, small changes in the input frequency will cause a change in the phase detector output voltage in the direction that drives the error signal back toward zero. The change in frequency will be the forward loop gain times the error voltage, or

$$\Delta f = K_d K_o \cos \theta_\epsilon$$

where $\theta_\epsilon$ is the phase error. This equation assumes that the signals are changing so slowly that there is no attenuation in the low-pass filter. Since $\cos \theta_\epsilon \leq 1$,

$$\Delta f \leq K_d K_o = K_v$$

That is, the maximum change in frequency for which a type I loop can remain locked, referred to as the *lock range* or *tracking range,* is less than or equal to the forward loop gain. Lock range refers to how far the input frequency can change (slowly) without the loop's losing frequency lock. The same analysis holds for digital phase detectors, except that $K_d$ is replaced by the detector output voltage in the expression for $K_v$. *Capture range* refers to how close the input frequency must come to the free-running frequency of the VCO before lockup can occur. When the frequency error is other than zero, the PD output voltage is attenuated by the low-pass filter before it reaches the VCO. The capture range is thus less than the lock range. The actual value depends on the type of low-pass filter used. A general expression for loop capture range is not available, as the system is highly nonlinear.

For type II loops, $K_v$ is theoretically equal to infinity. The dc value of the phase detector output is integrated over time and, if saturation does not occur, can become arbitrarily large. The tracking range of type II loops is limited by the dynamic range of the voltage-controlled oscillator. The loop capture range cannot exceed the frequency range of the VCO.

**Transient Analysis of the Linearized PLL**

In studying the transient response of phase-locked loops, one must consider the linear (small-signal) and nonlinear (large-signal) operation of the loop. We will consider both the linear and nonlinear responses of type I and type II PLLs. For the linear model, the error signal $\theta_\epsilon(s)$ is obtained from Eq. (9.1), and the output signal is

$$\theta_o(s) = \frac{\theta_r(s) K_o K_d F(s)/s}{1 + [K_o K_d F(s)/s]N} \tag{9.48}$$

In the following analysis it is assumed that the initial phase error is zero and that it remains sufficiently small that the loop operation remains linear. Also $N$ will be assumed equal to 1 to simplify the notation.

### Type I Systems: Phase Step Response

If at $t = 0$ the input phase is suddenly changed by an amount $\phi[\theta_r(s) = \phi/s]$, the error signal is

$$\theta_\epsilon(s) = \frac{\phi}{s\{1 + K_v/[sF(s)]\}}$$

If no low-pass filter is used in the loop [$F(s) = 1$], then

$$\theta_\epsilon(s) = \frac{\phi}{s + K_v}$$

and                                        $$\theta_\epsilon(t) = \phi e^{-K_v t} \qquad (9.49)$$

The error voltage decays exponentially toward zero with increasing time. The larger the loop gain, the faster the loop responds to the change in phase. If $F(s)$ is a simple low-pass filter, then $F(s) = (\tau s + 1)^{-1}$, and the error voltage will be

$$\theta_\epsilon(s) = \frac{\phi(\tau s + 1)}{s(\tau s + 1) + K_v} = \frac{\phi(s + 2\zeta\omega_n)}{s^2 + 2\zeta\omega_n s + \omega_n^2}$$

where                                      $$2\zeta\omega_n = \tau^{-1}$$

and                                        $$\omega_n^2 = \frac{K_v}{\tau}$$

For $\zeta > 1$ the response is overdamped, and the system has two real poles on the negative axis. For $\zeta < 1$, the case of greatest interest is

$$\theta_\epsilon(t) = \phi e^{-\zeta\omega_n t}\left[\cos\omega_n(1 - \zeta^2)^{1/2}t + \frac{\zeta}{(1 - \zeta^2)^{1/2}}\sin\omega_n(1 - \zeta^2)^{1/2}t\right] \qquad (9.50)$$

The plots of this equation for various values of $\zeta$ and constant $\omega_n$ are given in Fig. 9.36. The maximum error voltage $\theta_\epsilon(t)_{max} = \phi$ occurs at $t = 0$. Thus if the initial step input change $\phi$ is small enough for the small-angle approximation to be valid, the complete transient analysis is valid. From Eq. (9.50) and Fig. 9.36 it is seen that the larger $\zeta\omega_n = (2\tau)^{-1}$ is, the faster the error voltage settles to zero. The wider the bandwidth of the low-pass filter, the faster the response of the loop. The smaller the damping ratio

$$\zeta = \frac{1}{2}\left(\frac{2}{K_v}\right)^{1/2}$$

the larger will be the step-response overshoot. The overshoot is readily determined by using Eq. (9.17).

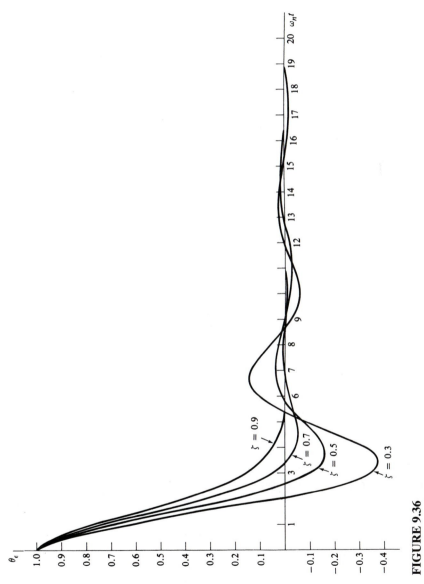

**FIGURE 9.36**
Error signal in response to a step input change in phase for a type I second-order PLL, given for selected damping ratios.

**Type I Systems: Frequency Step Response**

If at $t = 0$ the input frequency is suddenly changed by amount $\Delta\omega [\theta_r(s) = \Delta\omega/s^2]$, the error signal is

$$\theta_\epsilon(s) = \frac{\Delta\omega}{s^2[1 + K_v F(s)/s]} \qquad (9.51)$$

If $F(s) = 1$ (no filter),

$$\theta_\epsilon(s) = \frac{\Delta\omega}{s(s + K_v)}$$

and

$$\theta_\epsilon(t) = \frac{\Delta\omega}{K_v}(1 - e^{-K_v t}) \qquad (9.52)$$

If a simple low-pass filter $[F(s) = (\tau s + 1)^{-1}]$ is used in the loop, the phase error will be

$$\theta_\epsilon(t) = 2\zeta \frac{\Delta\omega}{\omega_n} + \frac{\Delta\omega}{\omega_n} e^{-\zeta\omega_n t}$$

$$\times \left[ \frac{1 - 2\zeta^2}{(1 - \zeta^2)^{1/2}} \sin \omega_n(1 - \zeta^2)^{1/2}t - 2\zeta \cos \omega_n(1 - \zeta^2)^{1/2}t \right] \qquad (9.53)$$

where

$$\zeta = \frac{1}{2}\left(\frac{\omega_L}{K_v}\right)^{1/2} \qquad \text{and} \qquad \omega_n = (K_v\omega_L)^{1/2}$$

The error signal in response to a unit ramp input is plotted in Fig. 9.37 for $\zeta = 0.3, 0.5, 0.7,$ and $0.9$ with a constant $K_v$ and in Fig. 9.38 with a constant $\omega_n$. In

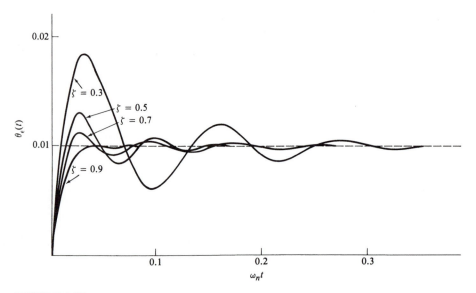

**FIGURE 9.37**
Error signal in response to a step input change in frequency for a type I second-order PLL, given for selected damping ratios; $K_v$ is constant.

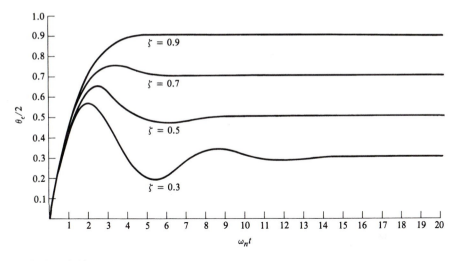

**FIGURE 9.38**
Error signal in response to a step input change in frequency for a type I second-order PLL, given for selected damping ratios; $\omega_n$ is constant.

Fig. 9.37 the damping ratio is varied by changing the filter bandwidth, and in Fig. 9.38 the product $2K_v$ is kept constant. An example of the input and output waveforms is given in Fig. 9.39. For the error expression given by Eq. (9.51), the steady-state error is

$$\epsilon_{ss}(t) = \lim_{t \to 0} \theta_\epsilon(t) = \lim_{s \to 0} s\theta_\epsilon(s) = \frac{\Delta\omega}{K_v} = \frac{2\zeta}{\omega_n}\Delta\omega \qquad (9.54)$$

provided the limit exists. Therefore, the larger the loop gain $K_v$, the smaller the steady-state error; but this implies a smaller damping $\zeta$ and hence more overshoot. If the steady-state error and transient performance are both important parameters, it may be necessary to use a type II loop.

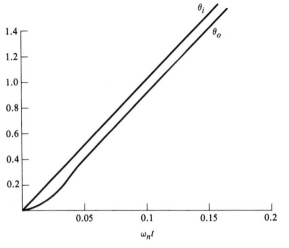

**FIGURE 9.39**
Input ($\theta_i$) and output ($\theta_o$) waveforms of a type I systems.

**Type II Loop**

For the type II second-order loop,

$$\theta_o(s) = \frac{\theta_r(s)(s/\omega_z + 1)}{s^2 + K_a(s/\omega_z + 1)} = \frac{\theta_r(s)(s/\omega_z + 1)}{s^2 + 2\zeta\omega_n s + \omega_n^2} \tag{9.55}$$

where

$$\omega_n = K_a^{1/2} \tag{9.56}$$

and

$$2\zeta = \left(\frac{K_a}{\omega_z}\right)^{1/2} \tag{9.57}$$

Also, the error response is

$$\theta_\epsilon(s) = \frac{\theta_r(s)s^2}{s^2 + 2\zeta\omega_n s + \omega_n^2} \tag{9.58}$$

where $\omega_n$ and $\zeta$ are defined by Eqs. (9.56) and (9.57), respectively.

If the damping ratio $\zeta < 1$, the output in response to a unit step input $[\theta_r(t) = 1]$ is given by

$$\theta_o(t) = 1 - e^{-\zeta\omega_n t}\left[\cos\omega_n(1 - \zeta^2)^{1/2}t - \frac{\zeta}{(1 - \zeta^2)^{1/2}}\sin\omega_n(1 - \zeta^2)^{1/2}t\right] \tag{9.59}$$

and the corresponding error signal is given by

$$\theta_\epsilon(t) = e^{-\zeta\omega_n t}\left[\cos\omega_n(1 - \zeta^2)^{1/2}t - \frac{\zeta}{(1 - \zeta^2)^{1/2}}\sin\omega_n(1 - \zeta^2)^{1/2}t\right] \tag{9.60}$$

The error and output signals of a type II system ($\zeta = 0.5$) in response to a unit step input are shown in Fig. 9.40. Error signals in response to a step input for

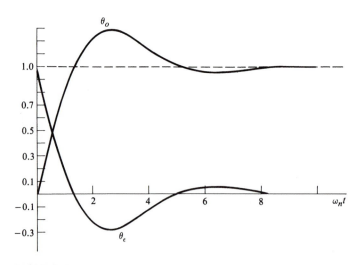

**FIGURE 9.40**
Error and output response of a type II PLL ($\zeta = 0.5$) to a step change in phase.

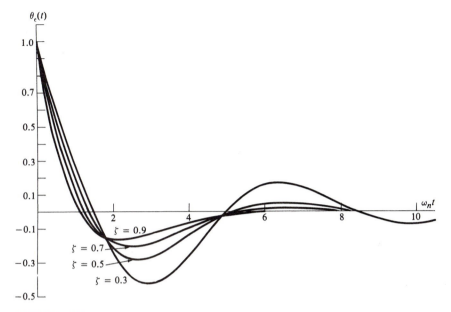

**FIGURE 9.41**
Error response of a type II PLL to a step change in phase, given for selected damping ratios.

$\zeta = 0.3, 0.5, 0.7$ and $0.9$ are given in Fig. 9.41. The steady-state error for the type II system in response to step inputs is zero.

### Ramp Inputs

For a step change in frequency $[\theta_r(t) = \Delta\omega\, t]$, the output of the second-order type II system is

$$\theta_o(t) = \Delta\omega\left[t - \frac{e^{-\zeta\omega_n t}}{\omega_n}\frac{\sin\omega_n(1-\zeta^2)^{1/2}t}{(1-\zeta^2)^{1/2}}\right] \tag{9.61}$$

The input and output waveforms of an underdamped system are illustrated in Fig. 9.42, and the error signals of a type II second-order system in response to a ramp input are illustrated in Fig. 9.43. It is seen that the steady-state error of type II systems in response to step changes in frequency decays to zero.

### Comparison of Type I and Type II Loops

When it is necessary to have zero steady-state error to ramp inputs, a type I system cannot be used, but either a type I or type II system can realize zero steady-state error to step inputs. For step inputs, a type II system will have a shorter rise time but also more overshoot than a type I system. Figure 9.44 compares the step response

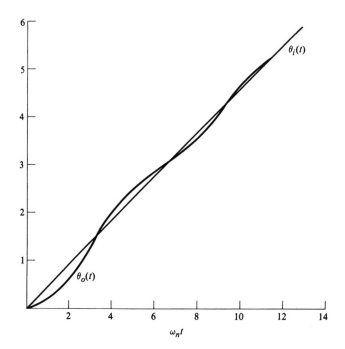

**FIGURE 9.42**
Input ($\theta_i$) and output ($\theta_o$) waveforms of a type II PLL.

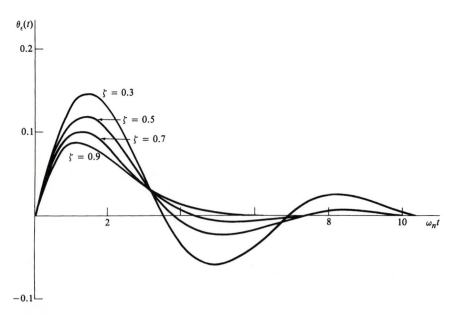

**FIGURE 9.43**
Error response of a type II PLL to a step change in frequency, given for selected
damping ratios.

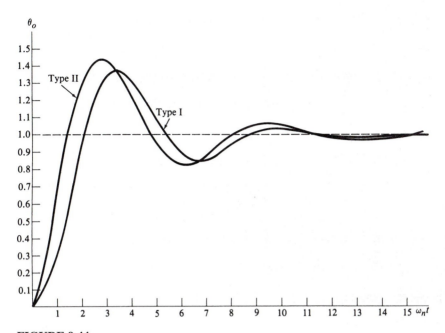

**FIGURE 9.44**

Step responses of second-order type I and type II PLL systems with identical $\zeta$ and $\omega_n$.

of type I and II systems with the same damping ratio and $\omega_n$. The type II system has a shorter rise time and more overshoot, as predicted from the stability analysis of the system. In many applications it is the settling time which is most important and not the rise time. Settling time in a phase-locked loop is usually defined as the time required for the phase error to settle below a specified time for which the frequency output is unusable while switching from one frequency to another. It is difficult to measure the instantaneous frequency, but relatively easy to measure the instantaneous phase, so the phase settling time is usually used for frequency synthesizer specifications. The phase error of a type II system in response to a step change frequency of $\Delta\omega$ rad/s is obtained from Eq. (9.61). At any time the phase error satisfies the inequality

$$|\theta_\epsilon(t)| \le \frac{\Delta\omega}{\omega_n} \frac{e^{-\zeta\omega_n t}}{(1 - \zeta^2)^{1/2}}$$

So, for a specified maximum steady-state error $\theta_m$,

$$t_s \le (\zeta\omega_n)^{-1} \ln\left\{ \frac{\Delta\omega}{\omega_n\theta_m}[(1 - \zeta^2)^{1/2}]^{-1} \right\} \tag{9.62}$$

The normalized settling time $\omega_n t_s$ is plotted in Fig. 9.45 for a change in frequency $(\Delta\omega/\omega_n = 0.5)$ and for three different values of steady-state error $\theta_m$: 0.01, 0.05, and 0.1. The settling time to a step change in frequency of $\Delta\omega/\omega_n = 0.1$ is plotted in Fig. 9.46.

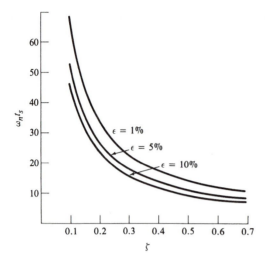

**FIGURE 9.45**
Normalized settling time $\omega_n t_s$ for $\Delta\omega/\omega_n = 0.5$.

## Large-Signal Behavior

When the error signal is large, the linear PLL model is no longer valid. The main source of the nonlinearity is the phase detector (the nonlinear characteristics being different for each type of phase detector), and the system's performance must be analyzed in the time domain. The following sections illustrate the analysis of the large-signal performance of phase-locked loops.

## Digital Phase Detectors

The output of a digital phase or phase-frequency detector is either a two- or three-amplitude level signal. As the phase error increases, the amplitude does not change; however, the duration of the output pulse does. The analysis and performance of

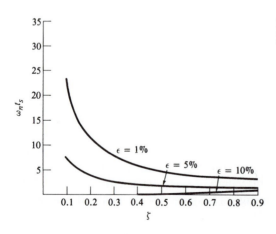

**FIGURE 9.46**
Normalized settling time for $\Delta\omega/\omega = 0.1$.

such systems will be illustrated by the analysis of type I phase-locked loops. The analysis provides insight into the design of modern phase-locked loops.

### First-Order Systems

It is assumed that the phase-frequency detector outputs a pulse of $+V$ V when the phase error is positive and $-V$ V when the phase error is negative. The width of the phase detector output pulse $\tau$ is equal to the time between the leading edges of the two waveforms if the output leads in phase. However, if the phase error is positive (reference leads in phase), a positive voltage will be applied to the VCO, the VCO will speed up, and the output phase will reset the phase detector in a shorter time than would elapse if the phase error were negative. To illustrate this operation, we will consider a type I loop, illustrated in Fig. 9.47, which uses no filtering. The PLL is a first-order system. If the input and output frequencies are equal, the phase error is defined as

$$\theta_\epsilon = \omega\tau \tag{9.63}$$

where $\tau$ is the time between the leading edges of the two waveforms and $\omega$ is the waveform frequency. The output frequency is

$$\omega_o = \omega_c + K_o V_c \tag{9.64}$$

where $\omega_c$ is the free-running frequency of VCO. Since the control voltage $V_c$ is either $\pm V$ or zero, the VCO frequency can take on only three distinct values. The output phase is

$$\theta_o(t) = \omega_c t + \int_0^t K_o V_c(t)\, dt + \phi_o \tag{9.65}$$

where $\phi_o$ is the phase at $t = 0$, or $\theta_o(0) = \phi_o$.

We will first consider the case in which the input leads in phase. It is assumed that the input frequency $\omega_r$ is equal to $\omega_c$. (If the input frequency is not equal to $\omega_c$, the output frequency cannot adjust to $\omega_r$, since this loop does not have the capability of continuously changing the frequency.) At $t = 0$, the phase of the reference is taken as zero. That is, $\theta_r(0) = 0$. Then for positive phase error, the output phase is negative, or $\phi_o(0) = -\phi_o$, and the leading edge of the reference frequency sets the phase detector. The leading edge of the output lags by $\tau$ s, so the phase error is

$$\theta_\epsilon(0) = \phi_o$$

where $\tau = \phi_o/\omega_c$.

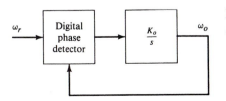

**FIGURE 9.47**
A fundamental type I PLL.

The phase detector outputs a pulse of $+V$ V until the leading edge of the output signal resets the phase detector at time $t_o$. At $t = 0$ the VCO frequency instantaneously increases to

$$\omega_o = \omega_c + K_o V \qquad \text{rad/s}$$

and the output phase is

$$\theta_o = \omega_c t + K_o \int_0^t V_c(t)\, dt - \phi_o$$

$$= \omega_c t + K_o V_c t - \phi_o \qquad 0 \le t \le t_o$$

The output phase becomes zero at the time

$$t_o = \frac{\phi_o}{\omega_c + K_o V} \tag{9.66}$$

at which time the phase detector resets and the VCO control voltage $V_c$ becomes zero. The phase error at $t_o$ is $\theta_\epsilon(t_o) = \omega_c t_o$; and since the VCO and the reference signal are again operating at the same frequency, this will also be the phase error at the time $T$ when the reference frequency again sets the phase detector and the VCO again increases in frequency. At the end of the $n$th reference period, the phase error will be

$$\theta_\epsilon(nT) = \frac{\phi_o}{(1 + K_o V / \omega_c)^n} \tag{9.67}$$

(Note that the larger $K_o V$ is, the smaller will be the phase error; but once the phase error is positive, it remains positive.)

If the initial phase error is negative (the VCO output leads the reference waveform by $\tau$ s), the phase detector output will be set to $-V$ V and will remain at this level until the leading edge of the reference waveform resets the phase detector after $\tau$ s. At this time the output phase will be

$$\theta_\epsilon(\tau) = -(\omega_c \tau - K_o V \tau) = -(\phi_o - K_o V \tau) \tag{9.68}$$

and if the loop gain $K_o V = \omega_c$, the phase error will be zero when the reference waveform resets the phase detector. The phase error will remain negative, provided $K_o V < \omega_c$. This will always be the case, since the maximum frequency deviation of a VCO below its free-running frequency $\omega_c$ is equal to $\omega_c$. The frequency deviation is also $K_o V_c$, so $K_o V / \omega_c$ cannot be larger than 1 for frequency deviations below the free-running frequency of the VCO. The phase error after $n$ output cycles will be

$$\theta_\epsilon(nT) = -\phi_o \left(1 - \frac{K_o V}{\omega_c}\right)^n \tag{9.69}$$

The analysis of this simple system illustrates that the large-signal response of PLL systems that utilize a digital phase detector is not symmetric, since it differs depending upon whether the initial phase error is positive or negative.

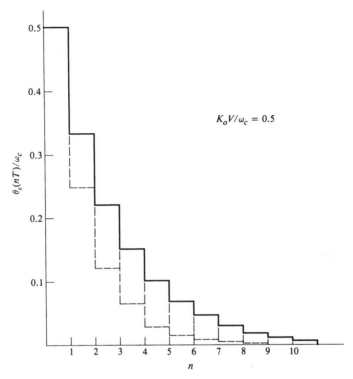

**FIGURE 9.48**
Step response of a fundamental type I PLL.

If $K_oV/\omega_c$ is close to 1, negative phase errors will be reduced faster than positive phase errors. This suggests that to reduce the settling time, the phase detector logic could be modified to provide a larger gain for positive phase errors. Figure 9.48 illustrates the phase-error transient response of a type I first-order loop to a step input of phase. The normalized loop gain $K_oV/\omega_c = 0.5$ in this example. The negative phase error diminishes faster than the positive phase error, but it must be kept in mind that the phase detector cycle time is longer for negative error signals.

The analysis of this simple system also illustrates some of the limitations of linearized analysis. For example, for this type I system, the final-value theorem predicts that the final frequency error in response to step changes $\Delta\omega$ in frequency will be

$$\lim_{t\to\infty}[f_r(t) - f_o(t)] = \lim_{s\to\infty} \frac{s(\Delta\omega/s^2)}{1 + K_v/s} = \frac{\Delta\omega}{K_v}$$

However, it is obvious from the large-signal analysis that the VCO cannot track arbitrary changes in the input frequency since the VCO can take on only three different values of frequency.

Another feature of loop performance brought out by this analysis is that the time for the phase detector to reset is different for positive and negative phase errors. If the output leads in phase and sets the phase detector, the phase detector is set for a time equal to the initial phase error; but if the input leads in phase and sets the phase detector, it is set for a time given by Eq. (9.66). In this case, time is a function of the loop parameters.

### Type I Loop Including a Low-Pass Filter

The preceding analysis has shown that when the loop does not contain a low-pass filter, the VCO frequency jumps discontinuously from $\omega_c$ to $\omega_c \pm K_oV$. A low-pass filter can be added to the loop to smooth out the frequency transitions and improve the spectral purity of the output signal. It will be shown that the filter can also reduce the system settling time. It is assumed that the low-pass filter is first-order, with a transfer function of

$$F(s) = \left(\frac{s}{a} + 1\right)^{-1}$$

If the charge pump phase detector and the filter illustrated in Fig. 9.14 are used, the filter is realized by selecting the capacitor $C$ such that $RC = a^{-1}$.

The system illustrated in Fig. 9.49 will now be analyzed. The case in which the reference leads in phase by $\phi_o$ rad will be considered first. The phase detector output voltage will consist of a pulse of width $t_o$ and amplitude $V$, so the voltage applied to the VCO is

$$V_c(s) = \frac{V(1 - e^{-t_o s})}{s(s/a + 1)} \tag{9.70}$$

or

$$V_c(t) = V(1 - e^{-at}) \qquad 0 \le t \le t_o \tag{9.71}$$

At time $t_o$ the phase detector resets to zero output, and

$$V_c(t) = Ve^{-at}(e^{at_o} - 1) \qquad t_o \le t \le T \tag{9.72}$$

The output phase during the time the phase detector output is $+V$ is

$$\theta_o(t) = \int_0^t [\omega_c + KV(1 - e^{-at})]\, dt - \phi_o$$

$$= \omega_c t + K_o V\left(t + \frac{e^{-at} - 1}{a}\right) - \phi_o \qquad 0 \le t \le t_o \tag{9.73}$$

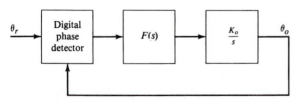

**FIGURE 9.49**
A type I PLL including a filter $F(s)$.

$\theta_r$ — Digital phase detector — $F(s)$ — $\dfrac{K_o}{s}$ — $\theta_o$

The output phase will be zero at $t_o$, so $t_o$ can be found by solving

$$0 = \omega_c t_o + K_o V \left( t_o + \frac{e^{-at_o} - 1}{a} \right) - \phi_o \tag{9.74}$$

The value of $t_o$ can be found by using numerical approximation techniques to solve Eq. (9.74). For $t_o \le t \le T$, the VCO control voltage is

$$V_c(t) = V(1 - e^{-at_o})e^{-a(t-t_o)} \qquad t_o \le t \le T \tag{9.75}$$

The output phase at time $T$ is

$$\theta_o(T) = \theta_o(t_o) + \int_{t_o}^{T} [\omega_c + K_o V(1 - e^{-at_o})e^{-a(t-t_o)}] \, dt \tag{9.76}$$

and since $\theta_o(t_o) = 0$,

$$\theta_o(T) = \omega_c(T - t_o) + \frac{K_o V}{a}[1 - e^{-at_o} + e^{-aT}(1 - e^{at_o})] \tag{9.77}$$

The phase error at time $T$ is

$$\theta_\epsilon(T) = \theta_\epsilon(t_o) - K_o V(1 - e^{-at_o}) \int_{t_o}^{T} e^{-a(t-t_o)} \, dt$$

$$= \omega_c t_o - \frac{K_o V}{a}[1 - e^{-at_o} + e^{-aT}(1 - e^{at_o})] \tag{9.78}$$

This derivation assumes that the phase error $\theta_\epsilon(T)$ remains positive. To see that it does, the expression $\theta_\epsilon(T)$ will be rewritten, substituting Eq. (9.74) for $\omega_c t_o$ in Eq. (9.78). That is,

$$\theta_\epsilon(T) = \phi_o - K_o V \left( t_o + \frac{e^{-at_o} - 1}{a} \right) - \frac{K_o V}{a}[1 - e^{-at_o} + e^{-aT}(1 - e^{at_o})]$$

$$= \phi_o - K_o V t_o - \frac{K_o V}{a}e^{-aT}(1 - e^{at_o})$$

$$= \phi_o - K_o V t_o + \alpha$$

where

$$\alpha = \frac{K_o V}{a}e^{-aT}(e^{at_o} - 1) > 0$$

Also $\phi_o = \omega_c \tau$ and $K_o V \le \omega_c$, so

$$\phi_o - K_o V t_o \ge \omega_c(\tau - t_o)$$

Since $\tau > t_o$, we have $\theta_\epsilon(T) > 0$, provided $\theta_\epsilon(0) > 0$. Therefore, as was the case when no filter was used, if in a type I loop containing a simple low-pass filter the phase error becomes positive, then it will remain positive in the absence of further external input. Note that this is a marked difference from the linear model, which predicts that the response is underdamped if the loop gain is sufficiently large.

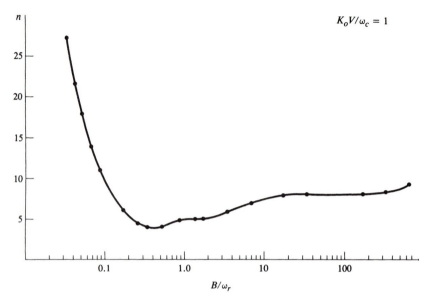

**FIGURE 9.50**
Settling time of a type I PLL as a function of filter bandwidth (for positive
phase error).

Figure 9.50 illustrates how the system settling time varies as a function of
the filter bandwidth for positive phase errors. For this example the loop gain
$K_oV/\omega_c = 1$ and the initial phase error is $180°$. The vertical axis represents how
many reference cycles it takes for the time difference between the two signals to be
less than $10^{-6}$ s ($\phi_\epsilon < 10^{-6}\omega_r$). If the filter bandwidth is too narrow, a large number
of reference cycles is needed before the phase error settles below the prescribed
limit. As the filter bandwidth increases, the settling time reaches a minimum at
$B = 0.5\omega_r$. As the bandwidth is further increased, the settling time again increases,
approaching the limit of 12 reference cycles for the case of infinite filter bandwidth.
In the case where the reference leads in phase, there is an optimum value of filter
bandwidth for minimizing the settling time. The minimum in Fig. 9.50 is relatively
broad. For positive phase errors, the bandwidth of this type I system should be
between the limits

$$0.25\,\omega_r \le B \le 2.5\,\omega_r \qquad \text{rad/s}$$

for the fastest response.

Figure 9.51 illustrates how the settling time varies as a function of the normalized loop gain $K_v/\omega_r$. The filter bandwidth is $2.5\omega_r$, and curves are plotted for two
different values of initial error. The larger the loop gain, the shorter the settling time.
Larger loop gain implies a larger frequency deviation of the VCO. The most important factor in reducing the loop settling time is the frequency deviation of the VCO.

If the output leads the reference in phase by $\tau$ s, the phase error is negative; for
this case the width of the phase-detector pulse is $\tau$ s because the phase detector

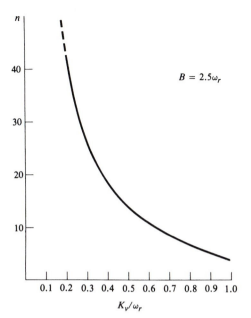

FIGURE 9.51
Settling time of a type I PLL as a function of loop gain (for positive phase error).

resets on the leading edge of the reference waveform, which is independent of the VCO waveform. Therefore,

$$\theta_o(\tau) = \int_o^\tau [\omega_c + K_o V_c(t)]\, dt$$

In this case the output phase is taken to be zero at $t = 0$,

$$\theta_o(0) = 0$$

and the phase error is

$$\theta_\epsilon(0) = -\omega_c \tau$$

The phase detector is set by the leading edge of the output waveform, and the control voltage is

$$V_c(t) = -V(1 - e^{-at}) \qquad 0 \le t \le \tau \tag{9.79}$$

Therefore, the output phase at time $\tau$ is

$$\theta_o(\tau) = \omega_c \tau - K_o V\left(\tau + \frac{e^{-a\tau} - 1}{a}\right) \tag{9.80}$$

After the phase detector is reset, the VCO control voltage decays toward zero:

$$V_c(t) = -K_o V e^{-at}(e^{a\tau} - 1) \qquad \tau \le t \le t_o \tag{9.81}$$

The output phase is

$$\theta_o(t) = \theta_o(\tau) - \frac{K_o V(e^{a\tau} - 1)(e^{-a\tau} - e^{-at})}{a} + \omega_c(t - \tau) \qquad \tau \le t \le t_o \tag{9.82}$$

If Eq. (9.80) is substituted for $\theta_o(\tau)$,

$$\theta_o(t) = \omega_c t - K_o V \left[ \tau + \frac{e^{-a\tau} - 1}{a} + \frac{(e^{a\tau} - 1)(e^{a\tau} - e^{-at})}{a} \right] \qquad (9.83)$$

The phase error at time $t_o$

$$\theta_\epsilon(t_o) = \theta_o(t_o) - \theta_i(t_o)$$
$$= \theta_o(t_o) - \omega_c(t_o - \tau) \qquad (9.84)$$

The problem now is to find the time $t_o$ at which the output phase is again 0 or $2\pi$. That is, $\theta_o(t_o) - 2\pi$. The time $t_o$ is readily found by numerical methods. (The following results were obtained by using the Newton-Raphson root-finding algorithm.)

How the settling time of a type I system varies as a function of normalized loop gain $K_v/\omega_r$ when the initial phase error is negative is illustrated in Fig. 9.52 for the case in which the low-pass filter bandwidth $B = 2.5\omega_r$. The curve is similar to the case in which the initial phase error is positive (Fig. 9.51). To minimize the loop

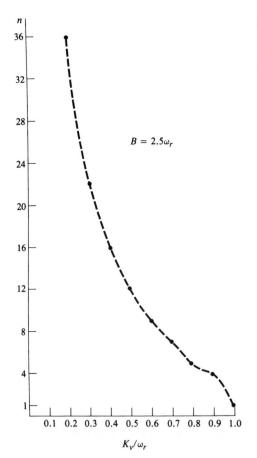

**FIGURE 9.52**
Settling time of a type I PLL as a function of loop gain (for negative phase error).

$B = 2.5\omega_r$

$K_v/\omega_r$

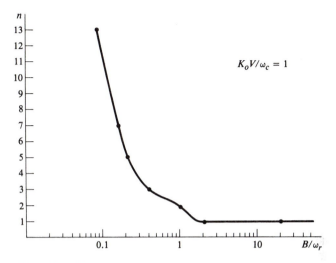

**FIGURE 9.53**
Settling time of a type I PLL as a function of loop bandwidth
(for negative phase error).

settling time, the normalized loop gain needs to be close to unity. How the settling time varies as a function of low-pass filter bandwidth is illustrated in Fig. 9.53 for a normalized loop gain of unity. The settling time reaches a minimum value of 1 when the filter bandwidth is approximately $2^{1}/_{2}$ times the reference frequency. The settling time remains at this value as the filter bandwidth is further increased. Figure 9.54 is similar to Fig. 9.53 except that the normalized loop gain $K_v/\omega_r = 0.5$.

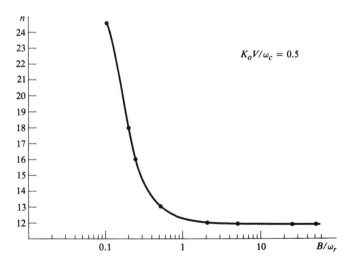

**FIGURE 9.54**
Settling time of a type I PLL as a function of loop bandwidth
(for negative phase error and reduced loop bandwidth).

There is little interaction between the minimum filter bandwidth and the normalized loop gain. A comparison of Figs. 9.50 and 9.53 shows that if only the loop settling time is considered in selecting the filter bandwidth, the bandwidth should be approximately $2^1/_2$ times the reference frequency.

### Sinusoidal Phase Detection

If a sinusoidal phase detector is used without a low-pass filter, and the upper-frequency term $\omega_r + \omega_c$ is ignored, the phase detector output voltage will be equal to the voltage applied to the VCO:

$$V_c = K_d \sin \theta_\epsilon \qquad (9.85)$$

The phase error is

$$\theta_\epsilon = \theta_r - \theta_o$$

so the frequency error is

$$\frac{d\theta_\epsilon}{dt} = \omega_r - \frac{d\theta_o}{dt}$$

If $\omega_c$ is the free-running frequency of the VCO, then

$$\theta_o(t) = \omega_c + K_o V_c(t)$$

so
$$\frac{d\theta_\epsilon}{dt} = \omega_r - \omega_c - K_o V_c(t) \qquad (9.86)$$

$$= \Delta\omega - K_o V_c(t)$$

where $\Delta\omega - \omega_r - \omega_c$ is the difference between the reference frequency and the free-running frequency of the VCO. If Eq. (9.85) is substituted in Eq. (9.86),

$$\frac{d\theta_\epsilon}{dt} = \Delta\omega - K_o K_d \sin \theta_\epsilon \qquad (9.87)$$

This is the nonlinear equation describing the phase error in a phase-locked loop containing a sinusoidal phase detector and no filter. Also, the high-frequency term has been ignored. For the sake of generality, assume the loop also contains an amplifier; then Eq. (9.87) can be rewritten as

$$\frac{d\theta_\epsilon}{dt} = \Delta\omega - K_v \sin \theta_\epsilon(t) \qquad (9.88)$$

where
$$K_v = K_o K_d K$$

If $\Delta\omega > K_v$, the loop cannot acquire phase lock since the maximum change of VCO frequency $(\Delta\omega)_{max} = K_v$ and it is assumed that $K_v \geq \Delta\omega$. Equation (9.88) can be rewritten as

$$\frac{d\theta_\epsilon}{\Delta\omega - K_v \sin\theta_\epsilon(t)} = dt$$

which has the solution

$$t - t_o = \frac{2}{[(\Delta\omega)^2 - K_v^2]^{1/2}} \tan h^{-1} \frac{\Delta\omega + K_v}{\Delta\omega - K_v} \tan\left(\frac{\pi}{4} - \frac{\theta_\epsilon}{2}\right) \qquad (9.89)$$

where the initial time $t_o$ will be assumed to be zero. This equation can be rearranged to express the phase error as a function of time.[2]

$$\theta_\epsilon(t) = \left[2\tan\left(\frac{K_v - \Delta\omega}{K_v + \Delta\omega}\right)^{1/2}\right]\frac{1 - \exp\{-[K_v^2 - (\Delta\omega)^2]^{1/2}t\}}{1 + \exp\{-[K_v^2 - (\Delta\omega)^2]^{1/2}t\}} + \frac{\pi}{2} \qquad (9.90)$$

Note that

$$\lim_{t\to\infty} \theta_\epsilon(t) = 2\tan^{-1}\left(\frac{K_v - \Delta\omega}{K_v + \Delta\omega}\right)^{1/2} + \frac{\pi}{2} \qquad (9.91)$$

## Numerical Example

Consider the simplest case where $\Delta\omega = 0$. Then Eq. (9.88) becomes

$$\frac{d\theta_\epsilon}{-K_v \sin\theta_\epsilon(t)} = dt$$

which is readily solved to give

$$\frac{1}{2}\ln\frac{1 - \cos\theta_\epsilon}{1 + \cos\theta_\epsilon} = -K_v(t - t_o) + C$$

or

$$\frac{1 - \cos\theta_\epsilon}{1 + \cos\theta_\epsilon} = Ae^{-2K_v(t-t_o)}$$

Without loss of generality, let the initial error be $\theta_o$ at $t = t_o = 0$. Then

$$\frac{1 - \cos\theta_o}{1 + \cos\theta_o} = A$$

and

$$\cos\theta_\epsilon = \frac{1 - Ae^{-2K_v(t)}}{1 + Ae^{-2K_v t}} = \frac{1 + \cos\theta_o - (1 - \cos\theta_o)e^{-2K_v t}}{1 + \cos\theta_o + (1 - \cos\theta_o)e^{-2K_v t}}$$

Several significant results for the large-signal behavior of PLLs containing a sinusoidal phase detector have also been obtained by using phase-plane techniques.[3] Since the emphasis here is on the response of loops containing digital phase detectors, those results will not be discussed, except to emphasize that the large-signal behavior of PLLs is markedly different for different types of phase detectors.

## 9.5

### PROBLEMS

**9.1** A simple phase-locked loop has $K_d K_o = 1000$ rad/s, $F(s) = 1$, and $N = 20$. What is the closed-loop bandwidth? What is the closed-loop bandwidth if the filter described by the transfer function

$$F(s) = \left(\frac{s}{50} + 1\right)^{-1}$$

is added inside the loop?

**9.2** The system of Prob. 9.1 (including the filter) is to be designed so that it will be able to track step changes in frequency ($\Delta\omega = 1$ rad/s) with less than 0.01-rad steady-state error. Is the loop gain adequate? If not, how much additional gain must be added to meet the steady-state error specification?

**9.3** Graph the open-loop frequency response, determine the crossover frequency, and calculate the phase margin of the PLL described in Prob. 9.1, both with and without the low-pass filter in the loop.

**9.4** Derive the formula [Eq. (9.16)] for the closed-loop bandwidth of a type I system.

**9.5** Calculate the overshoot of the system described in Prob. 9.1 in response to a $2°$ step change in the input phase.

**9.6** What is the minimum bandwidth that the filter of Prob. 9.1 can have if the system peak overshoot to step inputs in phase is to be less than 20 percent? Solve the problem by using both straight-line approximations and the analytic method.

**9.7** Modify the system of Prob. 9.1 so that it has less than 20 percent overshoot to step inputs and also has a rise time of less than 10 ms. Design the circuit to be added to the loop.

**9.8** The following open-loop system $G(s)$ has zero steady-state phase error to constant-frequency inputs. Determine the filter frequency $\omega_z$ so that the closed-loop system has approximately 20 percent overshoot to step inputs. What will the closed-loop bandwidth be?

$$G(s) = \frac{10^6(s/\omega_z + 1)}{s^2}$$

**9.9** Design a type II system that has a closed-loop bandwidth of $10^3$ rad/s and a 30 percent overshoot to a step input.

**9.10** For a type II system with $K_a = 10^6$ (rad/s)$^2$, determine a lead-lag network so that the closed-loop system has less than 30 percent overshoot to step changes in phase. The network pole location should be minimized in order to maximize the high-frequency filtering.

**9.11** Determine the transfer function of the amplifier shown in Fig. P9.11. The transconductance $g_m = 10^{-4}$ S. If this amplifier is connected as a unity-gain voltage-follower, what is the phase margin? How large a capacitor must be connected between the compensation terminals to obtain a 45° phase margin? The transfer function of the dependent source is

$$V_o = \frac{(-2 \times 10^3) V_a}{2 \times 10^{-5} s + 1}$$

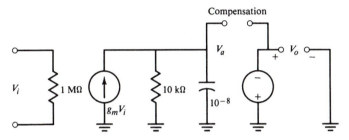

**FIGURE P9.11**
A two-stage amplifier.

**9.12** An open-loop transfer function is given by

$$G(s) = e^{-sT} \frac{10^4 (s/100 + 1)}{s^2}$$

For what value of time delay $T$ will the system become unstable?

**9.13** A type I system has the open-loop transfer function of

$$G(s) = \frac{K_v}{s(s/10^5 + 1)^2}$$

Select $K_v$ so that the closed-loop system has approximately 25 percent overshoot to a step input. Estimate the system's closed-loop bandwidth.

**9.14** A type II system has the open-loop transfer function of

$$G(s) = \frac{K_a(s/200 + 1)}{s^2(s/2000 + 1)}$$

Determine the value of $K_a$ that will minimize the overshoot to a step input. What will the minimum overshoot be?

**9.15** For the simple phase-locked loop consisting of a phase-frequency detector and a VCO, sketch the phase error as a function of time for the case where

$$\frac{K_o V}{\omega_c} = \frac{3}{4}$$

For what value of loop gain will the system become unstable?

**9.16** Design an FM demodulator using the CD4046 integrated-circuit PLL (see App. 7 for specifications). The carrier frequency is to be 100 kHz, and the frequency deviation is to be 4 kHz. Select the loop filter so that the closed-loop bandwidth is approximately 8 kHz.

**9.17** A PLL has a 25-kHz bandwidth. How much filtering does the PLL provide to the VCO output at 25 kHz from the center frequency? At 5 kHz?

**\*9.18** Use the phase margin specifications described in this chapter to meet the following design specifications.

  (*a*) Design a unity-gain noninverting amplifier with approximately 20 percent overshoot to a step input using an operational amplifier with a gain-bandwidth product of $10^7$ rad/s and a bandwidth of 10 rad/s (internally compensated). The op-amp output impedance is 200 $\Omega$ (resistive), and the load impedance consists of a 0.01-$\mu$F capacitor. Describe two compensating circuits. The closed-loop bandwidth should be as wide as possible. Verify your design by using computer simulation.

  (*b*) Use the same op amp to convert the output of a 10-$\mu$A current source to an output voltage of 1 V (magnitude). The current source has an output capacitance of 100 pF ($+20$ percent). The rise time should be as brief as possible subject to the design constraint of a maximum of 30 percent overshoot to a step input of current.

## ▨ 9.6
### REFERENCES

1. H. Bode, *Network Analysis and Feedback Amplifier Design,* Van Nostrand, New York, 1945.
2. V. F. Kroupa, *Frequency Synthesis,* Wiley, New York, 1973.
3. A. Viterbi, *Principles of Coherent Communications,* McGraw-Hill, New York, 1966.

## ▨ 9.7
### ADDITIONAL READING

Barab, S., and A. McBride: Uniform Sampling Analysis of a Hybrid Phase-Locked Loop with a Sample-and-Hold Phase Detector, *IEEE Trans. AES*-11, 1975, pp. 210–216.
Best, R. E.: *Phase-Locked Loops,* McGraw-Hill, New York, 1984.
Blanchard, A.: *Phase-Locked Loops,* Wiley, New York, 1976.
Egan, W. F.: *Frequency Synthesis by Phase Lock,* Wiley, New York, 1981.
Gardner, F. M.: *Phaselock Techniques,* 2d ed., Wiley, New York, 1979.
Gupta, S. C.: Phase-Locked Loops, *Proc. of the IEEE,* **63:** 291–306 (1975).
Rohde, U. L.: *Digital PLL Frequency Synthesizers,* Prentice-Hall, Englewood Cliffs, NJ, 1983.

# Frequency Synthesizers

## INTRODUCTION

A frequency synthesizer is a device that generates a large number of precise frequencies from a single reference frequency. The term *frequency synthesis* was first used by Finden[1] in 1943 for the generation of frequencies that were a harmonic of a submultiple of a reference frequency. Recent advances in integrated-circuit design include the development of inexpensive frequency synthesizers and their subsequent application in most communication receivers.

A frequency synthesizer can replace the expensive array of crystal resonators in a multichannel radio receiver. A single-crystal oscillator provides a reference frequency, and the frequency synthesizer generates the other frequencies. Because they are relatively inexpensive and because they can be easily controlled by digital circuitry, frequency synthesizers are being included in many new communication system designs.

The oldest synthesis method, first described by Finden, is referred to as *direct frequency synthesis;* it utilizes mixers, frequency multipliers, dividers, and bandpass filters. Direct synthesis has been superseded in almost all applications by indirect (coherent) synthesis, which utilizes a phase-locked loop that may be analog or digital. The newest method, *direct digital frequency synthesis (DDFS),* uses a digital computer and digital-to-analog (D/A) converter to generate the signals. Each of these methods has advantages as well as disadvantages; and if the specifications are sufficiently stringent, it may be necessary to incorporate all three methods into the synthesizer design. In this chapter, the three methods of frequency synthesis are described, and a design example that combines the different synthesis methods to meet the overall specifications is presented. Because one of the most demanding synthesizer specifications is output noise, the oscillator being a primary source of the random noise associated with frequency synthesis, a brief description of the random noise occurring in quality oscillators and phase-locked loops is included in this chapter so that the reader can better assess the noise performance of synthesizers. First we will examine the various methods of frequency synthesis.

## ◼ 10.2

### DIRECT FREQUENCY SYNTHESIS

Direct frequency synthesis is the oldest of the frequency synthesis methods. It synthesizes a specified frequency from one or more reference frequencies from a combination of harmonic generators, filters, multipliers, dividers, and frequency mixers. Bipolar transistors, because of the exponential base-to-emitter voltage characteristics, are well suited for use as harmonic generators.

One method of using a harmonic generator is shown in Fig. 10.1. The desired frequency is obtained with a filter tuned to the desired output frequency. Highly selective filters are required with this method. The multiple-oscillator approach is an alternative method. The oscillators are usually easier to realize than the bandpass filters. Figure 10.2 illustrates a method of generating 99 discrete frequencies from 18 crystal oscillators. One switch selects one of the nine oscillators that cover the frequency range 1 to 9 kHz in 1-kHz steps, and the other switch covers the frequency range 10 to 90 kHz in 10-kHz steps. The two signals are then combined in a frequency mixer, and the bandpass filter selects the higher of the two mixer output frequencies.

Direct frequency synthesis refers to the generation of new frequencies from one or more reference frequencies by using a combination of multipliers, dividers, bandpass filters, and mixers. A simple example of direct synthesis is shown in

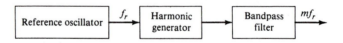

**FIGURE 10.1**
A direct frequency synthesizer.

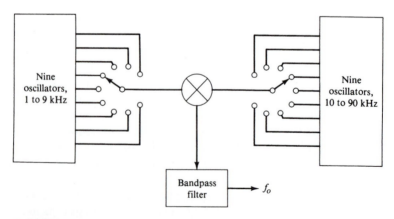

**FIGURE 10.2**
A two-decade direct frequency synthesizer.

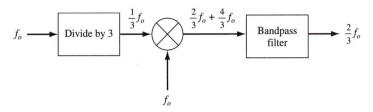

**FIGURE 10.3**
Example of direct synthesis.

Fig. 10.3. The new frequency $^2/_3 f_o$ is realized from $f_o$ by using a divide-by-3 circuit, a mixer, and a bandpass filter. In this example $^2/_3 f_o$ has been synthesized by operating directly on $f_o$.

One of the foremost considerations in the design of direct frequency synthesizers in the *mixing ratio*

$$r = \frac{f_1}{f_2} \tag{10.1}$$

where $f_1$ and $f_2$ are the two input frequencies to the mixer. If the mixing ratio is too large or too small, the two output frequencies will be too close together, and it will be difficult to remove one of the signals with filtering.

**EXAMPLE 10.1.** If the two mixer input frequencies are 100 and 1 MHz ($r = 100$), the mixer output frequencies will be 99 and 101 MHz. The removal of one of these frequencies would require an extremely complex filter.

The filter requirements can be reduced by using an offset frequency. This approach is utilized in the next direct synthesis method described.

Figure 10.4 illustrates a type of direct synthesis module frequently used in direct frequency synthesizers. The method is referred to as *double-mix-divide*. An input frequency $f_i$ is combined with a frequency $f_1$, and the upper frequency $f_1 + f_i$ is selected by the bandpass filter. This frequency is then mixed with a switch-selectable frequency $f_2 + f^*$. (In the following $f^*$ refers to any one of 10 switch-selectable frequencies.) Frequency $f_2 + f^*$ can be realized with one of the

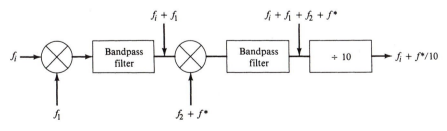

**FIGURE 10.4**
A double-mix-divide module.

methods illustrated in Figs. 10.1 and 10.2. The output of the second mixer consists of the two frequencies $f_i + f_1 + f_2 + f^*$ and $f_i + f_1 - f_2 - f^*$; only the higher frequency appears at the output of the bandpass filter. If frequencies $f_i$, $f_1$, and $f_2$ are selected so that $10f_i = f_i + f_1 + f_2$, then the frequency at the output of the divide-by-10 module will be

$$f_o = f_i + \frac{f^*}{10} \tag{10.2}$$

The double-mix-divide module has increased the input frequency by the switch-selectable frequency increment $f^*/10$. Double-mix-divide modules can be cascaded to form a frequency synthesizer with any degree of resolution. The double-mix-divide modular approach has the additional advantage that frequencies $f_1$, $f_2$, and $f_i$ can be the same in each module so that all modules can contain identical components.

Considered solely from a theoretical viewpoint, the double-mix-divide module appears unnecessarily complicated, since the output frequency $f_i + f^*/10$ could be realized by using one mixer and bandpass filter. The advantages of the approach shown in Fig. 10.4 are practical; it allows better mixing ratios (with relaxed filtering criteria) and allows for the same bandpass filters in each stage. The effect of deleting $f_2$ is illustrated after we discuss a three-digit synthesizer.

> **EXAMPLE 10.2.** A direct frequency synthesizer with three digits of resolution can be realized by using three double-mix-divide modules. Each decade switch selects one of 10 frequencies $f_2 + f^*$. In this example the output of the third module is taken before the decade divider. For example, it is possible to generate the frequencies between 10 and 19.99 MHz (in 10-kHz increments), using the three-module synthesizer, by selecting
>
> $$f_i = 1 \text{ MHz} \qquad f_1 = 3 \text{ MHz} \qquad f_2 = 6 \text{ MHz}$$
>
> Since
>
> $$f_i + f_1 + f_2 = 10f_i \tag{10.3}$$
>
> the output frequency before the last division by 10 will be
>
> $$f_o = 10f_i + f_3^* + \frac{f_2^*}{10} + \frac{f_1^*}{100} \tag{10.4}$$
>
> Since $f^*$ occurs in 1-MHz increments, $f_1^*/100$ will provide the desired 10-kHz frequency increments. The output is taken before the last decade divider as this provides a sine wave output. The divider output has a square waveform. If, for example, a frequency of 14.86 MHz is required, $f_1^*$ will be 6 MHz, $f_2^*$ will be 8 MHz, and $f_3^*$ will be 4 MHz.

Theoretically, either $f_1$ or $f_2$ could be eliminated, provided

$$f_i + f_1 \text{ (or } f_2) = 10f_i \tag{10.5}$$

but the additional frequency is used in practice to provide additional frequency separation at the mixer output. This frequency separation eases the bandpass filter requirements. For example, if $f_2$ is eliminated, $f_1 + f_i$ must equal $10f_i$, or 10 MHz in the preceding example. If an $f_1^*$ of 1 MHz is selected, the output of the first mixer will consist of the two frequencies 9 and 11 MHz. The lower of these closely spaced frequencies must be removed by the filter. The filter required would be extremely complex to achieve such selectivity. If, instead, a 5-MHz signal $f_2$ is also used so that $f_i + f_1 + f_2 = 10$ MHz, then the two frequencies at the first mixer input will be $f_i + f_1 = 5$ MHz and $f_2 + f_1^* = 6$ MHz. Therefore, the frequencies present at the mixer output (for an $f_1^*$ of 1 MHz) will be 1 and 11 MHz. In this case the two frequencies will be much easier to separate with a bandpass filter. The ancillary frequencies $f_1$ and $f_2$ can only be selected in each design after all possible frequency ratios at the mixer output have been considered.

Direct synthesis can produce fast frequency switching, almost arbitrarily fine frequency resolution, low phase noise, and the highest frequency of operation of any of the methods. Direct frequency synthesis requires considerably more hardware (oscillators, mixers, and bandpass filters) than the two other synthesis techniques to be described. The hardware requirements result in direct synthesizers being larger and more expensive to construct. Another disadvantage of the direct synthesis technique is that unwanted (spurious) frequencies can appear at the output. The wider the frequency range, the most likely it is that spurious components will appear in the output. These disadvantages must be weighed against the versatility, speed, and flexibility of direct synthesis.

## ■ 10.3

### FREQUENCY SYNTHESIS BY PHASE LOCK

The disadvantages associated with direct synthesis are greatly diminished with the frequency synthesis technique (often referred to as *indirect synthesis*) that employs a phase-locked loop (PLL). A simple PLL is illustrated in Fig. 10.5. A detailed analysis of PLL characteristics is given in Chaps. 8 and 9, but for the present discussion it is sufficient to state that when the PLL is functioning properly, the two phase-detector input frequencies are equal. That is,

$$f_r = f_d \qquad (10.6)$$

Frequency $f_d$ is obtained by dividing the voltage-controlled oscillator (VCO) output frequency $f_o$ by $N$:

$$f_d = \frac{f_o}{N} \qquad (10.7)$$

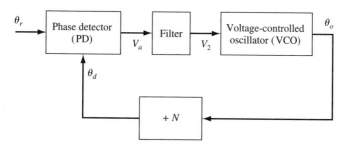

**FIGURE 10.5**
An indirect frequency synthesizer.

Therefore, the output frequency $f_o$ is an integer multiple of the reference frequency,

or
$$f_o = Nf_r \tag{10.8}$$

The PLL with a frequency divider in the loop thus provides a method for obtaining a large number of frequencies from a single reference frequency. If the divide ratio $N$ is realized by using a programmable divider, it is possible to easily change the output frequency in increments of $f_r$. The PLL with a programmable divider provides an easy method for synthesizing a large number of frequencies, all of which are an integer multiple of the reference frequency. There are, however, problems associated with this method. First the major difficulties will be discussed, and then some of the methods presently used to circumvent these problems will be described.

From Eq. (10.8) we note that the frequency resolution is equal to $f_r$. That is, the output frequency can be changed in increments as small as $f_r$; however, this is in conflict with the requirement of a short time interval for changing frequencies. Although an exact expression for the switching time has yet to be derived, a frequently used rule of thumb is that the switching time

$$t_s = \frac{25}{f_r} \tag{10.9}$$

It takes approximately 25 reference periods to switch frequencies. The frequency resolution is therefore inversely proportional to the switching speed. A contemporary specification for satellite communication systems, which use frequency hopping, is that the frequency resolution is equal to 10 Hz and the switching time is less than $10\,\mu s$! Since the above rule of thumb predicts a switching time of 2.5 s, it is clear that the simple PLL frequency synthesizer cannot meet both specifications. The choice of reference frequency dominates loop performance.

**Effects of Reference Frequency on Loop Performance**

The expression for the output frequency [Eq. (10.8)] shows that to obtain fine frequency resolution, the reference frequency must be small. This creates conflicting requirements. One problem is that to cover a broad frequency range requires a large variation in $N$. Even if the hardware problems can be overcome, some method will normally be needed to compensate for the variations in loop dynamics that occur for widely varying values of $N$. It is shown in Chap. 9 that the linearized loop transfer function is

$$\frac{f_o(s)}{f_i(s)} = \frac{\theta_o(s)}{\theta_i(s)} = \frac{K_v F(s)/s}{1 + K_v F(s)/(Ns)} \tag{10.10}$$

where $F(s)$ is the transfer function of the low-pass filter. If $N$ is to assume a large number of values, say, from 1 to 1000, then there will be a 60-dB variation in the open-loop gain and a correspondingly wide variation in the loop dynamics, unless some method (such as the use of a programmable amplifier) is employed to alter the loop gain for different $N$ values.

A second problem encountered with a low reference frequency is that the loop bandwidth must be less than or equal to the reference frequency, because the low-pass filter must filter out the reference frequency and its harmonics present at the phase-detector output. Thus, the filter bandwidth must be less than the reference frequency. It is explained in Sec. 9.3 (PLL stability analysis) that the loop bandwidth is normally less than the filter bandwidth for adequate stability. Therefore, a low reference frequency results in a frequency synthesizer that will be slow to change frequency.

Another problem introduced by a low reference frequency is the effect on noise introduced in the VCO. Figure 10.6 shows a linearized model of a PLL with the three main sources of noise. Here $\phi_{N_r}$ is the noise on the reference signal, and $\phi_{N_d}$ is the noise created in the phase detector. The largest phase-detector noise components are at the reference frequency and the harmonics of this frequency. And $\phi_{N_o}$ is the noise introduced by the VCO. Figure 10.7 illustrates a frequency spectrum typical of VCO noise. Most of the energy content of VCO noise is near the oscillator frequency; in the PLL model it can be interpreted as a low-frequency noise. The total noise of the closed-loop system at the VCO output $\phi_N$ is given by

$$\phi_N = \frac{(\phi_{N_r} + \phi_{N_d})K_v F(s)/s}{1 + K_v F(s)/(Ns)} + \frac{\phi_{N_o}}{1 + K_v F(s)Ns}$$

$$= G(s)(\phi_{N_r} + \phi_{N_d}) + G_r(s)\phi_{N_o}$$

(10.11)

Since $F(s)$ is either unity or a low-pass transfer function, $G(s)$ is a low-pass transfer function and $G_r(s)$ is a high-pass transfer function. The PLL functions as a low-pass filter for phase noise arising in the reference signal and phase detector, and it functions as a high-pass filter for phase noise originating in the VCO. Since the VCO noise is a low-frequency noise, the output noise due to $\phi_{N_o}$ is minimized by having the loop bandwidth as wide as possible. At the same time, the loop bandwidth should be less than the reference frequency in order to minimize the effect of $\phi_{N_d}$, which is dominated by spurious frequency components at the reference frequency and its harmonics.

Therefore, the desire to have a low reference frequency $f_r$ in order to obtain fine frequency resolution is offset by the need to have $f_r$ large in order to reduce the loop settling time and to minimize the noise contributed by the VCO.

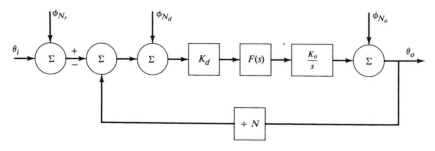

**FIGURE 10.6**
A PLL synthesizer including three noise sources.

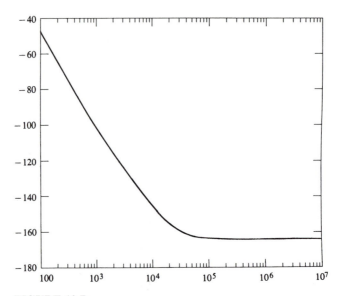

**FIGURE 10.7**
Frequency spectrum of VCO noise.

## Variable-Modulus Dividers

Another difficulty with the system illustrated in Fig. 10.5 is that the maximum operating speed of programmable dividers is slower than that required in many communication systems. The upper limit of a programmable divider realized from transistor-transistor logic (TTL) components is approximately 25 MHz, and that realized with complementary-symmetry metal-oxide semiconductor (CMOS) logic is about 4 MHz. So, for example, if one is to build a $2 \times 10^9$ Hz synthesizer for satellite communications, some other method must be used. There are various ways to overcome this problem. First we will discuss the problem of the relatively low operating speed of programmable dividers.

Programmable dividers are slower than fixed-modulus dividers (prescalers). In fact, prescalers are available that operate at gigahertz frequencies. Figure 10.8 illustrates an indirect synthesizer that contains both a prescaler and a programmable divider in the loop. The prescaler, which can operate frequencies into the gigahertz region, first reduces the output frequency by the factor $P$ before it is applied to the programmable divider. When the loop is in lock,

$$f_r = \frac{f_o}{PN} \qquad \text{or} \qquad f_o = N(Pf_r) \tag{10.12}$$

Although the use of the prescaler allows the loop to operate with higher output frequencies, the output frequency can be changed only in increments of $Pf_r$. Since the channel spacing is equal to the reference frequency, in order to obtain the same resolution, the reference frequency must be decreased by the prescaler factor $P$.

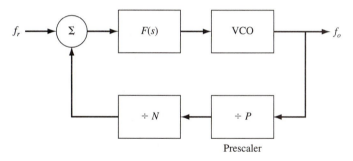

**FIGURE 10.8**
A PLL including a prescaler.


Another approach for obtaining good frequency resolution while operating at high output frequencies uses a method known as *variable-modulus prescaling*. Reconsidering Eq. (10.12), we see that the output frequency resolution could be improved if the value of $N$ were an integer plus a fraction. For example, if $N = N_o + AQ/P$ (where $A$ and $Q$ are integers), then the output frequency will be given by

$$f_o = P\left(N_o + \frac{AQ}{P}\right)f_r = PN_o f_r + AQf_r$$

and the resolution can be regained. This equation is not easily implemented, but if $\pm AP$ is added, the result is

$$f_o = (N_o P + AQ - AP + AP)f_r = [P(N_o - A) + (P + Q)A]f_r$$

From this equation, it is apparent that a dual-modulus counter that divides by $P + Q$ for $A$ cycles and by $P$ for $N_o - A$ cycles could be used to implement the function.

The dual-modulus prescaler system illustrated in Fig. 10.9 has a prescaler that divides by the modulus $P + Q$ when the modulus control is high and divides by $P$ when the modulus control is low. In this particular scheme, the output of the variable-modulus prescaler simultaneously drives the two programmable dividers 1 and 2. The programmable dividers operate at the input clock rate $f_i$ divided by $P$ or $P + Q$. The divide cycle begins with counter 1 preset to $A$, counter 2 preset to $N$, and the modulus control high so that the two-modulus prescaler output frequency is equal to the frequency divided by $P + Q$. The prescaler will divide-by-$P + Q$ until the $A$ counter reaches 0. At this point, the divide-by-$N$ counter is at a value equal to $N(\text{preset}) - A(\text{preset})$. Next, counter $A$ pulses the prescaler modulus control to low to change to the divide-by-$P$ mode. The prescaler then divides by $P$, $N - A$ times, until the $N$ counter reaches 0. Finally, the divide cycle is restarted by reloading the counters with their preset values and resetting the modulus control signal. The number of input cycles in one complete divide cycle is

$$D = (P + Q)A + P(N - A) = AQ + PN \tag{10.13}$$

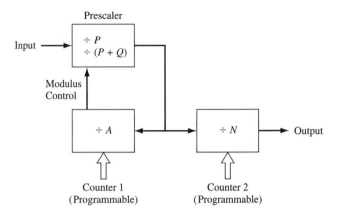

**FIGURE 10.9**
A programmable divider realized with a dual-modulus
prescaler.

(Note that $N$ must be greater than $A$ for the method to work.) If $Q = 1$, then the divide
ratio, even though it has a minimum value $D_{min} = PN$, can be implemented in unit
steps. A frequently used divide ratio is $P = 10$ and $P + Q = 11$. Then Eq. (10.13)
becomes

$$D = 10N + A \qquad (10.14)$$

which shows that the $10/11$ prescaler can be used to obtain division ratios with
increments of 1, provided $N > A$. Since $A_{max} = 9$, $N$ must be at least 10 and $D_{min}$
is 100. The minimum divide ratio is not usually a problem in frequency synthesizer
design.

> **EXAMPLE 10.3.** If it is desired to design a frequency synthesizer to cover the fre-
> quency range from 100 to 109 MHz in 1-MHz increments, a reference frequency of 1
> MHz is suitable (a higher reference frequency would not be). Since 100 MHz is too fast
> for a programmable divider, a $10/11$ variable-modulus prescaler will be considered.
> Now $A$ will vary from 0 to 9, so $N$ must be at least 10. The minimum value of $D$ will be,
> using the $10/11$ prescaler,
>
> $$D_{min} = \frac{100 \times 10^6}{10^6} = 100$$
>
> which will provide $(f_o)_{min} = 100$ MHz. Thus the desired division ratio can be obtained
> by using a $10/11$ variable-modulus prescaler together with the programmable dividers.

Other variable-modulus division ratios such as $5/6, 8/9, 32/33, 40/41, 64/65,$
$100/101,$ and $128/129$ are also frequently used. In Example 10.3, if it is necessary
to cover the frequency interval from 100 to 100.99 MHz in 10-kHz increments, a
maximum reference frequency of 10 kHz will be needed, and the minimum divide
ratio will be

$$D_{min} = \frac{100 \times 10^6}{10^4} = 10^4 = \frac{(f_o)_{min}}{f_r}$$

since $A_{max} = 99$ and $N_{min}$ is 100. Also, $D = AQ + PN$, so $Q$ must equal 1 for a frequency resolution of 10 kHz. It is possible to select $P = 10$ and $N = 10^3$, but it is better to select $P = 100$ and $N = 100$, since the maximum frequency to the two programmable dividers will then be 1.0099 MHz. This will allow for the use of low-noise CMOS logic for the programmable dividers. Therefore, a 100/101 variable-modulus divider is suited for this design. If $P = 10$ is selected, the maximum frequency to the programmable dividers would be 10.0A99 MHz.

**PLL Frequency Synthesizer ICs**

There are several types of PLL frequency synthesizer integrated circuits available today. Most contain digital phase detectors and presettable counters for the feedback divide. Some of the more advanced chips contain the control logic and counters necessary for an external dual-modulus prescaler and can be preset either serially or in parallel. One such chip is the Motorola MC145152-2. This parallel-input PLL frequency synthesizer has many useful features, as shown in the data sheets in Fig. 10.10.

> **EXAMPLE 10.4.** In the spread-spectrum technique known as *frequency hopping,* the carrier frequency is hopped over a bandwidth $B$. Each hop in frequency is located at a predetermined frequency *bin*. (For more information on spread spectrum, refer to Robert Dixon's book *Spread Spectrum Systems with Commercial Applications*). Suppose that we want to hop from 180.4 to 185.6 MHz [note that the fifth harmonic of these frequencies is located in the ISM (Industrial, Scientific and Medical) band] with at least 20 kHz between hops and at least 50 bins. Design a PLL frequency synthesizer circuit that will implement these requirements.
>
> *Solution.* The following components are used to design this circuit:
>
> - 2-MHz crystal
> - MC145152-2
> - $2^7 \times 16$ bit ROM
> - MC12017 (Motorola 64/65 Prescaler)
> - Addition digital control logic

To begin with the hopping-bin channel spacing requirement of at least 20 kHz, a 2-MHz crystal is connected to the MC14512-2 with the reference address inputs (pins 4, 5, 6) connected such that the crystal is divided by 64(RA2 = 0, RA1 = 0, RA0 = 1). This gives a reference frequency of 31.25 kHz, and the maximum number of hops is 5.2 MHz$/31.25$ kHz $= 166.4$ hops. For 185.6 MHz the values of $N$ and $A$ are found as follows: First, let $A = 0$, and solve for $N$.

$$PN + A = \frac{f_o}{f_r} = \frac{185.6 \text{ MHz}}{31.25 \text{ kHz}} = 5939.2 \approx 5939 \qquad \therefore N = 92$$

Second, for this value of $N$, find $A$ by

$$A = \frac{f_o}{f_r} - NP = 5939 - (92)(64) = 51$$

**MOTOROLA**
**SEMICONDUCTOR TECHNICAL DATA**

Order this document
by MC145151–2/D

# PLL Frequency Synthesizer Family
## CMOS

| |
|---|
| **MC145151-2** |
| **MC145152-2** |
| **MC145155-2** |
| **MC145156-2** |
| **MC145157-2** |
| **MC145158-2** |

The devices described in this document are typically used as low–power, phase–locked loop frequency synthesizers. When combined with an external low–pass filter and voltage–controlled oscillator, these devices can provide all the remaining functions for a PLL frequency synthesizer operating up to the device's frequency limit. For higher VCO frequency operation, a down mixer or a prescaler can be used between the VCO and the synthesizer IC.

These frequency synthesizer chips can be found in the following and other applications:

| | |
|---|---|
| CATV | TV Tuning |
| AM/FM Radios | Scanning Receivers |
| Two–Way Radios | Amateur Radio |

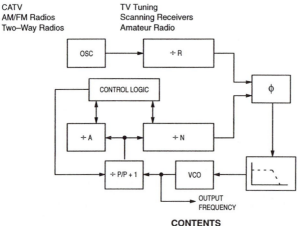

**CONTENTS**

REV 1
8/95

🅜 **MOTOROLA**

© Motorola, Inc. 1995

**FIGURE 10.10**

**MC145151–2 BLOCK DIAGRAM**

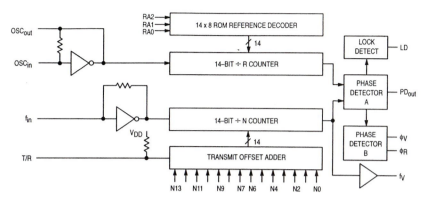

NOTE: N0 – N13 inputs and inputs RA0, RA1, and RA2 have pull–up resistors that are not shown.

## PIN DESCRIPTIONS

### INPUT PINS

#### $f_{in}$
**Frequency Input (Pin 1)**

Input to the ÷ N portion of the synthesizer. $f_{in}$ is typically derived from loop VCO and is ac coupled into the device. For larger amplitude signals (standard CMOS logic levels) dc coupling may be used.

#### RA0 – RA2
**Reference Address Inputs (Pins 5, 6, 7)**

These three inputs establish a code defining one of eight possible divide values for the total reference divider, as defined by the table below.

Pull–up resistors ensure that inputs left open remain at a logic 1 and require only a SPST switch to alter data to the zero state.

| Reference Address Code | | | Total Divide Value |
|---|---|---|---|
| **RA2** | **RA1** | **RA0** | |
| 0 | 0 | 0 | 8 |
| 0 | 0 | 1 | 128 |
| 0 | 1 | 0 | 256 |
| 0 | 1 | 1 | 512 |
| 1 | 0 | 0 | 1024 |
| 1 | 0 | 1 | 2048 |
| 1 | 1 | 0 | 2410 |
| 1 | 1 | 1 | 8192 |

#### N0 – N11
**N Counter Programming Inputs (Pins 11 – 20, 22 – 25)**

These inputs provide the data that is preset into the ÷ N counter when it reaches the count of zero. N0 is the least significant and N13 is the most significant. Pull–up resistors en-

sure that inputs left open remain at a logic 1 and require only an SPST switch to alter data to the zero state.

#### T/R
**Transmit/Receive Offset Adder Input (Pin 21)**

This input controls the offset added to the data provided at the N inputs. This is normally used for offsetting the VCO frequency by an amount equal to the IF frequency of the transceiver. This offset is fixed at 856 when T/R is low and gives no offset when T/R is high. A pull–up resistor ensures that no connection will appear as a logic 1 causing no offset addition.

#### $OSC_{in}$, $OSC_{out}$
**Reference Oscillator Input/Output (Pins 27, 26)**

These pins form an on–chip reference oscillator when connected to terminals of an external parallel resonant crystal. Frequency setting capacitors of appropriate value must be connected from $OSC_{in}$ to ground and $OSC_{out}$ to ground. $OSC_{in}$ may also serve as the input for an externally–generated reference signal. This signal is typically ac coupled to $OSC_{in}$, but for larger amplitude signals (standard CMOS logic levels) dc coupling may also be used. In the external reference mode, no connection is required to $OSC_{out}$.

### OUTPUT PINS

#### $PD_{out}$
**Phase Detector A Output (Pin 4)**

Three–state output of phase detector for use as loop–error signal. Double–ended outputs are also available for this purpose (see $\phi_V$ and $\phi_R$).

Frequency $f_V > f_R$ or $f_V$ Leading: Negative Pulses
Frequency $f_V < f_R$ or $f_V$ Lagging: Positive Pulses
Frequency $f_V = f_R$ and Phase Coincidence: High–Impedance State

**FIGURE 10.10** (*continued*)

**MOTOROLA**
SEMICONDUCTOR TECHNICAL DATA

# MC145151-2

# Parallel-Input PLL Frequency Synthesizer
## Interfaces with Single-Modulus Prescalers

The MC145151-2 is programmed by 14 parallel-input data lines for the N counter and three input lines for the R counter. The device features consist of a reference oscillator, selectable-reference divider, digital-phase detector, and 14-bit programmable divide-by-N counter.

The MC145151-2 is an improved-performance drop-in replacement for the MC145151-1. The power consumption has decreased and ESD and latch-up performance have improved.

- Operating Temperature Range: – 40 to 85°C
- Low Power Consumption Through Use of CMOS Technology
- 3.0 to 9.0 V Supply Range
- On- or Off-Chip Reference Oscillator Operation
- Lock Detect Signal
- ÷ N Counter Output Available
- Single Modulus/Parallel Programming
- 8 User-Selectable ÷ R Values: 8, 128, 256, 512, 1024, 2048, 2410, 8192
- ÷ N Range = 3 to 16383
- "Linearized" Digital Phase Detector Enhances Transfer Function Linearity
- Two Error Signal Options: Single-Ended (Three-State) or Double-Ended
- Chip Complexity: 8000 FETs or 2000 Equivalent Gates

**P SUFFIX**
PLASTIC DIP
CASE 710

28

1

**DW SUFFIX**
SOG PACKAGE
CASE 751F

28

1

**ORDERING INFORMATION**
MC145151P2    Plastic DIP
MC145151DW2   SOG Package

**PIN ASSIGNMENT**

| | | |
|---|---|---|
| $f_{in}$ | 1 ● | 28 | LD |
| $V_{SS}$ | 2 | 27 | $OSC_{in}$ |
| $V_{DD}$ | 3 | 26 | $OSC_{out}$ |
| $PD_{out}$ | 4 | 25 | N11 |
| RA0 | 5 | 24 | N10 |
| RA1 | 6 | 23 | N13 |
| RA2 | 7 | 22 | N12 |
| $\phi_R$ | 8 | 21 | T/R |
| $\phi_V$ | 9 | 20 | N9 |
| $f_V$ | 10 | 19 | N8 |
| N0 | 11 | 18 | N7 |
| N1 | 12 | 17 | N6 |
| N2 | 13 | 16 | N5 |
| N3 | 14 | 15 | N4 |

REV 1
8/95

© Motorola, Inc. 1995

Ⓜ **MOTOROLA**

**FIGURE 10.10 (*continued*)**

**MC145152–2 BLOCK DIAGRAM**

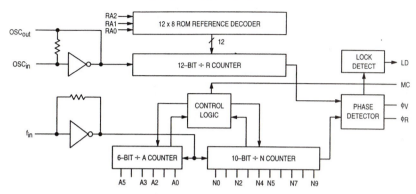

NOTE: N0 – N9, A0 – A5, and RA0 – RA2 have pull–up resistors that are not shown.

## PIN DESCRIPTIONS

### INPUT PINS

**f$_{in}$**
**Frequency Input (Pin 1)**

Input to the positive edge triggered ÷ N and ÷ A counters. f$_{in}$ is typically derived from a dual–modulus prescaler and is ac coupled into the device. For larger amplitude signals (standard CMOS logic levels) dc coupling may be used.

**RA0, RA1, RA2**
**Reference Address Inputs (Pins 4, 5, 6)**

These three inputs establish a code defining one of eight possible divide values for the total reference divider. The total reference divide values are as follows:

| Reference Address Code | | | Total Divide Value |
|---|---|---|---|
| **RA2** | **RA1** | **RA0** | |
| 0 | 0 | 0 | 8 |
| 0 | 0 | 1 | 64 |
| 0 | 1 | 0 | 128 |
| 0 | 1 | 1 | 256 |
| 1 | 0 | 0 | 512 |
| 1 | 0 | 1 | 1024 |
| 1 | 1 | 0 | 1160 |
| 1 | 1 | 1 | 2048 |

**N0 – N9**
**N Counter Programming Inputs (Pins 11 – 20)**

The N inputs provide the data that is preset into the ÷ N counter when it reaches the count of 0. N0 is the least significant digit and N9 is the most significant. Pull–up resistors ensure that inputs left open remain at a logic 1 and require only a SPST switch to alter data to the zero state.

**A0 – A5**
**A Counter Programming Inputs**
**(Pins 23, 21, 22, 24, 25, 10)**

The A inputs define the number of clock cycles of f$_{in}$ that require a logic 0 on the MC output (see **Dual–Modulus**

Prescaling section). The A inputs all have internal pull–up resistors that ensure that inputs left open will remain at a logic 1.

**OSC$_{in}$, OSC$_{out}$**
**Reference Oscillator Input/Output (Pins 27, 26)**

These pins form an on–chip reference oscillator when connected to terminals of an external parallel resonant crystal. Frequency setting capacitors of appropriate value must be connected from OSC$_{in}$ to ground and OSC$_{out}$ to ground. OSC$_{in}$ may also serve as the input for an externally–generated reference signal. This signal is typically ac coupled to OSC$_{in}$, but for larger amplitude signals (standard CMOS logic levels) dc coupling may also be used. In the external reference mode, no connection is required to OSC$_{out}$.

### OUTPUT PINS

**$\phi_R$, $\phi_V$**
**Phase Detector B Outputs (Pins 7, 8)**

These phase detector outputs can be combined externally for a loop–error signal.

If the frequency f$_V$ is greater than f$_R$ or if the phase of f$_V$ is leading, then error information is provided by $\phi_V$ pulsing low. $\phi_R$ remains essentially high.

If the frequency f$_V$ is less than f$_R$ or if the phase of f$_V$ is lagging, then error information is provided by $\phi_R$ pulsing low. $\phi_V$ remains essentially high.

If the frequency of f$_V$ = f$_R$ and both are in phase, then both $\phi_V$ and $\phi_R$ remain high except for a small minimum time period when both pulse low in phase.

**MC**
**Dual–Modulus Prescale Control Output (Pin 9)**

Signal generated by the on–chip control logic circuitry for controlling an external dual–modulus prescaler. The MC level will be low at the beginning of a count cycle and will remain low until the ÷ A counter has counted down from its programmed value. At this time, MC goes high and remains high until the ÷ N counter has counted the rest of the way down from its programmed value (N – A additional counts since both ÷ N and A are counting down during the first

**FIGURE 10.10** (*continued*)

portion of the cycle). MC is then set back low, the counters preset to their respective programmed values, and the above sequence repeated. This provides for a total programmable divide value ($N_T$) = N • P + A where P and P + 1 represent the dual–modulus prescaler divide values respectively for high and low MC levels, N the number programmed into the ÷ N counter, and A the number programmed into the ÷ A counter.

**LD**
**Lock Detector Output (Pin 28)**

Essentially a high level when loop is locked ($f_R$, $f_V$ of same phase and frequency). Pulses low when loop is out of lock.

**POWER SUPPLY**

**$V_{DD}$**
**Positive Power Supply (Pin 3)**

The positive power supply potential. This pin may range from + 3 to + 9 V with respect to $V_{SS}$.

**$V_{SS}$**
**Negative Power Supply (Pin 2)**

The most negative supply potential. This pin is usually ground.

## TYPICAL APPLICATIONS

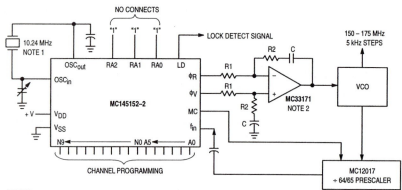

NOTES:
1. Off–chip oscillator optional.
2. The $\phi_R$ and $\phi_V$ outputs are fed to an external combiner/loop filter. See the Phase–Locked Loop — Low–Pass Filter Design page for additional information. The $\phi_R$ and $\phi_V$ outputs swing rail–to–rail. Therefore, the user should be careful not to exceed the common mode input range of the op amp used in the combiner/loop filter.

**Figure 1. Synthesizer for Land Mobile Radio VHF Bands**

**FIGURE 10.10** (*continued*)

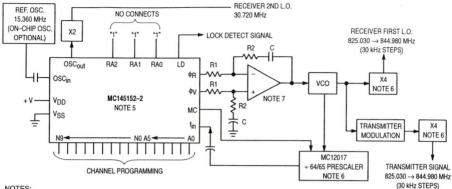

NOTES:
1. Receiver 1st I.F. = 45 MHz, low side injection; Receiver 2nd I.F. = 11.7 MHz, low side injection.
2. Duplex operation with 45 MHz receiver/transmit separation.
3. $f_R$ = 7.5 kHz; ÷ R = 2048.
4. $N_{total}$ = N • 64 + A = 27501 to 28166; N = 429 to 440; A = 0 to 63.
5. MC145158–2 may be used where serial data entry is desired.
6. High frequency prescalers (e.g., MC12018 [520 MHz] and MC12022 [1 GHz]) may be used for higher frequency VCO and $f_{ref}$ implementations.
7. The $\phi_R$ and $\phi_V$ outputs are fed to an external combiner/loop filter. See the Phase–Locked Loop — Low–Pass Filter Design page for additional information. The $\phi_R$ and $\phi_V$ outputs swing rail–to–rail. Therefore, the user should be careful not to exceed the common mode input range of the op amp used in the combiner/loop filter.

**Figure 2. 666–Channel, Computer–Controlled, Mobile Radiotelephone Synthesizer
for 800 MHz Cellular Radio Systems**

MC145152–2 Data Sheet Continued on Page 23

MC145151–2 through MC145158–2
8

MOTOROLA

**FIGURE 10.10** (*continued*)

In the same manner $N = 90$ and $A = 13$ for 180.4 MHz. Notice that the decimal values were rounded to the integer part. The decimal appears because the output frequency choice was not a multiple of the reference frequency. The divide-counter values are summarized below

$$N = 92 \qquad\qquad\qquad N = 90$$
$$A = 51 \qquad\qquad\qquad A = 12$$
$$P = 64 \qquad\qquad\qquad P = 64$$
$$f_o = 185,593,750 \text{ Hz} \qquad f_o = 180,406,250 \text{ Hz}$$

With these values there are 166 hops possible. Since only 50 hops are required and we will need a 16-bit read-only memory (ROM)(10 bits for the $N$ counter and 6 bits for the $A$ counter) to store these digital values, it is best to use 128 hops and a $2^7 \times 16$ bit ROM. The new divide-counter values are

$$N = 92 \qquad\qquad\qquad N = 90$$
$$A = 13 \qquad\qquad\qquad A = 12$$
$$P = 64 \qquad\qquad\qquad P = 64$$
$$f_o = 184,406,250 \text{ Hz} \qquad f_o = 180,406,250 \text{ Hz}$$

A low-pass filter (LPF) is designed with the techniques learned in Chaps. 8 and 9, and a VCO is designed with the techniques learned in Chap. 7. In addition, a timer circuit is required to clock the ROM. The time interval for which a hop is in a frequency bin is called the *dwell time.* For this application, the dwell time must be at least as long as the longest settling time of the PLL (note, as stated earlier in this chapter, that the different divide values change the open-loop gain and the loop dynamics). A block diagram of this implementation is given in Fig. 10.11.

**Down Conversion**

Another approach to circumventing the high-frequency limitation of the programmable dividers is to shift the output frequency down by mixing the output frequency with a local oscillator frequency. Figure 10.12 illustrates a single down-conversion synthesizer. The low-pass filter following the mixer is used to filter out the higher mixer output frequency $f_o + f_L$. The divider output frequency is

$$f_d = f_r = \frac{f_o - f_L}{N}$$

so

$$f_o = f_L + N f_r$$

The main disadvantages of this method are that the complexity and size are increased, the possibility of spurious components being introduced by the mixer is increased, and the phase lag of the filter used in the feedback path can degrade the loop performance.

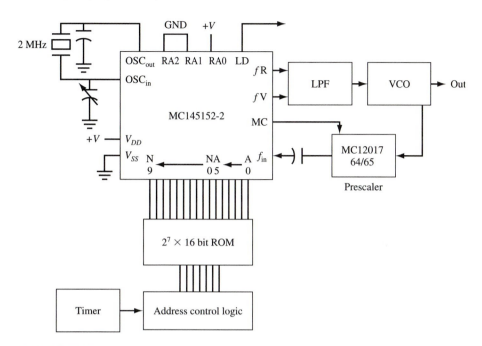

**FIGURE 10.11**
Implementation of PLL frequency synthesizer using MC145152-2.

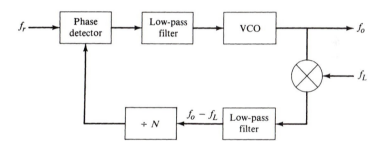

**FIGURE 10.12**
A PLL frequency synthesizer with down conversion.

## Methods for Reducing Switching Time and/or Widening the Loop Bandwidth

There are methods available for circumventing the conflict between the need for fine frequency resolution and the need to quickly change frequencies. A method of reducing the response time is to include a coarse steering signal. When the frequency is changed by altering the divide ratio $N$, a steering signal can be generated and applied immediately to direct the VCO to the new frequency (see Fig. 10.13). The steering signal can be obtained from a lookup table stored in memory with the D/A converter used to generate the analog steering signal. Another frequently used method is to incorporate multiple phase-locked loops in the synthesizer.

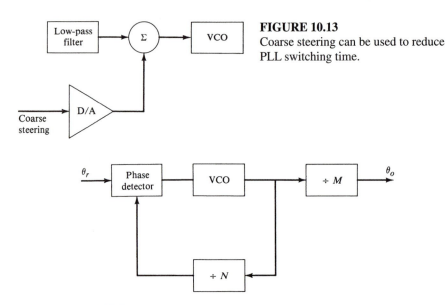

**FIGURE 10.13**
Coarse steering can be used to reduce PLL switching time.

**FIGURE 10.14**
A PLL frequency synthesizer with a postdivider for increased frequency resolution.

### Multiple-loop frequency synthesizers

One possible method of obtaining fine frequency resolution with a high reference frequency is illustrated in Fig. 10.14. The output frequency is obtained by dividing the VCO output frequency by $M$. That is,

$$f_o = \frac{Nf_r}{M}$$

and the frequency resolution is

$$f = \frac{f_r}{M}$$

Hence fine resolution is obtained by making sure that $M$ is sufficiently large. A problem inherent in this technique is that the loop frequency may become too large. The difficulty is illustrated by the following numerical example.

EXAMPLE 10.5. Consider the design of a frequency synthesizer to cover the frequency range from 10 to 10.1 MHz with 1-kHz resolution. The reference frequency is to be 100 kHz. To obtain the 1-kHz frequency resolution from the 100-kHz reference frequency requires that the VCO output frequency be divided by 100. An output frequency of 10 MHz will require that the VCO be operating at 1 GHz!

Although adding a postdivider is not a good solution in general, the concept does find practical application in multiple-loop synthesizers. A multiple-loop synthesizer uses one or more loops to obtain the fine frequency resolution and combines the outputs of these loops with that of another loop, which generates the high-frequency components of the desired output frequency. The principles involved can be easily understood by examining one such synthesizer.

**EXAMPLE 10.6.** Consider the design of a frequency synthesizer to cover the frequency range from 35.40 to 40.00 MHz in 1-kHz increments. If this is to be accomplished in a single-loop synthesizer, a reference frequency of 1 kHz will be required (with a response time of approximately 25 ms), together with the divide ratio $N$

$$35.40 \times 10^3 \leq N \leq 40.00 \times 10^3$$

An alternate design is shown in Fig. 10.15.[2] The synthesizer consists of three PLLs. PLLs A and B both use the 100-kHz reference frequency. Loop C locks the divided output of loop A $(f_A)$ to the difference between the output frequency $f_o$ and the output of loop B $(f_B)$. That is,

$$f_A = f_o - f_B \tag{10.15}$$

or

$$f_o = f_B + f_A \tag{10.16}$$

Phase-locked loop C serves as a mixer and filter for $f_A$ and $f_B$. If $f_A$ and $f_B$ are directly combined in a mixer, the sum and difference frequencies will be too close together to be adequately separated with a bandpass filter. The present technique of using a phase-locked loop for frequency mixing does accomplish good separation.

Since the reference frequency of loop A is 100 kHz, its output frequency $f_a$ can be varied in 100-kHz increments, and

$$f_A = \frac{f_a}{100} = N_a \times 10^3$$

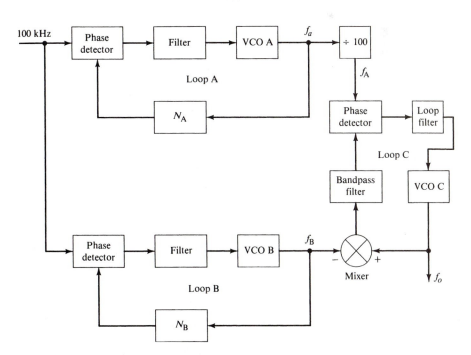

**FIGURE 10.15**
The three-loop frequency synthesizer discussed in Example 10.6.

varies in 1-kHz increments. Loop A is used to generate the 1- and 10-kHz increments of output frequency, and loop B the 0.1- and 1-MHz changes in output frequency. Frequency $f_A$ could be selected to vary between 0 and 99 kHz in 1-kHz increments, but $f_A$ also serves as the reference frequency for loop C. If, for example, $f_A = 1$ kHz, this will require that loop C be a relatively slow loop and will determine the overall response time of the synthesizer. To reduce the response time of loop C, $f_A$ is increased by 300 kHz so that

$$300 \text{ kHz} \le f_A \le 399 \text{ kHz}$$

Therefore,
$$3.0 \times 10^7 \le f_a \le 3.99 \times 10^7$$

and
$$300 \le N_a \le 399$$

Since $f_B = f_o - f_A$, frequency $f_B$ is reduced by 300 kHz so that

$$35.40 - 0.3 \le f_B \le 40 - 0.3 \text{ MHz}$$

and
$$351 \le N_B \le 397$$

The response time of the frequency synthesizer is determined by the response times of loops A and B, both of which have a reference frequency of 100 kHz. Hence the overall response time will be approximately $25 \times 10^{-2}$ ms, even though 1-kHz frequency increments are obtained. Frequency resolution down to 10 Hz could be obtained by first combining loop A with another loop and a divide-by-100 circuit (requiring two additional loops).

**Fractional-$N$ loops**

An alternative method of decreasing loop response time would be possible if $N$ could be made to take on fractional values. The output frequency could then be changed in fractional increments of the reference frequency. Although a digital divider cannot provide a fractional division ratio, ways can be found to effectively accomplish the same task. The technique was originally called *digiphase*,[3] and a commercial version[4] is referred to as *Fractional-N*. The most frequently used method is to divide the output frequency by $N + 1$ every $M$ cycles and to divide by $N$ the rest of the time. The effective division ratio is then $N + M^{-1}$, and the average output frequency is given by

$$f_o = (N + M^{-1})f_r \tag{10.17}$$

This expression shows that $f_o$ can be varied in fractional increments of the reference frequency by varying $M$. A simplified method for generating the fractional division is shown in Fig. 10.16. The divider divides the input frequency by $N$, and the counter counts the number of cycles of waveform output. Each time the counter reaches a count of $M - 1$, the counter output goes low for one input cycle and one input cycle does not reach the divider. Therefore, the divider requires $N + 1$ input cycles to change state.

The number of output cycles during one complete cycle of the $M$ counter is

$$f_o = f_d N(M - 1) + f_d(N + 1) = f_d(NM + 1)$$

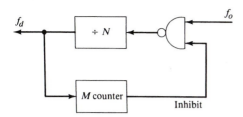

**FIGURE 10.16**
A simplified method of implementing fractional division.

and the average output frequency per cycle of the $M$ counter is

$$(f_o)_{av} = f_d(N + M^{-1})$$

The average frequency at the divider output is the output frequency divided by $N + M^{-1}$, so a form of fractional division has been realized. The method of implementing fractional-$N$ division shown in Fig. 10.16 will work as long as $M$ is an integer, but normally it will not be. A more general method of implementing fractional division can be obtained by using a phase accumulator. The phase accumulator approach is illustrated by the following example.

EXAMPLE 10.7. Consider the problem of generating 455 kHz by using a fractional-$N$ loop with a 100-kHz reference frequency. The integer part of the division is $N = 4$, and the fractional part is $M^{-1} = 0.55$, or $M = 1.8$. Clearly $M$ is not an integer; the VCO output is to be divided by $5(N + 1)$ every 1.8 cycles, or 55 times every 100 cycles. Although $M$ is not an integer, the fractional division can be easily implemented by adding the number 0.55 $M^{-1}$ to the contents of an accumulator every output cycle. Each time the accumulator overflows (the contents exceed 1), the divider divides by 5 rather than by 4. Only the fractional value of the addition is retained in the phase accumulator.

The phase accumulator realization of fractional division is illustrated in Fig. 10.17. Fine frequency resolution can be arbitrarily obtained by increasing the length of the phase accumulator. The previous example used a 100-kHz reference frequency. A resolution of $10^5/10^5 = 1$ Hz could be obtained by using a five-stage binary-coded decimal (BCD) accumulator. The performance of a fractional-$N$ synthesizer will be further illustrated with another numerical example.

EXAMPLE 10.8 Consider the problem of incrementing the output frequency of a 1-MHz synthesizer by 1000 Hz, the reference frequency being 10 kHz. Since

$$f_o = (N + M^{-1})f_r$$

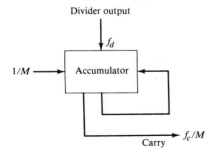

**FIGURE 10.17**
A phase accumulator used for fractional division.

$N = 100$; if $f_o$ is to be increased to $1.001 \times 10^6$ Hz, then $M = 10$. That is, at every 10 reference cycles ($10^3$ output cycles), the output frequency is divided by 101. The average output frequency is then $100.1 \times 10^4$ Hz, which is the desired frequency. While the reference signal goes through one period, the VCO signal goes through 100.1 cycles and the output of the divider ($\div 100$) goes through 1.001 cycles; its phase relative to the reference frequency advances by $0.001 \times 2\pi$ rad each reference cycle. After the 10 reference cycles, the divider reference output leads the reference signal by $0.01 \times 2\pi$ rad. At this time one VCO cycle is skipped; the skipping of one VCO cycle delays the divider output by $0.01 \times 2\pi$ rad, which is exactly how much the divider output had increased in phase.

Although the average output frequency was 1.001 MHz in the preceding example, the instantaneous output frequency changes with each reference cycle because of the increasing phase difference between the divider output and the reference signal. The timing diagram of Fig. 10.18 illustrates this point. If the divider output frequency is slightly faster than the reference frequency, the phase-detector output will consist of pulses of increasing width, and the dc value of these pulses will appear as shown in Fig. 10.19. This voltage will create fluctuations in the output frequency if the frequency is not eliminated before it reaches the VCO.

Figure 10.20 contains a simplified diagram of a fractional-$N$ synthesizer that eliminates the deterministic noise occurring at the phase-detector output by adding a signal equal in magnitude and opposite in sign to the deterministic voltage present at the detector output. The fraction register, adder, and phase register determine how often a pulse is to be removed from the VCO output. The phase register contains the

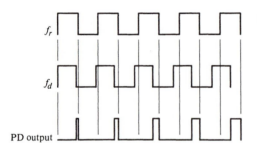

$f_r$

$f_d$

PD output

**FIGURE 10.18**
Timing signals in a PLL frequency synthesizer using fractional division.

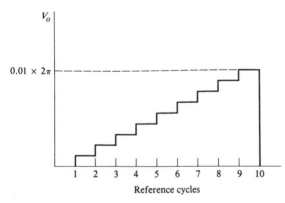

$V_o$

$0.01 \times 2\pi$

1  2  3  4  5  6  7  8  9  10
Reference cycles

**FIGURE 10.19**
Typical waveform of the average value of the phase-detector output of a fractional-$N$ synthesizer.

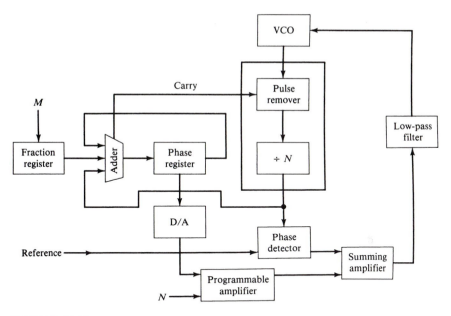

**FIGURE 10.20**
A complete fractional-$N$ frequency synthesizer.

fractional portion of the divisor, and this information is converted to an analog signal in the D/A converter. The analog signal is then used to reduce the phase noise.

One further feature of this analog noise-canceling signal is that it depends on both $M$ and $N$. If, for example, it is desired to synthesize a frequency of 2.001 MHz, then $M$ is again 10, and every 10 reference cycles the output frequency is divided by 201. During each reference cycle the VCO goes through 200.1 cycles. Therefore, the divider output phase relative to the reference frequency advances by $0.001/2 \times 2\pi$ rad each reference cycle. This phase increment is one-half of that which occurs for an output frequency of 1.001 MHz. In general, the amplitude of these steps is inversely proportional to frequency. Therefore, the D/A output amplitude must be adjusted by a programmable gain amplifier, with the gain inversely proportional to $N$. The analog signal is subtracted from the phase-detector output in order to provide a low-noise VCO control signal.

## ▓ 10.4
### DIRECT DIGITAL SYNTHESIS[5]

Direct digital frequency synthesis (DDFS) is achieved either by solving a digital recursion relationship using a general-purpose computer or microcomputer or by storing the sine wave values in a lookup table. Recent advances in microelectronics[6] make DDFS practical at frequencies up to approximately 150 MHz. Further scaling, the use of pipeline phase accumulators, and the use of parallel-architecture D/A converters may soon enable frequencies up to 500 MHz. The synthesizers can be small and low-power and can provide very fine frequency resolution (can be less than 1 Hz) with virtually instantaneous, phase-coherent switching of frequencies. In addition, they have a very fast settling time (usually measured in nanoseconds), experience minimal drift, and exhibit very little phase noise.

There are at least two problems with the method of solving a linear recursion relationship to generate the sine wave. The noise can increase until a limit cycle (nonlinear oscillation) occurs. Also, the finite word length used to represent the coefficients places a limitation on the frequency resolution. For these two reasons, the direct table lookup method is preferred today. One direct table lookup method outputs the same points for each cycle of the sine wave and changes the output frequency by adjusting the rate at which the data are output. It is relatively difficult to obtain fine frequency resolution with this approach, and a modified table lookup method is usually used if fine frequency resolution is desired. It is the latter method that will be described here. The basic idea is to store $N$ uniformly spaced samples of a sine wave in memory and then to output these samples at a uniform rate to a digital-to-analog converter, where they are converted to an analog signal. The lowest-output-frequency waveform then will contain $N$ distinct points. A waveform of twice the frequency can then be generated by using the same data output rate but outputting every other value stored in memory. A waveform $k$ times as fast is obtained by outputting every $k$th point at the same rate. The frequency resolution is the same as the lowest frequency $f_L$. There is an upper frequency limit that is determined by the number of points stored in memory. Theoretically, it is only necessary to output two samples of the sine wave and to recover the fundamental frequency with analog filtering on the output of the D/A converter. Normally four or more points are used in the highest frequency signal, as this somewhat eases the requirements of the analog filter at the output. The architecture of a complete DDFS is shown in Fig. 10.21. The system consists of a phase accumulator, which is simply a digital accumulator, a read-only memory, a reference oscillator, a D/A converter, and a low-pass filter. To generate the lowest frequency, the value 1 is added to the phase accumulator each reference cycle, and the next value from the lookup table is outputted. To output the frequency which is $k$ times as fast as the lowest frequency, the value $k$ is added to the phase accumulator each time and the corresponding value from the lookup table is outputted.

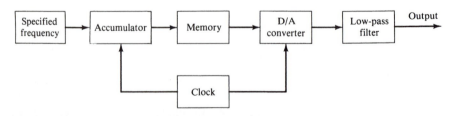

**FIGURE 10.21**
A direct digital frequency synthesizer (DDFS).

To determine the frequency resolution of a DDFS system, consider a $2^N$-bit phase accumulator and a reference clock $f_{clk}$. This phase accumulator can address up to $2^N$ different ROM locations (i.e., samples); and if it is incremented by 1 on each clock edge, then all $2^N$ samples are accessed. Assuming that all the samples are unique (usually the case), the frequency resolution is given by

$$f_r = \frac{f_{clk}}{2^N} \tag{10.18}$$

For example, with a 32-bit accumulator and a 10-MHz clock, the frequency resolution is 2.33 mHz!

If $P$ samples are used to represent the waveform at the highest output frequency $f_{max}$, then $N = (f_{max}/f_{min})$. $P$ samples are used in the lowest-frequency waveform. The number $N$ is limited by the amount of available memory, and $P$, which must be greater than 2, is determined by the output low-pass filtering requirements. For the period of the highest output frequency,

$$T_{max} = \frac{1}{f_{max}} = PT \quad \text{or} \quad f_{max} = \frac{1}{PT}$$

where $T$ is the reference clock's period. Therefore, the highest possible obtainable output frequency is determined by the fastest sampling rate possible. The single most important factor limiting the high-frequency performance of direct frequency synthesizers is the speed of the D/A converter. Not only does it limit the maximum output frequency, but also it introduces noise and harmonic distortion. For frequency synthesizers realized with a microprocessor, the upper frequency limit will be determined by the number of computer clock cycles required to do the phase accumulation and memory lookup transfer. For the new high-speed digital signal-processing integrated circuits, this time can be reduced to less than 20 ns. There is no lower limit on the lowest output frequency with this method. It will be subsequently shown that the lower frequency limit can be extended simply by extending the size of the phase accumulator.

To complete the DDFS, the memory size and length (number of bits) of each word must be determined. Word length is determined by the system noise requirements. The D/A output samples are those of an exact sinusoid corrupted with deterministic noise due to the truncation caused by the finite length of the digital words. It can be shown that if an $(n + 1)$-bit word length is used (including 1 bit as the sign bit), the worst-case noise power (relative to the signal) due to the truncation will be approximately

$$\sigma^2 = (2^n)^{-1} \quad \text{or} \quad \sigma^2 = -6n \quad \text{dB} \tag{10.19}$$

For each bit added to the word length, the spectral purity improves by 6 dB.

**EXAMPLE 10.9.** What word length will be required in a DDFS if the output spectral purity is to be at least 80 dB?

**Solution.** Since the noise power is $-6n$ dB, $n$ must be at least 14. One additional bit is needed for the sign; therefore, the minimum word length needed is 15 bits for an 80-dB signal-to-noise ratio.

For four output samples at the highest frequency, the memory size is determined from Eq. (10.18)

$$N = 4\frac{f_u}{f_L}$$

where $N$ is the number of points in the lowest-frequency sinusoid. Clearly $N$ words of memory would be sufficient for storing the data. However, the amount of memory required can usually be markedly reduced. First, it is only necessary to store the values for the first quadrant (0 to 90°) of the sine wave, since the values for the other three quadrants can be computed directly from these values; so a maximum of $N/4 = f_u/f_L$ memory points is required. The amount of memory can also be reduced by including one or more multipliers; but since multiplication is relatively slow, particularly with microprocessors, and memory is small and inexpensive, multiplication is rarely used to reduce the memory requirements. The amount of memory may still be reduced from that specified by Eq. (10.19) when the spectral purity requirements are not too severe. This point is illustrated in the following example.

**EXAMPLE 10.10.** Design a DDFS to cover the frequency range 0 to 10 kHz with a frequency resolution of at least 0.001 Hz. The spectral purity is to be at least 40 dB.

*Solution.* The use of 8-bit words, including the sign bit, will give a spectral purity of 42 dB $(6 \times 7)$, and this meets the noise specification. Since

$$N = 4\frac{f_u}{f_L} = \frac{4 \times 10^4}{0.001} = 4 \times 10^7 < 2^{26}$$

it appears at first inspection that a large amount of memory is required. However, only $2^8 = 256$ different words can be realized using 8-bit words, so 256 memory locations should suffice. The explanation of this apparent contradiction is that although $4 \times 10^7$ different points are specified, the phase increments $\Delta\theta = \omega T$ are so small that approximately $2^{26} \div 2^8 = 2^{18}$ increments are needed before a change is registered in the 8-bit word. (A 26-bit word would be required to represent all $2^{26}$ words.) The complete design is illustrated in Fig. 10.22. If four samples are used to represent the maximum frequency of 10 kHz, then a 40-kHz clock is required. From Eq. (10.18), for a resolution of at least 1 mHz, a 26-bit phase accumulator is required. However, only 8 bits are needed to address the ROM, and the remainder of the phase accumulator bits is unused.

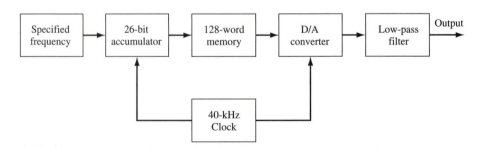

**FIGURE 10.22**
The direct digital frequency synthesizer discussed in Example 10.10.

This procedure is known as *phase accumulator truncation* and adds noise spurs to the output spectrum. The strength of these spurs and their location in the spectrum were first determined by Henry T. Nicholas, III, and Henry Samueli in their paper "An Analysis of the Output Spectrum of Direct Digital Frequency Synthesizers in the Presence of Phase-Accumulator Truncation."[7] The worst-case magnitude of a spur is given by

$$\delta = 2^{T-N} \frac{\pi(\text{FCW}, 2^T)/2^T}{\sin[\pi(\text{FCW}, 2^T)/2^T]}$$

where $T$ is the number of bits truncated, $N$ is the number of phase accumulator bits, FCW is the frequency control word (the amount that the phase accumulator is incremented), and $(\text{FCW}, 2^T)$ represents the greatest common divisor (GCD) of FCW and $2^T$. With a simple hardware modification,[6] the FCW is held relatively prime to $2^T$ and reduces the expression $(\text{FCW}, 2^T)$ to unity. In this case the worst-case spur amplitude is about $-48.2$ dBc for 18 bits of truncation, and the design requirements are still satisfied. Notice that greater frequency resolution could be obtained by increasing the number of phase accumulator bits, but at the cost of greater-magnitude noise spurs.

At the upper frequency limit $f_u$, the output waveform will consist of only four samples, and so it will not look much like a sine wave unless the harmonics of the fundamental frequency are removed by a low-pass filter. This filter should have a bandwidth slightly greater than $f_u$ and a steep attenuation rate outside the passband. Although the harmonic filtering will not be as great for lower-frequency waveforms, these waveforms contain more sample points, and hence the harmonic content will be less.

The main drawback of DDFS is that it is limited to relatively low frequencies. The upper frequency is restricted by the maximum possible clock frequency and the settling time of the D/A converter. In addition, at higher frequencies the power consumption can become large. (Combining DDFS for fine frequency resolution with other synthesis methods to obtain high-frequency performance is discussed in the next section.) DDFS is also spectrally noisier than the other methods. The spectral content of spurs due to phase accumulator truncation changes with each new FCW value, making the low-pass filtering requirements very stringent. More importantly, as seen by Eq. (10.19), the magnitude of the noise spurs at the output is directly affected by the bit resolution of the D/A converter (DAC).

Despite these disadvantages, DDFS systems are easily constructed with conventional components; they are flexible, have fast settling times, maintain phase coherence between frequency steps, and have very good frequency resolution. DDFS systems also offer easy digital modulation (on-off keyed and FM at the accumulator phase between the accumulator and ROM, and AM between the ROM and D/A converter. In addition, since the phase progresses in a linear manner, the phase linearity is the same as the reference clock's progression linearity. A DDFS system is essentially a clock divider, making the phase noise of the system normally less than that of clock (low-phase noise crystal).

Integrated-circuit versions of the DDFS system are available from many vendors including Harris Semiconductor and Analog Devices. For example, the Analog Devices AD9850 is a 28-pin DDFS with a 10-bit DAC and 32-bit phase accumulator, and it can be clocked up to 125 MHz. Current research efforts are focused on improving the settling time and resolution of the DAC, compressing the ROM samples, lowering the power consumption, and reducing the noise spurs at the system output. If these issues can be resolved, the DDFS may become the main frequency synthesis architecture in the future.

## 10.5
### SYNTHESIZER DESIGN EXAMPLE

Consider the design of a frequency synthesizer to cover the range of 198 to 200 MHz in 10-Hz increments. The frequency switching time should be as short as possible. These specifications are typical of those imposed on a synthesizer to be contained in a satellite communication system. Fine frequency resolution and short settling time are required in a system which uses *frequency hopping* as a means of preventing unauthorized reception of the data transmission. A switching time of 10 $\mu$s is realistic today.

There are many systems that can meet the frequency and resolution requirements. They include

1. A single-loop indirect synthesizer
2. A multiple-loop indirect synthesizer
3. A fractional-N synthesizer combined with one or more local oscillators for up conversion
4. A direct frequency synthesizer
5. A PLL-DDFS combination
6. A combination of DDFS, direct, and indirect synthesizers

The following discussion considers some of these possibilities.

The single-loop indirect (PLL) synthesizer is not a good choice for several reasons. First, $N$ must vary from $19.8 \times 10^6$ to $20 \times 10^6$, and programmable dividers that operate at 198 MHz are not yet available. However, a variable-modulus prescaler could be added to reduce the speed requirements of the programmable dividers. Also, the 10-Hz frequency resolution specification requires that the reference frequency be 10 Hz. For a 10-Hz reference frequency, the loop settling time would be on the order of 2.5 s, which is must too slow.

A two-loop synthesizer that can cover the specified frequency spectrum is shown in Fig. 10.23. The output frequency is the sum of the local oscillator frequency $f_L$ and the frequency of VCO 1 ($f_1$). That is,

$$f_o = f_L + f_1 \tag{10.20}$$

The output $f_1$ is found from

$$f_r = \frac{f_1 - f_2/1000}{M} \tag{10.21}$$

and

$$f_r = \frac{f_2 - f_L/2}{N} \tag{10.22}$$

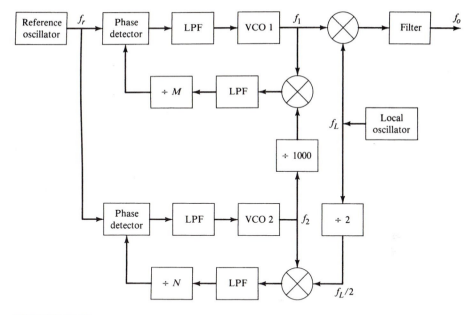

**FIGURE 10.23**
A hybrid frequency synthesizer.

Therefore,
$$f_o = f_L + M_r + \frac{f_2}{1000}$$

$$\text{(10.23)}$$

$$= f_L + Mf_r + N\frac{f_r}{1000} + \frac{f_L}{2000}$$

for a reference frequency $f_r = 10$ kHz, and

$$f_o = 1.0005 f_L + 10^4 M + 10 N$$

where $N$ could be a three-digit decimal number (1 to 999) to select the three least significant digits of the frequency, and $M$ could vary between 0 and 200 to select the three most significant digits. But then the output bandpass filter requirements would be too stringent. Therefore, it is better to place a minimum value on $M$ and reduce $f_L$ such that Eq. (10.23) is satisfied. For example, $M$ could vary from 800 to 1000; then

$$1.0005 f_L = 198 \times 10^6 - 800 \times 10^4$$

or
$$f_L = 189.90504 \times 10^6 \text{ Hz}$$

For this synthesizer the reference frequency for each loop is 10 kHz, so the settling time will be approximately $t_s = 25/10^4 = 2.5$ ms, which is a marked improvement over the single-loop system. If a shorter settling time is required, other alternatives need to be considered. Another possibility would be a three-loop synthesizer. The design of such a system is left as an exercise.

   A direct frequency synthesizer could be designed to meet the specifications, but the hardware would be complex. A direct synthesizer using double-mix-and-divide modules would require seven such modules. Currently, a DDFS cannot be used because of the high output frequency required, but a direct digital frequency synthesizer could be used to obtain the fine frequency resolution with a very short settling time. The DDFS could then be combined with a direct or indirect synthesizer to obtain the high output frequency.

Figure 10.24 illustrates one possible solution to the design problem. The DDFS realizes 1 to 3 MHz in 10-Hz increments. The lower frequency is offset to 1 MHz in order to make the first high-pass filter practical. An infinite variety of frequencies could be selected for the mixing frequencies, but the combination of $f_1 = 7$ MHz, $f_2 = 30$ MHz, and $f_3 = 160$ MHz is one possibility. As previously described, the memory requirements for the DDFS will be determined by the word length, which is determined by the noise specification.

Figure 10.25 illustrates another solution to the problem. The configuration is often referred to as a *direct digital/direct/indirect hybrid synthesizer.* In this system,

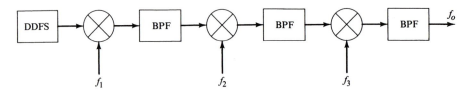

**FIGURE 10.24**
A combination direct/direct digital frequency synthesizer.

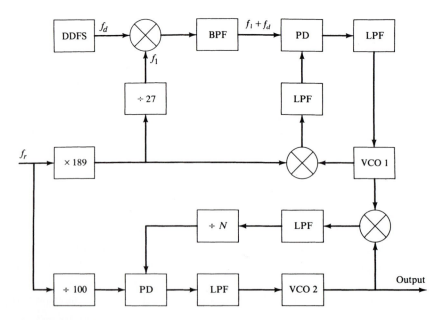

**FIGURE 10.25**
A hybrid frequency synthesizer.

the DDFS generates the frequencies 2.0 to 2.1 MHz in 10-Hz increments $(f_d)$. The minimum frequency of 2.0 MHz is selected to ease the requirements of the bandpass filter. The lower PLL uses a reference frequency of 100 kHz to generate the frequency increments of 100 kHz $(1 \le N \le 21)$. Since the DDFS responds almost instantaneously, the overall settling time is determined primarily by the loop with the 100-kHz reference frequency; the settling time is estimated to be 250 $\mu$s.

## 10.6

### PHASE NOISE

The preceding discussion of frequency synthesizers emphasized that the output noise is an important design consideration. The main sources of output noise in PLL synthesizers are spurious components at the reference frequency and at its harmonics (due to the phase detector) and the noise originating in the VCO. This noise creates a theoretical noise floor, which is a minimum against which actual systems can be compared. As shown earlier, the phase noise for a DDFS is actually less than that of the reference clock.

### A Model for Oscillator Phase Noise

If the power spectral density (power as a function of frequency) is measured at the output of an oscillator, a curve such as that of Fig. 10.7 is observed. Rather than all the power's being concentrated at the oscillator frequency, some of the power is distributed in frequency bands on both sides of the oscillator frequency. These unwanted frequency components are referred to as *oscillator noise*.

Oscillator noise will have a different impact on system performance depending upon the application. The noise of a synthesizer used in a transmitter is transmitted on frequencies above and below the desired frequency of transmission. A similar process occurs in a receiver. The local oscillator phase noise can mix with an unwanted signal to create an unwanted signal in the intermediate-frequency (IF) passband. This process is referred to as *reciprocal mixing*. The phase noise is one of the limiting factors in determining how closely spaced (in the frequency domain) two communication channels can be.

With a spectrum analyzer, it will not be possible to measure the noise characteristics of a signal unless the spectrum analyzer oscillators have substantially less noise than the signal to be measured. Leeson[8] developed a model that describes the origins of phase noise in oscillators, and since it closely fits experimental data, the model is widely used and will be described later in this chapter. First a relation between the observed power spectral density function and $\theta(t)$ will be developed.

The oscillator output $S(t)$ can be expressed by

$$S(t) = V(t) \cos[\omega_o t + \theta(t)] \tag{10.24}$$

where $V(t)$ describes the amplitude variation as a function of time, and $\theta(t)$ is the phase variation [$\theta(t)$ is referred to as *phase noise*.] A well-designed, high-quality oscillator is very amplitude-stable, and $V(t)$ can be considered constant. For a constant amplitude signal, all oscillator noise is due to $\theta(t)$.

A carrier signal of amplitude $V$ and frequency $f_o$, which is frequency-modulated by a sine wave of frequency $f_m$, can be represented by

$$S(t) = V \cos\left(\omega_o t + \frac{\Delta f}{f_m} \sin \omega_m t\right) \tag{10.25}$$

where $\Delta f$ is the peak frequency deviation and $\theta_p = \Delta f / f_m$ is the peak phase deviation, often referred to as the *modulation index* $\beta$. Equation (10.25) can be expanded as

$$S(t) = V[\cos(\omega_o t)\cos(\theta_P \sin \omega_m t) - \sin(\omega_o t)\sin(\theta_p \sin \omega_m t)] \tag{10.26}$$

If the peak phase deviation is much less than 1 $(\theta_p \ll 1)$, then

$$\cos(\theta_p \sin \omega_m t) \approx 1$$

and

$$\sin(\theta_p \sin \omega_m t) \approx \theta_p \sin \omega_m t$$

That is, for $\theta_p \ll 1$, the signal $S(t)$ is approximately equal to

$$S(t) = V[\cos \omega_o t - \sin \omega_o t\,(\theta_p \sin \omega_m t)]$$

$$= V\left\{\cos \omega_o t - \frac{\theta_p}{2}[\cos(\omega_o + \omega_m)t - \cos(\omega_o - \omega_m)t]\right\} \tag{10.27}$$

Equation (10.27) shows that if the peak phase deviation is small, the phase deviation results in frequency components on each side of the carrier of amplitude $V\theta_p/2$. This development has shown that a constant-amplitude signal of frequency $f_o$ phase-modulated with a signal of constant frequency $f_m$ and peak phase deviation $\theta_p$ results in frequency sidebands at the frequencies $f_o \pm f_m$. The ratio of the peak sideband voltage $V_n$ to the peak carrier voltage $V$ is

$$\frac{V_n}{V} = \frac{\theta_p}{2}$$

and the power ratio is

$$\left(\frac{V_N}{V}\right)^2 = \frac{\theta_p^2}{4} = \frac{\theta_{rms}^2}{2} \tag{10.28}$$

It is customary to extend this result to the interpretation of the power spectral density of a constant-amplitude signal. Consider the normalized power spectral density plot shown in Fig. 10.26. If the normalized power spectral density $P_\theta(f)$ is approximately constant over a unit bandwidth, then the power in that bandwidth $S_\theta'$ is

$$S_\theta'(f_m) = \int_{f_m - 1/2}^{f_m + 1/2} P_\theta(f)\,df = P_\theta(f_m) \tag{10.29}$$

Since $P_\theta(f)$ is symmetric about the carrier frequency $f_o$, the power in both sidebands is

$$S_\theta(f_m) \simeq 2P_\theta(f_m) \tag{10.30}$$

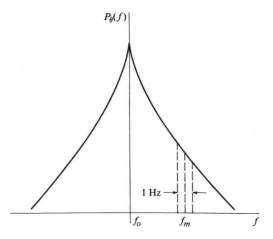

**FIGURE 10.26**
Oscillator-noise-power spectral density.

This noise power is interpreted as due to a phase-modulating noise at the frequency $f_m$:

$$S_\theta(f_m) = \theta_{rms}^2(f_m) = \frac{\theta_p^2(f_m)}{2} \tag{10.31}$$

where $\theta_p$ is the peak value of the phase modulation. Then $S_\theta(f)$ is the ratio of power in the unit bandwidth centered at $f_m$ to the carrier power. With this interpretation of the noise-power spectral density, the noise can now be described in terms of its origins.

We will assume that the oscillator is composed of an amplifier with gain $A$ and a high-$Q$ resonant circuit, as illustrated in block diagram form in Fig. 10.27. Since the gain of the resonant circuit has been normalized to unity at the resonant frequency $f_o$, the amplifier gain $A$ must also be unity in order for the circuit to oscillate. Let $S_\theta$ represent the amplifier-noise-power spectral density referenced to the amplifier input. The available white noise $N$ power per unit bandwidth at the amplifier input is given by

$$N = N_i + N_a = FkT \tag{10.32}$$

where $F$ is the amplifier noise figure. Therefore, the ratio of white noise power per unit bandwidth to signal power $P_s$ is $FkT/P_s$, which is a component of $S_\theta$. In

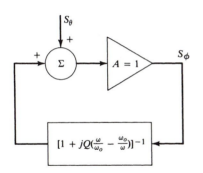

**FIGURE 10.27**
A model used to characterize oscillator noise.

addition, amplifiers generate an additional "flicker," or $1/f$ phase noise, about the carrier frequency, assumed probably to be due to carrier density fluctuations in the transistor.

A plot of a typical $S_\theta$ spectrum is contained in Fig. 10.28. For frequencies below $f_\alpha$, $S_\theta$ has a $1/f$ spectrum. At higher frequencies, the spectrum is flat and equal to $FkT/P_s$. The frequency $f_\alpha$ below which the spectrum has a $1/f$ shape depends upon the characteristic of the individual amplifier. For the circuit of Fig. 10.28 with positive feedback and an $A$ of 1, the closed-loop steady-state transfer function between the amplifier input and output is given by

$$\frac{C(j\omega)}{S(j\omega)} = [1 - H(j\omega)]^{-1} \tag{10.33}$$

where

$$H(j\omega) = \left[1 + jQ\left(\frac{\omega}{\omega_o} - \frac{\omega_o}{\omega}\right)\right]^{-1} \tag{10.34}$$

Since $H(j\omega)$ is a high-$Q$ filter and we are interested in describing the noise power distribution about the center frequency $\omega_o$, $H(j\omega)$ can be replaced by its low-pass equivalent.[9]

$$H_L(j\omega) = \left(1 + \frac{j\omega}{\omega_L}\right)^{-1} \tag{10.35}$$

where

$$\omega_L = \frac{\omega_o}{2Q} \tag{10.36}$$

is the equivalent bandwidth. The noise-power spectral density $S_o(\omega)$ at the output of a filter with a voltage transfer function $G(\omega)$ in terms of the spectral density $S_i(\omega)$ of the input noise, is given by

$$S_o(\omega) = S_i(\omega)|G(\omega)|^2 \tag{10.37}$$

Therefore, the ratio of the equivalent phase noise-to-signal power $S_\phi$ of the closed-loop system measured at the output of the unity-gain amplifier is

$$S_\phi = S_\theta[|1 - H(\omega)|^2]^{-1}$$

$$= \frac{S_\theta}{[1 - (1 + j\omega/\omega_L)^{-1}][1 - (1 - j\omega/\omega_L)^{-1}]}$$

$$= \frac{S_\theta(1 + \omega^2/\omega_L^2)}{\omega^2/\omega_L^2} \tag{10.38}$$

$$= S_\theta\left(1 + \frac{\omega_L^2}{\omega^2}\right)$$

which can be written [using Eqs. (10.28), (10.32), and (10.36)] as

$$S_\phi(\omega) = S_\theta(\omega)\left(1 + \frac{\omega_o^2}{4Q^2\omega^2}\right) \tag{10.39}$$

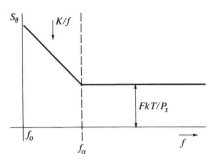

**FIGURE 10.28**
Amplifier-noise-power spectral density.

This is the expression proposed by Leeson[8] for describing the noise at the output of an oscillator.

For $S_\theta$ as depicted in Fig. 10.28, the output phase noise spectrum $S_\phi$ is as shown in Fig. 10.29a, provided the filter bandwidth is greater than the $f_\alpha$ of the amplifier. At frequencies close to the carrier frequency $f_o$, the noise power decreases with a $1/f^3$ (−18 dB per octave) slope; between $f_\alpha$ and $f_L$ the power spectral density (PSD) decreases as $1/f^2$ (−12 dB per octave); and for frequencies above the filter bandwidth, the output phase noise is white. If the filter bandwidth $f_L$ is less than $f_\alpha$, the noise PSD is as depicted in Fig. 10.29b. In this case the PSD decreases as $1/f$ (−6 dB per octave) for frequencies between $f_L$ and $f_\alpha$ and is independent of frequency for frequencies greater than $f_\alpha$. Equation (10.39) provides a quantitative measure for comparing an oscillator's noise performance to a theoretical minimum based on the amplifier's noise figure and $f_\alpha$.

At high frequencies, the oscillator noise floor is proportional to the noise figure of the amplifier used in realizing the oscillator. Since the minimum noise figure is 1, the minimum noise floor is $kT$, or −174 dB/Hz. At lower frequencies close to the oscillating frequency, the noise increases, but it is seen from Eq. (10.39) that the actual amplitude is inversely proportional to $Q^2$ of the resonator. So the higher the $Q$, the smaller will be the phase noise near the oscillating frequency.

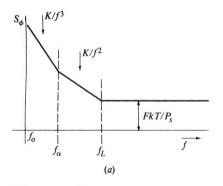

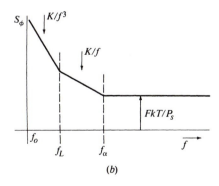

(a)          (b)

**FIGURE 10.29**
(a) Output noise of an oscillator with a low-Q resonator; (b) output noise of an oscillator with a high-Q resonator.

## Phase Noise in Phase-Locked Loops

The synthesizer output noise due to the noise generated in the VCO can be determined (assuming a linear model) with Eqs. (10.37) and (10.39). It is

$$
\begin{aligned}
S_{\phi_o} &= \frac{S_\theta(\omega)[1 + \omega_o^2/(4Q^2\omega^2)]}{|1 + K_v F(j\omega)/N(j\omega)|^2} \\
&= \frac{S_\phi(\omega)}{|1 + K_v F(j\omega)/(\omega N)|^2} \\
&= \frac{FkT/\{P_s(1 + \omega_\alpha/\omega)[1 + \omega_o^2/(4Q^2\omega^2)]\}}{|1 + K_v F(j\omega)/(j\omega N)|^2}
\end{aligned}
\tag{10.40}
$$

The denominator can be approximated by

$$
[|1 + G(j\omega)|^2]^{-1} \approx [|G(j\omega)|^2]^{-1} \qquad \text{for } |G(j\omega)| \gg 1
$$

(that is, inside the loop bandwidth). For frequencies above the loop bandwidth,

$$
[|1 + G(j\omega)|^2]^{-1} \approx 1 \qquad \text{for } |G(j\omega)| \ll 1
$$

The frequency at which the two approximations coincide is the open-loop crossover frequency, which is approximately the closed-loop bandwidth. Therefore, for frequencies higher than the open-loop crossover frequency, the closed-loop noise due to the VCO noise $S_{\phi_o}$ is approximately the same as the VCO noise, or

$$
S_{\phi_o} \approx S_\phi(\omega) \qquad \omega > \omega_c
\tag{10.41}
$$

and for lower frequencies

$$
S_{\phi_o} \approx S_\phi(\omega) \frac{\omega^2 N^2}{K_v^2 |F(j\omega)|^2}
\tag{10.42}
$$

If a type I loop is employed with a filter bandwidth greater than $\omega_c$ and the VCO output phase noise spectrum is as shown in Fig. 10.29a, then the synthesizer output noise spectrum due to the VCO noise will be as shown in Fig. 10.30. The open-loop noise power spectrum decreases at $-18$ dB per octave, so the closed-loop noise power spectrum decreases at a rate of $-6$ dB per octave until the frequency $\omega_\alpha$; then

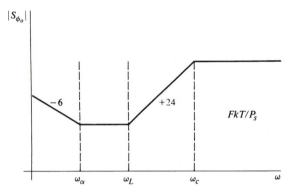

**FIGURE 10.30**
Output noise power spectrum of a type I PLL, due to VCO noise.

the noise power spectrum does not change with frequency until the PLL filter frequency is reached. The noise power density then increases at $+12$ dB per octave until the open-loop crossover frequency $\omega_c$ is reached. At higher frequencies, the power spectral density plot assumes the same shape as $S_\phi$.

For a type II loop, the open-loop transfer function is

$$G(j\omega) = \frac{K_v(j\omega/\omega_z + 1)}{N(j\omega)^2} \tag{10.43}$$

so the closed-loop noise spectrum (due to the VCO noise) will increase with a rate of $+6$ dB per octave at low frequencies. The shape of the normalized closed-loop noise power spectrum (due to the VCO noise only) is shown in Fig. 10.31. The slope increases from $+6$ to $+12$ dB per octave at $\omega_\alpha$ and to $+24$ dB per octave at $\omega_L$, then decreases to $+12$ dB per octave at $\omega_z$, and assumes the same shape at $S_\phi$ for frequencies above $\omega_c$. The type II system provides substantially more filtering of the VCO noise within the loop bandwidth.

**Effect of Frequency Division and Multiplication on Phase Noise**

Equation (10.25) states that the instantaneous phase $\theta_i(t)$ of a carrier frequency modulated by a sine wave of frequency $f_m$ is given by

$$\theta_i(t) = \omega_o t + \frac{\Delta f}{f_m} \sin \omega_m t$$

*Instantaneous frequency* is defined as the time rate of change of phase, or

$$\omega = \frac{d\theta_i}{dt} = \omega_o + \frac{\Delta f}{f_m}\omega_m \cos \omega_m t \leq \omega_o + \Delta\omega \tag{10.44}$$

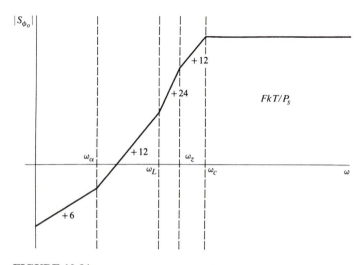

**FIGURE 10.31**
Output noise power spectrum of a type II PLL, due to VCO noise.

If this signal is passed through a frequency divider that divides the frequency by $N$, the output frequency $\omega^1$ will be given by

$$\omega^1 = \frac{\omega_o}{N} + \frac{\Delta\omega}{N} \cos \omega_m t$$

and the output phase by

$$\theta^1(t) = \frac{\omega_o t}{N} + \frac{\Delta f}{N f_m} \sin \omega_m t \qquad (10.45)$$

The divider reduces the carrier frequency by $N$, but does not change the frequency of the modulating signal. The peak phase deviation is reduced by the divide ratio $N$. Since it was shown in Eq. (10.28) that the ratio of the noise power to carrier power is

$$\left(\frac{V_n}{V}\right)^2 = \frac{\theta_p^2}{4}$$

frequency division by $N$ reduces the noise power by $N^2$.

**EXAMPLE 10.11.** The indirect frequency synthesizer shown in Fig. 10.32 is used to generate a 5-GHz ($5 \times 10^9$) signal. A 1-kHz reference signal is obtained from a 5-MHz reference oscillator ($M = 5000$) which is specified to have a signal sideband power of $-140$ dB/Hz at a frequency separation of 0.5 kHz from the oscillator's operating frequency. What will be the single sideband noise power (due to the input noise) at this frequency?

*Solution.* If the loop bandwidth is assumed to be approximately 1 kHz, then the noise from the reference oscillator 0.5 kHz from the carrier frequency will not be reduced by the low-pass filtering of the PLL. The approximate loop transfer function is

$$\theta_o = \frac{\theta_r K_v F(s)/s}{1 + K_v F(s)/(sN)} \simeq N\theta_r = \frac{N}{M}\theta_i \qquad (10.46)$$

for reference frequencies below the loop bandwidth of 1 kHz . Although the divider $M$ reduces the input noise power, the net effect is that the output noise power is the reference oscillator noise power multiplied by $(N/M)^2$. So $N$ must equal $5 \times 10^6$ in order to obtain the output frequency of $5 \times 10^9$ Hz, and the output noise power due to the

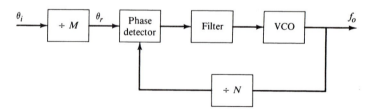

**FIGURE 10.32**
A PLL for synthesizing high-frequency signals.

reference oscillator is therefore

$$N_o = -140 \text{ dB/Hz} + 10 \log \left( \frac{5 \times 10^6}{5 \times 10^3} \right)^2 = -80 \text{ dB/Hz}$$

at a frequency offset of 0.5 kHz.

Actually, the noise would be much worse than predicted in the previous example. The reference oscillator is already a low-noise device, and the noise cannot be reduced below some noise floor by additional division. This illustrates a problem inherent in PLL frequency synthesizers used to generate an output frequency much higher than the reference oscillator frequency. Although the reference oscillator noise power may be small, the same noise power appears on the output signal amplified by the factor $N^2$, where $N$ is the ratio of the output frequency to the reference oscillator frequency.

## ▮ 10.7

### PROBLEMS

**10.1** A relationship was developed in Chap. 4 that expressed the filter attenuation of the $n$th multiple of the center frequency of a second-order tuned circuit in terms of the circuit $Q$. For the frequency synthesizer shown in Fig. 10.4, what must be the $Q$ of the first bandpass filter in order to obtain 80-dB suppression of the other mixer product?

**10.2** Design a direct frequency synthesizer to generate $15.8 \times 10^6$ Hz from a $1 \times 10^6$ Hz reference oscillator.

**10.3** Design a direct frequency synthesizer, using double-mix-and-divide modules, to cover the frequency spectrum of 25 to 29.999 MHz in 1-kHz increments. Specify all frequencies used and the maximum frequency mixing ratio of all mixers.

**10.4** Design a phase-locked loop synthesizer to meet the specifications of Prob. 10.3.

(*a*) What is the reference frequency?

(*b*) What is the range of the divide ratio $N$?

(*c*) The 25-MHz output frequency is too high for a programmable divider, and a variable-modulus divider should be used. Use a 10/11 divider, and determine the initial values of the counters required to synthesize 26.111 MHz.

**10.5** Could a 100/101 variable-modulus divider be used in Prob. 10.4? Explain your answer.

**10.6** For the three-loop frequency synthesizer illustrated in Fig. 10.15 determine $N_A$ and $N_B$ required to obtain an output frequency of $38.912 \times 10^6$ Hz.

**10.7** Design a multiple-PLL synthesizer to cover the frequency range of 35.4 to 40.0 MHz to 10-Hz increments. The reference frequency is to be 100 kHz. No loop should operate with a reference frequency below 100 kHz.

**10.8** Use a fractional-$N$ frequency synthesizer to synthesize a frequency of 2.33 kHz using a 1-kHz reference frequency.

(*a*) What must be the size of the phase accumulator?

    (*b*) Sketch the output of the phase detector ($N = 2, M = 3$).

    (*c*) Sketch the phase-detector output if the output frequency is to be 4.33 kHz ($N = 4, M = 3$).

    (*d*) What must the phase accumulator size be to realize a frequency resolution of 10 Hz?

**10.9** It is desired to design a direct digital frequency synthesizer with a maximum output frequency of 10 kHz and a step size of 10 Hz.

    (*a*) What must the clock frequency be if four samples are to be used in the highest-frequency waveform?

    (*b*) How many bits must the accumulator contain?

    (*c*) What number must be added to the accumulator at each cycle to generate an output frequency of 100 kHz?

    (*d*) What must the accumulator size be if the minimum step size is to be reduced to 1 Hz?

    (*e*) What word length is required for a 5-dB signal-to-noise ratio?

**10.10** Design a DDFS to cover the frequency range of 0 to 5 kHz in 0.01-Hz increments. The spectral purity is to be at least 50 dB. Specify the accumulator size, memory requirements, sampling rates, and characteristics of the output low-pass filter.

**10.11** Design a frequency synthesizer to cover the frequency range of 100 to 100.999 MHz in 10-Hz increments. The frequency switching time is to be less than 100 $\mu$s. Discuss different configurations which can be used to meet the specifications.

**10.12** Design a three-loop synthesizer to cover the frequency range of 198 to 200 MHz with a frequency resolution of 10 Hz. The loop frequency switching time should be as short as possible.

**10.13** Figure P10.13 illustrates another method for covering the 198- to 200-MHz frequency range with a frequency resolution of 10 Hz. Select frequency ranges for the

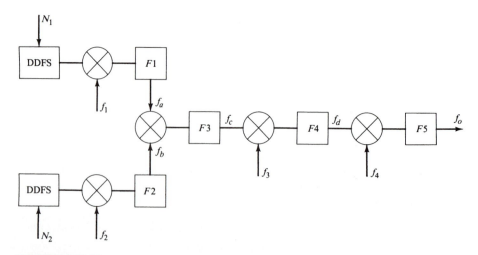

**FIGURE P10.13**
A hybrid direct/direct digital frequency synthesizer.

two direct digital frequency synthesizers, and specify the frequencies of the four oscillators used to mix the frequency up to the specified operating frequency.

**10.14** A 10-MHz oscillator has a noise figure of 10 dB, an $f_a$ of 20 kHz, and a $Q$ of 100.

    (*a*) What is the noise power (relative to the carrier) in a 1-Hz bandwidth 1 MHz above the carrier frequency?
    (*b*) What is the relative noise power 100 kHz above the carrier frequency?
    (*c*) What is the relative noise power 10 kHz above the carrier frequency?

**10.15** If $f_a$ in Prob. 10.14 had been 1 MHz above the carrier frequency, what would have been the answers to parts (*a*), (*b*), and (*c*) of the problem?

**10.16** A 10-MHz reference oscillator is to be used with a phase-locked loop to synthesize an output frequency of $10^9$ Hz. If the output noise is to be $-100$ dB/Hz relative to the carrier, what must be the output noise level of the reference oscillator?

**10.17** Discuss various methods of realizing a frequency synthesizer to cover the frequency spectrum of 10 Hz to 1 kHz in 10-Hz increments. The discussion should include the reference frequency used in each case, the synthesizer switching speed, and the advantages and disadvantages of each method.

**10.18** Design an indirect frequency synthesizer, using the Motorola MC145152-2 and MC12017 prescaler that generates a 50-MHz output from a 10-kHz reference input.

**10.19** Design a DDFS to cover the frequency range of 0 to 1 MHz with a frequency resolution of at least 0.001 Hz. The spectral purity is to be at least 55 dB.

## ▓ 10.8
### REFERENCES

1. H. J. Finden, The Frequency Synthesizer, *J. IEEE,* **90**(3): 165–180, 1943.
2. M. E. Peterson, The Design and Performance of an Ultra Low-Noise Digital Frequency Synthesizer for Use in VLF Receivers, *Proc. 26th Annual Symposium on Frequency Control.* U.S. Army Electronics Command, Fort Monmouth, N.J., Natl. Tech. Infor. Serv. Accession Nr. AD 771043, 1972, pp. 55–70.
3. G. C. Gillette, The Digiphase Synthesizer, *Proc. 23rd Annual Symposium on Frequency Control,* U.S. Army Electronics Command, Fort Monmouth, N.J., Natl. Tech. Infor. Serv. Accession Nr. 746209, 1969, pp. 201–210.
4. J. Gibbs, and R. Temple, Frequency Domain Yields Its Data to Phase-Locked Synthesizer, *Electronics,* April 27, 1978, pp. 107–113.
5. J. Tierney, C. M. Rader, and B. Gold, A Digital Frequency Synthesizer, *IEEE Trans. on Audio and Electroacoustics,* AU-19, 1971, pp. 48–57.
6. Nicholas, Henry T., III, and Henry Samueli: A 150-MHz Direct Digital Frequency Synthesizer in 1.25-$\mu$m CMOS with $-90$-dBc Spurious Performance, *IEEE Journal of Solid-State Circuits,* December 1991, Vol. 26, No. 12, pp. 1959–1969.
7. Nicholas, Henry T., III, and Henry Samueli: An Analysis of the Output Spectrum of Direct Digital Frequency Synthesizers in the Presence of Phase-Accumulator Truncation, *41$^{st}$ Annual Frequency Control Symposium,* 1987, pp. 495–502.

8. D. B. Leeson, A Simple Model of Feedback Oscillator Noise Spectrum, *Proc. IEEE,* **54:** 329–330, 1966.

9. E. K. Clark, and D. Hess, *Communication Circuits: Analysis and Design,* Addison-Wesley, Reading, Mass., 1971.

## ■ 10.9

### ADDITIONAL READING

### Frequency Synthesis

Apetz, B., B. Scheckel, and G. Weil: A 120 MHz AM/FM PLL-IC with Dual on-Chip Programmable Charge Pump/Filter Op-Amp, *IEEE Trans. on Cons. Elect.,* **27:** 234–242, 1981.

Beyers, B. E.: Frequency Synthesis Tuning Systems with Automatic Offset Tuning, *IEEE Trans. on Ce,* **CE-24** (3): 419–428, August 1978.

Bjerede, B. E., and G. Fisher: An Efficient Hardware Implementation for High Resolution Frequency Synthesis, *Proc. 31st Ann. Symp. on Frequency Control,* U.S. Army Electronics Command, Fort Monmouth, N.J., Natl. Tech. Infor. Serv. Accession Nr. AD 771043, 1977, pp. 318–321.

――― and ―――: A New Phase Accumulator Approach for Frequency Synthesis, *Proc. IEEE NAECON,* 76, May 1976, pp. 928–932.

Blachowicz, L. F.: Dial Any Channel to 500 MHz, *Electronics,* May 2, 1966, pp. 60–69.

d'Andrea, G., V. Libal, and G. Weil: Frequency Synthesis for Color TV-Receivers with a New Dedicated $\mu$ Computer, *IEEE Trans. on Cons. Elect.,* **27:** 272–283, 1981.

Dayoff, I., and B. Krischner: A Bulk CMOS 40-Channel CB Frequency Synthesizer, *IEEE Trans. on CE,* **CE-23** (4): 518–521, November 1977.

Egan, W. F.: *Frequency Synthesis by Phase Lock,* Wiley, New York, 1981.

Fukui, K.: A Portable All-Band Radio Receiver Using Microcomputer Controlled PLL Synthesizer, *IEEE Trans. on CE,* **CE-26:** 299–310, August 1980.

Furuno, K., S. Mitra, K. Hirano, and Y. Ito: Design of Digital Sinusoidal Oscillators with Absolute Periodicity, *IEEE Trans. AES,* **11:** 1286–1299, 1975.

Gorski-Popiel, J.: *Frequency Synthesis: Techniques and Applications,* IEEE, New York, 1975.

Ichinose, K.: One Chip AM/FM Digital Tuning System, *IEEE Trans. on CE,* **CE-26:** 282–288, August 1980.

Howald, Robert: Introduction to Analog and Direct Digital Frequency Synthesis, *RF Design,* January, 1995, p 78.

Kroupa, V. F.: *Frequency Synthesis: Theory, Design and Applications,* Wiley, New York, 1976.

Manassewitsch, V.: *Frequency Synthesizers: Theory and Design,* Wiley, New York, 1981.

Mills, T. B.: An AM/FM Digital Tuning System, *IEEE Trans. on CE,* **CE-24** (4): 507–513, November 1978.

Mueller, K. J., and C. P. Wu: A Monolithic : ECL/$I^2$L Phase-Locked Loop Frequency Synthesizer for AM/FM TV, *IEEE Trans. on CE,* **CE-25** (3): 670–676, August 1979.

Rhodes, R., W. Hutchinson, and B. Hutchinson: Frequency Agile Phase-Locked Loop Synthesizer for a Communications Satellite. *Proc. Natl. Telecommunications Conference,* 1980, pp. 22.3.1–22.3.6.

Rohde, U. L.: *Digital PLL Frequency Synthesizers,* Prentice-Hall, Englewood Cliffs, N.J., 1983.

Rohde, U. L.: Modern Design of Frequency Synthesizers, *Ham Radio,* July 1976, pp. 10–22.

Rzezewski, T., and T. Kawasaki: A Microcomputer Controlled Frequency Synthesizer for TV, *IEEE Trans. on CE,* **CE-24** (2): 145–153, May 1978.

Sample , L.: A Linear CB Synthesizer, *IEEE Trans. on CE,* **CE-23** (3): 200–206, August 1977.

Stinehelfer, J., and J. Nichols: A Digital Frequency Synthesizer for an AM and FM Receiver, *IEEE Trans.,* **BTR-15:** 235–243, 1969.

Tanaka, K., S. Ike Guichi, Y. Nakayama, and Osamuu Ikeda: New Digital Synthesizer LSI for FM/AM Receivers, *IEEE Trans. on Cons. Elect.,* **27:** 210–219, 1981.

Yamada, T.: A High Speed NMOS PLL-Synthesizer LSI with On-Chip Prescaler for AM/FM Receivers, *IEEE Trans. on CE,* **CE-26:** 289–298, August 1980.

Yuen, G. W. M.: An Analog-Tuned Digital Frequency Synthesizer Tuning System for FM/AM Tuner, *IEEE Trans. on CE,* **CE-23** (4): 507–513, November 1978.

**Phase Noise**

Baghdady, E. J., R. N. Lincoln, and B. D. Nelin: Short-Term Frequency Stability: Characterization, Theory, and Measurements, *Proc. IEEE,* **53:** 704–722, 1965.

Barnes, J. A., and R. C. Mockler: The Power Spectrum and Its Importance in Precise Frequency Measurements, *IRE Trans. on Instrumentation,* **9:** 149–155, 1960.

——— et al.: *Characterization of Frequency Stability,* U.S. Dept. of Comm., NBS Technical Note 394, 1970.

Cutler, L. S., and C. L. Searle: Some Aspects of the Theory and Measurement of Frequency Fluctuations in Frequency Standards, *Proc. IEEE,* **54:** 136–154, 1966.

Edson, W. A.: Noise in Oscillators, *Proc. IRE,* **48:** 1454–1466, 1960.

Hafner, Erich: The Effects of Noise in Oscillators, *Proc. IEEE,* **54:** 179–198, 1966.

Leeson, D. B.: Short-Term Stable Microwave Sources, *Microwave J.,* June 1970, pp. 59–69.

———: A Simple Model of Feedback Oscillator Noise Spectrum, *Proc. IEEE,* **54:** 136–154, 1966.

Lindsey, W. C., and C. M. Chie: Specification and Measurement of Oscillator Phase Noise Instability, *Proc. Ann. Frequency Control Symp.,* 1981, pp. 302–310.

Reynolds, Chuck: Measure Phase Noise, *Electronic Design,* February 15, 1977, pp. 106–108.

Rutman, Jaques: Characterization of Frequency Stability: A Transfer Function Approach and Its Application to Measure via Filtering of Phase Noise, *IEEE Trans. on Instrumentation and Measurement,* **22:** 40–48, 1974.

# 11

# Power Amplifiers

## INTRODUCTION

Power amplifiers are those amplifiers whose design concerns are based on a combination of output power, drive level, power dissipation, distortion, size, weight, and efficiency (power output divided by power supplied). Simultaneously, the transistors used in power amplifiers have requirements based on breakdown voltage, current limitations, and maximum power dissipation. The output power of power amplifiers can range from the milliwatt region for small, portable transistor amplifiers to the megawatt region for large broadcast stations.

The power amplifier is invariably the last stage in the amplifier chain because the power level is highest at this point; no intermediate-state amplifier would be operated with a power gain significantly less than 1. Because the signal level is the largest at this point, it results in the maximum amount of distortion due to the nonlinear characteristics of the device. These nonlinearities produce unwanted frequency components (harmonics) and intermodulation distortion (IMD) products. However, there are various methods of designing circuits, methods which lead to different levels of efficiency and create different amounts of distortion. Because various modulation techniques can tolerate different amounts of distortion, power amplifier design depends on the type of signal (modulation) to be amplified. In the least efficient design, the maximum amount of power is dissipated in the transistor, requiring larger and more expensive transistors than would otherwise be the case. However, good power amplifier design techniques can result in more economical and reliable electronics.

Power amplifiers are classified according to their mode of operation; the most frequently used classes are discussed in this chapter. The class of operation is determined by how the transistor is biased and the nature of the output circuit.

The original classification of operating modes included class A, B, AB, and C amplifiers. A class A amplifier is a linear amplifier. Theoretically, it will produce a sine wave output in response to a sine wave input. The output frequency will be the same as the input frequency, and the output amplitude will be a linear function of the input amplitude. If the amplifier output is a linear function of the input over 50 percent (180°) of the input waveform, it is categorized as class B. If the linear conduction angle is less than 50 percent, it is class C; and if the conduction angle is greater than 180° but less than 360°, it is referred to as class AB.

Today, additional classes of amplifiers exist, most of them using the transistor as a switch. The more popular forms of switching amplifiers are considered after the class A, B, and C amplifiers have been described.

## ▥ 11.2
### CLASS A AMPLIFIERS

Class A power amplifiers are no different in behavior from the linear amplifiers studied up to this point, except that their power and distortion levels are of primary importance. For class A operation, the output will be a sine wave in response to a sine wave input. The class of generation is determined by the input signal level and how the transistor is biased. Figure 11.1 describes an ac-coupled amplifier which can be biased for class A, B, or C.

For the amplifier shown in Fig. 11.1, the transistor quiescent voltage (no collector alternating current) is

$$V_{ce} = V_Q = V_{cc} - I_c R_E \tag{11.1}$$

The slope of the dc load line $dI_c/dV_{ce}$ is then $-1/R_E$, as illustrated in Fig. 11.2. In class A power amplifiers, the dc resistance $R_E$ will normally be much less than the ac resistance. Resistance $R_E$ is kept small to limit the dc power dissipated in the

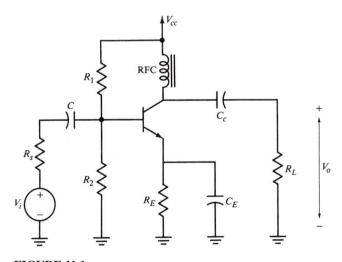

**FIGURE 11.1**
An ac-coupled amplifier.

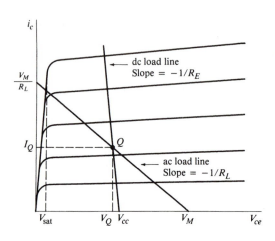

**FIGURE 11.2**
Transistor collector-emitter characteristics including the ac and dc load
lines.

resistor. In some designs $R_E$ is zero, but a small $R_E$ is often included for bias stabilization of bipolar transistor amplifiers and self-biasing of field-effect transistor
amplifiers. If $R_E$ is small, the $Q$ (quiescent) point voltage is

$$V_Q \approx V_{cc}$$

For ac operation, the coupling and emitter bypass capacitors act as short circuits, and the collector alternating current $i_c$ is given by

$$-i_c R_L = V_o$$

The output voltage is equal to the ac voltage drop across the transistor output:

$$V_{cc} - i_c R_E + V_{ce} \approx V_Q - i_c R_L \tag{11.2}$$

From the transistor characteristics of Fig. 11.2, it is seen that the collector-to-emitter
voltage is at a maximum when the collector current is zero. This occurs when the ac
component of collector current is equal in magnitude and opposite in direction to
the collector direct current ($i_c = -I_C$). Likewise, the collector-to-emitter voltage is
zero when the collector current is at a maximum (the ac component is equal in magnitude and direction to the dc component $I_C$). As the collector current decreases
from the $Q$ point due to the ac signal, the collector-to-emitter voltage increases
from $V_Q$ to $V_M$; and as the collector current increases, the collector voltage decreases from $V_Q$ to zero.

   For small-signal amplifiers, it is well known that the maximum power gain is
obtained by matching the load impedance to the transistor output impedance. In
power amplifiers, the objective is to obtain maximum output power, not maximum
gain. The amplifiers are operated at less-than-maximum gain. This requires a large
input drive signal but results in greater output power. For class A operation, the
maximum output power is obtained by selecting the ac load impedance such that the
maximum signal swing can be obtained from the device. The output must be symmetric to avoid distortion. In an ideal transistor $V_{sat} = 0$, the collector-to-emitter

voltage can decrease from $V_{cc}$ to zero; so for symmetric operation it can also increase to

$$V_M = 2V_Q = 2V_{cc}$$

Likewise, the output current can decrease from $I_Q$ to zero, so the maximum output current is

$$I_M = \frac{V_M}{R_L} = \frac{2V_{cc}}{R_L}$$

and

$$I_Q = \frac{V_{cc}}{R_L}$$

The slope of the ac load line is $-1/R_L$ and

$$-R_L^{-1} = -\frac{I_M}{V_M} = \frac{-I_Q}{V_{cc}}$$

This is the value of the load resistance that results in maximum output power.

A surprising characteristic of this amplifier is that the maximum voltage dropped across the transistor is twice the supply voltage, and the peak-to-peak output voltage is also $2V_{cc}$. How this is possible can be seen from the equivalent circuit shown in Fig. 11.3. Here the RF choke has been replaced by a constant current source since no alternating current flows through the device and the direct current is $I_Q$. Also, no direct current flows through the capacitor; therefore the dc voltage drop across the device is $V_{cc}$, and there is no ac voltage drop across the capacitor. Here the capacitor can be replaced by a battery $V_{cc}$. At the instant at which the load current is

$$i_L = -i_C = -I_Q = \frac{V_{cc}}{R_L}$$

the output voltage is

$$V_o = I_Q R_L = V_{ce}$$

If the emitter-resistor is sufficiently small,

$$V_{ce} = V_o + V_{cc}$$

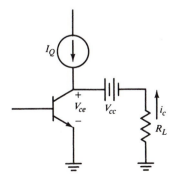

**FIGURE 11.3**
A small-signal equivalent circuit of the amplifier shown in Fig. 11.1.

so
$$(V_{ce})_{max} \approx 2V_{cc}$$

For a sinusoidal input signal, the collector current consists of a dc component $I_Q$, which does not flow through the load, and an ac component $i_L$, which does flow through the load. That is,

$$i_C = I_Q + i_L$$

where $i_L = I_p \sin \omega t$. Since the collector current cannot become negative, $I_p \leq I_Q$. The average power will be

$$(P_o)_{av} = \frac{I_p^2 R_L}{2} \leq \frac{I_Q^2 R_L}{2}$$

and the power supplied by the battery (neglecting the small amount of power dissipated in the base bias circuitry) will be

$$P_{cc} = V_{cc} I_Q = \frac{V_{cc}^2}{R_L}$$

so the efficiency is

$$\eta = \frac{P_o}{P_{cc}} = \frac{I_p^2 R_L^2}{2V_{cc}^2} \tag{11.3}$$

The maximum efficiency occurs for $I_p = I_Q$ and is

$$\eta_{max} = \frac{I_Q^2 R_L^2}{2V_{cc}^2} = 50\%$$

The maximum operating efficiency for a class A ac-coupled power amplifier is 50 percent and occurs with the maximum input signal. If the output signal decreases, $I_p$ decreases and so does the efficiency.

The power dissipated in the transistor is

$$P_T = P_{cc} - P_o = \frac{V_{cc}^2}{R_L} \left( 1 - \frac{I_p^2}{2I_Q^2} \right) \tag{11.4}$$

The maximum power dissipated in the transistor

$$(P_T)_{max} = \frac{V_{cc}^2}{R_L} = 2(P_o)_{max} \tag{11.5}$$

occurs when there is no input signal, and it is equal to twice the maximum power that can be delivered to the load. Also, as previously discussed, the maximum collector-to-emitter voltage is twice the supply voltage.

**EXAMPLE 11.1.** Design a class A amplifier to deliver 5 W to a 50-$\Omega$ load.

*Solution.* For 5-W power in a 50-$\Omega$ resistor, the peak value of the sinusoidal voltage across the resistor is

$$V_p = (2 \times 5 \times 50)^{1/2} = 22.4 \text{ V}$$

Since the peak voltage cannot exceed $V_{cc}$, a standard supply $V_{cc} = 24$ V is suitable. The corresponding peak value of the ac load is

$$I_p = \frac{22.4}{50} = 0.448 \text{ A}$$

Therefore, the transistor must be biased so that

$$I_Q \geq 0.448 \text{ A}$$

For a 24-V supply with no emitter (or source) resistance,

$$I_Q = \frac{V_{cc}}{50} = 0.48 \text{ A}$$

The power supplied is then

$$P_{cc} = V_{cc} I_Q = 11.52 \text{ W}$$

and the efficiency is

$$\eta = \frac{5}{11.52} = 43.4\%$$

The transistor that is selected must be able to dissipate 11.52 W in case the input power drops to zero, and the transistor collector-to-emitter breakdown voltage must be at least 48 V $(2V_{cc})$.

The effect of the saturation voltage is to create signal distortion and reduce efficiency. The maximum value of collector current that can be applied without $V_{ce}$ decreasing to $V_{\text{sat}}$ has the peak value

$$(I_p)_{\text{max}} = \frac{V_{cc} - V_{\text{sat}}}{R_L}$$

Therefore, when the saturation voltage is considered, the maximum efficiency is

$$\eta_{\text{max}} = \frac{1}{2} \left( 1 - \frac{V_{\text{sat}}}{V_{cc}} \right)^2 \tag{11.6}$$

## Transformer-Coupled Class A Amplifiers

If a load impedance is specified, a transformer can be used to improve the power gain by transforming the load impedance to that required for maximum output power. A class A transformer-coupled FET amplifier is shown in Fig. 11.4, and the load lines for the amplifier are shown in Fig. 11.5. The characteristic curves are similar to those of the bipolar transistor amplifier, except that the controlling signal for the field-effect transistor is the gate-to-source voltage.

If the transformer is ideal,

$$i_L = N i_d$$

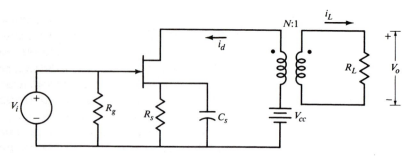

**FIGURE 11.4**
A transformer-coupled class A amplifier.

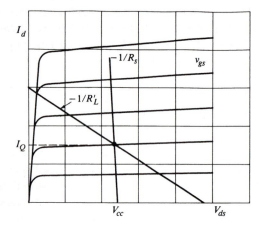

**FIGURE 11.5**
The ac and dc load lines of the amplifier shown in Fig. 11.4.

where $i_d$ is the drain alternating current. The ac voltage across the drain-to-source junction is

$$v_d = N v_o$$

Therefore, the ac load impedance seen by the transistor is

$$R'_L = N^2 R_L \tag{11.7}$$

The slope of the dc load line is $-1/R_S$, and if $R_S \ll R'_L$, the slope of the ac load line is $-1/R'_L$. The maximum signal swing is again $2V_{cc}$ (ignoring $V_{\text{sat}}$), and the peak drain current is

$$I_M = (i_d)_p = \frac{2V_{cc}}{R'_L} \tag{11.8}$$

Power calculations for the transformer-coupled load are the same as those for the capacitive-coupled load. The transformer provides greater flexibility for matching the load to the source, but the power dissipated in the transformer can significantly reduce the amplifier efficiency. Transformer- or inductor-coupled bipolar transistor amplifiers are subject to a phenomenon known as *thermal runaway.*[1] The heating up of the transistor causes more current to flow, which causes greater self-heating, which can cause the device to self-destruct. The problem rarely occurs with resistive loads since the increased current results in a reduced collector-to-emitter voltage, and eventually the circuit will reach equilibrium. Nor does the problem arise with FET amplifiers because as their temperatures increase, their output currents decrease.

### Class A Push-Pull Amplifiers

As mentioned in Chap. 2, push-pull amplifiers eliminate the even-harmonic distortion present in the amplifier output. This can provide a significant improvement in the performance of linear amplifiers. Also, push-pull operation reduces the power requirements of the individual transistors. Figure 11.6 illustrates a class A transformer-coupled push-pull stage. The ac equivalent circuit is shown in Fig. 11.7. In class A operation, both transistors continuously drive the output, and the transistor outputs, which are 180° out of phase, are combined in the center-tapped output transformer. The circuit of Fig. 11.7 can be redrawn as shown in Fig. 11.8 by using Thévenin's theorem; the results developed for center-tapped

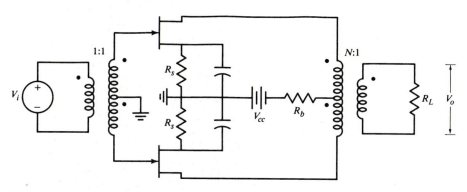

**FIGURE 11.6**
A class A transformer-coupled push-pull amplifier.

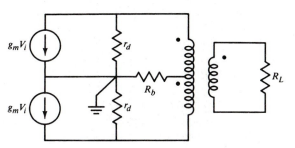

**FIGURE 11.7**
A small-signal equivalent circuit of the push-pull amplifier.

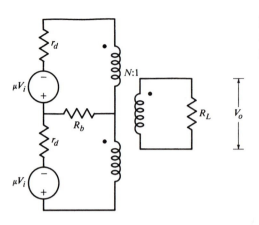

**FIGURE 11.8**
Another equivalent circuit of the push-pull amplifier shown in Fig. 11.6

transformers (Chap. 6) are now directly applicable. Here the amplification factor $\mu = g_m r_d$. If the dynamic drain resistances of the transistors are equal, then no current will flow through $R_b$ and the circuit can be redrawn as shown in Fig. 11.9. The output voltage $V_o$ is determined by

$$2\mu V_i - 2r_d i_d = -2NV_o$$

If the transformer is lossless

$$2NV_o = -i_d(2N)^2 R_L = -2R'_L i_d$$

where $R'_L = 2N^2 R_L$, the output voltage is

$$V_o = \frac{-N\mu R_L V_i}{r_d + R'_L} = \frac{-N g_m R_L V_i}{1 + R'_L/r_d}$$

Normally $r_d \gg R'_L$, so the voltage gain is

$$A_v = -g_m R_L N$$

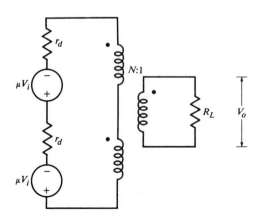

**FIGURE 11.9**
A simplified equivalent circuit of the push-pull amplifier.

since the load for each transistor is

$$R'_L = 2N^2R_L$$

For maximum output power (i.e., maximum signal swing) each transistor is biased so that

$$R'_L = \frac{V_{cc}}{I_d}$$

where $I_d$ is the drain direct current.

The maximum output power (from both transistors) is

$$P_o = \frac{2V_{cc}^2}{2R'_L} = \frac{V_{cc}^2}{2N^2RL}$$

The total output power of the class A push-pull amplifier is twice that of the single-ended class A amplifier; however, the maximum voltage drop across each transistor is the same as that of the single-ended amplifier. The power supplied to each transistor is

$$P_{cc} = I_d V_{cc} = \frac{V_{cc}^2}{R'_L}$$

The total power supplied is $2P_{cc}$, so the maximum efficiency of the class A push-pull amplifier is $\eta = 50$ percent, the same as the efficiency of the single-ended amplifier. Besides reducing even-harmonic distortion, the push-pull configuration can provide twice the output power of the single-ended design.

### Square-Wave Input

The efficiency of a class A amplifier depends on the input signal level and on the signal waveshape. Consider the case of a square-wave input. If the amplifier is class A, the collector current is also a square wave, as shown in Fig. 11.10. The dc value is $I_Q$, and the peak current is

$$I_p \le I_Q$$

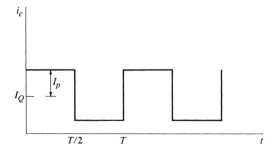

**FIGURE 11.10**
A collector-current waveform.

In this case

$$P_{cc} = I_Q V_{cc} = I_Q^2 R_L$$

and

$$P_o = I_p^2 R_L$$

so the efficiency

$$\eta = \frac{P_o}{P_{cc}} = \frac{I_p^2}{I_Q^2} \tag{11.9}$$

will approach 100 percent if $I_p$ approaches $I_Q$. The efficiency of the class A amplifier depends on the waveform of the input signal.

If a tuned circuit is used as the load, as is frequently the case in order to minimize distortion, then efficiency is reduced. With a square-wave input, the collector current will again be a square wave, as shown in Fig. 11.10, and the power supplied is again

$$P_{cc} = I_Q V_{cc}$$

If the tuned-circuit bandwidth is sufficiently narrow and tuned to the fundamental frequency, the output power is

$$P_o = \frac{I_1^2 R_L'}{2}$$

where $I_1$ is the amplitude of the fundamental frequency component of the output current

$$I_1 = \frac{4}{\pi} I_p$$

The maximum output power will be

$$(P_o)_{max} = \frac{8 I_p^2 R_L'}{\pi^2} = \frac{8 I_Q V_{cc}}{\pi^2}$$

and the maximum efficiency will be

$$\eta = \frac{(P_o)_{max}}{P_{cc}} = \frac{8}{\pi^2}$$

If the output is tuned to the $n$th harmonic ($n$ odd), the efficiency will be $\eta = 8/(n^2 \pi^2)$, since the amplitude of the $n$th (odd) harmonic of a square wave is $1/n$ times that of the fundamental frequency. A tuned-circuit load decreases the efficiency, but it is frequently used since the tuned circuit reduces the output harmonic distortion.

## Broadband Class A Power Amplifier

There is an increasing demand for linear, broadband power amplifiers in applications such as those in the CATV (Cable Television) industry. In many applications this demand is being met by combining several low-power modules in parallel, because it is possible to obtain more bandwidth by reducing the power gain of each module. Each module consists of a class A transformer-coupled push-pull amplifier such as the one shown in Fig. 11.6; four of these modules can then be combined, as shown in Fig. 11.10a, to obtain approximately 4 times as much output power as could be obtained from one module. High-power FETs are normally used, as they basically have no safe operating area limitation and no secondary breakdown and their high input impedance simplifies biasing. For broadband applications the combiners and splitters are usually obtained with broadband transformers, such as the transmission line transformer. A major broadband design is to keep the load seen by the transistors relatively constant over the specified bandwidth.

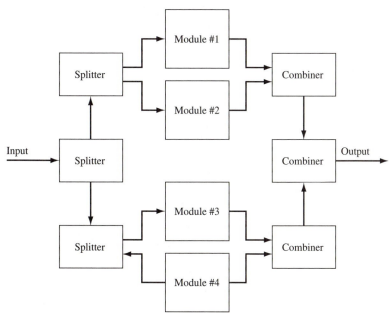

**FIGURE 11.10A**
A Class A Power Combiner.

## ▣ 11.3

### CLASS B AMPLIFIERS

A major disadvantage of the class A amplifier is that all the supply power is dissipated in the transistor when there is no input. It is usually advantageous to have no power supplied when there is no input signal, which is the case with a class B amplifier. A class B amplifier is biased as shown in Fig. 11.11. The quiescent collector current is zero, and the collector-to-emitter quiescent voltage is $V_{ce}$. It is biased just at the edge of the active region so that for a sine wave input the transistor will conduct over $180°$ of the input waveform, as illustrated in Fig. 11.12.

The output current is a highly distorted sine wave, but the distortion can be removed by using a narrowband tuned circuit for the load or, what is more frequently done, by operating two class B transistor amplifiers in push-pull, as illus-

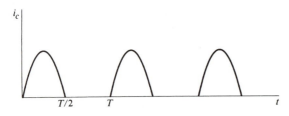

**FIGURE 11.11**
Collector-current waveform of a
class B amplifier.

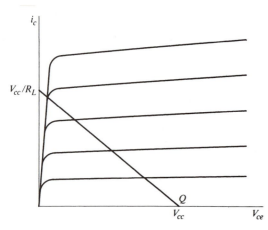

**FIGURE 11.12**
The $Q$-point biasing for a class B
amplifier.

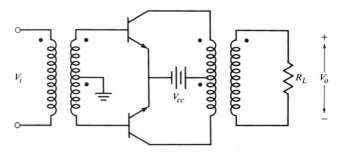

**FIGURE 11.13**
A class B push-pull amplifier.

trated in Fig. 11.13. Ideally, each transistor conducts over alternate $180°$ of the input
cycle, and the two outputs are summed so that an undistorted sine wave appears
across the load resistor $R_L$. At any time, one transistor is conducting and the other
is not, so the equivalent circuit can be drawn as shown in Fig. 11.14. The load seen
by each transistor

$$R_L' = N^2 R_L$$

is one-half that of the class A push-pull amplifier. The voltage gain is

$$\frac{V_o}{V_i} = \frac{-g_m r_o N R_L}{r_o + R_L'}$$

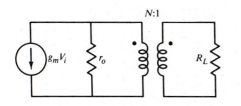

**FIGURE 11.14**
The small-signal equivalent circuit of the class B push-pull amplifier.

It is usually the case that $r_o \gg R'_L$, so $A_v = -g_m R'_L N$. The voltage gain of the class B stage is one-half that of the class A push-pull amplifier, so the drive requirements for the class B stage are twice as great. The increased drive requirement, however, is offset by the greater efficiency and power-handling capabilities of the class B amplifier.

The power-handling capability of each class B amplifier shown in Fig. 11.13 will now be evaluated by considering a sinusoidal input and a resistive load. The transistors are assumed to be ideal. Each transistor conducts when $V_i$ is greater than zero, and the collector current is zero when $V_i$ is negative. Since the load current conducts for 180° of a complete input cycle,

$$i_c(t) = \begin{cases} I_p \sin \omega t & 0 \le \omega t \le \pi \\ 0 & \pi \le \omega t \le 2\pi \end{cases}$$

The collector direct current is then

$$I_c = T^{-1} \int_0^{T/2} I_p \sin \omega t \, dt = \frac{I_p}{\pi} \tag{11.10}$$

The dc value of the base current is much less than $I_c$, so the dc power supplied by the supply voltage is

$$P_{cc} = I_c V_{cc} = \frac{I_p V_{cc}}{\pi}$$

The ac power delivered to the load is

$$P_o = \frac{R'_L}{T} \int_0^{T/2} (I_p \sin \omega t)^2 \, dt = \frac{I_p^2 R'_L}{4}$$

Since $I_p \le V_{cc}/R'_L$, the maximum output power from each transistor is

$$(P_o)_{\max} = \frac{V_{cc}^2}{4 R_L} \tag{11.11}$$

The power supplied by the input circuit will normally be much less than the dc power, so the efficiency at maximum output power is

$$\eta = \frac{V_{cc}^2}{4 R_L} \frac{\pi R_L}{V_{cc}^2} = \frac{\pi}{4} = 78.5\%$$

The efficiency of a class B amplifier is much higher than that of a class A stage. In addition, the class B stage consumes no power when an input signal is not present. This is a most significant advantage in many applications (the class A amplifier has maximum power dissipation when no signal is present). The power dissipated in the transistor of the class B amplifier is

$$P_T = P_{cc} - P_o = \frac{I_p V_{cc}}{\pi} - I_p^2 \frac{R_L'}{4} \tag{11.12}$$

The maximum power dissipated in the transistor is found by differentiating Eq. (11.12) with respect to $I_p$. It is found that the maximum dissipation occurs for

$$I_p = \frac{2V_{cc}}{\pi R_L'}$$

and

$$(P_T)_{max} = \frac{V_{cc}^2}{\pi^2 R_L'} = (P_o)_{max} \frac{4}{\pi^2} \tag{11.13}$$

Note, however, that the maximum transistor dissipation does not occur when the output power is a maximum. The class B stage also results in less transistor power dissipation than in a class A stage. This is an expected result of greater operating efficiency. Another important difference between the two is that the maximum voltage drop across the transistor in a class B amplifier is $V_{cc}$; it is $2V_{cc}$ for the class A amplifiers previously described. Class A amplifiers, therefore, require transistors with a higher collector-to-emitter breakdown voltage. Differences in class A and B power amplifiers are illustrated by the following example.

**EXAMPLE 11.2.** Design a class B push-pull amplifier to deliver 5 W (maximum) to a 50-$\Omega$ load.

**Solution.** Assume that a transformer-coupled amplifier such as the one illustrated in Fig. 11.13 is used with a 1:1 turns ratio. Since a push-pull amplifier is used, each class B amplifier will supply 2.5 W. The required supply voltage can be obtained from Eq. (11.11):

$$V_{cc}^2 = 4R_L(P_o)_{max} = 500 \text{ V}^2$$

A supply voltage $V_{cc}$ of 24 V would be a suitable choice. The power-handling requirements of the transistor can be determined from Eq. (11.13):

$$(P_T)_{max} = (P_o)_{max} \frac{4}{\pi^2} = 1 \text{ W}$$

The peak output current will be

$$I = \left(\frac{4P_o}{R_L}\right)^{1/2} = 0.45 \text{ A}$$

The voltage and/or current requirements can be modified by selecting a different turns ratio for the transformer.

Example 11.1 discussed using a class A amplifier to realize these same specifications. Note that the class A amplifier transistor dissipation is over 11 times greater than that of the class B transistor, so if a class A push-pull amplifier is used, the power-handling requirements of each transistor will be 5 1/2 times that of the transistors used in the class B amplifier. Also, the collector-to-emitter breakdown voltage of the class A transistors must be double that of the transistors used in class B amplifiers.

While the class B amplifier also has the significant advantage of no power's being dissipated when an input signal is not present, practical class A amplification does produce less distortion. All in all, if the distortion produced by class B amplification is acceptable, its benefits make it preferable to class A amplification. Large-signal distortion can be reduced with negative feedback.

### Push-Pull Amplification with Complementary Transistors

Transformerless push-pull amplifiers can be realized by using either bipolar or field-effect transistors with complementary symmetry. Figure 11.15 illustrates a push-pull amplifier using power MOSFETs. The current source and resistor $R_b$ are used to bias the transistors for class B operation, and the MOSFETs do not turn on until the input voltage exceeds the threshold voltage of the device.

Power MOSFETs have several advantages over bipolar transistors. The most important is that the drain current has a negative temperature coefficient that decreases with increases in temperature, since the positive temperature coefficient of bipolar transistors can result in self-destruction unless complicated biasing circuits are used. Bipolar transistors also have an undesirable characteristic when operated at high voltages. The collector current is no longer uniform, but tends to concentrate in small areas, causing very high peak temperatures, known as *hot spots*, in these areas. The average junction temperature, measured at the transistor case, does not show the presence of hot spots, which degrade the transistor performance. The drain current of power MOSFETs is distributed uniformly, preventing the development of hot spots. The high input impedance (and hence high power gain) of the MOSFETs is another advantage. The upper-frequency limits and power-handling capability of power MOSFETs have been continually increasing as manufacturing techniques have improved. Bipolar transistors are minority carrier devices that accumulate charge in the base region. This causes a problem in class B

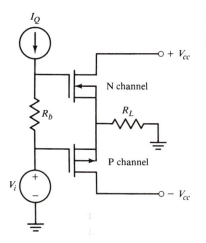

**FIGURE 11.15**
A class B push-pull amplifier using complementary MOSFETs.

and C operation because removing the charge takes time and energy. One consequence is increasing power dissipation with increasing frequency. FETs are majority carrier devices; the charge carriers are controlled by an electric field and not by injection of minority carriers into the active region. The gate regions, therefore, contain no stored charge, and switching between the on and off states is very fast.

The first field-effect transistors were useful only at lower (less than 1 W) power levels. Their major limitation was that the FET drain-to-source channel was parallel to the chip surface, so the current density was much less than that of bipolar transistors (which utilized vertical current flow). For a given current the FET chip had to be much larger than an equivalent bipolar junction transistor (BJT) chip, which meant a lower yield and a higher cost. Several new technologies have recently been developed that produce high-voltage, high-current FETs. The three most frequently used manufacturing technologies are VMOS (vertical MOS), VJFET (vertical JFET), and DMOS (double-diffusion MOS).[2]

### Power Relations in the Direct-Coupled Class B Push-Pull Amplifier

For the push-pull amplifier shown in Fig. 11.15, the maximum average ac output power in response to a sine wave input is

$$(P_o)_{av} = \frac{V_{cc}^2}{2R_L} \qquad (11.14)$$

The power delivered by each transistor is

$$(P_o)_{av} = \frac{V_{cc}^2}{4R_L} \qquad (11.15)$$

The power delivered by each supply is

$$P_{cc} = V_{cc}T^{-1} \int_0^{T/2} \frac{V_{cc}}{R_L} \sin \omega t \, dt = \frac{V_{cc}^2}{\pi R_L} \qquad (11.16)$$

so the efficiency of this circuit is the same as that of the transformer-coupled push-pull amplifier. The maximum power dissipated in each transistor is

$$(P_T)_{max} = \frac{V_{cc}^2}{\pi^2 R_L} \qquad (11.17)$$

the same as that for the transformer-coupled push-pull circuit. The only difference between the two circuits deduced from a power analysis is that the transformerless circuit requires two power supplies.

When the current requirements exceed the limitations of a single transistor, transistors can be operated in parallel. Figure 11.16 illustrates a class B amplifier containing two power MOSFETs.[3] It is usually easier to parallel FETs than bipolar transistors because of their larger input impedance. In this particular circuit, approximately 2 times as much output power could be obtained as from a class B circuit using a single transistor of the same type.

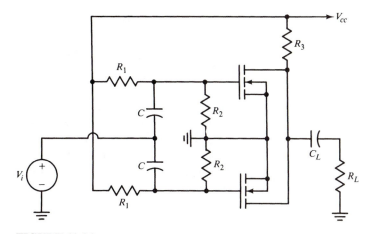

**FIGURE 11.16**
A class B amplifier using two MOSFETs in parallel.

Class B amplifiers all suffer from crossover distortion, which occurs because of the nonlinear behavior of small-signal levels. The silicon BJT requires that the base-to-emitter junction be forward-biased by approximately 0.7 V before the collector current will flow. FETs also exhibit a nonlinear behavior for low signal levels. For these reasons power amplifiers are usually operated with a slight forward bias to reduce crossover distortion. This operation is often referred to as *class AB,* but it is essentially the same in that the power levels and efficiency are only slightly reduced from those of class B amplifiers.

## 11.4

### CLASS C AMPLIFIERS

If the output current conduction angle is less than 180°, the amplifier operation is referred to as *class C*. This mode of operation can have a greater efficiency than class B, but it creates more distortion than class A or B amplifiers. The distortion is sometimes acceptable or, in the case of frequency multiplication, desirable. Class C is often used where there is no variation in signal amplitude and the output circuit contains a tuned circuit to filter out all the harmonics of the output current. In many applications, such as the amplification of FM signals, the signal frequency, and not the amplitude, is important. Class C power amplifiers are usually used for these applications. Figure 11.17 provides examples of FET and bipolar transistor class C amplifier circuits, and Fig. 11.18 illustrates the drain (or collector) current of a class C amplifier in which the conduction angle $(2\theta)$ is less than 180° and the drive level is sufficiently small that the output current does not saturate.

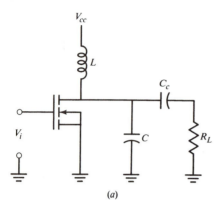

**FIGURE 11.17**
(a) A MOSFET class C amplifier;
(b) a class C amplifier realized with
a BJT.

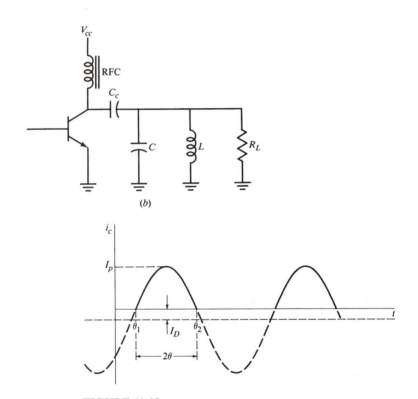

**FIGURE 11.18**
Collector-current waveform of a class C amplifier.

Several different models can be assumed for the current pulses. A relatively simple model is to assume that the pulses represent the tip of a sine wave. This is the model used here. That is (see Fig. 11.18),

$$i_c = \begin{cases} I_p \sin \omega t - I_D & \theta_1 \leq \omega t \leq \theta_2 \\ 0 & \text{otherwise} \end{cases} \tag{11.18}$$

where $$I_p > I_D$$

and $$I_D = I_p \sin \theta_1 \tag{11.19}$$

For this model the direct current is

$$I_c = T^{-1} \int_{\theta_1/\omega}^{\theta_2/\omega} (I_p \sin \omega t - I_D)\, dt$$
$$= \frac{2I_p \cos \theta_1 - I_D(\theta_2 - \theta_1)}{2\pi} \tag{11.20}$$

To simplify the notation, we will define the conduction angle as

$$2\theta = \theta_2 - \theta_1 \quad \text{or} \quad \theta = \frac{\pi}{2} - \theta_1 \tag{11.21}$$

Equation (11.20) can be rewritten as

$$I_c = \frac{I_p}{\pi}(\sin \theta - \theta \cos \theta) \tag{11.22}$$

The direct current determines the power supplied, since the base (or gate) direct current is much smaller than the output current. That is,

$$P_{cc} = V_{cc}I_c = \frac{V_{cc}}{\pi}I_p(\sin \theta - \theta \cos \theta) \tag{11.23}$$

If the output is a narrowband circuit tuned to the fundamental frequency of the current pulses, then the output power will be

$$P_o = \frac{I_1^2 R_L}{2} \tag{11.24}$$

where $I_1$ is the amplitude of the fundamental current component

$$I_1 = \frac{4}{T} \int_0^{\theta/\omega} (I_p \cos \omega t - I_D) \cos \omega t\, dt$$

Here the time origin has been shifted to the center of the current pulse to simplify the integration. The time shifting does not alter the amplitudes of the frequency components, only their phase. The amplitude of the fundamental frequency component is

$$I_1 = \frac{I_p}{2\pi}(2\theta - \sin 2\theta) \tag{11.25}$$

Since the conduction angle depends on the input amplitude, the fundamental current amplitude and thus the output voltage are a nonlinear function of the input signal amplitude.

For the class C FET amplifier shown in Fig. 11.17a, the maximum drain-to-source voltage is

$$(v_{ds})_{\max} = V_{cc} + (I_1)_{\max} R_L \leq 2V_{cc}$$

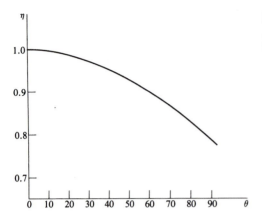

The efficiency at maximum output power is

$$\eta = \frac{P_o}{P_{cc}} = \frac{I_1^2 R_L}{V_{cc} I_c} = \frac{V_{cc} I_1}{V_{cc} I_c} = \frac{2\theta - \sin 2\theta}{4(\sin \theta - \theta \cos \theta)}$$

where $I_c$ is the dc value of the current [Eq. (11.20)]. The efficiency as a function of conduction angle is plotted in Fig. 11.19. The class C amplifier efficiency can be increased toward 100 percent (in an ideal amplifier) by decreasing the conduction angle toward zero. When the conduction angle $\theta$ is 90°, the operation is class B and the efficiency is 78.5 percent. The efficiency increases monotonically as the conduction angle decreases. This high efficiency is why class C amplifiers are often used for power amplification.

**Class C Power Amplifier Design**

For class C power amplifiers, as for all power amplifiers, the design parameters of greatest importance are the output power, transistor power dissipation, maximum collector-to-emitter (or drain-to-source) voltage, and maximum transistor output current $I_p$. For the BJT class C amplifier shown in Fig. 11.17b, the maximum collector-to-emitter voltage is

$$(v_{ce})_{max} = 2V_{cc}$$

The maximum collector current is, from Eq. (11.18),

$$I_M = (i_c)_{max} = I_p \sin \frac{\pi}{2} - I_D = I_p - I_p \sin \theta_1 \qquad (11.26)$$

and since $\theta = \pi/2 - \theta_1$,

$$I_M = I_p(1 - \cos \theta) \qquad (11.27)$$

The peak current is related to the amplitude $I_1$ of the fundamental frequency component [using Eq. (11.25)] by

$$I_M = \frac{2\pi I_1 (1 - \cos \theta)}{2\theta - \sin 2\theta} \tag{11.28}$$

The ac power output of the amplifier is approximately

$$P_o = \frac{I_1^2 R_L}{2} \tag{11.29}$$

provided the $Q$ of the tuned circuit is sufficiently high. The peak output current is a function of both collector current and output power. The maximum output power occurs for maximum $I_p$, and the average (maximum) output power is

$$P_o = \frac{I_1^2 R_L}{2} = \frac{V_{cc}^2}{2R_L} \tag{11.30}$$

since

$$I_1 R_L = V_{cc} \tag{11.31}$$

for maximum output power.

The power dissipated in the transistor is

$$P_T = P_{cc} - P_o = \frac{V_{cc} I_p}{\pi} (\sin \theta - \theta \cos \theta) - \frac{V_{cc}^2}{2R_L}$$
$$= \frac{V_{cc} I_p}{\pi} \frac{\sin \theta - \theta \cos \theta}{1 - \cos \theta} - \frac{V_{cc}^2}{2R_L} \tag{11.32}$$

For a specified load resistance, Eq. (11.30) determines the required supply voltage for a specified output power. The corresponding maximum current $I_M$ [from Eqs. (11.28) and (11.31)] is

$$I_M = \frac{2\pi V_{cc} (1 - \cos \theta)}{R_L (2\theta - \sin 2\theta)} \tag{11.33}$$

A normalized peak collector current is defined as

$$I'_M = \frac{I_M R_L}{2\pi V_{cc}} = \frac{1 - \cos \theta}{2\theta - \sin 2\theta} \tag{11.34}$$

The normalized peak collector current $I'_M$ as a function of conduction angle is plotted in Fig. 11.20. For a fixed level of output power, the peak value of the collector current increases as the conduction angle decreases.

The transistor dissipation for maximum output power can be expressed as a function of the output power and conduction angle [from Eqs. (11.32) and (11.33)]:

$$P_T = P_o \left[ \frac{4(\sin \theta - \theta \cos \theta)}{2\theta - \sin 2\theta} - 1 \right] \tag{11.35}$$

The normalized transistor dissipation $P_T / P_o$ is plotted as a function of the conduction angle in Fig. 11.21. As expected, the transistor dissipation increases with an increasing conduction angle. For a given maximum $P_T$ the conduction angle must be limited to a maximum value for a specified output power. The corresponding

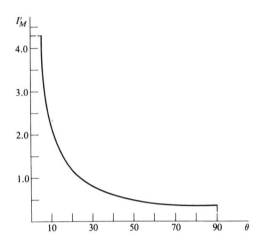

**FIGURE 11.20**
Normalized peak collector (or drain)
current as a function of conduction
angle $2\theta$.

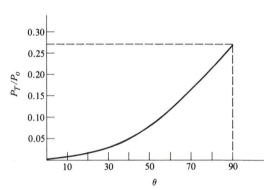

**FIGURE 11.21**
Normalized (by output power) transistor power dissipation as a function of conduction angle $2\theta$.

maximum value of transistor output current is then determined from Fig. 11.20. As
the conduction angle decreases, the transistor dissipation decreases but the peak
output current increases.

**EXAMPLE 11.3.** Design a class C amplifier that will deliver 5-W average power to a
50-$\Omega$ load at a frequency of 1 MHz using a transistor with a safe power dissipation rating of 0.5 W.

**Solution.** The average output power is

$$P_o = \frac{V_{cc}^2}{2R_L}$$

so

$$V_{cc}^2 = 500$$

and a supply voltage of 22.4 V is required. Since the allowable power dissipation is

$$(P_T)_{\max} = 0.1P_o$$

the maximum conduction angle is found from Fig. 11.21 to be 57.5°, and the corresponding peak value of normalized collector current is found from Fig. 11.20:

$$I_M' = 0.5$$

Therefore, the peak collector current is

$$I_M = \frac{2\pi \times 22.4}{50} 0.5 = 1.4 \, \text{A}$$

The selected transistor must be capable of handling this much current, and the collector circuit should be tuned to resonate at 1 MHz.

An alternate design procedure for class C amplifiers is to select the power supply and transistor and then determine the maximum output power possible without exceeding the ratings of the transistor. The transistor can then be driven to its maximum allowed value of output current. From Eq. (11.30) we know that the maximum output power will occur when $R_L$ is at a minimum. And $R_L$ can be expressed in terms of the supply voltage $V_{cc}$, and the peak transistor output current can be expressed [using Eqs. (11.28) and (11.31)] as

$$R_L = \frac{V_{cc}}{I_M} \times 2\pi \frac{1 - \cos\theta}{2\theta - \sin 2\theta}$$

For a specified $V_{cc}$ and $I_M$, it is seen from Fig. 11.20—a plot of $(1 - \cos\theta)/(2\theta - \sin 2\theta)$—that $R_L$ decreases with an increasing conduction angle. Actually, $R_L$ reaches a minimum for $\theta = 122.6°$ (class AB operation). This is the conduction angle for maximum output power constrained by the peak-current limitation. If the maximum transistor dissipation is exceeded at this conduction angle, the conduction angle must be reduced, and the output power will be decreased accordingly.

The transistor dissipation can also be written [from Eqs. (11.31) and (11.32)] as

$$P_T = \frac{V_{cc}I_M}{\pi} \frac{\sin\theta - \theta\cos\theta}{1 - \cos\theta} - V_{cc}\frac{V_{cc}}{2R_L}$$

$$= \frac{V_{cc}I_M}{\pi}\left(\frac{\sin\theta - \theta\cos\theta}{1 - \cos\theta} - \frac{V_{cc}I_1}{2}\right)$$

And $I_1$ is given by Eq. (11.28), so

$$P_T = \frac{V_{cc}I_M}{4\pi} \frac{4(\sin\theta - \theta\cos\theta) - (2\theta - \sin 2\theta)}{1 - \cos\theta} = \frac{V_{cc}I_M}{4\pi} f(\theta) \quad (11.36)$$

The normalized transistor power dissipation

$$P_T' = \frac{4\pi P_T}{V_{cc}I_M} = f(\theta)$$

is plotted as a function of the conduction angle in Fig. 11.22.

From Eq. (11.36) it is seen that for a given $V_{cc}$ and $I_M$ the maximum transistor dissipation occurs for maximum $f(\theta)$. For specified $P_T$, $V_{cc}$, and $I_M$, the value of $\theta$ that satisfies this equation, i.e., determines $(P_T)_{max}$, is the value of $\theta$ for maximum output power. Then $(P_o)_{max}$ is determined from Fig. 11.21.

**EXAMPLE 11.4.** Determine the maximum output power and the conduction angle of a class C amplifier using a transistor with a maximum power dissipation rating of 4 W and a maximum drain current of 1.5 A. The supply voltage is 48 V.

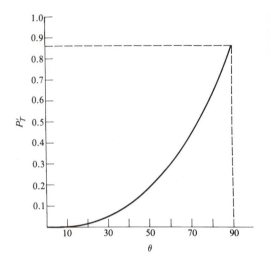

**FIGURE 11.22**

Normalized transistor power dissipation as a function of conduction angle $(2\theta)$.

*Solution.* The normalized maximum transistor dissipation is

$$P_T' = \frac{4\pi \times 4}{48 \times 1.5} = 0.7$$

From Fig. 11.21 it is found that the maximum possible conduction angle is $\theta = 80°$ without exceeding the maximum transistor dissipation. Figure 11.21 indicates that for this conduction angle the corresponding $P_T/P_o = 0.22$, so the output power is

$$P_o = \frac{4}{0.22} = 18.18 \text{ W}$$

The value of load resistance that results in this output power is determined from Eq. (11.30):

$$R_L = \frac{48^2}{2 \times 18.18} = 63.4 \ \Omega$$

## Frequency Multiplication

Since the current pulses of a class amplifier are rich in the harmonics of the input waveform, the class C amplifier can be used as a frequency multiplier by tuning the output circuit to the desired harmonic. The amplitude of the $n$th harmonic of the output current can be determined from a Fourier expansion of the current waveform. The collector current can be written [using Eqs. (11.18), (11.19), and (11.21)] as

$$i_c = I_p \cos \omega t - I_p \cos \theta$$

so

$$I_n = \frac{4}{T} \int_0^{\theta/\omega} (I_p \cos \omega t - I_p \cos \theta) \cos n\omega t$$

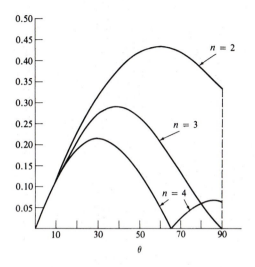

**FIGURE 11.23**
Amplitude of output current harmonics as a function of current conduction angle ($2\theta$).

$$= \frac{I_p}{\pi}\left[\frac{\sin(n+1)\theta}{n+1} - \frac{\sin(n-1)\theta}{n-1} - \frac{2\cos\theta\sin n\theta}{n}\right] \qquad n \geq 2 \quad (11.37)$$

$$= \frac{I_p}{\pi}\frac{\cos\theta\sin n\theta - n\sin\theta\cos n\theta}{n(n^2-1)}$$

The amplitudes of the harmonics as a function of the conduction angle are plotted in Fig. 11.23 (for $n = 2, 3,$ and 4).

**EXAMPLE 11.5.** A frequency quadrupler is to be designed. What should the conduction angle be to maximize the output signal voltage?

*Solution.* From Fig. 11.23 it is seen that the amplitude of the fourth harmonic has a maximum value for a conduction angle $2\theta$ of approximately $60°$. The output circuit would be tuned to the fourth harmonic of the input signal.

The analysis of class C operation assumed that the current pulses could be modeled as the tip of a sine wave. In many class C applications the transistor will saturate during part of the output cycle; however, a more detailed analysis of this behavior is possible but rarely necessary. Efficiency decreases as saturation increases, therefore, stable operation with maximum efficiency and output capability is achieved by driving the amplifier just hard enough to produce saturation of the transistor.

## ■ 11.5

### CLASS D AMPLIFIERS

The main source of power amplifier inefficiency is power that is dissipated in the transistor. A class A application is the poorest example, since current flows continuously through the device and the collector-to-emitter voltage is not zero. If the

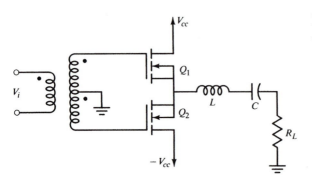

**FIGURE 11.24**
A class D amplifier.

collector-to-emitter (or drain-to-source) voltage is zero when the current flows, no power will be dissipated in the device, and the efficiency will approach 100 percent. This is the basic idea behind class D, E, and S power amplifiers. A class D amplifier is illustrated in Fig. 11.24. Transistors $Q_1$ and $Q_2$ operate as switches. When $Q_1$ is on, $Q_2$ is off, and vice versa. For an ideal transistor with zero saturation voltage, there will be no voltage drop across the transistor, and the circuit can be modeled as in Fig. 11.25. If the input $V_i$ is a square wave, the voltage $V_a$ at the input to the series tuned circuit will be as shown in Fig. 11.26. Since $V_a$ is a square wave, it can be expanded in a Fourier series, and the amplitude of the fundamental frequency component is

$$V_1 = \frac{4V_{cc}}{\pi}$$

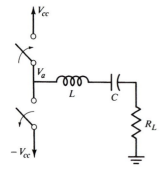

**FIGURE 11.25**
The equivalent circuit of a class D amplifier.

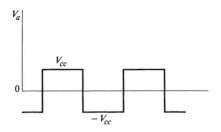

**FIGURE 11.26**
Tuned-circuit output voltage waveform for square-wave input to the class D amplifier.

If the output filter is relatively high $Q$ with a center frequency equal to the frequency of the input signal, the drain current in each transistor will be one-half of a sinusoid at the same frequency. Therefore, the direct current in each transistor is

$$I_D = (R_L T)^{-1} \int_0^{T/2} \frac{4V_{cc}}{\pi} \sin \omega t \, dt = \frac{4V_{cc}}{\pi^2 R_L} \tag{11.38}$$

and the total power supplied is

$$P_i = 2V_{cc}I_D = \frac{8V_{cc}^2}{\pi^2 R_L}$$

The output power will be

$$P_o = \left(\frac{4V_{cc}}{\pi}\right)^2 (2R_L)^{-1} = \frac{8V_{cc}^2}{\pi^2 R_L} \tag{11.39}$$

which is the same as the power supplied, so the theoretical efficiency of the ideal class D amplifier is $\eta = 100$ percent.

**EXAMPLE 11.6.**  Design a class D power amplifier to deliver 20 W to a 50-$\Omega$ load.

**Solution.**  Since the output power is

$$P_o = \frac{8V_{cc}^2}{\pi^2 R_L} = 20 \text{ W}$$

$$V_{cc} \approx 35.1 \text{ V}$$

The direct current in each transistor will be

$$I_{dc} = \frac{4V_{cc}}{R_L \pi^2} = 0.285 \text{ A}$$

and the maximum voltage drop across each transistor will be $2V_{cc}$, or 70.2 V. The load circuit would be tuned to resonate at the fundamental frequency of the input signal.

## Nonideal Performance

For actual transistors it is impossible to have a zero voltage drop across the device when it is saturated. Bipolar transistors are usually modeled by a saturation voltage $V_{sat}$, and FETs by an on resistance $R_{on}$. The class D FET amplifier can then be modeled as shown in Fig. 11.27. The voltage at the input to the tuned circuit is still a square wave, as illustrated in Fig. 11.26; only the amplitude is reduced. The output voltage will be (again assuming a narrowband filter)

$$V_o = \frac{4V_{cc}R_L}{\pi(R_L + R_{on})} \sin \omega t \tag{11.40}$$

and the direct current in each transistor will be

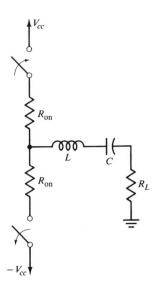

**FIGURE 11.27**
Circuit model of a class D FET amplifier.

$$I_d = \frac{4V_{cc}}{\pi^2(R_L + R_{on})} \tag{11.41}$$

The total power supplied is

$$P_{cc} = \frac{8V_{cc}^2}{\pi^2(R_L + R_{on})}$$

and the output power is

$$P_o = \frac{8V_{cc}^2}{\pi^2}\frac{R_L}{(R_L + R_{on})^2} = P_{o\text{ideal}}\left(\frac{R_L}{R_L + R_{on}}\right)^2 \tag{11.42}$$

so the efficiency of the nonideal class D FET amplifier is

$$\eta = \frac{R_L}{R_L + R_{on}}$$

**EXAMPLE 11.7.** Calculate the actual power output and efficiency of the class D amplifier used in the previous example if VMOS 2N6659s are used with a gate signal level of 8 V.

**Solution.** The specification sheet for the 2N6659 indicates that the on resistance is approximately $2\,\Omega$ for this drive level. Therefore, the output power is

$$P_o = 20\left(\frac{R_L}{R_L + R_{on}}\right)^2 = 18.5 \text{ W}$$

and the efficiency is

$$\eta = \frac{50}{52} = 96.2\%$$

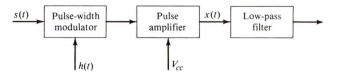

**FIGURE 11.28**
Block diagram of a class S amplifier.

## 11.6

### CLASS S POWER AMPLIFIERS

The high efficiency of the class D amplifier has encouraged the design of other switching amplifiers. Another possibility is to pulse-width-modulate the information, amplifying the pulses with a high-efficiency amplifier and then demodulating the amplified signal. This is the principle of class S amplification. The classification "class S" is not universal. Some authors interchange the class S and class D classifications, while others refer to class S as "broadband class D." A class S amplifier block diagram is illustrated in Fig. 11.28. The input signal is pulse-width-modulated, and the constant-amplitude pulses are increased by the pulse amplifier, which is a highly efficient switching amplifier. Pulse frequency modulation, which varies the rate of constant-width pulses, can also be used. It can be shown[4] that for pulse-width modulation (with fixed leading edges) of a sinusoidal signal $S(t) = A \cos \omega_s t$, the modulated signal is

$$X(t) = T^{-1} \left\{ \sum_{m=-\infty}^{\infty} -1^m \frac{e^{jm\omega_o t}}{jm\omega_o} - \sum_{\substack{m=-\infty \\ |m|+|n|\neq 0}}^{\infty} \sum_{n=-\infty}^{\infty} \frac{(-j)^n J_n[A(m\omega_o + n\omega_s)]}{j(m\omega_o + n\omega_s)} + \frac{1}{2} \right\}$$

(11.43)

The modulated signal contains a term at the input signal frequency ($m = 0$, $n = 1$). As long as the modulating frequency is much greater than the signal frequency, the other significant frequency components will be at much higher frequency and can be removed by low-pass filtering. The significant difference, then, between a class D and a class S power amplifier is that the output circuit of the class D amplifier is tuned to the fundamental frequency of the input signal, while the output circuit of the class S amplifier is a low-pass filter that recovers the input signal. It is important that the amplifier's fluctuations be large enough that they do not become a factor in determining the pulse width of the modulated signal.

The transistor-switching characteristics of a class S amplifier limit its high-frequency performance. A simplified class S amplifier is illustrated in Fig. 11.29. If the saturation resistance of the transistors is neglected, the efficiency will approach 100 percent; however, the nonzero switching time of the transistor can also affect its efficiency.

The class D amplifier is a nonlinear power amplifier that preserves the frequency (but not the amplitude) of the input signal. The purpose of the circuit is to

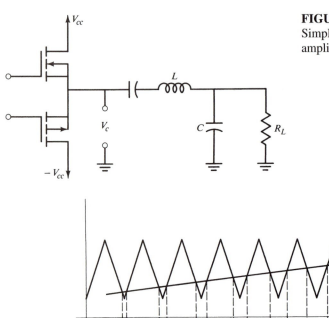

**FIGURE 11.29**
Simplified circuit for a class S
amplifier.

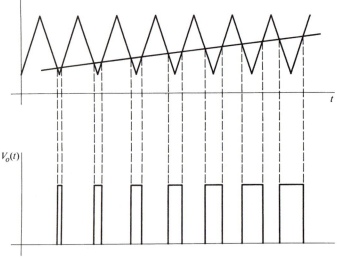

**FIGURE 11.30**
Input and output waveforms of a pulse-width modulator.

amplify digital data modulated on a carrier such as frequency-shift-keyed (FSK)
data.

## Pulse-Width Modulators

Class S amplifiers include a pulse-width modulator. A relatively simple method for
obtaining pulse-width modulation consists of using a comparator as a zero-crossing
detector to generate a pulse-width-modulated signal. One comparator input is a triangular wave $f(t)$, and the other input is the signal to be modulated $V_m(t)$. An
example of the modulated signal is shown in Fig. 11.30 for the case in which the
signal is a ramp waveform. The comparator output is high whenever $V_m(t) > f(t)$.
This particular technique modulates both the leading and trailing edges of the waveform. Single-edge modulation can also be easily implemented with integrated circuitry.

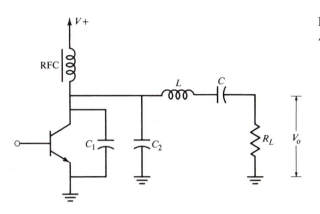

**FIGURE 11.31**
A class E amplifier.

## ◼ 11.7
### CLASS E TUNED POWER AMPLIFIERS[5]

The class D amplifier utilizes transistors as switches and is a power converter. The input signal waveform is not preserved, just the frequency. In class E operation the transistor acts as a switch, as it does in class D, but only one transistor is used. A simple class E amplifier is illustrated in Fig. 11.31. Class E further differs from class D in that the output tuned circuit is designed to realize certain collector-voltage and current waveform characteristics that are selected to minimize the power dissipated while the transistor is switching from on to off, or vice versa. The transistor switching time can occupy an appreciable fraction of the input signal period during which substantial power can be dissipated in the transistor, reducing amplifier efficiency. To prevent this problem, the output tuned circuit is designed so that (1) the rise of the collector voltage is delayed until after the transistor is turned off, (2) the collector voltage is reduced to zero when the transistor is turned on, and (3) the slope of the collector voltage is zero at the time of turn-on. The analysis of the circuit is as yet not well developed, and its main parameters must be further manipulated. An analysis of the idealized amplifier with design formulas is given in the literature.[6]

## ◼ 11.8
### PROBLEMS

**11.1** A transistor with a peak current rating of 5 A and a maximum drain-to-source voltage of 50 V is to be used in a class A power amplifier. What is the maximum output power? If a load resistance of 50 $\Omega$ is specified, what turns ratio for a matching transformer is required?

**11.2** Compare the designs of class A and class B power amplifiers to deliver 20 W to a 50-$\Omega$ load. The power supply is 24 V. Specify the maximum transistor dissipation, peak output voltage, and peak output current for each design.

**11.3** Calculate the efficiency of the direct-coupled class A power amplifier shown in Fig. P11.3. What is the maximum voltage across the transistor?

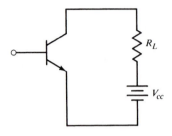

**FIGURE P11.3**
A direct-coupled class A amplifier.

**11.4** Design a transformerless class B power amplifier using complementary symmetry transistors to deliver 20 W to a 50-$\Omega$ load. Specify the maximum transistor dissipation, peak output voltage, and peak output current for each transistor.

**11.5** Figure P11.5 illustrates three different methods of biasing a class A amplifier. Compare the efficiency factor

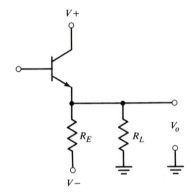

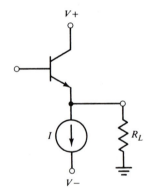

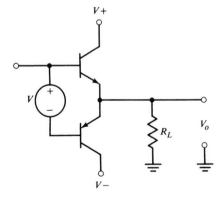

**FIGURE P11.5**
Three methods of class A biasing.

$$\frac{\text{Peak output power}}{\text{Transistor dissipation with no input signal}}$$

for the three configurations.

**11.6** Design a power amplifier using the DV1260T power MOSFET (see App. 1 for specification). The load resistance is 50-$\Omega$, and it is to be coupled to the transistor with a transformer (which is assumed to be 90 percent efficient). What is the maximum possible output power without exceeding the transistor maximum ratings? What are the required supply voltage and current?

**11.7** Design a class A push-pull amplifier to deliver 100 W to a 50-$\Omega$ load. Use DV1260T MOSPOWER FETs (App. 1). Specify the supply voltages, bias network, and turns ratios of any transformers used. The source resistance is 500 $\Omega$.

**11.8** Design a class C amplifier that will deliver an average power of 20 W at 1 MHz using the DV1260T MOSPOWER FET (see App. 1).

**11.9** Determine the maximum output power and conduction angle of a class C amplifier using the DV1260T MOSFET and a 24-V supply voltage (see App. 1).

**11.10** If a class C amplifier is to be used as a frequency tripler, what should the conduction angle be for maximum output power?

**11.11** Determine the maximum output power of a frequency tripler using the DV1260T MOSFET and a 24-V supply voltage (see App. 1).

**11.12** Design a complementary class D amplifier to deliver 100 W to a 50-$\Omega$ load using DV1260T transistors (see App. 1). Assume typical transistor values. A 24-V supply is to be used, and an impedance-matching transformer can be used for the output.

**11.13** What is the efficiency of the amplifier of Prob. 11.12? What will efficiency be if a DV1260T FET is used?

**11.14** What is the maximum output power possible using DV1260T transistors in a class D amplifier? (See App. 1.) What will be the maximum value of supply voltage?

**11.15** Describe a technique for class S amplification that uses single-edge pulse-width modulation.

*__**11.16**__ Design a class A power amplifier using MRF 177 transistors to deliver 20 W to a 50-$\Omega$ load. Calculate the maximum transistor dissipation, peak output voltage, and peak output current of each transistor; and compare with the values obtained from the simulation. Determine the bandwidth of the amplifier.

## ◼ 11.9

### REFERENCES

1. M. V. Joyce and K. Clarke, *Transistor Circuit Analysis,* Addison-Wesley, Reading, Mass., 1961.
2. A. D. Evans (ed.), *Designing with Field-Effect Transistors,* McGraw-Hill, New York, 1981.
3. H. O. Granberg, Four MOSFETs Deliver 600 W of RF Power, *Microwaves & RF,* **22:**89, January 1983.
4. H. E. Rowe, *Signals and Noise in Communication Systems,* Van Nostrand, New York, 1965, p. 277.
5. N. O. Sokal and A. D. Sokal, Class E—A New Class of High-Efficiency Tuned Single-Ended Switching Power Amplifiers, *IEEE J. Solid-State Circuits,* **10:** 168–176, 1975.
6. F. H. Raab, Idealized Operation of the Class E Tuned Power Amplifier, *IEEE Trans. Circuits and Systems,* **24:**725–735, 1977.

## ◼ 11.10

### ADDITIONAL READING

Clarke, K., and D. Hess: *Communication Circuits: Analysis and Design,* Addison-Wesley, Reading, Mass., 1971.

El-Hamamsy, S-A.: Design of High-Efficiency RF Class-D Power Amplifier, *IEEE Trans. Power Electronics,* **9:**297–507, 1994.

Furlan, J.: Preliminary Analysis of the Transistor Tuned Power Amplifier, *Proc. IEEE,* **52:** 311, 1964.

Gentzler, C.: Class A Design Techniques Create a New Broadband Amplifier, *RF Design,* May 1993, pp. 47–54.

Krauss, H. L., C. W. Bostian, and F. H. Raab: *Solid State Radio Engineering,* Wiley, New York, 1980.

Li, C-H., and Y-O. Lam: Maximum Frequency and Optimum Performance of Class E Power Amplifiers, *IEE Proc.-Circuits Devices, Syst.,* **141:**174–184, 1994.

Senak, P.: Amplitude Modulation of the Switched-Mode Tuned Power Amplifier, *Proc. IEEE,* **53:** 1658–1659, 1965.

# Modulators and Demodulators

## INTRODUCTION

Communication systems require circuits for frequency conversion, modulation, and detection. Modulation is the modification of a high-frequency carrier signal to include the information present in a relatively low frequency signal (the modulating signal). The information is modulated onto a higher-frequency signal because radio wave propagation is more efficient at higher frequencies and smaller antennas can be used. Also, a larger bandwidth can be obtained at the higher frequencies, enabling many information-containing signals to be multiplexed onto one carrier and sent simultaneously.

A practical illustration of this is the composite color-television signal used in the United States that was developed by the National Television Systems Committee (the NTSC system). This television signal includes a frequency-modulated sound signal with a resulting bandwidth of approximately ±25 kHz with 100 percent modulation. The sound signal is then shifted to a center frequency of 4.5 MHz so that it does not overlap with the video signal. The composite signal also includes the video information consisting of the picture signal and vertical and horizontal synchronization pulses with a bandwidth of 4 MHz. The color information, or chrominance signal, which is also contained in the composite signal, is modulated onto a 3.58-MHz subcarrier.

The total signal, then, includes time- and frequency-multiplexed components. This signal is then amplitude-modulated onto one of the standard broadcast channel carrier frequencies, which creates both upper- and lower-sideband frequency components. For standard TV broadcasting most of the lower sideband is filtered out before transmission. The removal of one sideband reduces the amount of power that must be transmitted and reduces the channel bandwidth, allowing for more channels in a given frequency spectrum. The frequency spectra of the transmitted signal are illustrated in Fig. 12.1. The signal bandwidth is 6 MHz, which includes 4.75 MHz above the carrier frequency and 1.25 MHz (the vestigial

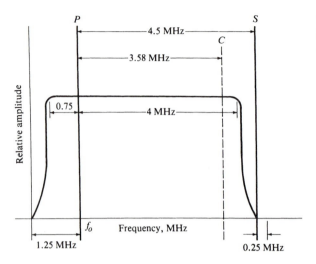

**FIGURE 12.1**
Frequency spectra of a standard broadcast TV channel.

sideband) below the carrier frequency. The color subcarrier $C$ is 3.58 MHz above the carrier frequency, and the center frequency of the sound signal is 4.5 MHz above the carrier frequency.

Demodulation, then separates the received information from the high-frequency carrier signal. Radio pioneers used the word *detector* for the device that detected the transmitted information. Today the terms *detector* and *demodulator* are often used interchangeably, but *detector* can have other connotations, such as when applied to an AGC detector or a frequency detector (for automatic frequency-control applications). In this chapter the terms *detector* and *demodulator* are used interchangeably, but it will be clear from the context which type of detector is being discussed.

In the following sections we will discuss the various types of mixers, modulators, and demodulators and their respective applications. The differences between time-varying and time-invariant circuits are also considered.

## ▪ 12.2

### FREQUENCY MIXERS

The most commonly used device for frequency modification is referred to as a *frequency mixer,* or simply *mixer*. The ideal mixer, represented by the schematic shown in Fig. 12.2, is a device which multiplies two input signals. If the inputs are sinusoids, the ideal mixer output is

$$V_o = (A_1 \sin \omega_1 t)(A_2 \sin \omega_2 t) = \frac{A_1 A_2}{2}[\cos (\omega_1 - \omega_2)t - \cos (\omega_1 + \omega_2)t] \quad (12.1)$$

The output consists of the sum and difference frequencies of the two input frequencies, one of which is the desired component. The other input frequency is removed with filtering—this combination of a mixer and filter to remove an output frequency is often referred to as a *single-sideband mixer*.

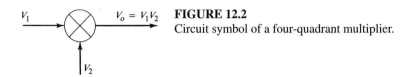

**FIGURE 12.2**
Circuit symbol of a four-quadrant multiplier.

Although the ideal mixer does not exist, there are many different circuits for approximating the ideal mixer. There are mixer circuits which provide gain (active mixers) and passive mixers which actually have a conversion loss. (The lowest-noise mixers are passive mixers.) Mixers can be classified in various ways, but in this text they are classified on the basis of the mode of operation, namely, nonlinear or switching-type mixers.

**Switching-Type Mixers**

In switching-type mixers, one or more switches, realized with diodes or transistors, function as time-varying circuit elements. The nonlinear or switching characteristics of diodes are often used for frequency mixing, particularly at high frequencies. Figure 12.3 provides an example of a simple switching-type mixer circuit containing diodes. If the center-tapped transformer is ideal, then the voltages will be as indicated in Fig. 12.4. The local oscillator $V_L$ is a constant-amplitude signal. The idea is for the local oscillator signal to be much larger than $V_i$ so that diode $D_1$ is on when $V_L$ is positive and diode $D_2$ is on when $V_L$ is negative. Therefore, the output voltage (ignoring the drop across the diode) will be

$$V_o = \begin{cases} V_i + V_L & V_L > 0 \\ -V_i + V_L & V_L < 0 \end{cases}$$

That is, the output consists of the local oscillator signal plus $V_i$ switched by 180° at the frequency of the local oscillator. If we consider this switched form of $V_i$ to be $V_i^*$, then

$$V_o = V_L + V_i^*$$

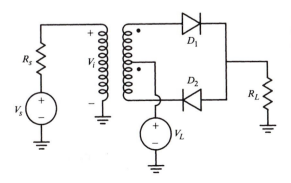

**FIGURE 12.3**
Circuit symbol of a simple two-diode switching-type mixer.

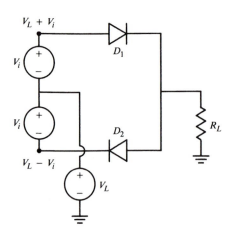

**FIGURE 12.4**
Simplified equivalent circuit of the two-diode mixer.

where
$$V_i^* = V_i P(t)$$

$$P(t) = \begin{cases} 1 & V_L > 0 \\ -1 & V_L < 0 \end{cases}$$

In Fig. 12.5, $P(t)$ is a square wave with a frequency equal to that of the local oscillator frequency $\omega_L$. It can be expanded in a Fourier series as

$$P(t) = \frac{4}{\pi} \sum_{n=0}^{\infty} \frac{\sin[(2n+1)\omega_L t]}{2n+1} \tag{12.2}$$

so
$$V_i^* = V_i \left[ \frac{4}{\pi} \sum_{n=0}^{\infty} \frac{\sin(2n+1)\omega_L t}{2n+1} \right]$$

If $V_i$ is a sine wave

$$V_i = V \sin \omega_i t$$

then
$$V_i^* = \frac{2V}{\pi} \sum_{n=0}^{\infty} \frac{\cos\left[(2n+1)\omega_L - \omega_i\right]t - \cos\left[(2n+1)\omega_L + \omega_i\right]t}{2n+1} \tag{12.3}$$

Since $V_o = V_L + V_i^*$, the mixer output consists of the local oscillator signal plus an infinite number of additional frequencies created in the mixer. The output frequencies

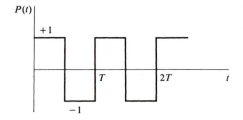

**FIGURE 12.5**
A local oscillator waveform.

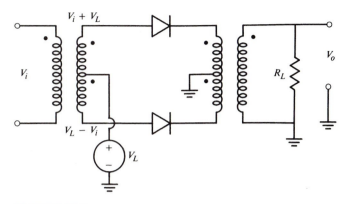

**FIGURE 12.6**
A two-diode mixer in which the load oscillator signal does not
appear in the output.

in addition to the upper and lower sidebands are called *spurious*. The desired mixing
component is selected from the mixer by filtering the output.

The preceding analysis assumed that the local oscillator signal was much larger than the input signal and sufficiently large to instantaneously switch the diodes.
Any deviation from these assumptions increases the distortion of the desired frequency component. Distortion is considered further in the discussion of double-
balanced mixers. A disadvantage of the mixer circuit shown in Fig. 12.3 is that the
local oscillator signal appears in the output. If the local oscillator frequency is much
larger than the input frequency, then the desired mixing product $\omega_L + \omega_i$ or $\omega_L - \omega_i$
may be close to the local oscillator frequency, and it will be difficult to separate the
frequencies by filtering.

The local oscillator signal does not appear in the output of the mixer circuit
shown in Fig. 12.6. If the center-tapped transformer is ideal, then the voltages will
be as shown in Fig. 12.7. If $V_L$ is positive and much larger than $V_i$, then both diodes
will be conducting, and $V_o = V_i$, since the local oscillator currents balance out in
the output transformer (assumed to have a unity turns ratio). If the local oscillator
signal is negative, the diodes will be open and the output signal will be equal to

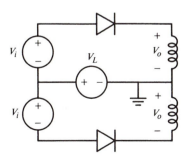

**FIGURE 12.7**
Simplified equivalent circuit of the diode
mixer shown in Fig. 12.6.

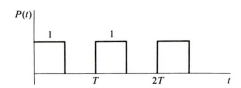

**FIGURE 12.8**
Output waveform of the mixer shown in Fig. 12.6.

zero. In general, the output voltage can be represented by

$$V_o = V_i P(t)$$

where

$$P(t) = \begin{cases} 1 & V_L > 0 \\ 0 & V_L \leq 0 \end{cases}$$

In this case $P(t)$ is a square wave (illustrated in Fig. 12.8) with a frequency equal to that of the local oscillator frequency, but it differs from the preceding case in that the dc value is no longer equal to zero. And $P(t)$ can be expanded in a Fourier series as

$$P(t) = \frac{1}{2} + \frac{2}{\pi} \sum_{n=0}^{\infty} \frac{\sin (2n + 1)\omega_L t}{2n + 1} \tag{12.4}$$

If $V_i$ is a sine wave

$$V_i = V \sin \omega_i t$$

then the output voltage is

$$V_o(t) = V \frac{\sin \omega_i t}{2} + \frac{V}{\pi} \sum_{n=0}^{\infty} \frac{\cos [(2n + 1)\omega_L - \omega_i]t - \cos [(2n + 1)\omega_L + \omega_i]t}{2n + 1} \tag{12.5}$$

The output of this mixer differs from that in Fig. 12.3 in that it does not contain the local oscillator signal, but it does contain a signal at the same frequency as the input signal. In some applications this can be a problem, as illustrated by the following example.

**EXAMPLE 12.1.** The input section of a general-coverage receiver with a 10-MHz intermediate-frequency (IF) filter is illustrated in Fig. 12.9. To tune the receiver to an input frequency of 20 MHz, the local oscillator frequency must be 10 MHz in a down-conversion receiver. If the mixer shown in Fig. 12.3 is used, then the local oscillator signal will be present at the IF output, and it will probably be so large that it will mask out the input signal. If the mixer is connected as illustrated in Fig. 12.6, the local oscillator signal will be removed from the output. Still, an unwanted 10-MHz signal present at the input will appear at the IF output unless it is removed by an input filter (preselector).

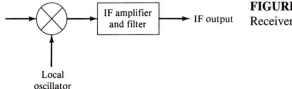

**FIGURE 12.9**
Receiver input section.

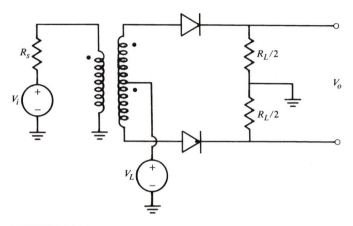

**FIGURE 12.10**
A double-balanced mixer.

A double-balanced mixer circuit, which can be used with a balanced load, is shown in Fig. 12.10. The behavior of this circuit is the same as that of the mixer illustrated in Fig. 12.6, and the output is given by Eq. (12.5) if one-to-one transformers are used.

A switching-type mixer containing four diodes in which neither the local oscillator signal nor the input signal appears at the output is shown in Fig. 12.11. If the local oscillator signal $V_L$ is positive, then diodes $D_2$ and $D_3$ will conduct and the equivalent circuit will be as shown in Fig. 12.12, with $r_d$ representing the equivalent diode on resistance. The circuit is more easily recognized if it is redrawn as shown in Fig. 12.13; the balanced output transformer eliminates the local oscillator signal from the load. The two loop equations are

$$V_i = (I_1 + I_2)R_L + I_1 r_d - V_L$$

and

$$V_i = (I_1 + I_2)R_L + I_2 r_d + V_L$$

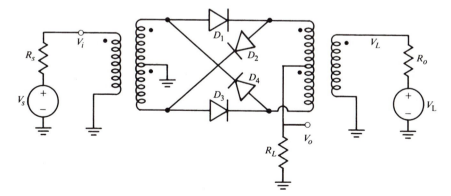

**FIGURE 12.11**
A four-diode switching-type mixer.

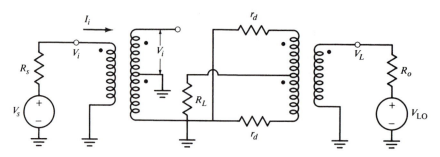

**FIGURE 12.12**
Equivalent circuit of Fig. 12.11 (for positive local oscillator voltage).

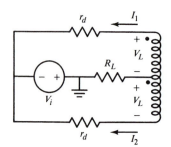

**FIGURE 12.13**
Circuit equivalent of that in Fig. 12.11.

If $V_L$ is eliminated,

$$I_1 + I_2 = \frac{V_i}{R_L + r_d/2} = -\frac{V_o}{R_L}$$

or

$$\frac{V_o}{V_i} = -\frac{R_L}{R_L + r_d/2}$$

If the local oscillator signal is negative, diodes $D_1$ and $D_4$ conduct, and the equivalent circuit is as shown in Figs. 12.14 and 12.15. The output voltage is then

$$V_o = \frac{V_i R_L}{R_L + r_d/2}$$

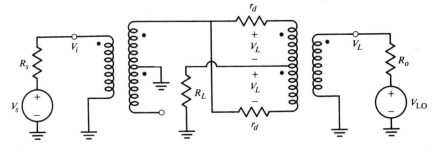

**FIGURE 12.14**
Equivalent circuit of Fig. 12.11 (for negative local oscillator voltage).

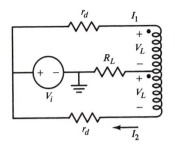

**FIGURE 12.15**
Circuit equivalent of that in Fig. 12.14.

In this mixer the output voltage is proportional to the input voltage and is switched at the local oscillator frequency; therefore,

$$V_o(t) = V_i P(t) \frac{R_L}{R_L + r_d/2}$$

where $P(t)$ is given by Eq. (12.4). If $V_i(t)$ is a sine wave

$$V_i = V \sin \omega_i t$$

then

$$V_o(t) = \frac{R_L}{R_L + r_d/2} \left\{ \frac{2V}{\pi} \sum_{n=0}^{\infty} \frac{\cos\left[(2n+1)\omega_L - \omega_i\right]t - \cos\left[(2n+1)\omega_L + \omega_i\right]t}{2n+1} \right\}$$

$$(12.6)$$

A double-balanced mixer with perfectly matched diodes and ideal transformer coupling will generate the upper and lower sidebands plus an infinite number of spurious frequencies centered on odd harmonics of the local oscillator frequency, but both the input and local oscillator signals are isolated from the output. Compact, inexpensive double-balanced mixers are commercially available that cover the frequency spectrum from the tens of kilohertz to the gigahertz region. Their excellent performance is due in a large part to modern fabrication techniques which enable one to construct closely matched diodes. High-frequency Schottky barrier diodes are invariably used today.

**Conversion Loss**

Mixer *conversion loss* is defined as the ratio of output power in one sideband to signal input power. It is a most important mixer parameter, particularly for the receiver input stage. To calculate the conversion loss, we will assume that the external impedances are adjusted for maximum power transfer. Consider first the double-balanced mixer circuit shown in Fig. 12.11. If the input transformer has a 1:1 turns ratio, then the equivalent circuit is as shown in Fig. 12.13, and the load impedance seen by $V_i$ is

$$\frac{V_i}{I_1 + I_2} = \frac{V_i}{I_i} = R_L + \frac{r_d}{2}$$

Normally $R_L \gg r_d$, so the input will be matched for maximum power transfer if $R_L = R_s$. Under this condition $V_i = V_s/2$ and

$$P_i = \frac{V_s^2}{4R_L}$$

From Eq. (12.6) we see that the output voltage in one sideband (assuming $R_L \gg r_d$) is

$$V_o|_{\omega_L \pm \omega_i} = \frac{V_i \times 2}{\pi} = \frac{V_s}{\pi}$$

and the output power is

$$P_o = \frac{V_s^2}{\pi^2 R_L}$$

So the conversion gain of the double-balanced mixer is

$$G = \frac{P_o}{P_i} = \frac{4R_L}{\pi^2 R_L} = \frac{4}{\pi^2} \tag{12.7}$$

which is less than 1. The mixer has a conversion loss of

$$L = 10 \log \frac{\pi^2}{4} \approx 4 \text{ dB}$$

For an ideal double-balanced mixer matched to the source impedance, and ignoring the power lost in the transformer and switching diodes, approximately 40 percent of the signal input power will be transferred to the output.

For the single-balanced mixer shown in Fig. 12.5, the output voltage of one sideband [from Eq. (12.5)] is

$$V_o|_{\omega_L + \omega_i} = \frac{V_i}{\pi}$$

If the input port is matched for maximum power transfer,

$$V_o = \frac{V_s}{2\pi}$$

$$P_i = \frac{V_s^2}{4R_L}$$

and

$$P_o = \frac{V_s^2}{4\pi^2 R_L}$$

The power gain is

$$G = \frac{P_o}{P_i} = (\pi^2)^{-1} \tag{12.8}$$

so the conversion loss is $L = 10 \log \pi^2 \approx 10$ dB, and the conversion loss of the single-balanced mixer is 4 times (6 dB) larger than that of the double-balanced mixer.

### Distortion

As the mixer input signal power increases, it will reach the level at which it is larger than the local oscillator power. The input signal then assumes the switching role, and the output power becomes proportional to the local oscillator power. Since the local oscillator power is constant, the output power will be constant. An idealized power transfer characteristic is shown in Fig. 12.16. At low input power levels, the power transfer is linear; but as the input power increases, distortion begins and the response becomes nonlinear. At high input levels, the output saturates at a level proportional to the local oscillator power. As the signal level further increases, the intermodulation distortion (IMD) also increases.

### Intermodulation Distortion in Diode-Ring Mixers [1]

Consider a diode-ring mixer with a resistance $R$ in series with each diode, as shown in Fig. 12.17. The purpose of the additional resistors will become clear once the IMD is determined. The transformers are assumed to be ideal, and the diodes all have identical characteristics. If the local oscillator power is sufficiently large, the circuit during either half-cycle is as shown in Fig. 12.18. The diode current then consists of a constant component $I$, due to the local oscillator, and a small component $i$, due to the input signal. The diode current is described by

$$i_D = I_s e^{V_d/V_T}$$

where $V_d$ is the voltage drop across the diode and $V_T = kT/q$.

The input signal $V_i$ causes a signal current $2i$ to flow through the load. Because of the circuit symmetry one-half flows through each diode. That is,

$$i_{D_1} = I - i$$

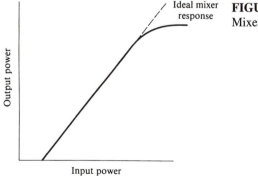

Ideal mixer response

**FIGURE 12.16**
Mixer power transfer characteristics.

Output power

Input power

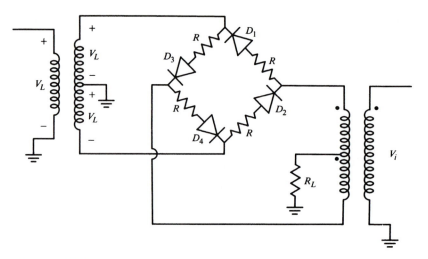

**FIGURE 12.17**
A four-diode mixer with linearizing resistors.

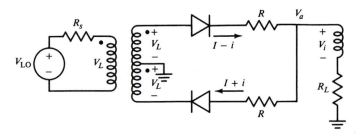

**FIGURE 12.18**
Equivalent circuit of mixer shown in Fig. 12.17.

and
$$i_{D_2} = I + i$$

The currents are shown in Fig. 12.18. The voltage equations are

$$V_L - V_a = V_{D_1} + i_{D_i} R$$

and
$$-V_L - V_a = -V_{D_2} - i_{D_2} R$$

or
$$V_L - V_a = V_T \ln \frac{I - i}{I_s} + (I - i)R$$

and
$$-V_L - V_a = -\left[ V_T \ln \frac{I + i}{I_s} + (I + i)R \right]$$

Adding the two equations, one obtains

$$-2V_a = V_T \ln \frac{I - i}{I + i} - 2iR$$

and since
$$2V_a = 2(V_i - 2iR_L)$$

the relation between input voltage and diode current is

$$V_i = i(2R_L + R) + \frac{V_T}{2} \ln \frac{I+i}{I-i}$$

This can be expanded for $i \ll I$ to

$$V_i = (2R_L + R)i + \frac{V_T}{2} \left[ \frac{i}{I} - \frac{1}{2}\left(\frac{i}{I}\right)^2 + \frac{1}{3}\left(\frac{i}{I}\right)^3 \left(\frac{i}{I}\right)^3 + \cdots \right]$$
$$- \left[ -\frac{i}{I} - \frac{1}{2}\left(\frac{i}{I}\right)^2 - \frac{1}{3}\left(\frac{i}{I}\right)^3 + \text{higher-order terms} \ldots \right]$$

The even-order terms cancel, so

$$V_i = \left( 2R_L + R + \frac{V_T}{I} \right) i + \frac{1}{3}\left(\frac{i}{I}\right)^3 + \text{odd higher-order terms}$$

Since the first term of the power series is not zero, the series can be inverted[2]:

$$i = \frac{V_i}{2R_L + R} - \frac{V_T}{3} \frac{V_i^3}{(2R_L + R)^4 I^3}$$

Comparing this equation with Eq. (3.64), we see that

$$k_3 = -\frac{V_T}{3(2R_L + R)^4 I^3} \tag{12.9}$$

The third-order intermodulation distortion is proportional to $k_3^2$ [Eq. (2.60)], so the smaller the magnitude of $k_3$, the smaller will be the IMD. The IMD is inversely proportional to the sixth power of the current $I_1$, so the larger the local oscillator drive level, the smaller will be the IMD. The addition of a resistor $R$ in series with the diode reduces the IMD. An additional diode can be used in place of the resistor. This permits higher local oscillator drive levels and a corresponding reduction in IMD.

Diode-ring mixers are classified by the different elements in each leg. Figure 12.19 illustrates three classes of diode-ring mixers. The class II mixers have the

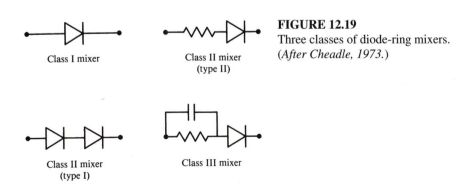

Class I mixer

Class II mixer
(type II)

Class II mixer
(type I)

Class III mixer

**FIGURE 12.19**
Three classes of diode-ring mixers.
(*After Cheadle, 1973.*)

highest power-handling capabilities, but because of the additional matched components, they are also the most expensive to manufacture. The capacitor in parallel
with the series resistor in the class III diodes acts to reduce the distortion that
occurs when the local oscillator signal switches state; it also reduces conversion
loss by acting as a high-frequency bypass for the signal voltage. At high frequencies the effects of finite switching time also cause distortion. It can be shown that
for the same power levels, a square-wave signal will cause less distortion than will
be the case if a sine wave signal is used for the local oscillator.[3]

### Square-Law Mixers

The square-law characteristic is approximated by several electronic devices. That a
square-law device can function as a mixer is readily seen by squaring the sum of
two sine waves:

$$(A_1 \sin \omega_1 t + A_2 \sin \omega_2 t)^2 = (A_1^2 \sin \omega_1 t)^2 + (A_2 \sin \omega_2 t)^2 + 2A_1 A_2 \sin \omega_1 t \sin \omega_2 t$$

$$= \frac{A_1^2(1 - \cos 2\omega_1 t)}{2} + \frac{A_2^2(1 - \cos 2\omega_2 t)}{2}$$

$$+ \frac{2A_1 A_2[\cos (\omega_1 - \omega_2)t - \cos (\omega_1 + \omega_2)t]}{2} \quad (12.10)$$

An ideal square-law device will provide the upper and lower sidebands, together
with a dc component and the second harmonic of both input waveforms. The circuit
is frequently used at microwave frequencies for down conversion to the lower sideband, which is at a lower frequency than either of the input signals. A simple square-
law mixer, which has been used since the earliest radio receivers were built, is
shown in Fig. 12.20. Crystal diodes were originally used, but today Schottky barrier diodes are the best choice.

At lower frequencies this form of the diode mixing is normally not used
because of the large conversion loss. Transistor mixers are preferred because they
can provide conversion gain. Transistors are often used to approximate the square-
law characteristic. The input and local oscillator signal voltages are applied to the
transistor so that they effectively add to the dc bias voltage to produce the total
gate-source or base-emitter voltage. The composite signal is then passed through
the device nonlinearity to create the sum and difference frequencies.

**FIGURE 12.20**
A simple square-law mixer-detector.

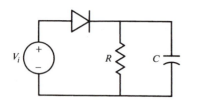

## BJT Mixers

Figure 12.21 illustrates a simplified bipolar junction transistor (BJT) mixer. For this circuit the base-to-emitter voltage is

$$V_{be} = V_{DC} + V_i - V_L$$

where $V_{DC}$ is the base-to-emitter bias voltage. The collector current in a bipolar transistor is described by $(V_{be} > 0)$

$$i_C = I_s e^{V_{be}/V_T}$$

and since $\qquad\qquad V_{be} = V_{DC} + V_i - V_L$

the current is $\qquad\qquad i_C = I_s e^{V_{DC}/V_T} e^{V_i/V_T} e^{-V_L/V_T}$

If $V_i = V_1 \cos \omega_i t$ and $V_L = V_L \cos \omega_o t$, then the current can be expanded in a series of modified Bessel functions[4] as

$$i_C(t) = I_s e^{V_{DC}/V_T} [I_o(y)I_o(x) - 2I_o(y)I_1(x) \cos \omega_o t + 2I_1(y)I_o(x) \cos \omega_i t$$

$$- 4I_1(y)I_1(x) \cos \omega_i t \cos \omega_o t + \text{higher-order terms}] \quad (12.11)$$

where $y = V_1/V_T$, $x = V_L/V_T$, and $I_n$ is the $n$th-order modified Bessel function.

The collector current consists of a dc component $I_C$, components at both the input and local oscillator frequencies, components at the frequencies $\omega_o \pm \omega_i$, and an infinite number of high-frequency components. The amplitude of either the upper- or lower-sideband component is

$$I = I_s e^{V_{DC}/V_T} 2I_1(y)I_1(x)$$

$$= 2I_C \frac{I_1(y)I_1(x)}{I_o(y)I_o(x)} \qquad\qquad (12.12)$$

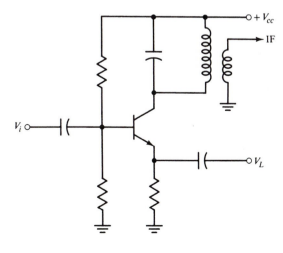

**FIGURE 12.21**
A bipolar junction transistor (BJT) mixer.

The local oscillator voltage amplitude is constant, and if $V_L \gg V_1$, then the collector direct current will not vary with changes in the amplitude of the input signal, since

$$\lim_{y \to 0} I_o(y) = 1$$

The mixer should have a linear response to changes in the input amplitude. The ratio is given as

$$\frac{I_1(y)}{I_o(y)} \approx \frac{y}{2}\left(1 - \frac{y^2}{8} + \frac{y^4}{16}\right)$$

So if the input amplitude is sufficiently small, the mixer upper- and lower-sideband outputs will be a linear function of the input signal. For $y \leq 0.4$ $(V_1 \leq 10.5 \text{ mV})$ the response will be within 2 percent of a linear response. The amplitude of the sideband current is

$$I = 2I_C \frac{I_1(y)}{I_o(y)} \frac{I_1(x)}{I_o(x)}$$

$$\approx I_{CY} \frac{I_1(x)}{I_o(x)} = g_m V_1 \frac{I_1(x)}{I_o(x)} \tag{12.13}$$

provided the input signal is sufficiently small. Note that $g_m$ is the transconductance $I_C/V_T$. The ratio $I_1(x)/I_o(x)$ rapidly approaches 1 as $x$ increases:

$$\lim_{x \to \infty} \frac{I_1(x)}{I_o(x)} = 1$$

and          $$\frac{I_1(x)}{I_o(x)} \approx 0.86 \qquad \text{for } x = 4$$

So if the local oscillator drive level is approximately 100 mV $(x \approx 4)$ or larger, the amplitude of the upper- and lower-sideband current components is $I \approx g_m V_1$. If the collector circuit is tuned to either of these frequencies, the mixer conversion gain $G = g_m R_L$, where $R_L$ is the equivalent load resistance at the frequency of interest. The BJT mixer has the advantage over diode mixers in that it provides conversion gain, although it will be much noisier than a properly designed diode-ring-type mixer. Another advantage of the BJT mixer over diode mixers is that the local oscillator drive level required is much smaller. This reduces oscillator design and system-shielding requirements.

### FET Mixers

If an FET is operated in its "constant-current" region, the idealized FET current transfer characteristic is the square-law relation

$$i_D = I_{DSS}\left(1 - \frac{V_{gs}}{V_p}\right)^2 \tag{12.14}$$

where $V_{gs}$ is the gate-to-source voltage and $V_p$ is the transistor pinch-off voltage. Because of the square-law characteristic, the FET will not generate any harmonics higher than second-order. Since the third-order intermodulation distortion arises from the cubic term in the transfer characteristic, an ideal FET mixer will not produce any third-order intermodulation distortion. However, in reality, the transfer characteristic deviates from the idealized version, and some intermodulation distortion will be produced. Still, a properly biased and operated FET mixer will produce much smaller high-order mixing products than a bipolar transistor. This is one reason why an FET mixer is usually preferred to a bipolar transistor mixer. The FET also provides at least 10 times as great an input voltage range as the BJT. Figure 12.22 illustrates an FET mixer circuit. The drain current is (in the constant-current region)

$$i_D = I_{DSS} \left( 1 - \frac{v_i - v_L + V_{DC}}{V_p} \right)^2 \tag{12.15}$$

where $V_{DC}$ is the gate-to-source bias voltage (or $V_{GS} - V_T$ for a MOSFET). If the applied signals are sine waves

$$v_i = V_i \sin \omega_i t$$

and

$$v_L = V_L \sin \omega_L t$$

then the output current [obtained by expanding Eq. (12.15)] is

$$i_D = I_{DC} + K_1(V_i \sin \omega_i t - V_L \sin \omega_L t) + K_2(V_i \sin 2\omega_i t + V_L \sin 2\omega_L t)$$
$$- K_3 \sin(\omega_i - \omega_L)t + K_3 \sin(\omega_i + \omega_L)t \tag{12.16}$$

The amplitude of the sum and difference frequencies is

$$K_3 = \frac{I_{DSS} V_i V_L}{V_p^2}$$

The term

$$\frac{K_3}{V_i} = \frac{I_{DSS} V_L}{V_p^2} \tag{12.17}$$

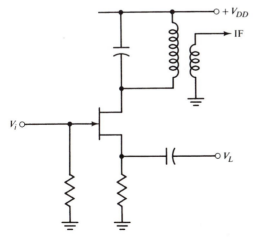

**FIGURE 12.22**
An FET mixer circuit.

is referred to as the *conversion transconductance* $g_c$. In general, the device with the lowest pinch-off voltage has the highest gain, and the conversion transconductance is directly proportional to the amplitude of the local oscillator signal. It would also appear that FETs with high $I_{DSS}$ are preferred, but $I_{DSS}$ and $V_p$ are related. It is usually the case that devices selected for high $I_{DSS}$ also have a high $V_p$ and a lower conversion transconductance than low-$I_{DSS}$ devices. Since the device is to be operated in the constant-current region, $V_L$ must be less than the magnitude of the pinch-off voltage. If

$$V_L = \frac{|V_p|}{2}$$

then

$$K_3 = \frac{I_{DSS}}{2V_p} V_i$$

and the sideband current is

$$i_D = K_3 \sin(\omega_i \pm \omega_2)t = V_1 \frac{I_{DSS}}{2V_p} \sin(\omega_i \pm \omega_1)t$$

Since for a JFET the transconductance is

$$g_m = \frac{\partial i_D}{\partial V_{gs}} = -2\frac{I_{DSS}}{V_p}\left(1 - \frac{V_{gs}}{V_p}\right)$$

$$= -2\frac{I_{DSS}}{V_p}\bigg|_{V_{gs}=0} \tag{12.18}$$

the conversion transconductance is one-fourth the small-signal transconductance evaluated at $V_{gs} = 0$ (provided $V_L = V_p/2$). For a MOSFET it can be shown that the conversion conductance cannot exceed one-half of the transconductance of the device when it is used as a small-signal amplifier.

**EXAMPLE 12.2.** An FET with $I_{DSS} = 40$ mA and transconductance $g_m = 14 \times 10^{-3}$ S at $V_{gs} = 0$ is to be used in a mixer. Estimate the conversion gain if a 50-$\Omega$ load is used.

***Solution.*** Since

$$g_m = \frac{-2I_{DSS}}{V_p}\left(1 - \frac{V_{gs}}{V_p}\right)$$

the transistor pinch-off voltage is

$$|V_p| = \frac{2 \times 40 \times 10^{-3}}{14 \times 10^{-3}} = 5.7 \text{ V}$$

The peak local oscillator voltage $V_L$ should be kept to approximately 50 percent of this value in order to minimize distortion. If $V_L \approx 2.85$ V, then the conversion transconductance is found from Eq. (12.18) to be

$$g_c = \frac{40 \times 10^{-3}}{2 \times 5.7} = 3.5 \times 10^{-3} \text{ S}$$

The magnitude of the voltage gain is

$$A_v \approx g_c R_L = 1.75$$

If the device is operated in the common-gate configuration, the current gain will be close to unity, so the voltage gain will also be equal to the power gain.

Although the conversion transconductance is smaller than the small-signal transconductance, it is large enough that the circuit can be operated as a mixer with power and voltage gains. This is an important difference from the diode-switching mixer. An FET mixer is capable of producing lower intermodulation and harmonic products than a comparable bipolar or diode mixer. Also, an FET mixer operating at a high level has a larger dynamic range and greater signal-handling capacity than does a diode mixer operated at the same local oscillator level. However, the noise figure of FET mixers is currently higher than that of diode mixers. The best intermodulation and cross-modulation performance is obtained with the FET operated in the common-gate configuration, where the input impedance is much lower than that for the common-source configuration. Figure 12.23 illustrates a double-balanced mixer in which the FET transistors are operated in the common-gate configuration. The push-pull output cancels the even-order output harmonics.

The dual-gate MOSFET is often used as a mixer. A typical dual-gate MOSFET mixer circuit is illustrated in Fig. 12.24. If the input signals are sinusoidal, the output will contain frequency components at both the sum and difference frequencies. Several other frequency components are also present in the output. The magnitude of either the sum or difference frequency is proportional to

$$A = K|V_{g_2}|$$

so the conversion gain is proportional to the magnitude of the local oscillator voltage. For maximum conversion gain, the local oscillator amplitude should be selected so that it drives the gate just to the point of transistor saturation.

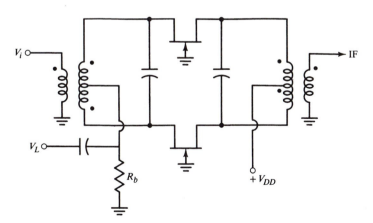

**FIGURE 12.23**
An FET double-balanced mixer.

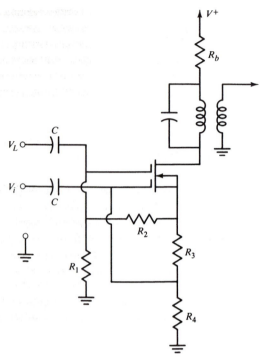

**FIGURE 12.24**
A dual-gate MOSFET mixer circuit.

The input signal is normally connected to the lower (closest to the ground) input gate terminal and the local oscillator signal to the upper gate. If the input is connected to the upper terminal, then the drain resistance of the lower transistor section appears as a source resistance to the input signal. It has been shown in Chap. 2 that the source resistance will reduce the voltage gain at the collector. Also, the connection has a larger drain-to-gate capacitance with a lower bandwidth than is attainable when the input signal is connected to the lower gate. The device is usually biased so that both transistors are operating in their triode (nonsaturated) region. In this region it can be shown[5] that the small-signal drain current is

$$i_d = g_{m_1} V_{g_1} + g_{m_2} V_{g_2}$$

where

$$g_{m_1} = a_o + a_1 V_{g_1} - a_2 V_{g_2}$$

and

$$g_{m_2} = b_o + b_1 V_{g_1} + b_2 V_{g_2}$$

The drain current can be written as

$$i_d = a_1 V_{g_1}^2 + a_o V_{g_1} + (a_2 + b_1) V_{g_1} V_{g_2} + b_o V_{g_2} + b_2 V_{g_2}^2 \qquad (12.19)$$

Since the drain current contains the product of the two signals, the dual-gate MOSFET can be used as a mixer when both transistors are operated in the nonsaturated region.

## ■ 12.3

### AMPLITUDE AND PHASE MODULATION AND DEMODULATION

Amplitude modulation (AM) is the process of varying the amplitude of a constant-frequency signal with a modulating signal. Continuous-wave (CW) modulations turn the carrier on and off in response to a modulating signal. It can be considered a special case of AM and also a special case of frequency modulation. In this section we concentrate on the circuit aspects of modulators and demodulators, but since the circuitry always depends on the frequency spectrum of the signal, the mathematical models for the various modulation methods are introduced in order to describe the frequency distributions of the modulated signals.

### Amplitude Modulation

An amplitude-modulated wave can be mathematically expressed as

$$S(t) = g(t) \sin \omega_c t$$

where $g(t)$ is the modulating signal (modulation) and $\omega_c$ is the carrier frequency. Normally the modulating signal varies slowly compared with the carrier signal frequency. For conventional AM the modulated signal is in the form of

$$S(t) = A[1 + mf(t)] \sin \omega_c t \qquad (12.20)$$

where $m$ is called the *modulation factor* and is normally less than 1; $100m$ is the *percentage of modulation*. Consider a simple modulating signal:

$$f(t) = \cos \omega_m t$$

then

$$S(t) = A \left\{ \sin \omega_c t + \frac{m}{2} [\sin (\omega_c + \omega_m)t + \sin (\omega_c - \omega_m)t] \right\} \qquad (12.21)$$

The frequency spectrum of the modulated signal is shown in Fig. 12.25. It consists of the carrier frequency plus upper- and lower-sideband components centered about the carrier frequency. This signal is often referred to as the *full-carrier double-sideband signal*. Figure 12.26 illustrates a sinusoidal modulating signal and the

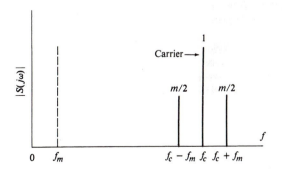

**FIGURE 12.25**
Frequency spectrum of an amplitude-modulated signal.

**FIGURE 12.26**
Amplitude-modulating and
amplitude-modulated signals.

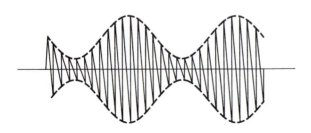

carrier amplitude modulated by the sinusoid. The signal envelope is of the same
form as the modulating signal ($m < 1$).

Equation (12.21) shows that for $m < 1$ the amplitude of the carrier is at least
twice as large as the amplitude of either sideband component, so at least two-thirds
of the signal power will be in the carrier and at most one-third in the two sidebands.
Because the carrier does not contain any modulating information, it is often removed
or suppressed, resulting in the signal

$$S(t) = \frac{Am}{2}[\sin(\omega_c + \omega_m)t + \sin(\omega_c - \omega_m)t] \qquad (12.22)$$

which is referred to as a *double-sideband (DSB) suppressed-carrier signal*. The carrier component is not present in the DSB signal. However, as the waveform gets
more efficient in terms of power-to-information content, the detection method gets
more complex. Some means of recovering the carrier component is needed for the
detector to recover the amplitude and frequency of the modulating signal. The DSB
signal, although more efficient in terms of transmitted power, still occupies the
same bandwidth as a normal AM signal. Since both sidebands contain the same
information, one sideband can be removed, resulting in a *single-sideband-signal
(SSB)*. SSB is the most efficient form of an amplitude-modulated signal. Circuitry
for the three types of AM will now be described.

### Amplitude Modulators: Standard AM

Full-carrier double-sideband amplitude modulation is achieved either by modulating the oscillator signal at a relatively low power level and amplifying the modulated signal with a cascade of amplifiers (including an output power amplifier) or
by using the modulating signal to control the supply voltage of the power amplifier. Both methods are illustrated in Fig. 12.27. The power requirements of the
modulator and modulating signal can be estimated by considering the power in an
amplitude-modulated waveform:

$$S(t) = A[1 + m(t)]\sin \omega_c t$$

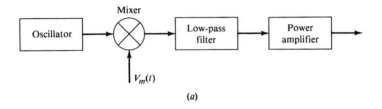

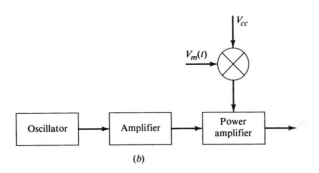

(b)

**FIGURE 12.27**
(a) A low-level amplitude-modulation circuit; (b) amplitude modulation at high power levels.

The peak output power is

$$(P_o)_{\text{peak}} = \frac{A^2}{2}(1+m)^2 \tag{12.23}$$

so if the maximum modulation index is unity,

$$(P_o)_{\text{peak}} = 2A^2 = 4(P_o)_{\text{car}}$$

The modulator must be designed to handle 4 times the average carrier power with 100 percent modulation; the output power will be 4 times the carrier power.

Several of the previously analyzed mixer circuits with output filtering can be used to realize the low-level modulation illustrated in Fig. 12.27a. For example, the output of the diode mixer shown in Fig. 12.3 is given by Eq. (12.3). If $V_L$ is a sine wave

$$V_L = V_1 \sin \omega_L t$$

and if a low-pass filter is added to the output with a bandwidth of

$$B = \omega_L + \omega_i$$

then the output will be

$$S(t) = V_1 \left(1 + \frac{4}{\pi}\frac{V}{V_1}\sin \omega_i t\right) \sin \omega_L t$$

Since the low-pass filter removes the higher-frequency component, the modulation index of the resulting AM waveform is $m = (4/\pi)V/V_1$. This particular amplitude modulator functions well only for low indices of modulation.

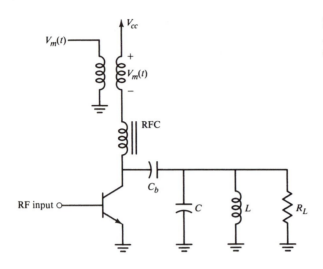

**FIGURE 12.28**
A collector-modulated circuit.

Both the FET and BJT mixers described earlier can function as amplitude modulators with a relatively high modulation index. The final amplifier will need to be linear. Linear class C can be obtained with the output circuit tuned to the carrier frequency. The output will then be linearly related to the input, provided the amplifier output circuit is not current-limited. (The peak value of the current pulses must be proportional to the peak value of the input drive current.)

The most frequently used method of amplitude modulation at high power levels is to modulate the supply voltage to the power amplifier, as shown in Fig. 12.27b. Figure 12.28 illustrates a collector-modulated circuit. The transistor can be operated as either class B or C; the output circuit is tuned to the carrier frequency and has a bandwidth equal to twice that of the modulating signal. The modulating signal is applied in series with the dc supply voltage, so the total low-frequency supply voltage for the transistor is

$$V = V_{cc} + V_m(t)$$
$$= V_{cc}(1 + m \cos \omega_m t)$$

provided           $V_m(t) = m V_{cc} \cos \omega_m t$

It was shown in the discussion of class C power amplifiers that the amplitude of the output signal under saturation-limited conditions equals the power supply voltage. Therefore, changing the transistor supply voltage modulates the output signal amplitude proportionally, and the output voltage becomes

$$V_o = V_{cc}(1 + m \cos \omega_m t) \cos \omega_c t$$

For 100 percent modulation the peak value of the voltage $V_m(t)$ must equal $V_{cc}$. At full modulation the sidebands each contain one-fourth the power of the carrier, so $P_o = \frac{3}{2} P_c$. The carrier signal has an amplitude $V_{cc}$, and the amplitude of each sideband component is one-half that of the carrier amplitude. The total output power is then

$$P_o = \frac{3}{2} \frac{V_{cc}^2}{2R_L}$$

The unmodulated carrier power is supplied by the power supply. The remaining power must be furnished by the modulator. One reason that output modulation has been the most frequently used method is that collector modulation results in less intermodulation distortion than existing low-level modulation circuits.

All the information in an AM wave is contained in one sideband. It is possible to eliminate the other sideband without loss of information; thus the required transmitter power is reduced to one-third of that previously required, since the power in each sideband is one-fourth the carrier power for sine wave modulation. Double-sideband suppressed-carrier modulation can be easily realized with the double-balanced mixers previously described. Although transmitter complexity is not markedly increased for DSB signals, the receiver is more complex. Single-sideband suppressed-carrier transmission results in an even more complex receiver, and the quality of the demodulated signal is generally not as good. However, SSB is extensively used because of the reduced power and bandwidth requirements. The signal quality is adequate for applications of voice transmission.

The simplest method of SSB generation is to generate the DSB signal using a double-balanced modulator and then remove one of the sidebands with a filter. A block diagram of this form of SSB generation is shown in Fig. 12.29. Another method of SSB generation, the *phasing method,* is illustrated in Fig. 12.30. Here both the modulating signal and the carrier signal are processed through phase splitters, which each generate two signals 90° out of phase with each other. The summing network output

$$S(t) = A \cos \omega_c t \sin \omega_o t + A \sin \omega_c t \cos \omega_o t$$

$$= A \sin (\omega_c + \omega_o)t$$

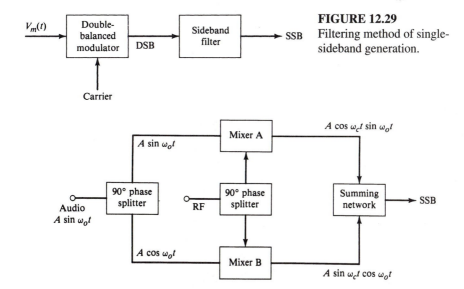

**FIGURE 12.29**
Filtering method of single-sideband generation.

**FIGURE 12.30**
Phasing method of single-sideband generation.

is the desired SSB signal. The phasing method has the advantage of not requiring the sharp cutoff filters of the filtering method of SSB generation, but it is difficult to realize a broadband phase-shifting network for the lower-frequency modulating signal. The television signal described at the beginning of the chapter uses vestigial sideband plus the low-frequency portion (the part closest to the carrier) of the other sideband. The technique is used primarily to conserve bandwidth. It is suitable for television because the low-frequency components are sufficient for reconstructing an acceptable picture.

## Demodulators

AM detection can be divided into synchronous and asynchronous detection. Synchronous detection employs a time-varying or nonlinear element synchronized with the incoming carrier frequency. Otherwise the detection is asynchronous. The simplest asynchronous detector, the average envelope detector, is described here.

### Average Envelope Detectors

A block diagram of the average envelope detector is shown in Fig. 12.31. The rectifier output

$$V_r(t) = \begin{cases} S(t) & S(t) > 0 \\ 0 & S(t) < 0 \end{cases}$$

can be written as

$$V_r(t) = S(t)P(t)$$

If $S(t)$ is periodic with a frequency $\omega_c$, then

$$P(t) = \begin{cases} 1 & \text{for } S(t) \geq 0 \\ 0 & \text{for } S(t) \leq 0 \end{cases}$$

And $P(t)$ is a rectangular wave (identical to that illustrated in Fig. 12.8) with the same frequency $\omega_c$, and so [from Eq. (12.4)]

$$P(t) = \frac{1}{2} + \frac{2}{\pi} \sum_{n=0}^{\infty} \frac{\sin (2n + 1)}{2n + 1} \omega_c t$$

If $S(t)$ is the AM wave described by Eq. (12.20), then

$$V_r(t) = A[1 + mf(t)] \left( \frac{\sin \omega_c t}{2} + \pi^{-1} + \frac{\cos 2\omega_c t}{\pi} + \text{higher harmonics of } \omega_c \right)$$

$$(12.24)$$

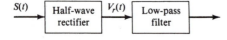

**FIGURE 12.31**
Block diagram of an average envelope detector.

If the low-pass filter bandwidth is chosen to filter out the component at $\omega_c$ and all higher harmonics, the output will be

$$V_o(t) = \frac{A[1 + mf(t)]}{\pi}$$

which is a dc term plus the modulating information.

Two additional points will be made to further describe the operation of the envelope detector. First, consider the case where

$$f(t) = \sin \omega_m t$$

Then Eq. (12.24) can be written as

$$V_r(t) = A \left[ \frac{\sin \omega_c t}{2} + \frac{m}{\pi} \sin \omega_m t + \pi^{-1} + \pi^{-1} \frac{\cos(\omega_c - \omega_m)t - \cos(\omega_c + \omega_m)t}{2} \right.$$

$$\left. + \text{higher-frequency terms} \right] \qquad (12.25)$$

The output will contain a term at the frequency $\omega_c - \omega_m$, which must also be removed by the low-pass filter. This is not possible if $\omega_m$ is close to $\omega_c$. To ensure that this distortion does not occur, the maximum modulating frequency should be constrained so that

$$(\omega_m)_{\max} \leq \frac{\omega_c}{2}$$

and the corresponding low-pass filter bandwidth $B$ must be selected so that

$$V_r(t) > 0 \quad \text{if} \quad S(t) > 0$$

This is only possible if $m$ is not greater than 1 (the maximum modulation is 100 percent) and the carrier term is present. Average envelope detection will only work for normal AM with a modulation index less than 1. However, if a large carrier component $A \cos \omega_c t$ is added to the SSB signal, the resultant signal can also be detected with an envelope detector (see Prob. 12.10).

A simple diode envelope detector circuit is shown in Fig. 12.32. It is assumed here that the input signal amplitude is large enough that the diode can be considered either on or off, depending upon the input signal polarity. The diode can then be replaced by an open circuit when it is reverse-biased and by a constant resistance

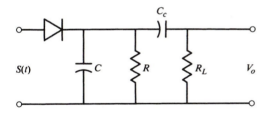

**FIGURE 12.32**
A diode envelope detector.

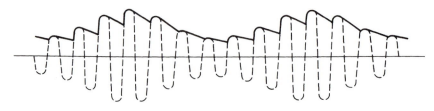

**FIGURE 12.33**
Envelope detector input (dashed line) and output waveforms.

when it is forward-biased. The series capacitor $C_c$ is included to remove the dc component [Eq. (12.24)]. The purpose of the load capacitor $C$ in the circuit is to eliminate the high-frequency component from the output and to increase the average value of the output voltage. The effect of the load capacitor can be seen from Fig. 12.33, which illustrates the input and output signal waveforms of a diode detector. As the input signal is applied, the capacitor charges up until the input waveform begins to decrease. At this time the diode becomes open-circuited, and the capacitor discharges through the load resistance $R_L$ as

$$V_L = V_p e^{-t/(R_L C)}$$

where $V_p$ is the peak value of the input signal, and the diode opens at time $t = 0$. The larger the value of capacitance used, the smaller will be the output ripple (which means the smaller will be the undesired high-frequency components and the larger will be the dc value of the output voltage). However, $C$ cannot be too large, or it will not be able to follow changes in the modulated signal. The time constant is often selected as

$$RC = [(\omega_m \omega_c)^{-1}]^{1/2}$$

If the highest modulation frequency approaches the carrier frequency, these two constraints become incompatible and some other form of detection, such as a synchronous detector, must be used.

**Synchronous Detection**

Envelope detectors will not detect some AM signals such as DSB. If it is possible to obtain a signal synchronized in frequency and phase with the original carrier, then it is possible to readily detect DSB signals. Some communications systems transmit a small pilot carrier signal synchronized with the original carrier signal. FM stereo uses this technique. If a local oscillator signal synchronized with the carrier signal is available, then demodulation can be accomplished, as shown in Fig. 12.34. Consider, for example, the DSB signal given by Eq. (12.22). If the local oscillator signal is

$$V_L = V \sin \omega_c t$$

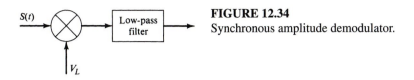

**FIGURE 12.34**
Synchronous amplitude demodulator.

then the output signal is

$$V_o = V_L S(t)$$

$$= \frac{Am}{4} V [2 \cos \omega_m t - \cos (2\omega_c + \omega_m)t - \cos (2\omega_c - \omega_m)t]$$

If this signal is low-pass-filtered ($\omega_m < B \le \omega_c$), the output is

$$V_o' = \frac{AVm}{2} \cos \omega_m t$$

which is proportional to the original modulating signal. The same form of synchronous detection can be used to detect normal AM and also SSB signals. Therefore, if a signal phase-synchronized with the carrier is available, any of the previously described mixer circuits together with a low-pass filter can be used as a detector. If the local oscillator signal is a square wave, the same results apply. As previously shown, the square wave will result in additional high-frequency mixing products, but these can all be removed by the low-pass filter. The mixer requirements for detection are far less stringent than those of the receiver input mixer. For detection the signal level is larger, and mixer noise is less of a problem. Also the mixer-detector usually follows a narrowband filter, and intermodulation distortion is not a significant problem. Phase-locked loops can also be used for AM detection and pilot carrier recovery. These applications are discussed in Chap. 8.

**Angle Modulation**

Amplitude modulation alters the amplitude of a sine wave carrier. Information can also be transmitted by modulating the phase and frequency. Such modulation is, of course, often used. Angle modulation has the disadvantage, compared with amplitude modulation, of occupying a wider bandwidth, but it can provide better discrimination against noise and other interfering signals.

An angle-modulated waveform can be written as

$$S(t) = A(t) \cos [\omega_c t + \theta(t)] \tag{12.26}$$

where $\omega_c$ is the carrier frequency and $A(t)$ is the amplitude modulation. Ideally, $A(t)$ will be constant, with the phase $\theta(t)$ representing the angle modulation. Angle modulation can be further subdivided into phase and frequency modulation, depending on whether it is the phase or the derivative of phase (frequency) that is modulated. Frequency modulation and phase modulation are not distinct, since

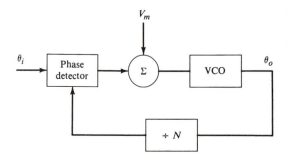

**FIGURE 12.35**
A phase-locked-loop frequency
demodulator.

changing the frequency will result in a change in phase and modulating the phase also modulates the frequency.

## Angle Modulators

Frequency modulation can be achieved directly by modulating a VCO (direct FM) or indirectly by phase-modulating the RF waveform by the integrated audio input signal (indirect FM). Another method of FM is to use a phase-locked loop, as shown in Fig. 12.35. The output in response to the modulating signal $V_M$ (assuming a linear model for the loop) is

$$\theta_o(s) = \frac{(K_o/s)V_M(s)}{1 + K_o K_d F(s)/(sN)} \tag{12.27}$$

where $K_d$ is the phase-detector gain constant and $K_o$ is the VCO sensitivity (hertz per volt). In the steady state, the output phase will be proportional to the modulating voltage, so the phase-locked loop can serve either as a phase modulator or, if $V_M$ is the integral of the modulating signal of interest, as a frequency modulator. The phase-locked loop bandwidth (described in Chap. 9) needs to be wider than the bandwidth of the modulating signal in order for the loop not to distort the signal.

## FM Demodulators

The same type of circuitry is used for detecting both types of angle modulation, and we will refer to either process as *FM detection*. In this section a few of the many different methods that have been used for the detection of frequency-modulated waves are evaluated. FM detector circuits are often referred to as *frequency discriminators*.

The ideal FM detector produces an output voltage that changes linearly with changes in the input frequency, as illustrated in Fig. 12.36. The output voltage will usually be zero at the carrier frequency. Any deviation from the linear characteristic distorts the detected waveform. Amplitude modulation caused by noise, signal fading, etc., can also cause the recovered signal to be distorted. Limiting circuitry is usually included in an FM detector to reduce the amount of amplitude modulation.

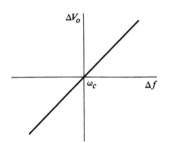

**FIGURE 12.36**
Ideal characteristics of a frequency-to-voltage converter.

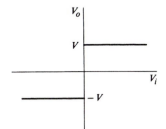

**FIGURE 12.37**
Transfer characteristics of an ideal limiter.

The transfer characteristic of an ideal limiter is illustrated in Fig. 12.37. The limiter output is restricted to the values that depend only on the sign of the input waveform. A single-stage differential-pair limiter is shown in Fig. 12.38. This circuit gives a close approximation to the ideal limiter characteristic. The transfer characteristic is symmetric (about $V_i = 0$), so there are no even harmonics (or direct current) in the output waveform. The input signal must be large enough to drive the differential amplifier into saturation. If the input signal is too small, several differential-pair stages can be cascaded in order for the output to be saturated. Integrated-circuit limiters frequently contain three cascaded differential sections.

**EXAMPLE 12.3.** A differential pair has a voltage gain of 60 dB into a 2-k$\Omega$ load and a differential input impedance of 2 k$\Omega$. If a 75-mV input voltage drives the stage into saturation, how many stages will be needed to limit a 5-$\mu$V input signal?

**Solution.** A voltage gain of 60 dB ($A_v = 1000$) means that the output of the first stage will be 5 mV, which is not sufficient to saturate the output of the second stage. Therefore, the complete limiter consists of three cascaded differential pairs.

## FM Detectors

An analytical basis of FM detection is obtained by considering the derivative of the FM signal:

$$\frac{d}{dt}\{A\cos\left[(\omega_c t + \theta(t)]\right\} = -\left(\omega_c + \frac{d\theta}{dt}\right)A\sin\left[\omega_c t + \theta(t)\right] \quad (12.28)$$

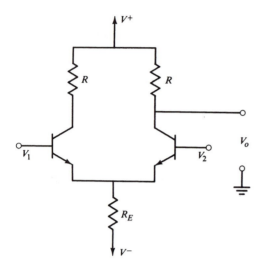

**FIGURE 12.38**
A differential-pair limiter.

The derivative of an angle-modulated signal is an amplitude-modulated FM wave-form; all the modulating information is contained in the amplitude of the differenti-ated waveform. Normally $\omega_c \gg d\theta/dt$. If so, the amplitude modulation can be removed with an envelope detector. Of course, any amplitude modulation on the original signal must first be removed by limiting.

The output of the envelope detector will be proportional to $\omega_c + d\theta/dt$, which is $\omega_c + KV_m(t)$ for a frequency-modulated waveform. If the output is then high-pass-filtered to remove the constant term $\omega_c$, the remainder will be proportional to the modulating signal. This approach does have the disadvantage that any dc com-ponent in the modulating signal is lost.

There are many circuits for realizing the differentiator, but the single-tuned cir-cuit is used most often. The frequency response of an ideal differentiator $H(j\omega) = j\omega K$ has a $+90°$ phase shift, and the magnitude increases with increasing fre-quency at 6 dB per octave. The frequency response of a simple tuned circuit will approximate this response at frequencies sufficiently below the circuit's resonant frequency.

The frequency response magnitude of the parallel tuned circuit is obtained from Eq. (4.7) and Table 4.1:

$$|A(j\omega)| = \frac{R}{[1 + Q^2(\omega/\omega_o - \omega_o/\omega)^2]^{1/2}}$$

Values for $Q$ and $\omega_o$ of the parallel tuned circuit are given in Table 4.1.

The magnitude of the frequency response of this circuit is plotted in Fig. 12.39. The carrier frequency must be sufficiently below the resonant frequency of the cir-cuit that the magnitude response is approximately linear. If the carrier frequency is too far below the resonant frequency, there will be too much attenuation, with a cor-responding degradation in the signal-to-noise ratio.

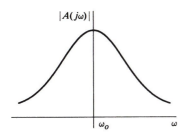

**FIGURE 12.39**
Tuned-circuit gain as a function of frequency.

At the frequency $\omega_c + \Delta\omega$,

$$|A(j\omega)| = \frac{R}{\{1 + Q^2[(\omega_c + \Delta\omega)/\omega_o - \omega_o/(\omega_c + \Delta\omega)]^2\}^{1/2}}$$

$$\approx \frac{R\omega_o(\Delta\omega + \omega_c)}{Q[\omega_o^2 - (\omega_c + \Delta\omega)^2]}$$

provided $\omega_c$ is close enough to $\omega_o$ that

$$Q\left[\frac{\omega_c + \Delta\omega - \omega_o^2}{\omega_o(\omega_c + \Delta\omega)}\right] \gg 1$$

Also, if

$$\omega_c + \Delta\omega \ll \omega_o$$

then

$$|A(j\omega)| \approx \frac{R(\omega_c + \Delta\omega)}{Q\omega_o}$$

The output consists of a constant term corresponding to $\omega_c$ plus a component proportional to the frequency deviation $\Delta\omega$. Balanced discriminators are often used to eliminate the constant term. A simplified balanced discriminator is illustrated in Fig. 12.40. The upper resonant circuit is tuned to the frequency $\omega_o - \omega_c$, and the output is proportional to $\omega_c + \Delta\omega$. The lower resonant circuit is tuned to $\omega_o + \omega_c$, and the output is proportional to $\omega_c - \Delta\omega$. The differential output is then

$$V_o = K[\omega_c + \Delta\omega - (\omega_c - \Delta\omega)] = 2K\,\Delta\omega$$

which is proportional to the frequency deviation from the carrier frequency.

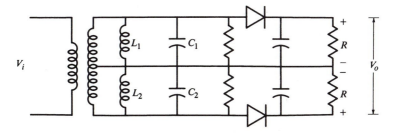

**FIGURE 12.40**
A balanced frequency discriminator.

**Pulse Discriminators**

A different approach to FM discrimination, one becoming increasingly popular as advances in digital integrated circuits continue, is shown in Fig. 12.41. The zero-crossing detector outputs a pulse on every other zero crossing of the input signal corresponding to a zero crossing for every full cycle of the input waveform. The period discriminator then determines the period between alternate zero crossings and converts the period information to an analog voltage with the period-to-voltage converter.

Figure 12.42 shows one method of implementing this frequency discriminator. The voltage comparator functions as an amplitude limiter. Whenever the input signal goes positive, the contents of the counter are strobed into the D/A converter and the counter is reset. The counter increments at a constant rate equal to the clock frequency. Therefore, the contents of the counter are proportional to the time lapse since the last positive-going zero crossing. A D/A converter connected to the counter output provides an analog signal proportional to the period or frequency. The circuit can also be implemented as shown in Fig. 12.43. For each trigger signal, the multivibrator outputs a positive pulse of constant duration. For this circuit the monostable multivibrator outputs a positive pulse of constant duration and a square wave when the input is triggered at a rate equal to the carrier frequency. Since the average value of square wave is zero, the output voltage will also be zero.

**FIGURE 12.41**
A pulse frequency discriminator.

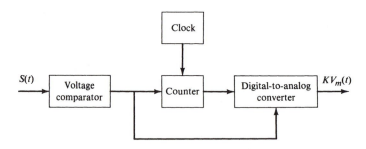

**FIGURE 12.42**
Block diagram implementation of a pulse frequency discriminator.

**FIGURE 12.43**
Another pulse frequency discriminator.

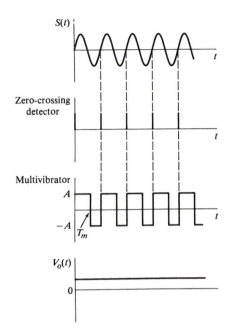

**FIGURE 12.44**
Pulse frequency discriminator waveforms.

If the input frequency increases, the multivibrator output will be as shown in Fig. 12.44. Since the pulse length is constant, the multivibrator output is no longer symmetric, and the average output voltage will be positive and proportional to the increase in frequency. If the input frequency decreases below the carrier frequency, the output voltage will become negative.

The pulse duration discriminator is very insensitive to amplitude modulation. Compared with ratio discriminators, it is primarily limited by the speed of the digital circuitry and the increased complexity of its hardware. Another popular FM discriminator, the phase-locked-loop frequency discriminator, is described in Chap. 8.

## ■ 12.4

### DIGITAL MODULATION

Today more and more information is being transmitted in digital form. This is true from low-frequency voice transmission to an all-digital television system to satellite communications at microwave frequencies. The many reasons for the increased use of digital modulation include the advances in modern technology, which made available small, compact, and inexpensive digital circuits; the ability to transmit at low signal-to-noise interference ratios; and the ease of time-multiplexing the data. Also, digital signals allow for the use of very efficient power amplifiers. Digital modulation can refer to the modulation of an analog signal by a digital waveform or the digital modulation of an analog signal. An analog signal is one that takes on a continuum of values, such as 0 to 10 V, and it can be either amplitude- or angle-modulated. A digital waveform assumes only the values corresponding to logical 0s and 1s. An analog signal can be modulated with digital data by varying either the amplitude (such as CW modulation), frequency (as in frequency-shift keying), or phase (as in phase-shift keying). Digital data can also be transmitted directly using various encoding methods, such as BCD, and data formats, such as nonreturn to zero (NRZ) or return to zero (RZ).

Figure 12.45 illustrates the components required to convert an analog-modulated signal $V_m(t)$ to a digitally encoded waveform. The signal is first sampled (see Chap. 8 for an analysis of the sampling process), providing a pulse-amplitude-modulated (PAM) signal, which can be transmitted directly but which is usually converted to a pulse-code-modulated (PCM) signal in the A/D converter. The Nyquist sampling criterion states that the sampling frequency $f_s$ must be greater than twice the bandwidth $B$ of the input signal in order to preserve all information. That is, $f_s > 2B$. If the sampling rate is less than twice the bandwidth, the frequency spectrum centered about the sampling frequency overlaps the original spectrum and cannot be separated from the original spectrum by filtering. (This is known as *foldover distortion,* or *aliasing.*)

Thus, a PAM system must include a band-limiting filter before the sampler to ensure that foldover distortion does not occur. The Nyquist sampling rate may need to be significantly exceeded if precise resolution of the signal is required in a short time. The Nyquist criterion states the minimum sampling rate required, but a large number of samples are required before the signal can be accurately reconstructed. Consider the problem of measuring the period $T$ of a sine wave (as illustrated in Fig. 12.46) from its PAM values. The signal period could be estimated by counting the number of samples between successive zero crossings or between zero

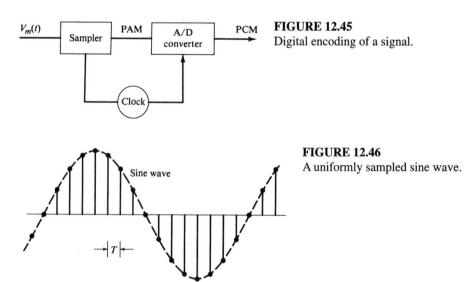

**FIGURE 12.45**
Digital encoding of a signal.

**FIGURE 12.46**
A uniformly sampled sine wave.

$\hat{f} = [(N + 1)T]^{-1}$     $f = (NT)^{-1}$

crossings of the same sign (corresponding to a full input cycle). If there are $N$ samples between zero crossings of the same sign, the estimated period of the signal is $\hat{T} = NT_s$, where $T_s$ is the sampling period. The maximum error in the estimation of the period is one sample period. That is,

$$NT_s \leq T \leq (N+1)T_s$$

The uncertainty in the frequency estimate is

$$\Delta f = (NT_s)^{-1} - [(N+1)T_s]^{-1} = [N(N+1)T_s]^{-1} \approx (N^2 T_s)^{-1}$$

Therefore, for small errors

$$\Delta f \approx \frac{T_s}{N^2 T_s^2} \approx \frac{f^2}{f_s}$$

or

$$f_s = \frac{f^2}{\Delta f} \qquad (12.29)$$

This equation provides a good estimate of the sampling frequency necessary to estimate the frequency of a sinusoidal waveform with an accuracy $\pm \Delta f$.

**EXAMPLE 12.4.** Determine the sampling rate required to estimate the frequency of a sine wave with a maximum error of 1 Hz. The maximum frequency of the unknown signal is 1 kHz.

*Solution.* If the estimate is to be made from one cycle of data, Eq. (12.29) can be used. The required sampling frequency is

$$f_s = \frac{(10^3)^2}{1} = 1 \text{ MHz}$$

The sampling rate requirements can be reduced if more samples (at the same rate) are used to estimate the period. For example, if $P$ full cycles of the sampled signal are used, the sampling uncertainty is still one sample period, so the required sampling frequency is

$$f_s = \frac{f^2}{P \Delta f} \qquad (12.30)$$

The PAM signal is usually converted to a PCM signal before transmission, using an analog-to-digital converter. In uniformly encoded PCM, all signals falling within a prescribed amplitude interval are represented by a single discrete value. The process is illustrated in Fig. 12.47, where eight equal-size quantization levels and a binary representation of each level are presented. If the signal level lies anywhere within the lowest level (and lowest 12.5 percent), it is assigned the binary code 000. If the signal lies in the highest level (the highest 12.5 percent), it is assigned the binary code 111. Signal amplitudes must be confined to the maximum range of the encoder, or else overload distortion will occur. The quantization process introduces quantization noise; the more quantization levels, the smaller will be the quantization noise. It was mentioned in Chap. 10 that for sampled data the

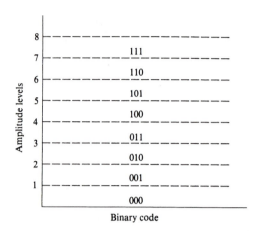

**FIGURE 12.47**
Amplitude levels of a PCM signal.

signal-to-noise ratio after uniform quantization of a noiseless signal is approximately

$$\frac{S}{N} = +6n \qquad \text{dB} \qquad (12.31)$$

where $n + 1$ is the number of bits in the digital word (including the sign bit). The uniform quantization process results in each level having the same amount of noise. Therefore, the small signals have a small signal-to-noise ratio, and the large signals have a large signal-to-noise ratio. In many applications such as voice transmission, the large signals are the least likely to occur; uniform PCM is not efficient for such applications. A more efficient encoding procedure results if smaller quantization levels are used for small signals and large quantization levels for large signals. The process is known as *companding;* it is possible to achieve the result with a linear A/D converter preceded by a nonlinear amplifier whose gain decreases with increasing signal level. Various nonlinear characteristics can be used to implement the amplifier (*compandor*). A widely used characteristic is the $u$ law, defined as

$$F_u(x) = \text{sgn}(x)\frac{\ln\left(1 + u|x|\right)}{\ln\left(1 + u\right)} \qquad (12.32)$$

where $x$ is the input signal amplitude $(-1 \le x \le 1)$, $\text{sgn}(x)$ is the polarity of $x$, and $u$ is a parameter used to define the amount of compression.[6] The gain compression functions are now included directly in the design of A/D converters (encoders), and the modulation is frequently referred to as *log-PCM*.

Many signals, such as voice signals, contain a large amount of sample-to-sample redundancy, which permits a reduction in the digital bit rate (and thus the bandwidth). If the sampling rate is fast enough, the next sample value is probably going to be close to the previous sample in amplitude. Differential pulse code modulation (DPCM) encodes the difference between samples. As in PCM, the analog-to-digital conversion process can be conventional or companded. DPCM is ideally

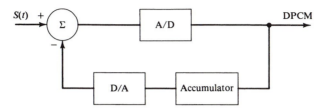

**FIGURE 12.48**
A differential pulse code modulator.

suited for digital implementation and large-scale integration. Figure 12.48 contains a block diagram representation of one system for DPCM implementation. The error signal is digitized, and the estimate of the input signal is constructed by accumulating the error signal. This estimate is then reconverted to analog form to obtain the next error estimate. There are many implementations of this technique, including many which do not reconstruct the analog signal but rather generate the error estimate in the digital domain. DPCM provides approximately a 1-bit reduction in word length for the digital modulation of speech signals.

Delta modulation (DM) is another method of digital modulation of analog signals. DM is a special case of DPCM. It uses only 1 bit per sample to represent the difference signal. The signal bit specifies whether the signal has increased or decreased since the last sample. Delta modulation requires an appreciably higher sampling rate than PCM or DPCM, but the hardware required is simpler. Figure 12.49 illustrates a delta modulator consisting of a clock, comparator, pulse generator, and integrator (to reconstruct an estimate of the input signal). The comparator output consists of digital bits, and the pulse generator output consists of two equal-amplitude pulses of opposite polarity. The polarity depends on the output bit.

**Digital Modulation and Demodulation of Analog Data**

The digitized data, often referred to as the *baseband data,* can be easily modulated onto a radio-frequency carrier since the baseband has only two amplitude levels. The circuit components previously discussed are directly applicable to this problem. The following are the most frequently used techniques.

**Amplitude Modulation**

For digital or continuous-wave modulation, the oscillator can be directly keyed on an "off," but this can create frequency distortion referred to as *chirps.* A better method is to operate the oscillator continuously and then switch the output on and off with a buffer amplifier (switch). On/off keying is a special case of amplitude or frequency modulation. It is the most economical method of digital communications, particularly if envelope detection is used, but it gives the poorest error performance. Some form of angle modulation gives better performance.

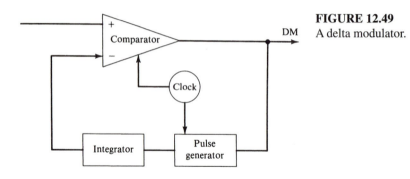

**FIGURE 12.49**

DM    A delta modulator.

Another type of amplitude modulation used for transmission of binary data is quadrature AM (QAM), which uses two carriers in quadrature (offset by 90°). If a carrier is amplitude-modulated by a signal that assumes only the values ±1, then the modulated signal is

$$S(t) = A\frac{1 \pm m}{2} \sin \omega_o t$$

The envelope will have an amplitude of $1 + m/2$ or $1 - m/2$. The digital data contribute a portion that is either in phase or 180° out of phase with respect to the carrier. QAM uses two carriers 90° out of phase; both are modulated by digital data, and then the modulated signals are summed. The relative phase of the signal portion is then 45, 135, 225, and 315°. QAM is able to transmit twice as much data as quadrature AM modulation.

### Frequency Modulation

Frequency-shift keying (FSK) is a special case of frequency modulation that can be achieved by pulling a VCO plus or minus a standard frequency deviation $\Delta\omega$. FSK is often used for low-cost, low-data-rate communication over analog telephone networks. The channel capacity can be increased by using multiple frequencies. An FSK system having four frequencies is denoted as $(FSK)_4$. FSK detection is readily achieved with phase-locked loops, as described in Chap. 8.

### Phase Modulation

It can be shown that for two-level line coding the maximum use of transmitted power is achieved when one signal is the negative of the other. Phase-shift keying (PSK) has become the most popular digital modulation method for high-performance applications. A relatively simple method of PSK modulation is to generate the carrier signals 180° apart in phase and then select between the two signals, depending on the data values. The data rate can be increased by transmitting multiple-phase signals. $(PSK)_8$, for example, will assume one of eight values spaced 45° apart.

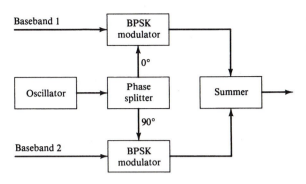

**FIGURE 12.50**
Block diagram of a QPSK
modulator.

Quadrature-phase-shift keying [QPSK or (PSK)$_4$] uses four phase values spaced 90° apart. QPSK is achieved by using two binary phase-shift-keyed (BPSK) modulators with the two carriers 90° apart in phase, as shown in Fig. 12.50. The two BPSK signals are then summed. The phase splitting can be achieved with a PLL, as described in Chap. 8. It can be shown that an ideal QPSK modem can transmit 2 bits/s per unit bandwidth. To transmit $50 \times 10^6$ bits/s, for example, will require a transmission bandwidth of 25 MHz if QPSK is used. Coherent demodulation is normally used for QPSK demodulation by recovering the carrier as described in Chap. 8. Many other phase modulation methods are possible,[7] such as staggered QPSK (SQPSK) and minimal shift keying (MSK), and can be realized with the basic circuitry described in this test.

## ■ 12.5

### INTEGRATED-CIRCUIT TECHNIQUES [8]

Most integrated-circuit modulators and demodulators use some form of a variable transconductance multiplier that relies on the dependence of the transistor's transconductance upon its bias current. The technique is illustrated by the differential amplifier in Fig. 12.51. It was shown in Chap. 2 that for small values of differential input voltage, the differential output voltage is

$$V_o \approx g_m R_L V_1$$

and the transconductance is

$$g_m = \frac{I_E}{V_T} \approx \frac{V_2}{R_E V_T}$$

Therefore, the output

$$V_o \approx \frac{R_L V_1 V_2}{R_E V_T} \tag{12.33}$$

is proportional to the product of the two input signals.

One difficulty with this circuit is that since the total current $I_E$ varies directly as a function of $V_2$, a large common-mode voltage swing will occur in the circuit. This common-mode swing is eliminated by the circuit shown in Fig. 12.52, which

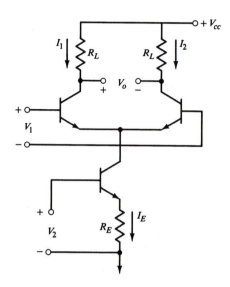

**FIGURE 12.51**
Balanced modulator stage commonly used in integrated circuits.

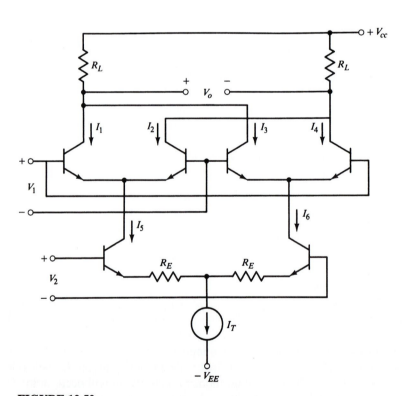

**FIGURE 12.52**
An improved integrated-circuit balanced modulator.

consists of two differential stages in parallel. This circuit will now be analyzed with the assumptions that the transistors are well matched and have a large current gain $(B \gg 1)$. Under these conditions,

$$I_1 + I_2 = I_5$$

$$I_3 + I_4 = I_6$$

$$I_5 + I_6 = I_T$$

If the differential input voltage is small,

$$I_1 - I_2 = g_{m_1} V_1$$

and

$$I_3 - I_4 = g_{m_3} V_1$$

where $g_{m_1}$ is the transconductance of the transistor pair $Q_1$ and $Q_2$ and $g_{m_3}$ is the transconductance of transistor pair $Q_3$ and $Q_4$. Also,

$$g_{m_1} = \frac{I_5}{V_T}$$

and

$$g_{m_3} = \frac{I_6}{V_T}$$

The output voltage is

$$V_o = R_L(I_1 - I_2 + I_3 - I_4)$$

$$= R_L(g_{m_1} - g_{m_3})V_1$$

$$= \frac{V_1 R_L}{V_T}(I_5 - I_6)$$

Also

$$V_2 = V_{be_5} + I_5 R_e - I_6 R_E - V_{be_6}$$

$$= V_T \ln \frac{I_5}{I_6} + R_E(I_5 - I_6) \approx R_E(I_5 - I_6)$$

provided $R_E$ is sufficiently large. Under these conditions

$$V_o = \frac{R_L}{R_E V_T} V_1 V_2 \tag{12.34}$$

This circuit operates as a multiplier, provided the input voltages are sufficiently small. If the voltages are comparable to, or larger than, the thermal voltage $V_T$, the transistor pairs function as synchronous switches, and the circuit functions as a balanced modulator. Improved circuits, which provide four-quadrant multiplication, have been developed, but the balanced modulator circuit suffices for most communications applications. In integrated-circuit terminology, a balanced modulator is a multiplier circuit in which the output is a linear function of only one of the inputs. This input is known as the *modulating input,* and the other input is referred to as the *carrier input.* When differential inputs are used (such as shown in Fig. 12.52), the carrier signal appearing at the output is greatly suppressed, and the circuit is referred to as a *balanced modulator.*

## 12.6

## PROBLEMS

**12.1** Design a passive double-balanced mixer (including the specification of the transformer turns ratio) that will provide maximum power transfer between a 300-$\Omega$ source resistance and a 50-$\Omega$ load resistance.

**12.2** If a passive double-balanced mixer is used to realize the mixing product $n\omega_L - \omega_i$, what is the conversion loss?

**12.3** In the circuit illustrated in Fig. P12.3, the local oscillator signal is fed to the primary of the center-tapped transformer. Derive an expression for the output voltage as a function of the input voltage, assuming the diodes and transformer are ideal.

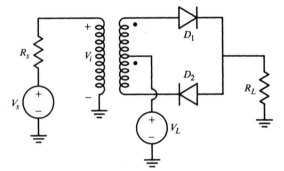

**FIGURE P12.3**
A two-diode mixer circuit.

**12.4** In synchronous detection of the DSB signal given by Eq. (12.22), what happens if the local oscillator signal is of the form

$$V_L = V \cos \omega_c t$$

What does this imply in terms of phase synchronization for the receiver?

**12.5** Show that synchronous detection can be used for SSB signal detection. What must be the phase relation between the carrier and local oscillator signal?

**12.6** Determine the sampling rate required to measure the period of a sine wave using the successive zero crossings to estimate the half-cycle period of the signal. The maximum signal frequency is 1000 Hz, and the maximum tolerable error is 1 Hz. What is the required rate if five consecutive zero crossings are used to estimate the period?

**12.7** Verify Eq. (12.30).

**12.8** Calculate the conversion loss of the simple double-balanced mixer shown in Fig. 12.10. The diode on resistance is much less than the load resistance $R_L$. What value of $R_L$ is required for maximum power transfer if the circuit employs a 1:1 transformer?

**12.9** If a 2:1 voltage transformer is used on the input of the double-balanced mixer shown in Fig. P12.9, what is the optimum value of $R_L$, in terms of $R_s$, for maximum power transfer?

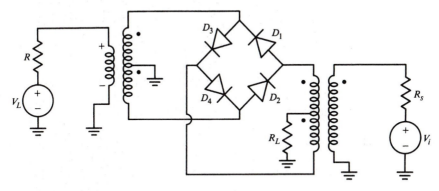

**FIGURE P12.9**
Double-balanced mixer circuit.

**12.10** Show that an amplitude-modulated wave consisting of one sideband plus a carrier component can be demodulated with envelope detection, provided the carrier amplitude is sufficiently large.

**12.11** Calculate the conversion loss of the single-balanced mixer shown in Fig. 12.3 for the condition $R_L = 2R_s$ and for the case $R_s = 2R_L$.

**12.12** Determine the output voltage of the circuit shown in Fig. P12.12. Compare the performance of this circuit with that of the circuit illustrated in Fig. 12.10.

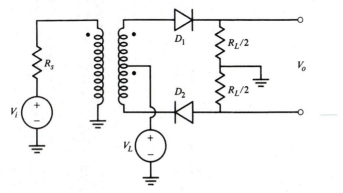

**FIGURE P12.12**
Mixer with balanced load.

**12.13** Verify that the differential output of the balanced discriminator illustrated in Fig. 12.40 is $2K\Delta\omega$, provided one tuned circuit is tuned to $\omega_c + \Delta\omega$ and the other is tuned to $\omega_c - \Delta\omega$.

**12.14** Design a demodulator for QAM signals using phase-locked-loop techniques. Describe the system in block diagram form.

**12.15** Design a $(FSK)_4$ demodulator using PLL techniques. Include circuitry to indicate which signal is present at the input.

**12.16** A system is to measure the period of a sine wave. Describe how the sampling rate can be reduced if single-point data interpolation midway between the sampling points is used.

**12.17** Describe a block diagram for QPSK detection.

**12.18** Calculate the common-mode voltage swing of the double-balanced modulator shown in Fig. 12.52 as a function of the input signal $V_2$.

**\*12.19** Design a class A power amplifier using MRF 177 transistors to deliver 20 W to a 50 ohm load. Calculate the maximum transistor dissipation, peak output voltage and peak output current of each transistor and compare with the values obtained from the simulation. Determine the bandwidth of the amplifier.

## ▨ 12.7
### REFERENCES

1. H. P. Walker, Sources of Intermodulation in Diode-Ring Mixers, *Radio and Electronic Engineer,* **46:**247–255, 1976.
2. W. Kaplan, *Advanced Calculus,* Addison-Wesley, Reading, Mass., 1952, p. 362.
3. D. Cheadle, Selecting Mixers for Best Intermod Performance, *Microwaves,* pp. 48–52, 1973.
4. K. K. Clarke, and D. T. Hess, *Communication Circuits: Analysis and Design,* Addison-Wesley, Reading, Mass., 1971.
5. B. G. Ulfers, "Characteristics and Applications of Dual Gate, Insulated Gate Field Effect Transistors." M.S. thesis, University of Florida, Gainesville, 1969.
6. J. Bellamy, *Digital Telephony,* Wiley, New York, 1982.
7. J. J. Spilker, Jr., *Digital Communications by Satellite,* Prentice-Hall, Englewood Cliffs, N.J., 1977.
8. A. B. Grebene, *Analog Integrated Circuit Design,* Van Nostrand Reinhold, New York, 1972.

## ▨ 12.8
### ADDITIONAL READING

El-Hamamsy, S–A, "Design of High-Efficiency RF Class-D Power Amplifier," *IEEE Trans. Power Electronics*, 9, 297–507, 1994.

Gentzler, C., "Class A Design Techniques Create a New Broadband Amplifier," RF Design, 47–54, May 1993.

Gray, P. R., and R. G. Meyer: *Analysis and Design of Analog Integrated Circuits,* 3d ed., Wiley, New York, 1993.

Hayward, W. H.: *Introduction to Radio Frequency Design,* Prentice-Hall, Englewood Cliffs, N.J., 1982

H. L. Krauss, C. W. Bostian, and F. H. Raab, Solid State Radio Engineering, New York: John Wiley and Sons, 1980.

Jayant, N. S.: Digital Coding of Speech Waveforms, *Proc. IEEE,* 1974, pp. 611–632.

Li, C-H, and Lam, Y-O, "Maximum Frequency and optimum performance of class E power amplifiers," *IEE Proc.-Circuits Devices, Syst., 141,* 174184, 1994.

Pappenfus, E. W., W. B. Brune, and E. O. Schoenke: *Single Sideband Principles and Circuits,* McGraw-Hill, New York, 1964.

Rohde, U., and T. T. N. Bucher: *Communication Receivers,* McGraw-Hill, New York, 1988.

Wheeler, H. A.: Design Formulas for Diode Detectors, *Proc. IRE,* **26:**745–780, 1938.

## Mixers

Abdullah, F., and F. M. Clayton: A Large-Signal Theory for Broad-Band Frequency Converters Using Abrupt Junction Varactor Diodes, *IEEE Trans. on Microwave Theory and Techniques,* MTT-25, **101:**117–136, 1977.

Barber, M. R.: Noise Figure and Conversion Loss of the Schottky Barrier Mixer Diode, *IEEE Trans. on Microwave Theory and Techniques,* **MIT-15:**629–635, 1967.

Gardiner, J. G.: The Relationship between Cross-Modulation and Intermodulation Distortions in the Double-Balanced Modulator, *Proc. IEEE,* **56:**2069–2971, 1968.

Howson, D. P., and J. G. Gardiner: High-Frequency Mixers Using Square-Law Diodes, *Radio and Electronic Engineer,* November 1968, pp. 311–316.

Huang, M. Y., R. L. Buskirk, and D. E. Carlile: Select Mixer Frequencies Painlessly, *Electronic Design* **8:**104–109, April 12, 1976.

Kurtz, S. R.: Specifying Mixers as Phase Detectors, *Microwaves,* January 1978, pp. 80–89.

Mouw, R. B., and S. M. Fukuchi: Broadband Double Balanced Mixer/Modulators, Part I, *Microwave J.,* March 1969, pp. 131–134.

——— and ———: Broadband Double Balanced Mixer/Modulators, Part II, *Microwave J.,* May 1969, pp. 71–76.

Stevens, R. T.: Linear Scales Show Mixer Harmonics, *Electronics,* **10:**37–39, 1964.

# A P P E N D I C E S

# Motorola NPN Transistor P2N2222A

## Amplifier Transistors
### NPN Silicon

COLLECTOR
1

2
BASE

3
EMITTER

**P2N2222A**

**CASE 29–04, STYLE 17**
**TO–92 (TO–226AA)**

### MAXIMUM RATINGS

| Rating | Symbol | Value | Unit |
|---|---|---|---|
| Collector–Emitter Voltage | $V_{CEO}$ | 40 | Vdc |
| Collector–Base Voltage | $V_{CBO}$ | 75 | Vdc |
| Emitter–Base Voltage | $V_{EBO}$ | 6.0 | Vdc |
| Collector Current — Continuous | $I_C$ | 600 | mAdc |
| Total Device Dissipation @ $T_A = 25°C$<br>Derate above 25°C | $P_D$ | 625<br>5.0 | mW<br>mW/°C |
| Total Device Dissipation @ $T_C = 25°C$<br>Derate above 25°C | $P_D$ | 1.5<br>12 | Watts<br>mW/°C |
| Operating and Storage Junction<br>Temperature Range | $T_J, T_{stg}$ | −55 to +150 | °C |

### THERMAL CHARACTERISTICS

| Characteristic | Symbol | Max | Unit |
|---|---|---|---|
| Thermal Resistance, Junction to Ambient | $R_{\theta JA}$ | 200 | °C/W |
| Thermal Resistance, Junction to Case | $R_{\theta JC}$ | 83.3 | °C/W |

### ELECTRICAL CHARACTERISTICS ($T_A = 25°C$ unless otherwise noted)

| Characteristic | Symbol | Min | Max | Unit |
|---|---|---|---|---|
| **OFF CHARACTERISTICS** | | | | |
| Collector–Emitter Breakdown Voltage<br>($I_C = 10$ mAdc, $I_B = 0$) | $V_{(BR)CEO}$ | 40 | — | Vdc |
| Collector–Base Breakdown Voltage<br>($I_C = 10$ µAdc, $I_E = 0$) | $V_{(BR)CBO}$ | 75 | — | Vdc |
| Emitter–Base Breakdown Voltage<br>($I_E = 10$ µAdc, $I_C = 0$) | $V_{(BR)EBO}$ | 6.0 | — | Vdc |
| Collector Cutoff Current<br>($V_{CE} = 60$ Vdc, $V_{EB(off)} = 3.0$ Vdc) | $I_{CEX}$ | — | 10 | nAdc |
| Collector Cutoff Current<br>($V_{CB} = 60$ Vdc, $I_E = 0$)<br>($V_{CB} = 60$ Vdc, $I_E = 0$, $T_A = 150°C$) | $I_{CBO}$ | —<br>— | 0.01<br>10 | µAdc |
| Emitter Cutoff Current<br>($V_{EB} = 3.0$ Vdc, $I_C = 0$) | $I_{EBO}$ | — | 10 | nAdc |
| Collector Cutoff Current<br>($V_{CE} = 10$ V) | $I_{CEO}$ | — | 10 | nAdc |
| Base Cutoff Current<br>($V_{CE} = 60$ Vdc, $V_{EB(off)} = 3.0$ Vdc) | $I_{BEX}$ | — | 20 | nAdc |

**P2N2222A**

**ELECTRICAL CHARACTERISTICS** ($T_A$ = 25°C unless otherwise noted) (Continued)

| Characteristic | Symbol | Min | Max | Unit |
|---|---|---|---|---|
| **ON CHARACTERISTICS** | | | | |
| DC Current Gain | $h_{FE}$ | | | — |
| ($I_C$ = 0.1 mAdc, $V_{CE}$ = 10 Vdc) | | 35 | — | |
| ($I_C$ = 1.0 mAdc, $V_{CE}$ = 10 Vdc) | | 50 | — | |
| ($I_C$ = 10 mAdc, $V_{CE}$ = 10 Vdc) | | 75 | — | |
| ($I_C$ = 10 mAdc, $V_{CE}$ = 10 Vdc, $T_A$ = –55°C) | | 35 | — | |
| ($I_C$ = 150 mAdc, $V_{CE}$ = 10 Vdc)[1] | | 100 | 300 | |
| ($I_C$ = 150 mAdc, $V_{CE}$ = 1.0 Vdc)[1] | | 50 | — | |
| ($I_C$ = 500 mAdc, $V_{CE}$ = 10 Vdc)[1] | | 40 | — | |
| Collector–Emitter Saturation Voltage[1] | $V_{CE(sat)}$ | | | Vdc |
| ($I_C$ = 150 mAdc, $I_B$ = 15 mAdc) | | — | 0.3 | |
| ($I_C$ = 500 mAdc, $I_B$ = 50 mAdc) | | — | 1.0 | |
| Base–Emitter Saturation Voltage[1] | $V_{BE(sat)}$ | | | Vdc |
| ($I_C$ = 150 mAdc, $I_B$ = 15 mAdc) | | 0.6 | 1.2 | |
| ($I_C$ = 500 mAdc, $I_B$ = 50 mAdc) | | — | 2.0 | |
| **SMALL–SIGNAL CHARACTERISTICS** | | | | |
| Current–Gain — Bandwidth Product[2] | $f_T$ | 300 | — | MHz |
| ($I_C$ = 20 mAdc, $V_{CE}$ = 20 Vdc, f = 100 MHz) | | | | |
| Output Capacitance | $C_{obo}$ | — | 8.0 | pF |
| ($V_{CB}$ = 10 Vdc, $I_E$ = 0, f = 1.0 MHz) | | | | |
| Input Capacitance | $C_{ibo}$ | — | 25 | pF |
| ($V_{EB}$ = 0.5 Vdc, $I_C$ = 0, f = 1.0 MHz) | | | | |
| Input Impedance | $h_{ie}$ | | | kΩ |
| ($I_C$ = 1.0 mAdc, $V_{CE}$ = 10 Vdc, f = 1.0 kHz) | | 2.0 | 8.0 | |
| ($I_C$ = 10 mAdc, $V_{CE}$ = 10 Vdc, f = 1.0 kHz) | | 0.25 | 1.25 | |
| Voltage Feedback Ratio | $h_{re}$ | | | $\times 10^{-4}$ |
| ($I_C$ = 1.0 mAdc, $V_{CE}$ = 10 Vdc, f = 1.0 kHz) | | — | 8.0 | |
| ($I_C$ = 10 mAdc, $V_{CE}$ = 10 Vdc, f = 1.0 kHz) | | — | 4.0 | |
| Small–Signal Current Gain | $h_{fe}$ | | | — |
| ($I_C$ = 1.0 mAdc, $V_{CE}$ = 10 Vdc, f = 1.0 kHz) | | 50 | 300 | |
| ($I_C$ = 10 mAdc, $V_{CE}$ = 10 Vdc, f = 1.0 kHz) | | 75 | 375 | |
| Output Admittance | $h_{oe}$ | | | µmhos |
| ($I_C$ = 1.0 mAdc, $V_{CE}$ = 10 Vdc, f = 1.0 kHz) | | 5.0 | 35 | |
| ($I_C$ = 10 mAdc, $V_{CE}$ = 10 Vdc, f = 1.0 kHz) | | 25 | 200 | |
| Collector Base Time Constant | $rb'C_C$ | — | 150 | ps |
| ($I_E$ = 20 mAdc, $V_{CB}$ = 20 Vdc, f = 31.8 MHz) | | | | |
| Noise Figure | $N_F$ | — | 4.0 | dB |
| ($I_C$ = 100 µAdc, $V_{CE}$ = 10 Vdc, $R_S$ = 1.0 kΩ, f = 1.0 kHz) | | | | |

**SWITCHING CHARACTERISTICS**

| | | | | | |
|---|---|---|---|---|---|
| Delay Time | ($V_{CC}$ = 30 Vdc, $V_{BE(off)}$ = –2.0 Vdc, $I_C$ = 150 mAdc, $I_{B1}$ = 15 mAdc) (Figure 1) | $t_d$ | — | 10 | ns |
| Rise Time | | $t_r$ | — | 25 | ns |
| Storage Time | ($V_{CC}$ = 30 Vdc, $I_C$ = 150 mAdc, $I_{B1}$ = $I_{B2}$ = 15 mAdc) (Figure 2) | $t_s$ | — | 225 | ns |
| Fall Time | | $t_f$ | — | 60 | ns |

1. Pulse Test: Pulse Width ≤ 300 µs, Duty Cycle ≤ 2.0%.
2. $f_T$ is defined as the frequency at which $|h_{fe}|$ extrapolates to unity.

**P2N2222**

## SWITCHING TIME EQUIVALENT TEST CIRCUITS

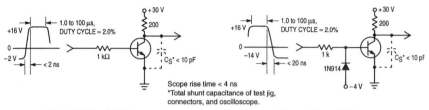

Scope rise time < 4 ns
*Total shunt capacitance of test jig,
connectors, and oscilloscope.

**Figure 1. Turn–On Time**          **Figure 2. Turn–Off Time**

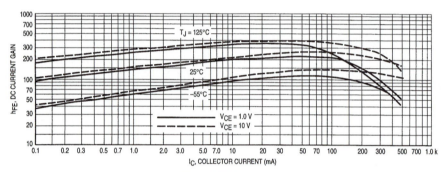

**Figure 3. DC Current Gain**

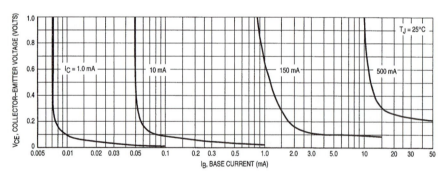

**Figure 4. Collector Saturation Region**

**P2N2222A**

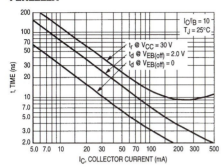

Figure 5. Turn–On Time

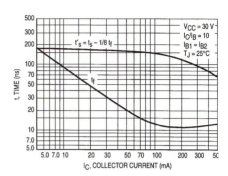

Figure 6. Turn–Off Time

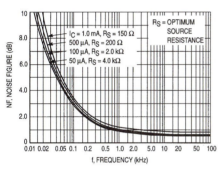

Figure 7. Frequency Effects

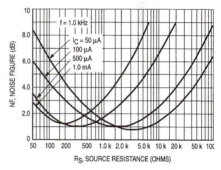

Figure 8. Source Resistance Effects

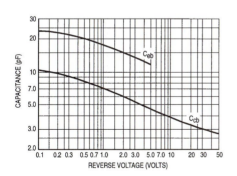

Figure 9. Capacitances

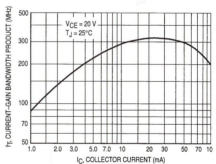

Figure 10. Current–Gain Bandwidth Product

**P2N2222A**

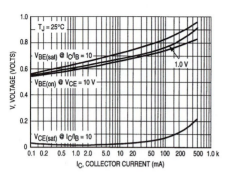

Figure 11. "On" Voltages

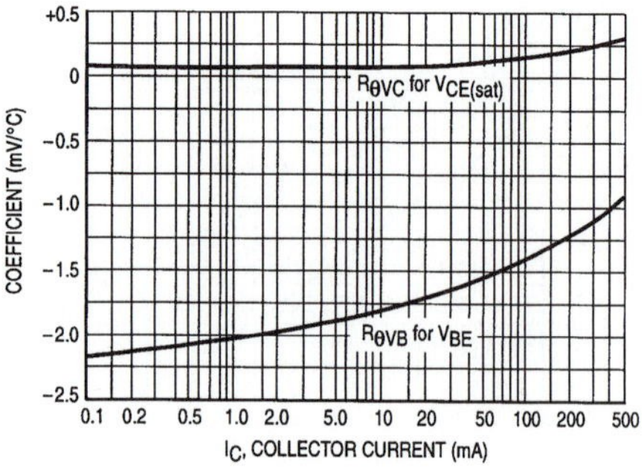

Figure 12. Temperature Coefficients

# U310 JFET Siliconix

## n-channel JFETs
## designed for . . .

**⬛ VHF Amplifiers**

**⬛ Front End High Sensitivity Amplifiers**

**⬛ Oscillators**

**⬛ Mixers**

**Siliconix**

**BENEFITS**
- Industry Standard
- High Power Gain
  16 dB at 105 MHz, Common-Gate
  11 dB at 450 MHz, Common-Gate
- Low Noise
  2.7 dB Noise Figure at 450 MHz
- Wide Dynamic Range
  Greater than 100 dB
- 75 Ω Input Match Common Gate

### ABSOLUTE MAXIMUM RATINGS (25°C)

Gate-Drain or Gate-Source Voltage . . . . . . . . . . . . . . –25 V
Gate Current . . . . . . . . . . . . . . . . . . . . . . . . . . . . 20 mA
Total Power Dissipation at $T_A$ = 25°C . . . . . . . . . 500 mW
Power Derating to 150°C . . . . . . . . . . . . . . . . . 4.0 mW/°C
Storage Temperature Range . . . . . . . . . . . . . . –65 to +200°C
Lead Temperature
  (1/16" from case for 10 seconds) . . . . . . . . . . . . . 300°C

TO-52
See Section 6

### ELECTRICAL CHARACTERISTICS (25°C unless otherwise noted)

| | | Characteristic | U308 Min | U308 Typ | U308 Max | U309 Min | U309 Typ | U309 Max | U310 Min | U310 Typ | U310 Max | Unit | Test Conditions | |
|---|---|---|---|---|---|---|---|---|---|---|---|---|---|---|
| 1 | | $I_{GSS}$ Gate Reverse Current | | | –150 | | | –150 | | | –150 | pA | $V_{GS}$ = –15 V, | |
| 2 | | | | | –150 | | | –150 | | | –150 | nA | $V_{GS}$ = 0 | $T_A$ = 125°C |
| 3 | S T A T I C | $BV_{GSS}$ Gate-Source Breakdown Voltage | –25 | | | –25 | | | –25 | | | V | $I_G$ = –1 μA, $V_{DS}$ = 0 | |
| 4 | | $V_{GS(off)}$ Gate-Source Cutoff Voltage | –1.0 | | –6.0 | –1.0 | | –4.0 | –2.5 | | –6.0 | | $V_{DS}$ = 10 V, $I_D$ = 1 nA | |
| 5 | | $I_{DSS}$ Saturation Drain Current (Note 1) | 12 | | 60 | 12 | | 30 | 24 | | 60 | mA | $V_{DS}$ = 10 V, $V_{GS}$ = 0 | |
| 6 | | $V_{GS(f)}$ Gate-Source Forward Voltage | | | 1.0 | | | 1.0 | | | 1.0 | V | $I_G$ = 10 mA, $V_{DS}$ = 0 | |
| 7 | D Y N A M I C | $g_{fg}$ Common-Gate Forward Transconductance (Note 1) | 10 | 17 | | 10 | 17 | | 10 | 17 | | mmho | $V_{DS}$ = 10 V, $I_D$ = 10 mA | f = 1 kHz |
| 8 | | $g_{og}$ Common-Gate Output Conductance | | | 250 | | | 250 | | | 250 | μmho | | |
| 9 | | $C_{gd}$ Drain-Gate Capacitance | | | 2.5 | | | 2.5 | | | 2.5 | pF | $V_{GS}$ = –10 V, $V_{DS}$ = 10 V | f = 1 MHz |
| 10 | | $C_{gs}$ Gate-Source Capacitance | | | 5.0 | | | 5.0 | | | 5.0 | | | |
| 11 | | $\overline{e_n}$ Equivalent Short Circuit Input Noise Voltage | | 10 | | | 10 | | | 10 | | $\frac{nV}{\sqrt{Hz}}$ | $V_{DS}$ = 10 V, $I_D$ = 10 mA | f = 100 Hz |
| 12 | H I F R E Q | $g_{fg}$ Common-Gate Forward Transconductance | | 15 | | | 15 | | | 15 | | | | f = 105 MHz |
| 13 | | | | 14 | | | 14 | | | 14 | | | | f = 450 MHz |
| 14 | | $g_{og}$ Common-Gate Output Conductance | | 0.18 | | | 0.18 | | | 0.18 | | mmho | $V_{DS}$ = 10 V, $I_D$ = 10 mA | f = 105 MHz |
| 15 | | | | 0.32 | | | 0.32 | | | 0.32 | | | | f = 450 MHz |
| 16 | | $G_{pg}$ Common-Gate Power Gain (Note 2) | 14 | 16 | | 14 | 16 | | 14 | 16 | | | | f = 105 MHz |
| 17 | | | 10 | 11 | | 10 | 11 | | 10 | 11 | | | | f = 450 MHz |
| 18 | | NF Noise Figure | | 1.5 | 2.0 | | 1.5 | 2.0 | | 1.5 | 2.0 | dB | | f = 105 MHz |
| 19 | | | | 2.7 | 3.5 | | 2.7 | 3.5 | | 2.7 | 3.5 | | | f = 450 MHz |

**NOTES:**
1. Pulse test duration ≈ 2 ms.
2. Gain ($G_{pg}$) measured at optimum input noise n

NZA

# Motorola MRF177 RF Power FET

## The RF MOSFET Line
## RF Power
## Field Effect Transistors
### N–Channel Enhancement Mode MOSFETs

Designed for broadband commercial and military applications up to 400 MHz frequency range. Primarily used as drivers or output amplifiers in push–pull configurations. Can be used in manual gain control, ALC and modulation circuits.

- Typical Performance at 400 MHz, 28 V:
    - Output Power — 100 W
    - Gain — 12 dB
    - Efficiency — 60%
- Low Thermal Resistance
- Low $C_{rss}$ — 10 pF Typ @ $V_{DS}$ = 28 Volts
- Ruggedness Tested at Rated Output Power
- Nitride Passivated Die for Enhanced Reliability
- Excellent Thermal Stability; Suited for Class A Operation
- Circuit board photomaster available upon request by contacting RF Tactical Marketing in Phoenix, AZ.

**MRF177**
**MRF177M**

100 W, 28 V, 400 MHz
N–CHANNEL
BROADBAND
RF POWER MOSFETs

CASE 744A–01

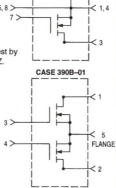

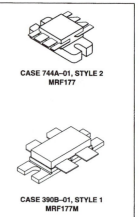

CASE 744A–01, STYLE 2
MRF177

CASE 390B–01

CASE 390B–01, STYLE 1
MRF177M

## MAXIMUM RATINGS

| Rating | Symbol | Value | Unit |
|---|---|---|---|
| Drain–Source Voltage | $V_{DSS}$ | 65 | Vdc |
| Drain–Gate Voltage ($R_{GS}$ = 1.0 MΩ) | $V_{DGR}$ | 65 | Vdc |
| Gate–Source Voltage | $V_{GS}$ | ±40 | Vdc |
| Drain Current — Continuous | $I_D$ | 16 | Adc |
| Total Device Dissipation @ $T_C$ = 25°C (1)<br>Derate above 25°C | $P_D$ | 270<br>1.54 | Watts<br>W/°C |
| Storage Temperature Range | $T_{stg}$ | −65 to +150 | °C |
| Operating Temperature Range | $T_J$ | 200 | °C |

## THERMAL CHARACTERISTICS

| Characteristic | Symbol | Max | Unit |
|---|---|---|---|
| Thermal Resistance, Junction–to–Case | $R_{θJC}$ | 0.65 | °C/W |

NOTE:
1. Total device dissipation rating applies only when the device is operated as an RF push–pull amplifier.

NOTE — **CAUTION** — MOS devices are susceptible to damage from electrostatic charge. Reasonable precautions in handling and packaging MOS devices should be observed.

**ELECTRICAL CHARACTERISTICS** ($T_C$ = 25°C unless otherwise noted)

| Characteristic (2) | Symbol | Min | Typ | Max | Unit |
|---|---|---|---|---|---|
| **OFF CHARACTERISTICS** | | | | | |
| Drain–Source Breakdown Voltage<br>($V_{GS}$ = 0, $I_D$ = 50 mA) | $V_{(BR)DSS}$ | 65 | — | — | Vdc |
| Zero Gate Voltage Drain Current<br>($V_{DS}$ = 28 V, $V_{GS}$ = 0) | $I_{DSS}$ | — | — | 2.0 | mAdc |
| Gate–Source Leakage Current<br>($V_{GS}$ = 20 V, $V_{DS}$ = 0) | $I_{GSS}$ | — | — | 1.0 | μAdc |
| **ON CHARACTERISTICS** (2) | | | | | |
| Gate Threshold Voltage<br>($V_{DS}$ = 10 V, $I_D$ = 50 mA) | $V_{GS(th)}$ | 1.0 | 3.0 | 6.0 | Vdc |
| Drain–Source On–Voltage<br>($V_{GS}$ = 10 V, $I_D$ = 3.0 A) | $V_{DS(on)}$ | — | — | 1.4 | Vdc |
| Forward Transconductance<br>($V_{DS}$ = 10 V, $I_D$ = 2.0 A) | $g_{fs}$ | 1.8 | 2.2 | — | mhos |
| **DYNAMIC CHARACTERISTICS** (2) | | | | | |
| Input Capacitance<br>($V_{DS}$ = 28 V, $V_{GS}$ = 0, f = 1.0 MHz) | $C_{iss}$ | — | 110 | — | pF |
| Output Capacitance<br>($V_{DS}$ = 28 V, $V_{GS}$ = 0, f = 1.0 MHz) | $C_{oss}$ | — | 105 | — | pF |
| Reverse Transfer Capacitance<br>($V_{DS}$ = 28 V, $V_{GS}$ = 0, f = 1.0 MHz) | $C_{rss}$ | — | 10 | — | pF |
| **FUNCTIONAL CHARACTERISTICS** (Figures 7 & 8) (4) | | | | | |
| Common Source Power Gain (3)<br>($V_{DD}$ = 28 Vdc, $P_{out}$ = 100 W, f = 400 MHz, $I_{DQ}$ = 200 mA) | $G_{PS}$ | 10 | 12 | — | dB |
| Drain Efficiency (3)<br>($V_{DD}$ = 28 Vdc, $P_{out}$ = 100 W, f = 400 MHz, $I_{DQ}$ = 200 mA) | η | 55 | 60 | — | % |
| Electrical Ruggedness (3)<br>($V_{DD}$ = 28 Vdc, $P_{out}$ = 100 W, f = 400 MHz, $I_{DQ}$ = 200 mA,<br>Load VSWR = 30:1, All Phase Angles At Frequency of Test) | ψ | No Degradation<br>in Output Power<br>Before & After Test | | | |
| **TYPICAL INPUT/OUTPUT DEVICE IMPEDANCES**<br>**MRF177** | | | | | |
| Series Equivalent Input Impedance<br>($V_{DD}$ = 28 V, $I_{DQ}$ = 200 mA, $P_{out}$ = 100 W, f = 400 MHz) | $Z_{in}$ | — | 2.35 + j0.4 | — | Ohms |
| Series Equivalent Output Impedance<br>($V_{DD}$ = 28 V, $I_{DQ}$ = 200 mA, $P_{out}$ = 100 W, f = 400 MHz) | $Z_{out}$ | — | 3.2 – j1.38 | — | Ohms |
| **MRF177M** | | | | | |
| Series Equivalent Input Impedance<br>($V_{DD}$ = 28 V, $I_{DQ}$ = 200 mA, $P_{out}$ = 100 W, f = 400 MHz) | $Z_{in}$ | — | 2.64 + j1.64 | — | Ohms |
| Series Equivalent Output Impedance<br>($V_{DD}$ = 28 V, $I_{DQ}$ = 200 mA, $P_{out}$ = 100 W, f = 400 MHz) | $Z_{out}$ | — | 3.15 + j0.05 | — | Ohms |

NOTES:
1. Note each transistor chip measured separately
2. Both transistor chips operating in push–pull amplifier
3. RF functional specification is the same for MRF177 & MRF177M

**TYPICAL CHARACTERISTICS**

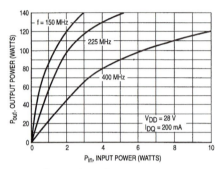

Figure 1. Output Power versus Input Power

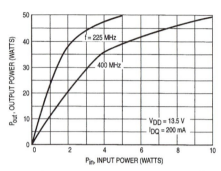

Figure 2. Output Power versus Input Power

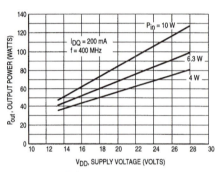

Figure 3. Output Power versus Supply Voltage

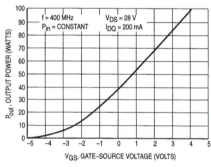

Figure 4. Output Power versus Gate Voltage

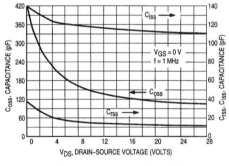

Figure 5. Capacitance versus Drain Voltage

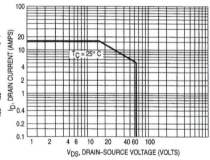

Figure 6. DC Safe Operating Area

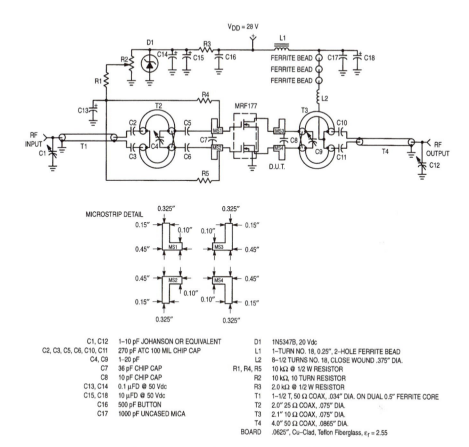

**MICROSTRIP DETAIL**

| C1, C12 | 1–10 pF JOHANSON OR EQUIVALENT | D1 | 1N5347B, 20 Vdc |
|---|---|---|---|
| C2, C3, C5, C6, C10, C11 | 270 pF ATC 100 MIL CHIP CAP | L1 | 1–TURN NO. 18, 0.25″, 2–HOLE FERRITE BEAD |
| C4, C9 | 1–20 pF | L2 | 8–1/2 TURNS NO. 18, CLOSE WOUND .375″ DIA. |
| C7 | 36 pF CHIP CAP | R1, R4, R5 | 10 kΩ @ 1/2 W RESISTOR |
| C8 | 10 pF CHIP CAP | R2 | 10 kΩ, 10 TURN RESISTOR |
| C13, C14 | 0.1 μFD @ 50 Vdc | R3 | 2.0 kΩ @ 1/2 W RESISTOR |
| C15, C18 | 10 μFD @ 50 Vdc | T1 | 1–1/2 T, 50 Ω COAX, .034″ DIA. ON DUAL 0.5″ FERRITE CORE |
| C16 | 500 pF BUTTON | T2 | 2.0″ 25 Ω COAX, .075″ DIA. |
| C17 | 1000 pF UNCASED MICA | T3 | 2.1″ 10 Ω COAX, .075″ DIA. |
| | | T4 | 4.0″ 50 Ω COAX, .0865″ DIA. |
| | | BOARD | .0625″, Cu–Clad, Teflon Fiberglass, $\varepsilon_r$ = 2.55 |

**Figure 7. Test Circuit Electrical Schematic — MRF177**

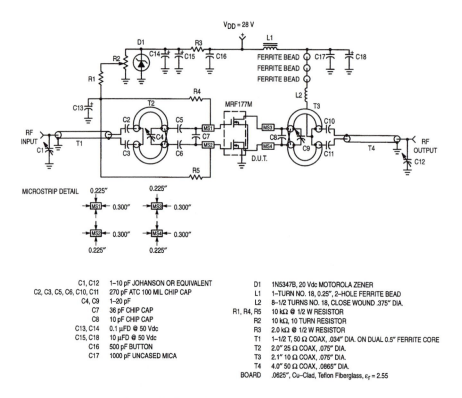

| | |
|---|---|
| C1, C12 | 1–10 pF JOHANSON OR EQUIVALENT |
| C2, C3, C5, C6, C10, C11 | 270 pF ATC 100 MIL CHIP CAP |
| C4, C9 | 1–20 pF |
| C7 | 36 pF CHIP CAP |
| C8 | 10 pF CHIP CAP |
| C13, C14 | 0.1 μFD @ 50 Vdc |
| C15, C18 | 10 μFD @ 50 Vdc |
| C16 | 500 pF BUTTON |
| C17 | 1000 pF UNCASED MICA |

| | |
|---|---|
| D1 | 1N5347B, 20 Vdc MOTOROLA ZENER |
| L1 | 1–TURN NO. 18, 0.25″, 2–HOLE FERRITE BEAD |
| L2 | 8–1/2 TURNS NO. 18, CLOSE WOUND .375″ DIA. |
| R1, R4, R5 | 10 kΩ @ 1/2 W RESISTOR |
| R2 | 10 kΩ, 10 TURN RESISTOR |
| R3 | 2.0 kΩ @ 1/2 W RESISTOR |
| T1 | 1–1/2 T, 50 Ω COAX, .034″ DIA. ON DUAL 0.5″ FERRITE CORE |
| T2 | 2.0″ 25 Ω COAX, .075″ DIA. |
| T3 | 2.1″ 10 Ω COAX, .075″ DIA. |
| T4 | 4.0″ 50 Ω COAX, .0865″ DIA. |
| BOARD | .0625″, Cu–Clad, Teflon Fiberglass, $\varepsilon_r$ = 2.55 |

**Figure 8. Test Fixture Electrical Schematic — MRF177M**

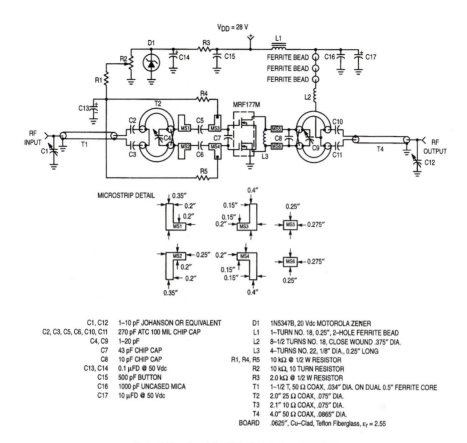

MICROSTRIP DETAIL

| C1, C12 | 1–10 pF JOHANSON OR EQUIVALENT |
| C2, C3, C5, C6, C10, C11 | 270 pF ATC 100 MIL CHIP CAP |
| C4, C9 | 1–20 pF |
| C7 | 43 pF CHIP CAP |
| C8 | 10 pF CHIP CAP |
| C13, C14 | 0.1 μFD @ 50 Vdc |
| C15 | 500 pF BUTTON |
| C16 | 1000 pF UNCASED MICA |
| C17 | 10 μFD @ 50 Vdc |

| D1 | 1N5347B, 20 Vdc MOTOROLA ZENER |
| L1 | 1–TURN NO. 18, 0.25″, 2–HOLE FERRITE BEAD |
| L2 | 8–1/2 TURNS NO. 18, CLOSE WOUND .375″ DIA. |
| L3 | 4–TURNS NO. 22, 1/8″ DIA., 0.25″ LONG |
| R1, R4, R5 | 10 kΩ @ 1/2 W RESISTOR |
| R2 | 10 kΩ, 10 TURN RESISTOR |
| R3 | 2.0 kΩ @ 1/2 W RESISTOR |
| T1 | 1–1/2 T, 50 Ω COAX, .034″ DIA. ON DUAL 0.5″ FERRITE CORE |
| T2 | 2.0″ 25 Ω COAX, .075″ DIA. |
| T3 | 2.1″ 10 Ω COAX, .075″ DIA. |
| T4 | 4.0″ 50 Ω COAX, .0865″ DIA. |
| BOARD | .0625″, Cu–Clad, Teflon Fiberglass, $\varepsilon_r$ = 2.55 |

**Figure 9. Broadband Amplifier Schematic — MRF177M**

# Comlinear CLC 430 Current Feedback Op-amp

## Comlinear CLC430
## General Purpose 100MHz Op Amp with Disable

### General Description

The ComlinearCLC430 is a low-cost, wideband monolithic amplifier for general purpose applications. The CLC430 utilizes Comlinear's patented current feedback circuit topology to provide an op amp with a slew rate of 2000V/μs, 100MHz unity-gain bandwidth and fast output disable function. Like all current feedback op amps, the CLC430 allows the frequency response to be optimized (or adjusted) by the selection of the feedback resistor. For demanding video applications, the 0.1dB bandwidth to 20MHz and differential gain/phase of 0.03%/0.05° make the CLC430 the preferred component for broadcast quality NTSC and PAL video systems.

The large voltage swing (28V$_{pp}$), continuous output current (85mA) and slew rate (2000V/μs) provide high-fidelity signal conditioning for applications such as CCDs, transmission lines and low impedance circuits. Even driving loads of 100Ω, the CLC430 provides very low 2nd and 3rd harmonic distortion at 1MHz (-76/-82dBc).

Video distribution, multimedia and general purpose applications will benefit from the CLC430's wide bandwidth and disable feature. Power is reduced and the output becomes a high impedance when disabled. The wide gain range of the CLC430 makes this general purpose op amp an improved solution for circuits such as active filters, differential-to-single-ended drivers, DAC transimpedance amplifiers and MOSFET drivers.

### Features

- 0.1dB gain flatness to 20MHz (A$_v$=+2)
- 100MHz bandwidth (A$_v$=+1)
- 2000V/μs slew rate
- 0.03%/0.05° differential gain/phase
- ±5V, ±15V or single supplies
- 100ns disable to high-impedance output
- Wide gain range
- Low cost

### Applications

- Video distribution
- CCD clock driver
- Multimedia systems
- DAC output buffers
- Imaging systems

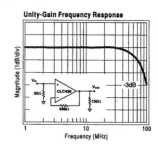

**Unity-Gain Frequency Response**

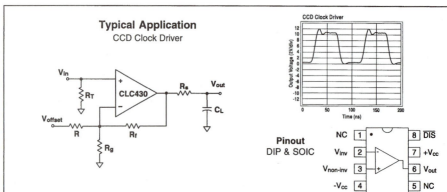

**Typical Application**
CCD Clock Driver

**Pinout**
DIP & SOIC

| | | | | |
|---|---|---|---|---|
| NC | 1 | • | 8 | $\overline{DIS}$ |
| V$_{inv}$ | 2 | | 7 | +V$_{cc}$ |
| V$_{non-inv}$ | 3 | | 6 | V$_{out}$ |
| -V$_{cc}$ | 4 | | 5 | NC |

## CLC430 Electrical Characteristics (V$_{cc}$ = ±15V; A$_v$ = +2V/V; R$_f$ =604Ω; R$_L$ = 100Ω; unless noted)

| PARAMETERS | CONDITIONS | V$_{cc}$ | TYP | MIN/MAX RATINGS | | | UNITS | NOTES |
|---|---|---|---|---|---|---|---|---|
| Ambient Temperature | CLC430 | | 25°C | 25°C | 0 to 70°C | -40 to 85°C | | |
| **FREQUENCY DOMAIN RESPONSE** | | | | | | | | |
| unity-gain bandwidth | V$_{out}$ < 1.0V$_{pp}$ | ±15 | 100 | | | | MHz | |
| small-signal bandwidth | V$_{out}$ < 1.0V$_{pp}$ | ±15 | 75 | 50 | 45 | 42 | MHz | |
| | V$_{out}$ < 1.0V$_{pp}$ | ±5 | 55 | 35 | | | MHz | |
| 0.1dB bandwidth | V$_{out}$ < 1.0V$_{pp}$ | ±15 | 20 | 7 | | | MHz | |
| | V$_{out}$ < 1.0V$_{pp}$ | ±5 | 16 | | | | MHz | |
| large-signal bandwidth | V$_{out}$ = 10V$_{pp}$ | | 30 | 22 | 20 | 19 | MHz | |
| gain flatness | V$_{out}$ < 1.0V$_{pp}$ | | | | | | | |
| peaking | DC to 10MHz | | 0.0 | 0.1 | 0.2 | 0.2 | dB | |
| rolloff | DC to 20MHz | | 0.1 | 0.7 | 1.0 | 1.2 | dB | |
| linear phase deviation | DC to 20MHz | | 0.5 | 1.8 | 2.0 | 2.1 | ° | |
| differential gain | 4.43MHz, R$_L$=150Ω | ±15 | 0.03 | 0.05 | 0.06 | 0.06 | % | |
| | 4.43MHz, R$_L$=150Ω | ±5 | 0.03 | 0.05 | | | % | |
| differential phase | 4.43MHz, R$_L$=150Ω | ±15 | 0.05 | 0.09 | 0.12 | 0.13 | ° | |
| | 4.43MHz, R$_L$=150Ω | ±5 | 0.09 | 0.19 | | | ° | |
| **TIME DOMAIN RESPONSE** | | | | | | | | |
| rise and fall time | 2V step | | 5 | 7 | 7 | 7 | ns | |
| | 10V step | | 10 | 14 | 14 | 14 | ns | |
| settling time to 0.05% | 2V step | | 35 | 50 | 55 | 55 | ns | |
| overshoot | 2V step | | 5 | 15 | 15 | 15 | % | |
| slew rate | 20V step | | 2000 | 1500 | 1450 | 1450 | V/μs | |
| **DISTORTION AND NOISE RESPONSE** | | | | | | | | |
| 2$^{nd}$ harmonic distortion | 1V$_{pp}$,1MHz, R$_L$=500 | | -89 | | | | dBc | |
| 3$^{rd}$ harmonic distortion | 1V$_{pp}$,1MHz, R$_L$=500 | | -92 | | | | dBc | |
| input voltage noise | >1MHz | | 3.0 | 3.5 | 3.7 | 3.8 | nV/√Hz | |
| non-inverting input current noise | >1MHz | | 3.2 | 6.0 | 6.3 | 6.8 | pA/√Hz | |
| inverting input current noise | >1MHz | | 15 | 18 | 20 | 21 | pA/√Hz | |
| **DC PERFORMANCE** | | | | | | | | |
| input offset voltage | | ±15 | 1.0 | 7.5 | 9.0 | 10.0 | mV | A |
| average drift | | | 25 | --- | 50 | 50 | μV/ C | |
| input bias current | non-inverting | ±15,±5 | 3 | 14 | 16 | 20 | μA | A |
| average drift | | | 10 | --- | 100 | 100 | nA/°C | |
| input bias current | inverting | ±15,±5 | 3 | 14 | 15 | 17 | μA | A |
| average drift | | | 10 | --- | 60 | 90 | nA/°C | |
| power-supply rejection ratio | DC | | 62 | 56 | 54 | 53 | dB | B |
| common-mode rejection ratio | DC | | 62 | 54 | 53 | 52 | dB | |
| supply current | R$_L$ = ∞ | ±15,±5 | 11, 8.5 | 12 | 13 | 14.5 | mA | A |
| disabled | R$_L$ = ∞ | ±15,±5 | 1.5 | 2.0 | 2.2 | 2.4 | mA | A |
| **SWITCHING PERFORMANCE** | | | | | | | | |
| turn on time | | | 200 | 300 | 320 | 340 | ns | |
| turn off time | (Note 2) | | 100 | 200 | 200 | 200 | ns | |
| off isolation | 10MHz | | 59 | 56 | 56 | 56 | dB | |
| high input voltage | V$_{IH}$ | ±15 | 11.8 | 12.5 | 12.7 | | V | |
| | | ±5 | 1.8 | 2.5 | 2.7 | | V | |
| low input voltage | V$_{IL}$ | ±15 | 10.8 | 10.5 | 10.0 | | V | |
| | | ±5 | 0.8 | 0.6 | 0.1 | | V | |
| **MISCELLANEOUS PERFORMANCE** | | | | | | | | |
| Non-inverting input resistance | | | 8.0 | 3.0 | 2.5 | 1.7 | MΩ | |
| Non-inverting input capacitance | | | 0.5 | 1.0 | 1.0 | 1.0 | pF | |
| input voltage range | common mode | ±15 | ±12.5 | ±12.3 | ±12.1 | ±11.8 | V | |
| | common mode | ±5 | ±2.5 | ±2.3 | ±2.2 | ±1.9 | V | |
| output voltage range | R$_L$ = ∞ | ±15 | ±14 | ±13.7 | ±13.7 | ±13.6 | V | |
| | R$_L$ = ∞ | ±5 | ±4.0 | ±3.9 | ±3.8 | ±3.7 | V | |
| output current | | | ±85 | ±60 | ±50 | ±45 | mA | |

Min/max ratings are based on product characterization and simulation. Individual parameters are tested as noted. Outgoing quality levels are determined from tested parameters.

### Absolute Maximum Ratings

| | |
|---|---|
| supply voltage | ±16.5V |
| short circuit current | (note 1) |
| common-mode input voltage | ±V$_{cc}$ |
| maximum junction temperature | +200°C |
| storage temperature | -65°C to +150°C |
| lead temperature (soldering 10 sec) | +300°C |

### Notes

A) J-level: spec is 100% tested at +25°C, sample tested at +85°C.
L-level: spec is 100% wafer probed at 25°C.
B) J-level: spec is sample tested at 25°C.
1) Output is short circuit protected to ground, however maximum reliability is obtained if output current does not exceed 125mA.
2) To >50dB attenuation @ 10MHz.

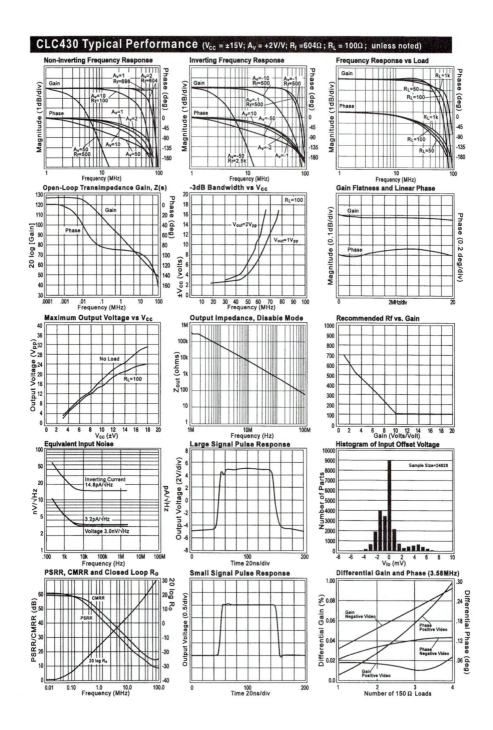

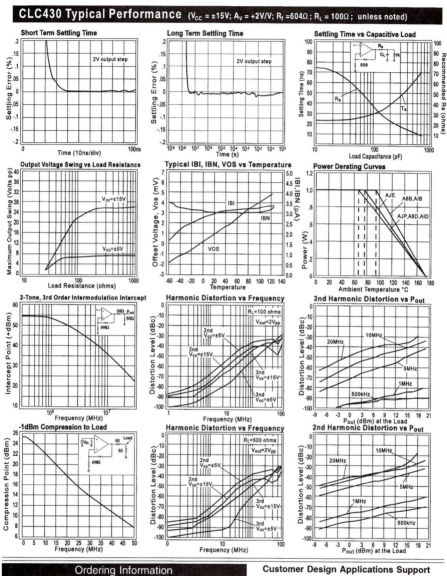

## CLC430 Typical Performance ($V_{CC} = \pm15V$; $A_v = +2V/V$; $R_f = 604\Omega$; $R_L = 100\Omega$; unless noted)

### Ordering Information

| Model | Temperature Range | Description |
|---|---|---|
| CLC430AJP | -40°C to +85°C | 8-pin PDIP |
| CLC430AJE | -40°C to +85°C | 8-pin SOIC |
| CLC430ALC | -40°C to +85°C | dice |
| CLC430A8B | -55°C to +125°C | 8-pin CERDIP, MIL-STD-883 |
| CLC430A8L-2 | -55°C to +125°C | 20-pin LCC, MIL-STD-883 |
| CLC430AMC | -55°C to +125°C | dice, MIL-STD-883 |
| CLC430A8D | -55°C to +125°C | 8-pin sidebrazed ceramic DIP, MIL-STD 833, Level B |

DESC SMD number: 5962-92030.

**Customer Design Applications Support**
National Semiconductor is committed to design excellence. For sales, literature and technical support call the National Semiconductor Customer Response Group at **1-800-272-9959** or fax **1-800-737-7018.**

## General Design Considerations

The CLC430 is a general purpose current-feedback amplifier for use in a variety of small- and large-signal applications. Use the feedback resistor to fine tune the gain flatness and -3dB bandwidth for any gain setting. Comlinear provides information for the performance at a gain of +2 for small and large signal bandwidths. The plots show feedback resistor values for selected gains.

## Gain

Use the following equations to set the CLC430's non-inverting or inverting gain:

$$\text{Non-Inverting Gain} = 1 + \frac{R_f}{R_g}$$

$$\text{Inverting Gain} = -\frac{R_f}{R_g}$$

Choose the resistor values for non-inverting or inverting gain by the following steps.

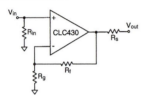

Fig. 0 Component Identification

1) Select the recommended feedback resistor $R_f$ (refer to plot in the plot section entitled $R_f$ vs Gain).
2) Choose the value of $R_g$ to set gain.
3) Select $R_s$ to set the circuit output impedance.
4) Select $R_{in}$ for input impedance and input bias.

## High Gains

Current feedback closed-loop bandwidth is independent of gain-bandwidth-product for small gain changes. For larger gain changes the optimum feedback register $R_f$ is derived by the following:

$$R_f = 724\Omega - 60\Omega \cdot (A_v)$$

As gain is increased, the feedback resistor allows bandwidth to be held constant over a wide gain range. For a more complete explanation refer to application note OA-25 Stability Analysis of Current-Feedback Amplifiers.

Resistors have varying parasitics that affect circuit performance in high-speed design. For best results, use leaded metal-film resistors or surface mount resistors. A SPICE model for the CLC430 is available to simulate overall circuit performance.

## Enable / Disable Function

The CLC430 amplifier features an enable/disable function that changes the output and inverting input from low to high impedance. The pin 8 enable/disable logic levels

are as follows:

| | ±15V | ±5V |
|---|---|---|
| $V_{cc}$ | ±15V | ±5V |
| Enable | >12.7V | >2.7V |
| Disable | <10.0V | <0.8V |

The amplifier is enabled with pin 8 left open due to the 2kΩ pull-up resistor, shown in Fig. 1.

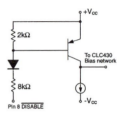

Fig. 1 Pin 8 Equivalent Disable Circuit

Open-collector or CMOS interfaces are recommended to drive pin 8. The turn-on and off time depends on the speed of the digital interface.

The equivalent output impedance when disabled is shown in Fig. 2. With $R_g$ connected to ground, the sum of $R_f$ and $R_g$ dominates and reduces the disabled output impedance. To raise the output impedance in the disabled state, connect the CLC430 as a unity-gain voltage follower by removing $R_g$. Current-feedback op-amps need the recommended $R_f$ in a unity-gain follower circuit. For high density circuit layouts consider using the dual CLC431 (with disable) or the dual CLC432 (without disable).

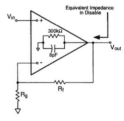

Fig. 2 Equivalent Disabled Output Impedance

## 2nd and 3rd Harmonic Distortion

To meet low distortion requirements, recognize the effect of the feedback resistor. Increasing the feedback resistor will decrease the loop gain and increase distortion. Decreasing the load impedance increases 3rd harmonic distortion more than 2nd.

## Differential Gain and Differential Phase

The CLC430 has low DG and DP errors for video applications. Add an external pulldown resistor to the CLC430's output to improve DG and DP as seen in Fig.3. A 604Ω $R_P$ will improve DG and DP to 0.01% and 0.02°.

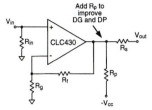

Fig. 3 Improved DG and DP Video Amplifier

## Printed Circuit Layout

To get the best amplifier performance careful placement of the amplifier, components and printed circuit traces must be observed. Place the 0.1µF ceramic decoupling capacitors less than 0.1" (3mm) from the power supply pins. Place the 6.8µF tantalum capacitors less than 0.75" (20mm) from the power supply pins. Shorten traces between the inverting pin and components to less than 0.25" (6mm). Clear ground plane 0.1" (3mm) away from pads and traces that connect to the inverting, non-inverting and output pins. Do not place ground or power plane beneath the op-amp package. Comlinear provides literature and evaluation boards 730013 DIP or 730027 SOIC illustrating the recommended op-amp layout.

## Applications Circuits

### Level Shifting

The circuit shown in Fig. 4 implements level shifting by AC coupling the input signal and summing a DC voltage. The resistor $R_{in}$ and the capacitor C set the high-pass break frequency. The amplifier closed-loop bandwidth is fixed by the selection of $R_f$. The DC and AC gains for circuit of Fig. 4 are different. The AC gain is set by the ratio of $R_f$ and $R_g$. And the DC gain is set by the parallel combination of $R_g$ and $R_2$.

$$V_{out} = V_{in_{ac}}\left(1+\left(\frac{R_f}{R_g\|R_2}\right)\right) - V_{in_{DC}}\left(\frac{R_f}{R_2}\right)$$

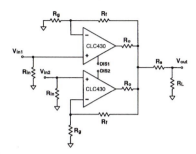

Fig. 4 Level Shifting Circuit

### Multiplexing

Multiple signal switching is easily handled with the disable function of the CLC430. Board trace capacitance at the output pin will affect the frequency response and switching transients. To lessen the effects of output capacitance place a resistor ($R_o$) within the feedback loop to isolate the outputs as shown in Fig. 5. To match the mux output impedance to a transmission line, add a resistor ($R_s$) in series with the output.

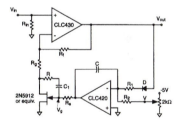

Fig. 5 Output Connection

### Automatic Gain Control

Current-feedback amplifiers can implement very fast automatic-gain control circuits. The circuit shown in Fig. 6 shows an AGC circuit using the CLC430, a half-wave rectifier, an integrator and a FET. The CLC430 current-feedback amplifier maintains constant bandwidth and linear phase over AGC's gain range. This circuit effectively controls the output level for continuous signals.

Fig. 6 AGC Circuit

The bandwidth of the CLC430 AGC is limited by $R_f$, the feedback resistor. The FET gate voltage is limited to a range of:

$$-2.5 < V_g < -1$$

R of 750Ω and C1 of 1.0µF gives a useful $R_{ds}$ range of approximately 150 to 2K ohms. Scaling the integrator gain or adding attenuation before the diode D accommodates large signal swings. Determine the overall gain by:

$$1+\frac{R_f}{R_g+R_{ds}}$$

The integrator sets the loop time constant.

# Burr Brown 3554 1.7 GHz op amp 8 pages

**3554**

---

## Wideband - Fast-Settling
## OPERATIONAL AMPLIFIER

### FEATURES
- SLEW RATE, 1000V/µsec
- FAST SETTLING, 150nsec, max (to ±.05%)
- GAIN-BANDWIDTH PRODUCT, 1.7GHz
- FULL DIFFERENTIAL INPUT

### APPLICATIONS
- PULSE AMPLIFIERS
- TEST EQUIPMENT
- WAVEFORM GENERATORS
- FAST D/A CONVERTERS

### DESCRIPTION
The 3554 is a full differential input, wideband operational amplifier. It is designed specifically for the amplification or conditioning of wideband data signals and fast pulses. It features an unbeatable combination of gain-bandwidth product, settling time and slew rate. It uses hybrid construction. On the beryllia substrate are matched input FETs, thin-film resistors and high speed silicon dice. Active laser trimming and complete testing provide superior performance at a very moderate price.

The 3554 has a slew rate of 1000V/µsec and will output ±10V and ±100mA. When used as a fast

settling amplifier, the 3554 will settle to ±0.05% of the final value within 150nsec. A single external compensation capacitor allows the user to optimize the bandwidth, slew rate or settling time in the particular application.

The 3554 is reliable and rugged and addresses almost any application when speed and bandwidth are serious considerations. It is particularly a good choice for use in fast settling circuits, fast D/A converters, multiplexer buffers, comparators, waveform generators, integrators, and fast current amplifiers. It is available in several grades to allow selection of just the performance required.

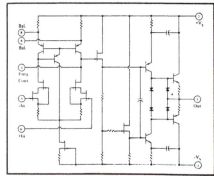

International Airport Industrial Park · P.O. Box 11400 · Tucson, Arizona 85734 · Tel. (602) 746-1111 · Twx: 910-952-1111 · Cable: BBRCORP · Telex: 66-6491

# TYPICAL CIRCUITS

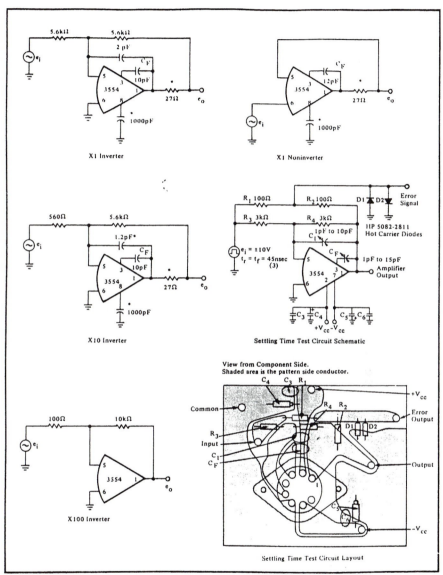

X1 Inverter

X1 Noninverter

X10 Inverter

Settling Time Test Circuit Schematic

X100 Inverter

View from Component Side.
Shaded area is the pattern side conductor.

Settling Time Test Circuit Layout

**NOTES:**

1. These circuits are optimized for driving large capacitive loads (to 470pF).
2. The 3554 is stable at gains of greater than 55 ($C_L \leq 100pF$) without any frequency compensation.
3. 45nsec is optimum. Very fast rise times (10-20nsec) may saturate the input stage causing less than optimum settling time performance.

*Indicates component that may be eliminated when large capacitive loads are not being driven by the device.

# ELECTRICAL SPECIFICATIONS

At T$_{CASE}$ = 25°C and ±15VDC, unless otherwise noted

| PARAMETERS | CONDITIONS | 3554AM | | | 3554BM | | | 3554SM | | | UNITS |
|---|---|---|---|---|---|---|---|---|---|---|---|
| | | MIN | TYP | MAX | MIN | TYP | MAX | MIN | TYP | MAX | |
| **OPEN-LOOP GAIN, DC** | | | | | * | * | * | * | * | * | |
| No Load | | 100 | 106 | | | | | | | | dB |
| Rated Load | R$_L$ = 100Ω | 90 | 96 | | | | | | | | dB |
| **RATED OUTPUT** | | | | | * | * | * | * | * | * | |
| Voltage | I$_o$ = ±100mA | ±10 | ±11 | | | | | | | | V |
| Current | V$_o$ = ±10V | ±100 | ±125 | | | | | | | | mA |
| Output Resistance, open loop | f = 10MHz | | 20 | | | | | | | | •Ω |
| **DYNAMIC RESPONSE** | | | | | * | * | * | * | * | * | |
| Bandwidth (0dB, small signal) | C$_f$ = 0 | 70† | 90 | | | | | | | | MHz |
| Gain-bandwidth Product | C$_f$ = 0, G = 10 V/V | 150 | 225 | | | | | | | | MHz |
| | C$_f$ = 0, G = 100 V/V | 425 | 725 | | | | | | | | MHz |
| | C$_f$ = 0, G = 1000 V/V | 1000 | 1700 | | | | | | | | MHz |
| Full Power Bandwidth | C$_f$ = 0, V$_o$ = 20V p-p, R$_L$ = 100Ω | 16 | 19 | | | | | | | | MHz |
| Slew Rate | C$_f$ = 0, V$_o$ = 20V p-p, R$_L$ = 100Ω | 1000 | 1200 | | | | | | | | V/µsec |
| Settling Time   to ±1% | A = 1 | | 60 | | | | | | | | nsec |
|    to ±.1% | A = 1 | | 120 | | | | | | | | nsec |
|    to ±.05% | A = 1 | | 140 | 150 | | | | | | | nsec |
|    to ±.01% | A = 1 | | 200 | 250 | | | | | | | nsec |
| **INPUT OFFSET VOLTAGE** | | | | | | | | | | | |
| Initial offset, T$_A$ = 25°C | | | ±0.5 | ±2 | | ±0.2 | ±1 | | ±0.2 | ±1 | mV |
| vs. Temp (T$_A$ = -25°C to +85°C) | | | ±20 | ±50 | | ±8 | ±15 | | | | µV/°C |
| vs. Temp (T$_A$ = -55°C to +125°C) | | | | | | | | | ±12 | ±25 | µV/°C |
| vs. Supply Voltage | | | ±80 | ±300 | | * | * | | * | * | µV/V |
| **INPUT BIAS CURRENT** | | | | | * | * | * | * | * | * | |
| Initial bias, 25°C | | 0 | -10 | -50 | | | | | | | pA |
| vs. Temp | | | ** | | | | | | | | |
| vs. Supply Voltage | | | ±1 | | | | | | | | pA/V |
| **INPUT DIFFERENCE CURRENT** | | | | | * | * | * | * | * | * | |
| Initial difference, 25°C | | | ±2 | ±10 | | | | | | | pA |
| **INPUT IMPEDANCE** | | | | | * | | | * | | | |
| Differential | | | 10¹¹ ‖ 2 | | | | | | | | Ω ‖ pF |
| Common-mode | | | 10¹¹ ‖ 2 | | | | | | | | Ω ‖ pF |
| **INPUT NOISE** | | | | | * | * | * | * | * | * | |
| Voltage, f$_o$ = 1Hz | R$_s$ = 100Ω | | 125 | 450† | | | | | | | nV/√Hz |
| f$_o$ = 10Hz | R$_s$ = 100Ω | | 50 | 160† | | | | | | | nV/√Hz |
| f$_o$ = 100Hz | R$_s$ = 100Ω | | 25 | 90† | | | | | | | nV/√Hz |
| f$_o$ = 1kHz | R$_s$ = 100Ω | | 15 | 50† | | | | | | | nV/√Hz |
| f$_o$ = 10kHz | R$_s$ = 100Ω | | 10 | 35† | | | | | | | nV/√Hz |
| f$_o$ = 100kHz | R$_s$ = 100Ω | | 8 | 25† | | | | | | | nV/√Hz |
| f$_o$ = 1MHz | R$_s$ = 100Ω | | 7 | 25† | | | | | | | nV/√Hz |
| f$_B$ = .3Hz to 10Hz | R$_s$ = 100Ω | | 2 | 7† | | | | | | | µV, p-p |
| f$_B$ = 10Hz to 1MHz | R$_s$ = 100Ω | | 8 | 25 | | | | | | | µV, rms |
| Current, f$_B$ = .3Hz to 10Hz | R$_s$ = 100Ω | | 45 | | | | | | | | fA, p-p |
| f$_B$ = 10Hz to 1MHz | R$_s$ = 100Ω | | 2 | | | | | | | | pA, rms |
| **INPUT VOLTAGE RANGE** | | | | | * | * | * | * | * | * | |
| Common-mode Voltage Range | Linear Operation | | ±(\|V$_{cc}$\|-4) | | | | | | | | V |
| Common-mode Rejection | f = DC, V$_{CM}$ = +7V, -10V | 44 | 78 | | | | | | | | dB |
| Max. Safe Input Voltage | | | ±Supply | | | | | | | | V |
| **POWER SUPPLY** | | | | | * | * | * | * | * | * | |
| Rated Voltage | | | ±15 | | | | | | | | VDC |
| Voltage Range, derated performance | | ±5 | | ±18 | | | | | | | VDC |
| Current, quiescent | | ±17 | ±35 | ±45 | | | | | | | mA |
| **TEMPERATURE RANGE** (ambient) | | | | | | | | | | | |
| Specification | | -25 | | +85 | -25 | | +85 | -55 | | +125 | °C |
| Operating, derated performance | | -55 | | +125 | -55 | | +125 | -55 | | +125 | °C |
| Storage | | -65 | | +150 | -65 | | +150 | -65 | | +150 | °C |
| θ junction-case | | | 15 | | | 15 | | | 15 | | °C/W |
| θ junction-ambient | | | 45 | | | 45 | | | 45 | | °C/W |

* Specifications same as for 3554AM
** Doubles every +10°C
† This parameter is untested and is not guaranteed. This specification is established to a 90% confidence level.
The information in this publication has been carefully checked and is believed to be reliable; however, no responsibility is assumed for possible inaccuracies or omissions. Prices and specifications are subject to change without notice. No patent rights are granted to any of the circuits described herein.

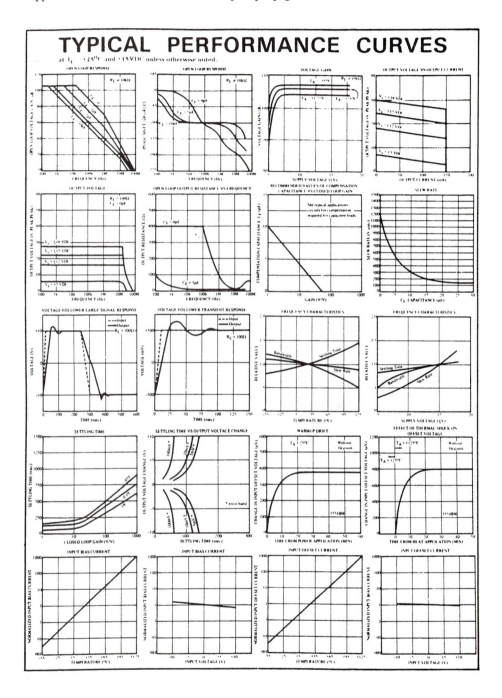

TYPICAL PERFORMANCE CURVES

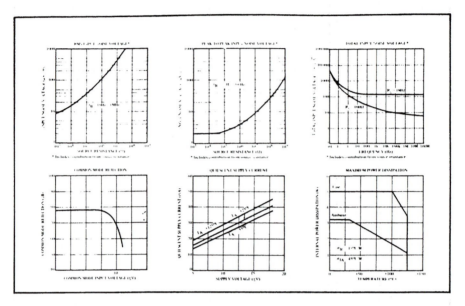

## MECHANICAL

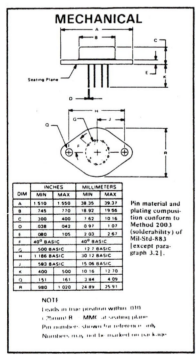

| DIM | INCHES | | MILLIMETERS | |
|-----|--------|--------|-------------|--------|
| | MIN | MAX | MIN | MAX |
| A | 1.510 | 1.550 | 38.35 | 39.37 |
| B | .745 | .770 | 18.92 | 19.56 |
| C | .300 | .400 | 7.62 | 10.16 |
| D | .038 | .042 | 0.97 | 1.07 |
| E | .080 | .105 | 2.03 | 2.67 |
| F | 40° BASIC | | 40° BASIC | |
| G | .500 BASIC | | 12.7 BASIC | |
| H | 1.186 BASIC | | 30.12 BASIC | |
| J | .593 BASIC | | 15.06 BASIC | |
| K | .400 | .500 | 10.16 | 12.70 |
| Q | .151 | .161 | 3.84 | 4.09 |
| R | .980 | 1.020 | 24.89 | 25.91 |

Pin material and plating composition conform to Method 2003 (solderability) of Mil-Std-883 [except paragraph 3.2].

NOTE

Leads in true position within .010 (.25mm) R MMC at seating plane.
Pin numbers shown for reference only.
Numbers may not be marked on package.

## AMPLIFIER CONNECTIONS

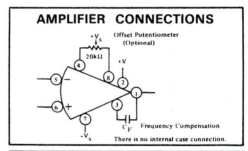

Offset Potentiometer (Optional)

$+V_S$

20kΩ

$+V$

Frequency Compensation $C_F$

There is no internal case connection.

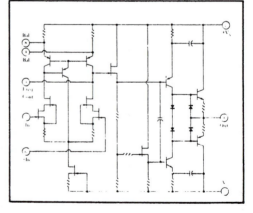

# APPLICATIONS INFORMATION

## WIRING PRECAUTIONS

The 3554 is a wideband, high frequency operational amplifier that has a gain-bandwidth product exceeding 1 Gigahertz. The full performance capability of this amplifier will be realized by observing a few wiring precautions and high frequency techniques.

Of all the wiring precautions, grounding is the most important and is described in an individual section. The mechanical circuit layout also is very important. All circuit element leads should be as short as possible. All printed circuit board conductors should be wide to provide low resistance, low inductance connections and should be as short as possible. In general, the entire physical circuit should be as small as practical. Stray capacitances should be minimized especially at high impedance nodes such as the input terminals of the amplifier. Pin 5, the inverting input, is especially sensitive and all associated connections must be short. Stray signal coupling from the output to the input or to pin 8 should be minimized. A recommended printed circuit board layout is shown with the "Typical Circuits." It also may be used for test purposes as described below.

When designing high frequency circuits low resistor values should be used; resistor values less than 5.6kΩ are recommended. This practice will give the best circuit performance as the time constants formed with the circuit capacitances will not limit the performance of the amplifier.

## GROUNDING

As with all high frequency circuits a ground plane and good grounding techniques should be used. The ground plane should connect all areas of the pattern side of the printed circuit board that are not otherwise used. The ground plane provides a low resistance, low inductance common return path for all signal and power returns. The ground plane also reduces stray signal pick up. An example of an adequate ground plane and good high frequency techniques is the Settling Time Test Circuit Layout shown with the "Typical Circuits."

Each power supply lead should be bypassed to ground as near as possible to the amplifier pins. A combination of a 1μF tantalum capacitor in parallel with a 470pF ceramic capacitor is a suitable bypass.

In inverting applications it is recommended that pin 6, the noninverting input, be grounded rather than being connected to a bias current compensating resistor. This assures a good signal ground at the noninverting input. A slight offset error will result; however, because the resistor values normally used in high frequency circuits are small and the bias current is small, the offset error will be minimal.

If point-to-point wiring is used or a ground plane is not, single point grounding should be used. The input signal return and the load signal return and the power supply common should all be connected at the same physical point. This will eliminate any common current paths or ground loops which could cause signal modulation or unwanted feedback.

It is recommended that the case of the 3554 not be grounded during use (it may, if desired). A grounded case will add a slight capacitance to each pin. To an already functional circuit, grounding the case will probably require slight compensation readjustment and the compensation capacitor values will be slightly different from those recommended in the typical performance curves. There is no internal connection to the case.

Proper grounding is the single most important aspect of high frequency circuitry.

## GUARDING

The input terminals of the 3554 may be surrounded by a guard ring to divert leakage currents from the input terminals. This technique is particularly important in low bias current and high input impedance applications. The guard, a conductive path that completely surrounds the two amplifier inputs, should be connected to a low impedance point which is at the input signal potential. It blocks unwanted printed circuit board leakage currents from reaching the input terminals. The guard also will reduce stray signal coupling to the input.

In high frequency applications guarding may not be desirable as it increases the input capacitance and can degrade performance. The effects of input capacitance, however, can be compensated by a small capacitor placed across the feedback resistor. This is described further in the following section.

## COMPENSATION

The 3554 uses external frequency compensation so that the user may optimize the bandwidth or slew rate or settling time for his particular application. Several typical performance curves are provided to aid in the selection of the correct compensation capacitance value. In addition several typical circuits show recommended compensation in different applications.

The primary compensation capacitor, $C_F$, is connected between pins 1 and 3. As the performance curves show, larger closed-loop gain configurations require less capacitance and an improved gain-bandwidth product will be realized. Note that no compensation capacitor is required for closed-loop gains above 55V/V and when the load capacitance is less than 100pF.

When driving large capacitive loads, 470pF and greater,

an additional capacitor, $C_x$, is connected between pin 8 and ground. This capacitor is typically 1000pF. It is particularly necessary in low closed loop voltage gain configurations. The value may be varied to optimize performance and will depend upon the load capacitance value. In addition, the performance may be optimized by connecting a small resistance in series with the output and a small capacitor from pin 1 to 5. See the "Typical Circuits" for the X10 Inverter.

The flat high frequency response of the 3554 may be preserved and any high frequency peaking avoided by connecting a small capacitor in parallel with the feedback resistor. This capacitor will compensate for the closed-loop, high frequency, transfer function zero that results from the time constant formed by the input capacitance of the amplifier, typically 2pF, and the input and feedback resistors. Using small resistor values will keep the break frequency of this zero sufficiently high, avoiding peaking and preserving the phase margin. Resistor values less than 5.6kΩ are recommended. The selected compensation capacitor may be a trimmer, a fixed capacitor or a planned PC board capacitance. The capacitance value is strongly dependent on circuit layout and closed-loop gain. It will typically be 2pF for a clean layout using low resistances (1kΩ) and up to 10pF for circuits using larger resistances.

## SETTLING TIME

Settling time is truly a complete dynamic measure of the 3554's total performance. It includes the slew rate time, a large signal dynamic parameter, and the time to accurately reach the final value, a small signal parameter that is a function of bandwidth and open loop-gain. The settling time may be optimized for the particular application by selection of the closed-loop gain and the compensation capacitance. The best settling time is observed in low closed-loop gain circuits. A performance curve shows the settling time to three different error bands.

Settling time is defined as the total time required, from the signal input step, for the output to settle to within the specified error band around the final value. This error band is expressed as a percentage of the magnitude of the output transition.

## SLEW RATE

Slew rate is primarily an output, large signal parameter. It has virtually no dependence upon the closed-loop gain or the bandwidth, per se. It is dependent upon compensation. Decreasing the compensation capacitor value will increase the available slew rate as shown in the performance curve. Stray capacitances may appear to the amplifier as compensation. To avoid limiting the slew rate performance, stray capacitances should be minimized.

## CAPACITIVE LOADS

The 3554 will drive large capacitive loads (up to 1000pF) when properly compensated. See the section on "Compensation." The effect of a capacitive load is to decrease the phase margin of the amplifier. With compensation the amplifier will provide stable operation even with large capacitive loads.

The 3554 is particularly well suited for driving 50Ω loads connected via coaxial cables due to its ±100mA output drive capability. The capacitance of the coaxial cable, 29pF/foot of length for RG-58, does not load the amplifier when the coaxial cable or transmission line is terminated in the characteristic impedance of the transmission line.

## OFFSET VOLTAGE ADJUSTMENT

The offset voltage of the 3554 may be adjusted to zero by connecting a 20kΩ linear potentiometer between pins 4 and 8 with the wiper connected to the positive supply. A small, noninductive potentiometer is recommended. The leads connecting the potentiometer to pins 4 and 8 should be no longer than 6 inches to avoid stray capacitance and stray signal pickup. Stray coupling from the output, pin 1, to pin 4 (negative feedback) or to pin 8 (positive feedback) should be avoided.

The potentiometer is optional and may be omitted when the guaranteed offset voltage is considered sufficiently low for the particular application.

For each microvolt of offset voltage adjusted, the offset voltage temperature drift will change by $\pm 0.004 \mu V/°C$.

## HEAT SINKING

The 3554 does not require a heat sink for operation in most environments. The use of a heat sink, however, will reduce the internal thermal rise and will result in cooler operating temperatures. At extreme temperature and under full load conditions a heat sink will be necessary as indicated in the "Maximum Power Dissipation" curve. A heat sink with 8 holes for the 8 amplifier pins should be used. Burr-Brown has heat sinks available in three sizes - 3°C/W, 4.2°C/W and 12°C/W. A separate product data sheet is available upon request.

When heat sinking the 3554, it is recommended that the heat sink be connected to the amplifier case and the combination not connected to the ground plane. For a single-sided printed circuit board, the heat sink may be mounted between the 3554 and the nonconductive side of the PC board, and insulating washers, etc., will not be required. The addition of a heat sink to an already functional circuit will probably require slight compensation readjustment for optimum performance due to the change in stray capacitances. The added stray capacitance from the heat sink to each pin will depend on the thickness and type of heat sink used.

## SHORT CIRCUIT PROTECTION

The 3554 is short circuit protected for continuous output shorts to common. Output shorts to either supply will destroy the device, even for momentary connections. Output shorts to other potential sources are not recommended as they may cause permanent damage.

## TESTING

The 3554 may be tested in conventional operational amplifier test circuits; however, to realize the full performance capabilities of the 3554, the test fixture must not limit the full dynamic performance capability of the amplifier. High frequency techniques must be employed. The most critical dynamic test is for settling time. The 3554 Settling Time Test Circuit Schematic and a test circuit layout is shown with the "Typical Circuits." The input pulse generator must have a flat topped, fast settling pulse to measure the true settling time of the amplifier. The layout exemplifies the high frequency considerations that must be observed. The layout also may be used as a guide for other test circuits. Good grounding, truly square drive signals, minimum stray coupling and small physical size are important.

Every 3554 is thoroughly tested prior to shipment assuring the user that all parameters equal or exceed their specifications.

# Texas Instrument's SN54LS297 Digital Phase-Locked-Loop Filters

- **Digital Design Avoids Analog Compensation Errors**

- **Easily Cascadable for Higher Order Loops**

- **Useful Frequency from DC to:**
  **50 MHz Typical (K Clock)**
  **35 MHz Typical (I/D Clock)**

## description

The SN54LS297 and SN74LS297 devices are designed to provide a simple, cost-effective solution to high-accuracy, digital, phase-locked-loop applications. These devices contain all the necessary circuits, with the exception of the divide-by-N counter, to build first order phase-locked loops as described in Figure 1.

Both exclusive-OR (XORPD) and edge-controlled (ECPD) phase detectors are provided for maximum flexibility.

Proper partitioning of the loop function, with many of the building blocks external to the package, makes it easy for the designer to incorporate ripple cancellation or to cascade to higher order phase-locked loops.

The length of the up/down K counter is digitally programmable according to the K counter function table. With A, B, C, and D all low, the K counter is disabled. With A high and B, C, and D low, the K counter is only

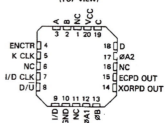

three stages long, which widens the bandwidth or capture range and shortens the lock time of the loop. When A, B, C, and D are all programmed high, the K counter becomes seventeen stages long, which narrows the bandwidth or capture range and lengthens the lock time. Real-time control of loop bandwidth by manipulating the A through D inputs can maximize the overall performance of the digital phase-locked loop.

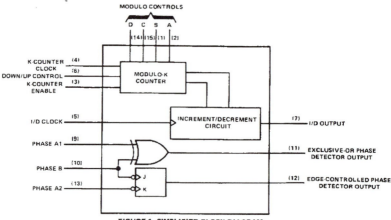

**FIGURE 1—SIMPLIFIED BLOCK DIAGRAM**
Pin numbers shown are for J, N and W packages.

## SN54LS297, SN74LS297
## DIGITAL PHASE-LOCKED-LOOP FILTERS

### description (continued)

The 'LS297 can perform the classic first-order phase-locked loop function without using analog components. The accuracy of the digital phase-locked loop (DPLL) is not affected by $V_{CC}$ and temperature variations, but depends solely on accuracies of the K clock, I/D clock, and loop propagation delays. The I/D clock frequency and the divide-by-N modulos will determine the center frequency of the DPLL. The center frequency is defined by the relationship $f_C = I/D$ Clock/2N (Hz).

### logic diagram (positive logic)

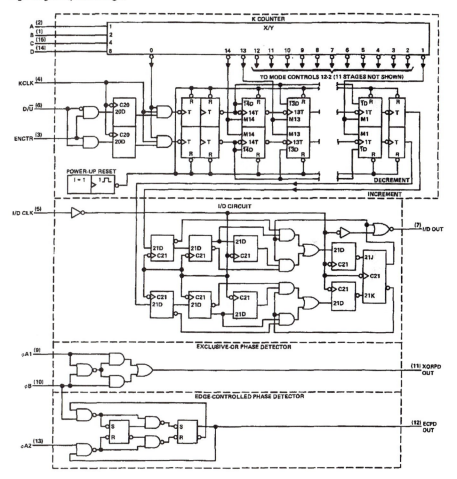

Pin numbers shown are for J, N, and W packages.

**SN54LS297, SN74LS297**
**DIGITAL PHASE-LOCKED-LOOP FILTERS**

| | | | | K COUNTER FUNCTION TABLE (DIGITAL CONTROL) |
|---|---|---|---|---|
| **D** | **C** | **B** | **A** | **MODULO (K)** |
| L | L | L | L | Inhibited |
| L | L | L | H | $2^3$ |
| L | L | H | L | $2^4$ |
| L | L | H | H | $2^5$ |
| L | H | L | L | $2^6$ |
| L | H | L | H | $2^7$ |
| L | H | H | L | $2^8$ |
| L | H | H | H | $2^9$ |
| H | L | L | L | $2^{10}$ |
| H | L | L | H | $2^{11}$ |
| H | L | H | L | $2^{12}$ |
| H | L | H | H | $2^{13}$ |
| H | H | L | L | $2^{14}$ |
| H | H | L | H | $2^{15}$ |
| H | H | H | L | $2^{16}$ |
| H | H | H | H | $2^{17}$ |

**FUNCTION TABLE**
**EXCLUSIVE-OR PHASE DETECTOR**

| $\phi$A1 | $\phi$B | XORPD OUT |
|---|---|---|
| L | L | L |
| L | H | H |
| H | L | H |
| H | H | L |

**FUNCTION TABLE**
**EDGE-CONTROLLED PHASE DETECTOR**

| $\phi$A2 | $\phi$B | ECPD OUT |
|---|---|---|
| H or L | ↓ | H |
| ↓ | H or L | L |
| H or L | ↑ | No change |
| ↑ | H or L | No change |

H = steady-state high level
L = steady-state low level
↓ = transition from high to low
↑ = transition from low to high

**schematics of inputs and outputs**

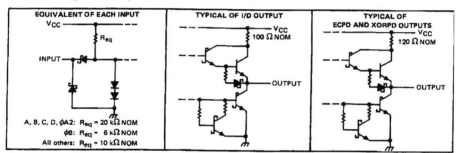

EQUIVALENT OF EACH INPUT
$V_{CC}$
$R_{eq}$
INPUT
A, 8, C, D, $\phi$A2: $R_{eq}$ = 20 kΩ NOM
$\phi$B: $R_{eq}$ = 6 kΩ NOM
All others: $R_{eq}$ = 10 kΩ NOM

TYPICAL OF I/D OUTPUT
$V_{CC}$
100 Ω NOM
OUTPUT

TYPICAL OF
ECPD AND XORPD OUTPUTS
$V_{CC}$
120 Ω NOM
OUTPUT

**operation**

The phase detector generates an error signal waveform that, at zero phase error, is a 50% duty cycle square wave. At the limits of linear operation, the phase detector output will be either high or low all of the time, depending on the direction of the phase error ($\phi_{in} - \phi_{out}$). Within these limits, the phase detector output varies linearly with the input phase error according to the gain $k_d$, which is expressed in terms of phase detector output per cycle of phase error. The phase detector output can be defined to vary between ±1 according to the relation:

$$\text{PD Output} = \frac{\% \text{ high} - \% \text{ low}}{100} \qquad (1)$$

The output of the phase detector will be $k_d \phi_e$, where the phase error $\phi_e = \phi_{in} - \phi_{out}$.

## SN54LS297, SN74LS297
## DIGITAL PHASE-LOCKED-LOOP FILTERS

---

Exclusive-OR phase detectors (XORPD) and edge-controlled phase detectors (ECPD) are commonly used digital types. The ECPD is more complex than the XORPD logic function, but can be described generally as a circuit that changes states on one of the transitions of its inputs. $k_d$ for an XORPD is 4 because its output remains high (PD output = 1) for a phase error of 1/4 cycle. Similarly, $k_d$ for the ECPD is 2 since its output remains high for a phase error of 1/2 cycle. The type of phase detector will determine the zero-phase-error point, i.e., the phase separation of the phase detector inputs for $\phi_e$ defined to be zero. For the basic DPLL system of Figure 2, $\phi_e = 0$ when the phase detector output is a square wave. The XORPD inputs are 1/4 cycle out of phase for zero phase error. For the ECPD, $\phi_e = 0$ when the inputs are 1/2 cycle out of phase.

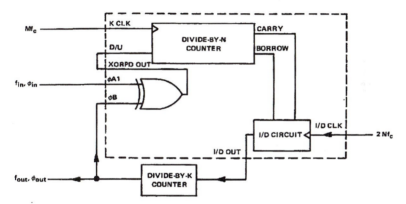

**FIGURE 2—DPLL USING EXCLUSIVE-OR PHASE DETECTION**

The phase detector output controls the up/down input to the K counter. The counter is clocked by input frequency $Mf_c$, which is a multiple M of the loop center frequency $f_c$. When the K counter recycles up, it generates a carry pulse. Recycling while counting down generates a borrow pulse. If the carry and borrow outputs are conceptually combined into one output that is positive for a carry and negative for a borrow, and if the K counter is considered as a frequency divider with the ratio $Mf_c/K$, the output of the K counter will equal the input frequency multiplied by the division ratio. Thus the output from the K counter is $(k_d \phi_e Mf_c)/K$.

The carry and borrow pulses go to the increment/decrement (I/D) circuit, which, in the absence of any carry or borrow pulse, has an output that is 1/2 of the input clock I/D CLK. The input clock is just a multiple, 2N, of the loop center frequency. In response to a carry or borrow pulse, the I/D circuit will either add or delete a pulse at I/D OUT. Thus the output of the I/D circuit will be $Nf_c + (k_d \phi_e Mf_c)/2K$.

The output of the N counter (or the output of the phase-locked loop) is thus:

$$f_o = f_c + (k_d \phi_e Mf_c)/2KN$$

If this result is compared to the equation for a first-order analog phase-locked loop, the digital equivalent of the gain of the VCO is just $Mf_c/2KN$ or $f_c/K$ for M = 2N.

Thus the simple first-order phase-locked loop with an adjustable K counter is the equivalent of an analog phase-locked loop with a programmable VCO gain.

### SN54LS297, SN74LS297
### DIGITAL PHASE-LOCKED-LOOP FILTERS

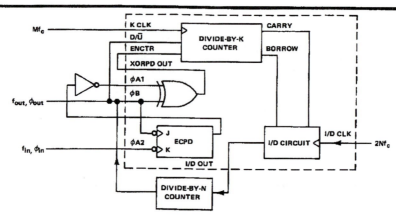

FIGURE 3—DPLL USING BOTH PHASE DETECTORS IN A RIPPLE-CANCELLATION SCHEME

**absolute maximum rating over operating free-air temperature range (unless otherwise noted)**

Supply voltage, $V_{CC}$ (see Note 1) . . . . . . . . . . . . . . . . . . . . . . . . . . . . . . . . . . . . . . . . . . . . . . . . . . 7 V
Input voltage . . . . . . . . . . . . . . . . . . . . . . . . . . . . . . . . . . . . . . . . . . . . . . . . . . . . . . . . . . . . . . . . . . . . . 7 V
Operating free-air temperature range:  SN54LS297 . . . . . . . . . . . . . . . . . . . . . . . . . . . . . . . . . . . . . −55°C to 125°C
              SN74LS297 . . . . . . . . . . . . . . . . . . . . . . . . . . . . . . . . . . . . . 0°C to 70°C
Storage temperature range . . . . . . . . . . . . . . . . . . . . . . . . . . . . . . . . . . . . . . . . . . . . . . . . . . . . . . −65°C to 150°C

NOTE 1:  Voltage values are with respect to network ground terminal.

**recommended operating conditions**

| | | | SN54LS297 | | | SN74LS297 | | | UNIT |
|---|---|---|---|---|---|---|---|---|---|
| | | | MIN | NOM | MAX | MIN | NOM | MAX | |
| $V_{CC}$ | Supply voltage | | 4.5 | 5 | 5.5 | 4.75 | 5 | 5.25 | V |
| $I_{OH}$ | High-level output current | I/D OUT | | | −1.2 | | | −1.2 | mA |
| | | EXOR, ECPD | | | −400 | | | −400 | µA |
| $I_{OL}$ | Low-level output current | I/D OUT | | | 12 | | | 24 | mA |
| | | XOR, ECPD | | | 4 | | | 8 | mA |
| $f_{clock}$ | Clock frequency | K Clock | 0 | | 32 | 0 | | 32 | MHz |
| | | I/D Clock | 0 | | 16 | 0 | | 16 | MHz |
| $t_w$ | Width of clock input pulse | K Clock | 16 | | | 16 | | | ns |
| | | I/D Clock | 33 | | | 33 | | | ns |
| $t_{su}$, to K | Setup time to K Clock ↑ | U/D̄, ENCTR | 30 | | | 30 | | | ns |
| $t_h$ | Hold time from K Clock ↑ | U/D̄, ENCTR | 0 | | | 0 | | | ns |
| $T_A$ | Operating free-air temperature | | −55 | | 125 | 0 | | 70 | °C |

## SN54LS297, SN74LS297
## DIGITAL PHASE-LOCKED-LOOP FILTERS

**electrical characteristics over recommended operating free-air temperature range (unless otherwise noted)**

| PARAMETER | | | TEST CONDITIONS† | | SN54LS297 MIN | SN54LS297 TYP‡ | SN54LS297 MAX | SN74LS297 MIN | SN74LS297 TYP‡ | SN74LS297 MAX | UNIT |
|---|---|---|---|---|---|---|---|---|---|---|---|
| $V_{IH}$ | High-level input voltage | | | | 2 | | | 2 | | | V |
| $V_{IL}$ | Low-level input voltage | | | | | | 0.7 | | | 0.8 | V |
| $V_{IK}$ | Input clamp voltage | | $V_{CC}$ = MIN, | $I_I$ = −18 mA | | | −1.5 | | | −1.5 | V |
| $V_{OH}$ | High-level output voltage | I/D OUT | $V_{CC}$ = MIN, $V_{IH}$ = 2 V, | $I_{OH}$ = MAX | 2.4 | | | 2.4 | | | V |
| | | Others | $V_{IL}$ = $V_{IL}$ max | $I_{OH}$ = MAX | 2.5 | | | 2.7 | | | |
| $V_{OL}$ | Low-level output voltage | I/D OUT | $V_{CC}$ = MIN, $V_{IH}$ = 2 V, | $I_{OL}$ = 12 mA | | 0.25 | 0.4 | | 0.25 | 0.4 | V |
| | | | | $I_{OL}$ = 24 mA | | | | | 0.35 | 0.5 | |
| | | Others | $V_{IL}$ = $V_{IL}$ max | $I_{OL}$ = 4 mA | | 0.25 | 0.4 | | 0.25 | 0.4 | |
| | | | | $I_{OL}$ = 8 mA | | | | | 0.35 | 0.5 | |
| $I_I$ | Input current at maximum input voltage | | $V_{CC}$ = MAX, | $V_I$ = 7 V | | | 0.1 | | | 0.1 | mA |
| $I_{IH}$ | High-level input current | U/D̄, EN, φA1 | $V_{CC}$ = MAX, | $V_I$ = 2.7 V | | | 40 | | | 40 | µA |
| | | φB | | | | | 60 | | | 60 | |
| | | All others | | | | | 20 | | | 20 | |
| $I_{IL}$ | Low-level input current | U/D̄, EN, φA1 | $V_{CC}$ = MAX, | $V_I$ = 0.4 V | | | −0.8 | | | −0.8 | mA |
| | | φB | $V_I$ = 0.4 V | | | | −1.2 | | | −1.2 | |
| | | All others | | | | | −0.4 | | | −0.4 | |
| $I_{OS}$ | Short-circuit output current § | I/D OUT | $V_{CC}$ = MAX | | −30 | | −130 | −30 | | −130 | mA |
| | | Others | | | −20 | | −100 | −20 | | −100 | |
| $I_{CC}$ | Supply current | | $V_{CC}$ = MAX, All inputs grounded, All outputs open | | | 75 | 120 | | 75 | 120 | mA |

†For conditions shown as MIN or MAX, use the appropriate value specified under recommended operating conditions.
‡All typical values are of $V_{CC}$ = 5 V, $T_A$ = 25°C.
§ Not more than one output should be shorted at a time and the duration of the short-circuit should not exceed one second.

**switching characteristics, $V_{CC}$ = 5 V, $T_A$ = 25°C**

| PARAMETER¶ | FROM (INPUT) | | TO (OUTPUT) | TEST CONDITIONS | MIN | TYP | MAX | UNIT |
|---|---|---|---|---|---|---|---|---|
| $f_{max}$ | KCLK | | I/D OUT | $R_L$ = 667 Ω, $C_L$ = 45 pF, See Note 2 | 32 | 50 | | MHz |
| | I/D CLK | | I/D OUT | | 16 | 35 | | |
| $t_{PLH}$ | I/D CLK ↑ | | I/D OUT | | | 15 | 25 | ns |
| $t_{PHL}$ | I/D CLK ↑ | | I/D OUT | | | 22 | 35 | ns |
| $t_{PLH}$ | φA1 or φB | Other input low | XOR OUT | | | 10 | 15 | ns |
| | φA1 or φB | Other input high | XOR OUT | | | 17 | 25 | |
| $t_{PHL}$ | φA1 or φB | Other input low | XOR OUT | $R_L$ = 2 kΩ, $C_L$ = 45 pF, See Note 2 | | 15 | 25 | ns |
| | φA1 or φB | Other input high | XOR OUT | | | 17 | 25 | |
| $t_{PLH}$ | φB ↓ | | ECPD OUT | | | 20 | 30 | ns |
| $t_{PHL}$ | φA2 ↓ | | ECPD OUT | | | 20 | 30 | ns |

¶$t_{PLH}$ = propagation delay time, low-to-high-level output
$t_{PHL}$ = propagation delay time, high-to-low-level output
NOTE 2: Load circuits and voltage waveforms are shown in Section 1.

# Motorola MC14046B Phase Locked Loop

**MOTOROLA**
**SEMICONDUCTOR TECHNICAL DATA**

## Phase Locked Loop

The MC14046B phase locked loop contains two phase comparators, a voltage–controlled oscillator (VCO), source follower, and zener diode. The comparators have two common signal inputs, PCA$_{in}$ and PCB$_{in}$. Input PCA$_{in}$ can be used directly coupled to large voltage signals, or indirectly coupled (with a series capacitor) to small voltage signals. The self–bias circuit adjusts small voltage signals in the linear region of the amplifier. Phase comparator 1 (an exclusive OR gate) provides a digital error signal PC1$_{out}$, and maintains 90° phase shift at the center frequency between PCA$_{in}$ and PCB$_{in}$ signals (both at 50% duty cycle). Phase comparator 2 (with leading edge sensing logic) provides digital error signals, PC2$_{out}$ and LD, and maintains a 0° phase shift between PCA$_{in}$ and PCB$_{in}$ signals (duty cycle is immaterial). The linear VCO produces an output signal VCO$_{out}$ whose frequency is determined by the voltage of input VCO$_{in}$ and the capacitor and resistors connected to pins C1$_A$, C1$_B$, R1, and R2. The source–follower output SF$_{out}$ with an external resistor is used where the VCO$_{in}$ signal is needed but no loading can be tolerated. The inhibit input Inh, when high, disables the VCO and source follower to minimize standby power consumption. The zener diode can be used to assist in power supply regulation.

Applications include FM and FSK modulation and demodulation, frequency synthesis and multiplication, frequency discrimination, tone decoding, data synchronization and conditioning, voltage–to–frequency conversion and motor speed control.

- Buffered Outputs Compatible with MHTL and Low–Power TTL
- Diode Protection on All Inputs
- Supply Voltage Range = 3.0 to 18 V
- Pin–for–Pin Replacement for CD4046B
- Phase Comparator 1 is an Exclusive Or Gate and is Duty Cycle Limited
- Phase Comparator 2 switches on Rising Edges and is not Duty Cycle Limited

## MC14046B

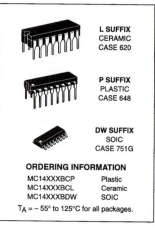

**L SUFFIX**
CERAMIC
CASE 620

**P SUFFIX**
PLASTIC
CASE 648

**DW SUFFIX**
SOIC
CASE 751G

**ORDERING INFORMATION**

| | |
|---|---|
| MC14XXXBCP | Plastic |
| MC14XXXBCL | Ceramic |
| MC14XXXBDW | SOIC |

T$_A$ = – 55° to 125°C for all packages.

**PIN ASSIGNMENT**

| | | | | |
|---|---|---|---|---|
| LD | 1 | | 16 | V$_{DD}$ |
| PC1$_{out}$ | 2 | | 15 | ZENER |
| PCB$_{in}$ | 3 | | 14 | PCA$_{in}$ |
| VCO$_{out}$ | 4 | | 13 | PC2$_{out}$ |
| INH | 5 | | 12 | R2 |
| C1$_A$ | 6 | | 11 | R1 |
| C1$_B$ | 7 | | 10 | SF$_{out}$ |
| V$_{SS}$ | 8 | | 9 | VCO$_{in}$ |

**BLOCK DIAGRAM**

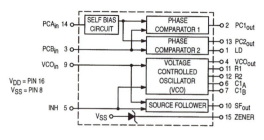

**MAXIMUM RATINGS\*** (Voltages Referenced to $V_{SS}$)

| Rating | Symbol | Value | Unit |
|---|---|---|---|
| DC Supply Voltage | $V_{DD}$ | $-0.5$ to $+18$ | Vdc |
| Input Voltage, All Inputs | $V_{in}$ | $-0.5$ to $V_{DD} + 0.5$ | Vdc |
| DC Input Current, per Pin | $I_{in}$ | $\pm 10$ | mAdc |
| Power Dissipation, per Package† | $P_D$ | 500 | mW |
| Operating Temperature Range | $T_A$ | $-55$ to $+125$ | °C |
| Storage Temperature Range | $T_{stg}$ | $-65$ to $+150$ | °C |

\* Maximum Ratings are those values beyond which damage to the device may occur.
†Temperature Derating:
    Plastic "P and D/DW" Packages: $-7.0$ mW/°C From 65°C To 125°C
    Ceramic "L" Packages: $-12$ mW/°C From 100°C To 125°C

**ELECTRICAL CHARACTERISTICS** (Voltages Referenced to $V_{SS}$)

| Characteristic | | Symbol | $V_{DD}$ Vdc | $-55$°C Min | $-55$°C Max | 25°C Min | 25°C Typ | 25°C Max | 125°C Min | 125°C Max | Unit |
|---|---|---|---|---|---|---|---|---|---|---|---|
| Output Voltage<br>$V_{in} = V_{DD}$ or 0 | "0" Level | $V_{OL}$ | 5.0<br>10<br>15 | —<br>—<br>— | 0.05<br>0.05<br>0.05 | —<br>—<br>— | 0<br>0<br>0 | 0.05<br>0.05<br>0.05 | —<br>—<br>— | 0.05<br>0.05<br>0.05 | Vdc |
| $V_{in} = 0$ or $V_{DD}$ | "1" Level | $V_{OH}$ | 5.0<br>10<br>15 | 4.95<br>9.95<br>14.95 | —<br>—<br>— | 4.95<br>9.95<br>14.95 | 5.0<br>10<br>15 | —<br>—<br>— | 4.95<br>9.95<br>14.95 | —<br>—<br>— | Vdc |
| Input Voltage #<br>($V_O = 4.5$ or 0.5 Vdc)<br>($V_O = 9.0$ or 1.0 Vdc)<br>($V_O = 13.5$ or 1.5 Vdc) | "0" Level | $V_{IL}$ | <br>5.0<br>10<br>15 | <br>—<br>—<br>— | <br>1.5<br>3.0<br>4.0 | <br>—<br>—<br>— | <br>2.25<br>4.50<br>6.75 | <br>1.5<br>3.0<br>4.0 | <br>—<br>—<br>— | <br>1.5<br>3.0<br>4.0 | Vdc |
| ($V_O = 0.5$ or 4.5 Vdc)<br>($V_O = 1.0$ or 9.0 Vdc)<br>($V_O = 1.5$ or 13.5 Vdc) | "1" Level | $V_{IH}$ | 5.0<br>10<br>15 | 3.5<br>7.0<br>11 | —<br>—<br>— | 3.5<br>7.0<br>11 | 2.75<br>5.50<br>8.25 | —<br>—<br>— | 3.5<br>7.0<br>11 | —<br>—<br>— | Vdc |
| Output Drive Current<br>($V_{OH} = 2.5$ Vdc)<br>($V_{OH} = 4.6$ Vdc)<br>($V_{OH} = 9.5$ Vdc)<br>($V_{OH} = 13.5$ Vdc) | Source | $I_{OH}$ | 5.0<br>5.0<br>10<br>15 | $-1.2$<br>$-0.25$<br>$-0.62$<br>$-1.8$ | —<br>—<br>—<br>— | $-1.0$<br>$-0.2$<br>$-0.5$<br>$-1.5$ | $-1.7$<br>$-0.36$<br>$-0.9$<br>$-3.5$ | —<br>—<br>—<br>— | $-0.7$<br>$-0.14$<br>$-0.35$<br>$-1.1$ | —<br>—<br>—<br>— | mAdc |
| ($V_{OL} = 0.4$ Vdc)<br>($V_{OL} = 0.5$ Vdc)<br>($V_{OL} = 1.5$ Vdc) | Sink | $I_{OL}$ | 5.0<br>10<br>15 | 0.64<br>1.6<br>4.2 | —<br>—<br>— | 0.51<br>1.3<br>3.4 | 0.88<br>2.25<br>8.8 | —<br>—<br>— | 0.36<br>0.9<br>2.4 | —<br>—<br>— | mAdc |
| Input Current | | $I_{in}$ | 15 | — | $\pm 0.1$ | — | $\pm 0.00001$ | $\pm 0.1$ | — | $\pm 1.0$ | µAdc |
| Input Capacitance | | $C_{in}$ | — | — | — | — | 5.0 | 7.5 | — | — | pF |
| Quiescent Current<br>(Per Package) Inh = PCA$_{in}$ = $V_{DD}$,<br>Zener = VCO$_{in}$ = 0 V, PCB$_{in}$ = $V_{DD}$<br>or 0 V, $I_{out}$ = 0 µA | | $I_{DD}$ | 5.0<br>10<br>15 | —<br>—<br>— | 5.0<br>10<br>20 | —<br>—<br>— | 0.005<br>0.010<br>0.015 | 5.0<br>10<br>20 | —<br>—<br>— | 150<br>300<br>600 | µAdc |
| Total Supply Current†<br>(Inh = "0", $f_0$ = 10 kHz, $C_L$ = 50 pF,<br>R1 = 1.0 MΩ, R2 = ∞ $R_{SF}$ = ∞,<br>and 50% Duty Cycle) | | $I_T$ | 5.0<br>10<br>15 | $I_T = (1.46$ µA/kHz$)$ f + $I_{DD}$<br>$I_T = (2.91$ µA/kHz$)$ f + $I_{DD}$<br>$I_T = (4.37$ µA/kHz$)$ f + $I_{DD}$ | | | | | | | mAdc |

\#Noise immunity specified for worst–case input combination.
  Noise Margin for both "1" and "0" level = 1.0 Vdc min @ $V_{DD}$ = 5.0 Vdc
                               2.0 Vdc min @ $V_{DD}$ = 10 Vdc
                               2.5 Vdc min @ $V_{DD}$ = 15 Vdc

†To Calculate Total Current in General:

$$I_T \approx 2.2 \times V_{DD}\left(\frac{VCO_{in} - 1.65}{R1} + \frac{V_{DD} - 1.35}{R2}\right)^{3/4} + 1.6 \times \left(\frac{VCO_{in} - 1.65}{R_{SF}}\right)^{3/4} + 1 \times 10^{-3}\,(C_L + 9)\,V_{DD}\,f +$$

$$1 \times 10^{-1}\,V_{DD}^2\left(\frac{100\% \text{ Duty Cycle of } PCA_{in}}{100}\right) + I_Q \quad \text{where: } I_T \text{ in µA, } C_L \text{ in pF, } VCO_{in}, V_{DD} \text{ in Vdc, f in kHz, and}$$
$$\text{R1, R2, } R_{SF} \text{ in MΩ, } C_L \text{ on } VCO_{out}.$$

**ELECTRICAL CHARACTERISTICS\*** ($C_L$ = 50 pF, $T_A$ = 25°C)

| Characteristic | Symbol | $V_{DD}$ Vdc | Minimum Device | Typical | Maximum Device | Units |
|---|---|---|---|---|---|---|
| Output Rise Time | $t_{TLH}$ | | | | | ns |
| $t_{TLH}$ = (3.0 ns/pF) $C_L$ + 30 ns | | 5.0 | — | 180 | 350 | |
| $t_{TLH}$ = (1.5 ns/pF) $C_L$ + 15 ns | | 10 | — | 90 | 150 | |
| $t_{TLH}$ = (1.1 ns/pF) $C_L$ + 10 ns | | 15 | — | 65 | 110 | |
| Output Fall Time | $t_{THL}$ | | | | | ns |
| $t_{THL}$ = (1.5 ns/pF) $C_L$ + 25 ns | | 5.0 | — | 100 | 175 | |
| $t_{THL}$ = (0.75 ns/pF) $C_L$ + 12.5 ns | | 10 | — | 50 | 75 | |
| $t_{THL}$ = (0.55 ns/pF) $C_L$ + 9.5 ns | | 15 | — | 37 | 55 | |
| **PHASE COMPARATORS 1 and 2** | | | | | | |
| Input Resistance — $PCA_{in}$ | $R_{in}$ | 5.0 | 1.0 | 2.0 | — | MΩ |
| | | 10 | 0.2 | 0.4 | — | |
| | | 15 | 0.1 | 0.2 | — | |
| — $PCB_{in}$ | $R_{in}$ | 15 | 150 | 1500 | — | MΩ |
| Minimum Input Sensitivity | $V_{in}$ | 5.0 | — | 200 | 300 | mV p–p |
| AC Coupled — $PCA_{in}$ | | 10 | — | 400 | 600 | |
| C series = 1000 pF, f = 50 kHz | | 15 | — | 700 | 1050 | |
| DC Coupled — $PCA_{in}$, $PCB_{in}$ | — | 5 to 15 | See Noise Immunity | | | |
| **VOLTAGE CONTROLLED OSCILLATOR (VCO)** | | | | | | |
| Maximum Frequency | $f_{max}$ | 5.0 | 0.5 | 0.7 | — | MHz |
| ($VCO_{in}$ = $V_{DD}$, C1 = 50 pF | | 10 | 1.0 | 1.4 | — | |
| R1 = 5.0 kΩ, and R2 = ∞) | | 15 | 1.4 | 1.9 | — | |
| Temperature — Frequency Stability | — | 5.0 | — | 0.12 | — | %/°C |
| (R2 = ∞ ) | | 10 | — | 0.04 | — | |
| | | 15 | — | 0.015 | — | |
| Linearity (R2 = ∞ ) | — | | | | | % |
| ($VCO_{in}$ = 2.5 V ± 0.3 V, R1 > 10 kΩ) | | 5.0 | — | 1.0 | — | |
| ($VCO_{in}$ = 5.0 V ± 2.5 V, R1 > 400 kΩ) | | 10 | — | 1.0 | — | |
| ($VCO_{in}$ = 7.5 V ± 5.0 V, R1 ≥ 1000 kΩ) | | 15 | — | 1.0 | — | |
| Output Duty Cycle | — | 5 to 15 | — | 50 | — | % |
| Input Resistance — $VCO_{in}$ | $R_{in}$ | 15 | 150 | 1500 | — | MΩ |
| **SOURCE–FOLLOWER** | | | | | | |
| Offset Voltage | — | 5.0 | — | 1.65 | 2.2 | V |
| ($VCO_{in}$ minus $SF_{out}$, RSF > 500 kΩ) | | 10 | — | 1.65 | 2.2 | |
| | | 15 | — | 1.65 | 2.2 | |
| Linearity | — | | | | | % |
| ($VCO_{in}$ = 2.5 V ± 0.3 V, $R_{SF}$ > 50 kΩ) | | 5.0 | — | 0.1 | — | |
| ($VCO_{in}$ = 5.0 V ± 2.5 V, $R_{SF}$ > 50 kΩ) | | 10 | — | 0.6 | — | |
| ($VCO_{in}$ = 7.5 V ± 5.0 V, $R_{SF}$ > 50 kΩ) | | 15 | — | 0.8 | — | |
| **ZENER DIODE** | | | | | | |
| Zener Voltage ($I_z$ = 50 µA) | $V_Z$ | — | 6.7 | 7.0 | 7.3 | V |
| Dynamic Resistance ($I_z$ = 1.0 mA) | $R_Z$ | — | — | 100 | — | Ω |

\* The formula given is for the typical characteristics only.

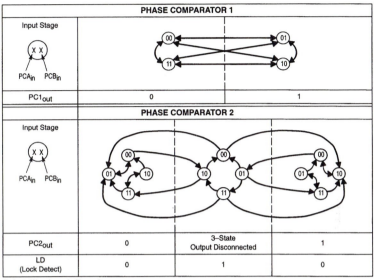

**Figure 1. Phase Comparators State Diagrams**

| Characteristic | Using Phase Comparator 1 | Using Phase Comparator 2 |
|---|---|---|
| No signal on input $PCA_{in}$. | VCO in PLL system adjusts to center frequency ($f_0$). | VCO in PLL system adjusts to minimum frequency ($f_{min}$). |
| Phase angle between $PCA_{in}$ and $PCB_{in}$. | 90° at center frequency ($f_0$), approaching 0° and 180° at ends of lock range ($2f_L$) | Always 0° in lock (positive rising edges). |
| Locks on harmonics of center frequency. | Yes | No |
| Signal input noise rejection. | High | Low |
| Lock frequency range ($2f_L$). | The frequency range of the input signal on which the loop will stay locked if it was initially in lock; $2f_L$ = full VCO frequency range = $f_{max} - f_{min}$. | |
| Capture frequency range ($2f_C$). | The frequency range of the input signal on which the loop will lock if it was initially out of lock. | |
| | Depends on low–pass filter characteristics (see Figure 3). $f_C \leq f_L$ | $f_C = f_L$ |
| Center frequency ($f_0$). | The frequency of $VCO_{out}$, when $VCO_{in}$ = 1/2 $V_{DD}$ | |
| VCO output frequency (f). <br><br> Note: These equations are intended to be a design guide. Since calculated component values may be in error by as much as a factor of 4, laboratory experimentation may be required for fixed designs. Part to part frequency variation with identical passive components is typically less than ± 20%. | $f_{min} = \dfrac{1}{R_2(C_1 + 32 \text{ pF})}$  ($V_{CO}$ input = $V_{SS}$) <br><br> $f_{max} = \dfrac{1}{R_1(C_1 + 32 \text{ pF})} + f_{min}$  ($V_{CO}$ input = $V_{DD}$) <br><br> Where:  $10K \leq R_1 \leq 1 M$ <br> $10K \leq R_2 \leq 1 M$ <br> $100pF \leq C_1 \leq .01 \mu F$ | |

**Figure 2. Design Information**

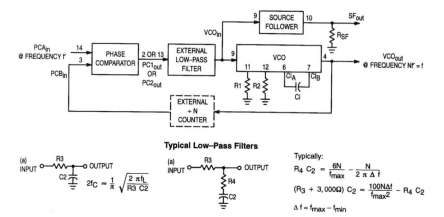

### Typical Low–Pass Filters

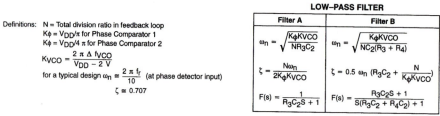

(a) INPUT —R3— OUTPUT

$2f_C \approx \dfrac{1}{\pi}\sqrt{\dfrac{2\,\pi f_L}{R3\;C2}}$

(a) INPUT —R3— OUTPUT

Typically:

$R_4\,C_2 = \dfrac{6N}{f_{max}} - \dfrac{N}{2\,\pi\,\Delta\,f}$

$(R_3 + 3{,}000\Omega)\,C_2 = \dfrac{100N\Delta f}{f_{max}{}^2} - R_4\,C_2$

$\Delta\,f = f_{max} - f_{min}$

NOTE: Sometimes R3 is split into two series resistors each R3 ÷ 2. A capacitor $C_C$ is then placed from the midpoint to ground. The value for $C_C$ should be such that the corner frequency of this network does not significantly affect $\omega_n$. In Figure B, the ratio of R3 to R4 sets the damping, R4 $\cong$ (0.1)(R3) for optimum results.

Definitions: N = Total division ratio in feedback loop

$K\phi = V_{DD}/\pi$ for Phase Comparator 1
$K\phi = V_{DD}/4\,\pi$ for Phase Comparator 2

$K_{VCO} = \dfrac{2\,\pi\,\Delta\,f_{VCO}}{V_{DD} - 2\,V}$

for a typical design $\omega_n \cong \dfrac{2\,\pi\,f_r}{10}$ (at phase detector input)

$\zeta \cong 0.707$

### LOW–PASS FILTER

| | Filter A | Filter B |
|---|---|---|
| $\omega_n =$ | $\sqrt{\dfrac{K_\phi K_{VCO}}{N R_3 C_2}}$ | $\sqrt{\dfrac{K_\phi K_{VCO}}{N C_2 (R_3 + R_4)}}$ |
| $\zeta =$ | $\dfrac{N\omega_n}{2K_\phi K_{VCO}}$ | $0.5\,\omega_n\left(R_3 C_2 + \dfrac{N}{K_\phi K_{VCO}}\right)$ |
| $F(s) =$ | $\dfrac{1}{R_3 C_2 S + 1}$ | $\dfrac{R_3 C_2 S + 1}{S(R_3 C_2 + R_4 C_2) + 1}$ |

### Waveforms

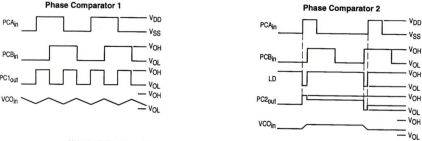

**Phase Comparator 1**

PCA_in
PCB_in
PC1_out
VCO_in

**Phase Comparator 2**

PCA_in
PCB_in
LD
PC2_out
VCO_in

Note: for further information, see:
(1) F. Gardner, "Phase–Lock Techniques", John Wiley and Son, New York, 1966.
(2) G. S. Moschytz, "Miniature RC Filters Using Phase–Locked Loop", BSTJ, May, 1965.
(3) Garth Nash, "Phase–Lock Loop Design Fundamentals", AN–535, Motorola Inc.
(4) A. B. Przedpelski, "Phase–Locked Loop Design Articles", AR254, reprinted by Motorola Inc.

**Figure 3. General Phase–Locked Loop Connections and Waveforms**

# Index